Pollution:
Causes, Effects, and Control
Third Edition

# Pollution: Causes, Effects and Control

## Third Edition

Edited by
**Roy M. Harrison**
*The University of Birmingham*

THE ROYAL
SOCIETY OF
CHEMISTRY
Information
Services

ISBN 0-85404-534-1

A catalogue record of this book is available from the British Library.

First published 1983

Second Edition 1990
Reprinted 1992, 1993, 1995

Third Edition © The Royal Society of Chemistry 1996

Published by The Royal Society of Chemistry,
Thomas Graham House, Science Park, Milton Road, Cambridge, CB4 4WF, UK

Typeset by Paston Press Ltd, Loddon, Norfolk, NR14 6JD
Printed by Hartnolls Ltd, Bodmin, Cornwall, UK

# Preface

The first edition of this book, published in 1983, arose from collation of the course notes from a Residential School held at the University of Lancaster in 1982, supplemented with additional chapters to give a more complete overview of the field. The aim was to provide a basic textbook covering the more important concepts. When the second edition was published in 1990, it was considerably expanded both by including totally new subject areas and by giving each author a greater length in which to cover his/her topic. The level of treatment was similar to the first edition, being essentially introductory, although some more advanced aspects were covered. Wholly new chapters were introduced dealing with radioactive pollution and the chemistry and pollution of the stratosphere. The second edition has proved enormously successful, going through a number of reprints, and sales have stayed remarkably strong despite the passage of time. While some aspects of the subject matter change little with the years, others rapidly become dated and this, the third edition, has been produced largely to update the contents in line with the current position in a fast moving field. I have, however, taken the opportunity to amend the content somewhat so as to remove some of the more specialized chapters which have been replaced with other more general topics which had been omitted previously. Notable examples of subjects now included are the marine environment, soils and contaminated land, solid waste disposal and toxic organic chemicals. Also, given the massive current public interest in the subject of air pollution and health, a wholly new chapter addresses this subject.

Once again, the chapter authors have been selected on the basis of their established reputation in the field and their ability to write with clarity of presentation. A high proportion of them wrote for the second edition and have had the opportunity to update their contribution to this third edition. A number of such authors, recognizing the fast development of their fields, have chosen to rewrite their contribution totally.

Scientific activity in the pollution field has intensified greatly since 1982 when the first edition was written. The knowledge base has expanded greatly and the legislation and regulatory environment has intensified in response to mounting public concern over environmental issues. The environment remains high on the political agenda, heightening the need for authoritative scientific information, and I hope that this book goes some way towards meeting that need.

Roy M. Harrison
Birmingham

v

# Contents

# Contributors

**B. J. Alloway,** *Department of Soil Science, University of Reading, London Road, Reading RG1 5AQ, UK*

**D. A. Arthur,** *Environmental Resources Ltd, Eaton House, Wallbrook Court, North Hinksey Lane, Oxford OX2 0QS, UK*

**J. Ayres,** *Department of Respiratory Medicine, East Birmingham Hospital NHS Trust, Bordesley Green East, Birmingham B9 5ST, UK*

**R. Chester,** *Oceanography Laboratories, Department of Earth Sciences, University of Liverpool, Bedford Street North, PO Box 147, Liverpool L69 3BX, UK*

**I. Colbeck,** *Department of Chemistry, University of Essex, Wivenhoe Park, Colchester CO4 3SQ, UK*

**B. Crathorne,** *Water Research Centre plc, Henley Road, Medmenham, Marlow, Bucks. SL7 2HD, UK*

**A. J. Dobbs,** *Water Research Centre plc, Henley Road, Medmenham, Marlow, Bucks. SL7 2HD, UK*

**G. H. Eduljee,** *Environmental Resources Ltd, Eaton House, Wallbrook Court, North Hinksey Lane, Oxford OX2 0QS, UK*

**S. J. Harrad,** *Institute of Public and Environmental Health, School of Chemistry, Edgbaston, Birmingham B15 2TT, UK*

**P. T. C. Harrison,** *Institute for Environment and Health, University of Leicester, PO Box 138, Lancaster Road, Leicester LE1 9HN, UK*

**R. M. Harrison,** *Institute of Public and Environmental Health, School of Chemistry, University of Birmingham, Edgbaston, Birmingham B15 2TT, UK*

**C. N. Hewitt,** *Division of Environmental Sciences, Institute of Environmental and Biological Sciences, University of Lancaster, Bailrigg, Lancaster LA1 4YQ, UK*

**C. Holman,** *Brook Cottage, Camp Lane, Elberton, Bristol BS9 2AU, UK*

**A. James,** *Department of Civil Engineering, University of Newcastle upon Tyne, Newcastle upon Tyne, NE1 7RU, UK*

**J. N. Lester,** *Department of Civil Engineering, Imperial College of Science and Technology and Medicine, London SW7 2BU, UK*

**P. W. Lucas,** *Division of Biological Sciences, Institute of Environmental and Biological Sciences, University of Lancaster, Bailrigg, Lancaster LA1 4YQ, UK*

**A. R. Mackenzie,** *Department of Chemistry, University of Cambridge, Lensfield Road, Cambridge CB2 1EW, UK*

**F. R. McDougall,** *Department of Civil Engineering, University of Newcastle upon Tyne, Newcastle upon Tyne, NE1 7RU, UK*

**R. Macrory** *Centre for Environmental Technology, Imperial College of Science, Technology and Medicine, 48 Prince's Gardens, London SW7 2PE, UK*

**T. A. Mansfield,** *Division of Biological Sciences, Institute of Environmental and Biological Sciences, University of Lancaster, Bailrigg, Lancaster LA1 4YQ, UK*

**C. F. Mason,** *Department of Biology, University of Essex, Wivenhoe Park, Colchester CO4 3SQ, UK*

**R. F. Packham,** *22 Harwood Road, Marlow, Bucks. SL7 2AS, UK*

**M. R. Preston,** *Oceanography Laboratories, Department of Earth Sciences, University of Liverpool, Liverpool L69 3BX, UK*

**Y. Rees,** *Water Research Centre plc, Henley Road, Medmenham, Marlow, Bucks. SL7 2HD, UK*

**D. H. Slater,** *Environment Agency, Rio House, Aztec West, Bristol BS12 4UD, UK*

**S. Walters,** *Department of Public Health and Epidemiology, Medical School, University of Birmingham, Edgbaston, Birmingham B15 2TT, UK*

**M. L. Williams,** *Air Quality Division, Department of the Environment, Romney House, 43 Marsham Street, London SW1P 3PY, UK*

CHAPTER 1

# Chemical Pollution of the Aquatic Environment by Priority Pollutants and its Control

B. CRATHORNE, A. J. DOBBS and Y. REES

## 1.1 INTRODUCTION

It is worth starting by introducing the distinction between pollution and contamination. Although these terms tend to be used in similar ways in everyday speech and journalism, in scientific areas there is a broad consensus that the term 'contamination' should be used where a chemical is present in a given sample with no evidence of harm and 'pollution' used in cases where the presence of the chemical is causing harm. Pollutants therefore are chemicals causing environmental harm. The effects of water pollution can be summarized as:[1]

- aesthetic: visual nuisance caused, *e.g.* litter, discoloration and smells
- temperature: usually heat
- deoxygenation: lack of oxygen in the water
- toxicity: acute or chronic toxicity causing damage to aquatic or human life
- acidity/alkalinity: disturbance of the pH regime
- eutrophication: nutrients giving rise to excessive growths of some organisms.

Any chemical can become a pollutant in water causing one or more of these effects if it is present at a high enough concentration. For example, serious pollution incidents result from spills of sugar and milk. Despite the fact that any chemical can be a pollutant, certain chemicals have been identified in regulation or by international agreement as being 'priority chemicals for control'. Such chemicals have generally been selected based on the following criteria:

- the chemicals are frequently found by monitoring programmes

---

[1] A. M. C. Edwards, 'The Implications of Water Regulations for Industry', Stanley Thornes, Cheltenham, UK, 1994.

- they are toxic at low concentrations
- they bioaccumulate
- they are persistent
- they are carcinogens.

Chemicals showing these characteristics constitute a rather large group, often loosely referred to as 'priority pollutants' although dangerous or hazardous substances would probably be more accurate terms. While a few years ago it would have been possible to quote the List I substances in the Dangerous Substances Directive (see Section 1.3.1) as being *the* priority pollutants, it is no longer possible to identify a single list. Different chemicals are priority pollutants in different contexts.

Priority pollutants are important but the majority of pollution incidents in the UK continue to be due to gross organic pollution. In 1994 sewage accounted for 23% of the 25 000 substantiated pollution incidents in England and Wales (*cf.* 3000 in Scotland) with oil being the single most frequent cause, accounting for 27%.[2] Effluents containing biodegradable organic chemicals generally act as pollutants, not because they contain chemicals at concentrations that are toxic, but rather the reverse. They contain chemicals that provide food for microorganisms which multiply rapidly as a result of the increased food input. The microorganisms, in the process of growing and oxidizing the organic chemical foodstuff, use up the dissolved oxygen rapidly which in some cases leads to the death of higher organisms, like fish.

Consents for discharge of effluents have historically been set in terms of the gross pollution parameters identified in the Royal Commission Report of 1912. This Report identified suspended solids concentrations and five-day biochemical oxygen demand. The current approach to discharge consents and compliance taken by the NRA (now Environment Agency) has been published.[3] With a much increased and growing array of synthetic organic chemicals being made and used, these two simple measurements cannot provide sufficient environmental protection in all cases. The use of ecotoxicity tests for effluent consenting and monitoring is increasingly being seen as the answer to the problems posed by complex effluents.

Most of the chemicals discussed in this chapter had been in production and use for many years before the various chemical approval regulations now in force were introduced. These regulations contain specific requirements relating to environmental effects. For example, the Directive 91/325/EEC concerned with the packaging and labelling of chemicals has specific guidance on rules for classifying chemicals as 'dangerous to the environment'. Directive 93/79/EEC on the 'Evaluation and Control of the Risks of Existing Substances' provides a mechanism for extending the evaluation of new substances to identify those chemicals already on the market (*i.e.* existing substances) that may be hazardous. The European Commission is in the process of drawing up a priority list of

---

[2]'Water Quality Series 25', HMSO, London, 1995.
[3]'Water Quality Series 17', HMSO, London, 1994.

chemicals that need full environmental risk assessments based on the data collected under this regulation.

## 1.2  POLLUTION CONTROL PHILOSOPHY

When discussing pollution control, it is necessary to recognize two broad types of chemical release; direct discharge, *i.e.* specific inputs from an industrial site or sewage works, *etc.*, and diffuse or non-point sources. In the last decade or so the emphasis on priority pollutant control has shifted away from control of point sources towards control of diffuse sources. It is also now recognized that control of pollution by priority pollutants is not a single environmental media or single industry issue and integrated pollution control and cross-industry reviews are now routine aspects of the control philosophy. A good recent example of the complex interactions that can occur, at least in theory, is the report of the potential accumulation of HCFC breakdown products in seasonal wetland areas.[4] The HCFCs (hydrochlorofluorohydrocarbons) are expected to break down in the atmosphere fairly rapidly. This is the reason they are preferred over the CFCs (chlorofluorohydrocarbons). However trifluoroacetic acid, which some HCFCs are expected to produce, is very stable and will wash out from the atmosphere in rain. In areas like seasonal wetlands which have high evapotranspiration rates the trifluoroacetic may concentrate to levels which may damage plants. At present, this is based on theory and modelling but it does illustrate the complexity of environmental processes.[4]

The main technical reason for the emphasis shifting from point to diffuse sources, is the success achieved by control of point source pollution. A combination of factors has caused this, principally:

- new technology enabling priority pollutant use to be avoided, *e.g.* new pesticide discovery and new production processes
- improved pollution control technology.

Given the success in this area, the relative contribution of diffuse sources has risen and attention has come to focus on these areas to facilitate further improvements. By their nature, diffuse sources require different types of regulatory control from those for point sources.

The growing internationalization of priority pollutant control is mostly driven by the recognition that pollution does not recognize national borders and particularly that protection of marine waters can only be accomplished by international action.

Government, regulatory agencies and industrial initiatives have recognized a hierarchy of approaches to priority pollutant control:

- replace: use another, more environmentally friendly chemical
- reduce: use as little of the priority pollutants as possible

[4]T. L. Tromp, M. K. W. Ko, J. M. Rodriquez and N. D. Sze, *Nature*, 1995, **376**, 327.

– manage: use in a carefully managed way to minimize accidental or adventitious loss and waste.

Banning a chemical from specific or all uses has been the primary regulatory process for pursuing replacement. However, increasingly more market-based instruments are being used, such as Eco-labelling and charging. In the case of reduction, internationally agreed targets for reduction of loads and the collection and publication of a toxic release inventory provide incentives. In the management areas the introduction of Eco-Management and Audit Scheme (EMAS) regulations and waste minimization programmes are examples of initiatives designed to improve environmental performance which can be expected to affect primary pollutants. All these activities are described in more detail later. The important points to emphasize here are that the regulation and control of priority pollutants:

– is not just a case of tightening up on discharge consents as diffuse sources can be important and can dominate in some cases
– is an international issue, and
– requires a portfolio of complementary activities based on a hierarchy of replace–reduce–manage.

## 1.3  DIRECT DISCHARGES

Two general approaches for controlling direct discharges to water are recognized; the use of Environmental Quality Objectives/Environmental Quality Standards (EQO/EQS) and Uniform Emission Standards (UES). This difference lies at the heart of a long running debate in Europe with the UK favouring the EQO/EQS approach and most other countries favouring UESs.

The EQO/EQS approach seeks to define the use that is to be made of a given water body, examples would be 'use for drinking water abstraction' or 'use for the support of salmonid fish populations', which defines an Environmental Quality Objective (EQO). In order to secure the objective in the presence of dangerous chemicals, Environmental Quality Standards (EQS) are needed. These are concentrations of the chemicals concerned below which there is expected to be no impact on the EQO. Emission limits for effluents can then be established by taking account of dilution capacity within the receiving waters on the basis that the EQS limits must not be exceeded outside the immediate impact zone (also called the mixing zone).

The UES, or limit value, approach sets limits for the concentration of dangerous substance in the effluent, without taking specific account of the available dilution capacity or the presence of other inputs to the same water body. The UES limits are usually expressed as effluent concentrations (monthly flow-weighted averages) with additional limits on daily values – the daily values being usually a factor of two to four higher. Values are also expressed as a total amount of substances per unit of production or use, *e.g.* 40 g $CCl_4$ per tonne of production.

At a technical level there has been a growing recognition that the most sensible approach to pollution control is a fusion of the two approaches as proposed in the 'Red List' Discussion Document and adopted in the Integrated Pollution Control system (see Sections 1.3.2 and 1.3.3).

### 1.3.1 Dangerous Substances Directive

The Dangerous Substances Directive (76/464/EEC) was adopted in 1976 to provide a framework for eliminating or reducing pollution of inland waters by particularly dangerous substances. In this Directive chemicals are either placed on List I, which has come to be known as the 'Black' List, or on List II, the 'Grey' List. Different control procedures are applied to chemicals on these lists. Those on List I have limit values and EQSs agreed at Community level. These appear in daughter Directives, *e.g.* Directive of 26 September 1993 on limit values and quality objectives for cadmium discharges (93/513/EEC). (Confusingly, in Directives the Commission uses 'quality objective' to mean the same as EQS in the discussion above!) The List I chemical categories are given in Table 1.1 and specific chemicals agreed for control as List I chemicals are provided in Table 1.2. In 1982 the Commission also published a list of 129 potential List I chemicals selected by the Commission on the basis of production volume and estimates of toxicity, persistence and bioaccumulation.

List II chemicals are to be controlled by using the EQO approach using quality standards set nationally. Member States are also required by the Directive to establish programmes to reduce pollution by these substances. The families of chemicals identified as List II are given in Table 1.3 and those for which National Quality Standards have been set are given in Table 1.4.

**Table 1.1**  *Categories of substances on List I*

---

List I contains certain individual substances which belong to the following families and groups of substances, selected mainly on the basis of their toxicity, persistence and bioaccumulation, with the exception of those which are biologically harmless or which are rapidly converted into substances which are biologically harmless:
1 organohalogen compounds and substances which may form such compounds in the aquatic environment.
2 organophosphorus compounds.
3 organotin compounds.
4 substances in respect of which it has been proved that they possess carcinogenic properties in or *via* the aquatic environment.
5 mercury and its compounds.
6 cadmium and its compounds.
7 persistent mineral oils and hydrocarbons of petroleum origin.
8 persistent synthetic substances which may float, remain in suspension or sink and which may interfere with any use of the waters.

---

**Table 1.2**   *Substances on List I and other priority lists*

| Priority substance | EU List I | UK Red List | UK prescribed substances | North sea conference |
|---|---|---|---|---|
| Mercury | X | X | X | X[1] |
| Cadmium | X | X | X | X[1] |
| Copper | | | | X[1] |
| Zinc | | | | X[1] |
| Lead | | | | X[1] |
| Arsenic | | | | X[1] |
| Chromium | | | | X[1] |
| Nickel | | | | X[1] |
| Aldrin | X | X | X | X |
| Dieldrin | X | X | X | X |
| Endrin | X | X | X | X |
| Isodrin | X | | | X |
| Hexachlorocyclohexane | X | X[2] | X | X[1] |
| DDT | X | X[3] | X | X |
| Pentachlorophenol | X | X | X[5] | X[1] |
| Hexachlorobenzene | X | X | X | X[1] |
| Hexachlorobutadiene | X | X | X | X |
| Carbon tetrachloride | X | | | X[1] |
| Chloroform | X | | | X |
| Trifluralin | | X | X | X |
| Endosulfan | | X | X | X |
| Simazine | | X | X | X |
| Atrazine | | X | X | X |
| Tributyltin compounds | | X[4] | X | X |
| Triphenyltin compounds | | X[4] | X | X |
| Azinphos-ethyl | | | | X |
| Azinphos-methyl | | X | X | X |
| Fenitrothion | | X | X | X |
| Fenthion | | | | X |
| Malathion | | X | X | X |
| Parathion | | | | X |
| Dichlorvos | | X | X | X |
| Trichloroethylene | X | | | X[1] |
| Tetrachloroethylene | X | | | X[1] |
| Trichlorobenzene | X | X | X | X[1] |
| 1,2-Dichloroethane | X | X | X | X |
| 1,1,1-Trichloroethane | | | | X[1] |
| Dioxins | | | | X[1] |
| PCBs | | X | X | |

[1]Substances for which a 50% reduction in atmospheric emissions is also expected to be achieved
[2]Gamma isomer only
[3]Including metabolites DDD and DDE
[4]All triorganotin compounds
[5]And its compounds

**Table 1.3**   *Categories of substances on List II*

*List II contains*

– substances belonging to the families and groups of substances in List I for which the limit values referred to in Article 6 of the Directive have not been determined.
– certain individual substances and categories of substances belonging to the families and groups of substances listed below and which have a deleterious effect on the aquatic environment, which can, however, be confined to a given area and which depend on the characteristics and location of the water into which they are discharged.

*Families and groups of substances referred to*

1 The following metalloids and metals and their compounds:

| | | | |
|---|---|---|---|
| 1) Zinc | 6) Selenium | 11) Tin | 16) Vanadium |
| 2) Copper | 7) Arsenic | 12) Barium | 17) Cobalt |
| 3) Nickel | 8) Antimony | 13) Beryllium | 18) Thallium |
| 4) Chromium | 9) Molybdenum | 14) Boron | 19) Tellurium |
| 5) Lead | 10) Titanium | 15) Uranium | 20) Silver |

2 Biocides and their derivatives not appearing in List I.
3 Substances which have a deleterious effect on the taste and/or smell of the products for human consumption derived from the aquatic environment, and compounds liable to give rise to such substances in water.
4 Toxic or persistent organic compounds of silicon and substances which may give rise to such compounds in water, excluding those which are biologically harmless or are rapidly converted in water into harmless substances.
5 Inorganic compounds of phosphorus and elemental phosphorus.
6 Non-persistent mineral oils and hydrocarbons of petroleum origin.
7 Cyanides, fluorides.
8 Substances which have an adverse effect on the oxygen balance, particularly ammonia, nitrites.

**Table 1.4**   *List II chemicals for which UK National Standards have been set*

Lead
Chromium
Zinc
Copper
Nickel
Arsenic
Boron
Iron*
pH*
Vanadium
Tributyltin compounds
Triphenyltin compounds
Cyfluthrin
Sulcofuron
Flucofuron
Permethrin
Polychlorochloromethylsulfonamidodiphenylether mothproofers

*Neither of these falls within the scope of the substances listed in Table 1.3 but DoE Circular 7/89 requires that they be treated in the same way as List III substances.

### 1.3.2 The North Sea Conferences and Other International Conventions

The North Sea Conferences were set up by governments of countries bordering the North Sea as a forum for agreeing policies aimed at reducing pollution from dangerous substances. At the second conference, in London in 1987, participating countries agreed to reduce inputs of dangerous substances to rivers and estuaries by 50% over the period 1985–1995. In response, the Department of the Environment and the Welsh Office issued a consultation paper entitled 'Inputs of Dangerous Substances to Water: Proposals for a Unified System of Control' in July 1988. This paper set out the Government's proposal for tightening controls over the input of dangerous substances to water. Dangerous substances were defined as those which represented the greatest threat to the aquatic environment due to their persistence, toxicity, their ability to bioaccumulate and their likely presence in the aquatic environment. The aim of the proposals in the consultation paper were stated as:

(i)   To reduce inputs to the aquatic environment of those substances which represent the greatest potential hazards.

(ii)  To improve the scientific basis of the system for identifying the most dangerous substances.

(iii) To develop a more integrated approach to pollution control.

(iv)  In controlling dangerous substances in water, to take full account not only of 'point source' discharges from production plants, but also of 'diffuse sources' of entry into the aquatic environment.

A further important objective was to propose a unified system for controlling the discharge of dangerous substances and reconcile the approach favoured in the UK, the EQO/EQS system, and the approach favoured by other states in the EC, that of uniform emission standards (UES), the more stringent of the two being applied as a control measure in any given case.

With these considerations in mind, the Government set out proposals for a unified approach to controlling the discharge of dangerous substances. The essential points of the proposals were as follows:

(i)   Identification of a limited range of the most dangerous substances selected according to clear scientific criteria, whose discharge to water should be minimized as far as possible – the 'Red List'.

(ii)  The setting of strict environmental quality standards for all Red List substances.

(iii) The introduction of a system of 'scheduling' of industrial processes discharging significant amounts of Red List substances, and the progressive application of technology-based emission standards, based on the concept of Best Available Technology Not Entailing Excessive Costs (BATNEEC).

(iv)  Measures designed, where possible, to reduce inputs of Red List substances from diffuse sources.

The control regime for the Red List chemicals is more stringent than that for Black List substances and marks a change in philosophy by becoming more 'precautionary' and requiring 'BATNEEC' (see Section 1.3.3) for all Red List substances. The proposals also imply that because there is always some uncertainty about the long-term environmental effects of Red List chemicals we should aim to reduce their environmental concentrations as much as is possible – regardless of the evidence that they are causing environmental damage.

After considerable consultation, the DoE published an initial priority Red List of 23 substances which is shown in Table 1.2. The Government has initiated a monitoring programme for these chemicals and the need for specific action will be considered on an individual basis.

Other participating countries produced their own national priority lists of hazardous substances different to those on the UK Red List. At the third conference in the Hague in 1990, the individual priority lists were amalgamated and redefined to produce a common list of 32 substances (see Table 1.2) for which the reduction target of 50% applied. A further target was set at the fourth conference in Esjberg, Denmark, in 1995 based on the objective of ensuring a sustainable, sound and healthy North Sea ecosystem. It was agreed to reduce continuously discharges of hazardous substances thereby moving towards the ultimate aim of reducing concentrations in the environment of natural hazardous substances to near background levels and of synthetic substances to zero within 25 years.

### 1.3.3 Integrated Pollution Control (see also Chapter 20)

In the Environmental Protection Act (1990), the Government introduced a new system, Integrated Pollution Control (IPC). The system was designed to implement the commitment made earlier to use the precautionary approach for control of the most dangerous substances and also moved to a more integrated system by considering all three media (air, water and land) together. IPC applies to emissions from the 'most polluting processes' (defined in Regulations) and is designed to tighten requirements on them by introducing a number of measures:

  – BATNEEC (Best Available Techniques Not Entailing Excessive Costs) must be applied to emissions of 'the most polluting substances' (also defined in Regulations[5] and, for water, virtually identical to the Red List (see Table 1.2) to prevent or, if that is not possible, minimize emissions. Other emissions will be minimized and rendered harmless by the use of BATNEEC (*i.e.* they must meet the EQS and tighter standards should be imposed where technology allows). The use of the term 'techniques' emphasizes that thought must be given to all aspects of the process including the nature of raw materials, the technology of the process, treatment of wastes and training of operators.

[5]'The Environmental Protection (Prescribed Processes and Substances) Regulations SI No. 472', HMSO, London, 1991.

– BPEO (Best Practicable Environmental Option) will have to be identified if emissions are to more than one medium in order to minimize the effect of emissions to the environment as a whole.

The EU Council of Ministers recently agreed the text for a proposal for an Integrated Pollution Prevention and Control (IPPC) Directive (COM(95)88) which will require the implementation of systems similar to that operated in the UK across the whole of Europe. The substances subject to the most stringent controls will, in this case, be the EU priority list, *i.e.* List I substances.

## 1.4  DIFFUSE SOURCES

Pollution from diffuse sources is normally less immediately apparent and more difficult to control than that from point sources. However, for many pollutants diffuse sources are of equal or greater importance, for example pesticides and nitrates from agriculture, urban run-off and deposition from the atmosphere all make significant contributions. The differing nature of the sources means that several approaches are required to control diffuse inputs to water.

### 1.4.1  Product Controls

A major contribution to controlling diffuse sources is the use of 'product controls'. This type of approach shifts the emphasis from 'end of pipe', to focus on the manufacture and use of the chemicals at a stage before they become wastes. The general aims of these controls are either to make the product more environmentally acceptable or to restrict or prohibit the use of certain substances in product formulations.

*1.4.1.1  Dangerous Substances and Preparations.*   Recognizing the need to reduce diffuse inputs of dangerous substances as well as point sources, various Member States introduced measures to totally ban or restrict the use of some substances. For example, the UK restricted the use of benzene in toys and in France, restrictions were introduced for the use of polychlorinated biphenyls (PCBs). Controls at a European level were introduced to prevent distortion in trade through a framework Directive (76/769/EEC) to ban or restrict the marketing and use of certain dangerous substances and preparations. Since the original Directive, restrictions on other substances continue to be introduced through daughter Directives (see Table 1.5). The Directives provide a mechanism to implement a precautionary approach in a quick and consistent manner across the whole of the European Union.

*1.4.1.2  Pesticides.*   The way in which pesticides are released to the environment is largely through their use. This is a diffuse source which is controlled through product registration. Recognition of the need to protect water is a key element in the process and approvals can be reviewed or revoked at any time. The Commission is currently introducing a system to harmonize approval procedures throughout the EU. The Directive concerning the placing of plant protection

**Table 1.5**  *Dangerous substances subject to restrictions on marketing and use (Directive 76/769/EEC and amendments)*

| Substance | Directive |
|---|---|
| PCBs | 76/769/EEC, 85/467/EEC, 89/677/EEC |
| PCTs | 76/769/EEC, 82/828/EEC, 85/467/EEC, 89/677/EEC |
| Chloro-1-ethylene (monomer vinyl chloride) | 76/769/EEC |
| Liquids in the following categories according to Directive 67/548/EEC, highly toxic, toxic, harmful, corrosive, explosive, extremely flammable, highly flammable, flammable and any liquid with a flashpoint <55°C | 79/663/EEC, 89/677/EEC |
| Tris (2,3-dibromopropyl) phosphate | 79/663/EEC |
| Benzene | 82/806/EEC, 89/677/EEC |
| Tris-(aziridinyl)-phosphinoxide | 83/264/EEC |
| PCBs | 83/264/EEC |
| Soap bark powder and its derivatives containing saponines | 83/264/EEC |
| Powder of the roots of Helleborus viridis and Helleborus niger | 83/264/EEC |
| Powder of the roots of Veratrum album and Veratrum nigrum | 83/264/EEC |
| Benzidine and/or its derivatives | 83/264/EEC |
| o-Nitrobenzaldehyde | 83/264/EEC |
| Wood powder | 83/264/EEC |
| Ammonium sulfide or ammonium hydrogen sulfide | 83/264/EEC |
| Ammonium polysulfide | 83/264/EEC |
| Methyl bromoacetate | 83/264/EEC |
| Ethyl bromoacetate | 83/264/EEC |
| Propyl bromoacetate | 83/264/EEC |
| Butyl bromoacetate | 83/264/EEC |
| Asbestos | 83/478/EEC, 85/610/EEC, 91/659/EEC |
| 2-Naphthylamine | 89/677/EEC |
| Benzidine and its salts | 89/677/EEC |
| 4-Nitrophenyl | 89/677/EEC |
| 4-Aminobiphenyl and its salts | 89/677/EEC |
| Neutral anhydrous lead carbonate | 89/677/EEC |
| Lead hydrocarbonate | 89/677/EEC |
| Lead sulfates | 89/677/EEC |
| Mercury | 89/677/EEC |
| Arsenic | 89/677/EEC |
| Organostannic compounds | 89/677/EEC |
| Di-$\mu$-oxo-di-n-butylstanniohydroxyborane | 89/677/EEC |
| Pentachlorophenol | 91/173/EEC |
| Cadmium | 91/338/EEC |
| Monomethyltetrachlorodiphenyl methane | 91/339/EEC |
| Monomethyldichlorodiphenyl methane | 91/339/EEC |
| Monomethyldibromodiphenyl methane | 91/339/EEC |
| Nickel | 94/27/EC |
| Chlorinated solvents | Proposal (COM 1992c) |
| Wood preservatives | Proposal (COM 1992c) |
| Carcinogens | Proposal (COM 1992c) |
| Mutagens | Proposal (COM 1992c) |
| Teratogens | Proposal (COM 1992c) |
| Flammable Substances | Proposal (COM 1993) |
| Hexachloroethane | Draft Proposal expected soon. |

products on the market (91/414/EEC) contains a 'positive' list of active ingredients that may be used in the formulation of plant protection products. Inclusion on the list is dependent upon the active ingredient satisfying a number of conditions, including the assessment of its impact on human health and the environment.

Another Directive (79/117/EEC) restricts the marketing and use of certain pesticides and lists the substances that may not be present in pesticide formulations. The list currently bans several mercury and persistent organochlorine compounds, for example DDT, aldrin, endrin, dieldrin and chlordane, as well as other compounds such as nitrofen and ethylene oxide.

Pesticides that are marketed must conform with various classification, packaging and labelling requirements (defined in Directive 78/631/EEC). Amongst other things, these specify application methods, timing and rates, and disposal methods to ensure that pollution of the environment is avoided. In the UK, those applying pesticides must meet certain training requirements and guidance is given by the Ministry of Agriculture, Fisheries and Food in a Code of Practice for the use and storage of pesticides.[6]

*1.4.1.3 Biocides.* Biocides are a diverse range of chemical additives, commonly used to control growth of microorganisms in many industrial processes. Increasing concerns about possible effects on non-target organisms, including man, has led the Commission to submit a proposal for a Directive to introduce a Community-wide scheme for the Authorization of Biocides (COM(93)440), very similar to that already in operation for pesticides. The proposal seeks to harmonize existing national regulations.

*1.4.1.4 Detergents.* In Western Europe in the 1960s, one of the most visible examples of water pollution was foaming in rivers due to the use of 'hard' detergents. These are poorly biodegradable, both naturally and in sewage treatment works. Foaming is not only an aesthetic problem but also impairs photosynthesis and oxygenation in rivers as well as reducing the operating efficiency of sewage treatment plants.

In an attempt to alleviate the problem, the Community developed the Detergent Directive (73/404/EEC) to restrict the sale of detergents with a poor biodegradability. The Directive banned the marketing of detergents based upon surface active agents (surfactants) with an average biodegradability less than 90%. The Directive applies to all types of surfactant – anionic, nonionic, cationic and amphoteric. More detailed 'daughter' directives stipulate the test methods that should be used to assess whether anionic and nonionic surfactants comply with the requirements.

### 1.4.2 Controlling Land Use

Most land uses or activities, whether industrial, urban or agricultural, have an impact on water quality. Whilst it is clearly impracticable to prohibit all such

---

[6]'Code of Practice for the Safe Use of Pesticides on Farms and Holdings', MAFF, London, 1990.

activities, it is also important to protect and enhance the environment. Increasingly, controls are being introduced to ensure that the environmental impact of land uses are taken into account both for new developments and those already in existence. The general aims of land use controls are to reduce the risk of pollution from priority activities such as the storage of large amounts of hazardous substances by regulating the way in which the activity is carried out (for example, requiring certain storage measures, or controlling fertilizer use). Such controls can be applied on a national basis, as with planning controls in the UK, or may be used to protect particularly sensitive areas or catchments, for example in water pollution zones or nitrate sensitive areas.

*1.4.2.1 Planning.* Planning permission is required for any development of land in the UK under the Town and Country Planning Act 1990 and the Planning and Compensation Act 1991. Through this system, Local Authorities are able to control the location of 'building, engineering, mining or other activities or other operations in, on, over or under land, or the making of any material change in the use of any buildings or other land'. Planning controls provide a primary opportunity for implementing the precautionary principle. They are the key to balancing the need for development whilst protecting the environment thereby achieving 'Sustainable Development' which the UK Government is determined to make the basis of its environmental policies for the future.

Increasingly, planners are required to take into account the environmental impact of proposed developments prior to providing planning permission. A number of procedures and guidance documents have been introduced. Most notably the DoE recently issued a planning policy guidance note on planning and pollution control[7] which is intended to provide guidance on industrial and waste disposal related developments. The National Rivers Authority (NRA) has also published guidance for Local Planning Authorities on methods for protecting the water environment.[8] In addition, the Environment Agency must be consulted for developments which may affect water quality and for certain projects, developers are required to carry out an environmental impact assessment as a part of an application for planning permission.

A new system enabling Local Authorities to control the storage of hazardous substances was introduced in the Planning (Hazardous Substances) Act 1990. The subsequent Regulations[9] list 71 substances, selected on the basis of their explosive or inflammable properties, for which consent will be needed if they are to be stored on a site in quantities above that prescribed in an Annex to the Regulations.

[7]'Planning Policy Guidance: Planning and Pollution Control', PPG23, HMSO, London, 1994.
[8]'Guidance Notes for Local Planning Authorities on the Methods of Protecting the Water Environment through Development Plans', NRA, 1994.
[9]'The Planning (Hazardous Substances) Regulations', S.I. No. 656, HMSO, London, 1992.

*1.4.2.2  Controls on Industrial Sites.*  The Control of Industrial Major Accident Hazards (CIMAH) in the UK (1984)[10] amended in 1990[11] and 1994[12] are concerned with reducing the hazard posed to man and the environment from accidents at industrial plants which involve 'one or more dangerous substances'. The Regulations were introduced as a result of the Seveso Directive (82/501/EC), adopted by the Community after the accident at Seveso in Italy. A proposal to replace this Directive, referred to as COMAH, (COM(94)4) was agreed by the Environment Council of Ministers in June 1995. This has a wider scope than the Seveso Directive and aims to:

- simplify the application system and to encourage more consistent implementation
- require the development of land use policies around the COMAH sites
- provide improved freedom of access to information and encourage greater public participation, and
- ensure an increased emphasis on safety management.

*1.4.2.3  Water Protection Zones.*  For certain areas requiring extra protection, the Secretary of State for the Environment may prohibit or restrict activities carried out by designating water protection zones (WPZs). This approach has proved to be effective at, for example, reducing the pollution of groundwater from diffuse sources in other European countries. In the UK, several voluntary water protection zones have been established and their effectiveness is currently being assessed. Recently the NRA announced proposals to designate the first statutory WPZ around the river Dee. The proposed order would prohibit, without the consent of the Environment Agency, the keeping or use within the catchment area of 'controlled substances' above defined threshold amounts. The 'controlled substances' are listed in Table 1.6.

The Environment Agency would be able to grant protection zone consents, either unconditionally or subject to conditions, likely to be based on BATNEEC, or to refuse consent.

*1.4.2.4  Nitrate Sensitive Areas and Vulnerable Zones.*  The Water Act 1989 introduced powers for the Secretary of State to establish nitrate sensitive areas (NSAs). Farmers in these areas are compensated if they adopt 'environmentally friendly' farming practices that result in a decrease in nitrate application to land. The initial scheme, in which ten areas were designated, is now being extended to a further 22 areas. Further areas, to be termed Nitrate Vulnerable Zones, where farming practices will have to be modified to reduce the inputs of nitrate, will be required in the near future to ensure compliance with the Nitrates Directive (91/692/EEC). Unlike NSAs, landowners in these areas will be bound to conform to

[10]'The Control of Industrial Major Accident Hazards Regulations', S.I. No. 1902, HMSO, London, 1984.
[11]'The Control of Industrial Major Accident Hazards (Amendment) Regulations', S.I. No. 2325, HMSO, London, 1990.
[12]'The Control of Industrial Major Accident Hazards (Amendment) Regulations', S.I. No. 118, HMSO, London, 1994.

**Table 1.6**  *Controlled Substances according to the NRA proposals for the Dee Water Protection Zone Designation Order*[13]

---

'Controlled substance' means any substance which is –

1  a dangerous substance;
2  a fuel, lubricant or industrial spirit or solvent;
3  a medicinal product;
4  food which is a liquid under normal conditions;
5  feeding stuff which is a liquid under normal conditions;
6  an inorganic fertiliser;
7  a cosmetic product;
8  a substance identified by its manufacturer as being toxic, harmful, corrosive or irritant, but does not include:
  – controlled waste;
  – radioactive waste;
  – any fuel, whether kept within a site and used exclusively for the production of heat or power;
  – any substance contained in an exempt pipe-line;
  – any substance at a site for a period of 24 hours or less;
  – any substance which is a gas or vapour under normal conditions.

The minimum quantities subject to control are:

  – in the case of food and feeding stuffs other than defined dangerous substances, an amount in excess of 500 litres;
  – in other cases, an amount equal to or in excess of, 50 litres when the substance is present in a single container but otherwise 200 litres.

---

specified agricultural practices and are unlikely to receive compensation payments.

*1.4.2.5  Groundwater Protection.*   In order to control groundwater contamination outside of areas designated as WPZs or NSAs, the NRA published a Groundwater Policy[14] as guidance for use by planning and waste regulatory authorities. Two of the central principles of the policy are the classification of groundwaters according to their vulnerability and the definition of three-tier 'Source Protection Zones' which will form the basis for controls on activities posing a potential threat to groundwater.

## 1.5  ALTERNATIVE CONTROL PROCEDURES

Regulation in the sense of allowing something to happen – a discharge to occur, a landfill site to be established – has been recognized as being a rather blunt instrument with which to achieve continuous improvement. Specifically in relation to priority pollutant control it has been difficult to find ways of encouraging replacements of priority chemicals by other, more benign substitutes. It has also been recognized that if the regulation can work with the 'grain'

---

[13]'Proposed Water Protection Zone (River Dee Catchment) Designation Order', NRA Welsh Region, 1994.
[14]'Policy and Practice for the Protection of Groundwater', NRA, 1992.

of the market, *i.e.* the normal market forces – supply and demand, consumer choice *etc.* – it will be more effective. Many of the newer regulatory initiatives can be considered as market-based instruments. They are roughly aggregated together into the subsections titled Regulations, Information and Initiatives.

### 1.5.1 Regulations

The introduction of an EC Ecolabelling scheme was agreed by the Council of Ministers in 1991. It was designed to enable consumers to select goods and services based on their environmental impact, the assumption being that given a choice consumers will select environmentally friendly goods which will encourage manufacturers in this direction. The UK Regulations came into force in November 1992. They created an Ecolabelling Board which can award an Ecolabel to a product, which thereby alerts consumers to the more environmentally friendly alternatives. The Ecolabel criteria are still being established, but priority pollutants are considered as a significant component, for example the German Federal Environmental Agency is advocating a negative list of substances for detergents. More specifically 'only traces' of mercury will be permitted in light bulbs awarded an Ecolabel.

The Eco-Management and Audit Scheme (EMAS) has a similar objective to Ecolabelling. EMAS was set up by the EC under Regulation 1836/93/EC 'allowing voluntary participation by companies to a Community Eco-Management and Audit Scheme'. Like the Ecolabelling scheme, EMAS is voluntary and is also providing a means by which consumers can choose to buy from, or deal with, industries that have better environmental performance. EMAS applies to sites, not companies, and in order to comply with the Scheme the following environmental components must be verified by an independent audit of each site: an environmental policy, programme, management system, review, audit procedure and statement.

In the UK there are few current examples of economic instruments being applied to environmental regulation of priority substances in water. The case of differential tax on unleaded and leaded petrol is a good example applied to priority pollutants in the atmosphere. The NRA in England and Wales and River Protection Boards in Scotland (now Environment Agency and Scottish Environmental Protection Agency respectively) are only allowed to use charges for discharge consents to cover their administration costs. So the scope for using consent charges as incentives is limited. However, the analytical costs associated with priority substances in effluents can be significant and the differential between charges for effluents containing these substances and effluents not containing them can be seen as a financial incentive, albeit possibly not substantial enough to cause a company to modify its process.

### 1.5.2 Information

Most activities we have grouped together in this section relate to total loads of priority pollutants. HMIP recently started a Chemical Release Inventory which will provide an annual estimate of releases of priority substances from Prescribed Processes. As explained in Section 1.3.2 the UK committed itself at the North Sea Conference in 1987 to reduce loads of dangerous substances from rivers and estuaries to the North Sea by around 50% by 1995, based on inputs for 1985. Its performance in relation to these targets were published at the 1995 North Sea Conference.[15]

Individual companies have also initiated schemes to provide annual reports on their environmental performance, alongside those relating to their financial performance. In some cases these reports show aggregate releases of priority substances and comparisons with previous years' releases and future targets.

All these activities help to provide pressure from public, shareholders and pressure groups to reduce discharges of dangerous substances at both a local and national level.

### 1.5.3 Initiatives

The activities dealt with in this section are concerned with research or technology transfer and most have been initiated by government departments, but also require industrial input and contribution. The initiatives are concerned with environmental protection in general, but often have priority chemicals as one of the focus areas.

A report in 1991[16] identified that industry is, in general, unaware of the benefits that could result from a systematic approach to reduction of emissions and the introduction of cleaner technology. A number of case study projects have now been undertaken which show the dual benefits of cost reduction in industry and pollution load reduction for the environment. In some cases the savings are quite staggering, for example in the Aire and Calder study 51 opportunities to reduce waste were identified, resulting in estimated savings of over £400 000.[1] Over 68% had payback periods of less than twelve months and over 89% of less than two years. In order to encourage technology transfer DTI and DoE launched the Environmental Technology Best Practice Programme in 1994. The initial objectives relate to waste minimization and cleaner technology, both of which can be expected to impact on priority pollutants in water.

## 1.6  CASE STUDIES

Certain chemicals or groups of chemicals have been chosen in this section to illustrate the wide range of problems which arise in considering the effects of chemicals on the environment. The examples illustrate the varying sources of pollutants, their fate and analysis and factors which can influence what type of control procedures have been or may need to be adopted.

[15]'Water Quality Series 24', HMSO, London, 1995.
[16]Centre for Exploitation of Science and Technology, 1991.

### 1.6.1  Polybrominated Diphenyl Ethers (PBDEs)

The general chemical formula of brominated diphenyl ethers is:

A large number of congeners exist (209), but the main interest is in the production of penta-, octa- and deca- varieties, used commercially as flame retardants. In this respect they are very effective and European consumption of PBDEs has increased considerably over the last ten years. For example, it has been estimated that production and imported quantities of PBDE amounted to around 8500 tonnes in 1986, rising to around 11 000 in 1991.[17] As the use of one of the major alternative fire retardants (polybrominated biphenyls, PBBs) has been reduced, the use of PBDEs has increased and worldwide production was estimated at 40 000 tonnes in 1994.[18]

In general terms PBDEs are not considered to be highly toxic although there are few data available; they exhibit low acute oral toxicity to rats, and there is no evidence that they give rise to mutations. Recent data from Sweden has, however, indicated that PBDEs can be detected in many fish species and their predators, such as the osprey and grey seal.[18] There is some evidence of carcinogenicity in rats, but they are not thought to present a carcinogenic risk to humans.[17]

There are various potential sources of release and exposure to PBDEs, although release from products is considered to be low. Production sites appear to be a source of environmental exposure with release possible in the form of dust or particles. No studies are available on the fate of PBDE-containing products in landfill and it is possible that leaching in the long-term may occur. Few data are available on the fate of PBDEs in the environment although some studies have been carried out on the commercially produced isomers. Data suggest that their solubility in water is low, that the higher brominated congeners are persistent and that they are likely to bind to particles and sediment (log $K_{ow}$ values ranging from around 4.3 to 10).[17] The presence of PBDE congeners in the environment has been reported, with highest values found in some sediments near to production sites[17] (up to 1 g $kg^{-1}$ has been reported although more typical levels appear to be below 100 $\mu g$ $kg^{-1}$ away from production sites). Concentrations in aqueous samples are much lower and the higher brominated congeners are frequently undetectable. PBDEs have, however, been regularly detected at low concentrations in the fatty tissue of a wide range of aquatic life, animals and humans. The analysis of PBDEs in

[17]'Environmental Health Criteria, 162', International Programme on Chemical Safety, WHO, Geneva, 1994.
[18]KEMI, The Swedish National Chemicals Inspectorate, Solna, Sweden, 1994.

environmental samples is relatively straightforward and several methods based on solvent extraction, clean-up and separation/detection using GLC with electron capture detection or GC-MS have been published. The main concern behind the proposals to restrict the use of PBDEs is, in fact, the possibility that they may contain, or form brominated dioxins and furans during combustion. Numerous studies have been conducted on this and the results vary according to the products tested, temperatures applied and other conditions such as the presence of oxygen or catalyst (see also Chapter 17).

The way in which PBDEs are controlled as environmental contaminants, presents regulators (and ultimately society itself) with a difficult dilemma. Concern for fire safety is a high priority and each year many people die directly from burns or indirectly through inhalation of fumes. There is no question that the use of fire retardants can, and does, save lives. Many fire regulations actually specify a certain level of fire resistance which currently could not be met for many materials without the use of a fire retardant.

As mentioned previously, PBBs are now being phased out of use due to concern over their persistence in the environment and their toxicity to a range of animals and possibly man. Questions have now been raised over the use of PBDEs, which leaves the materials/products industry short of obvious alternative products. Some alternatives are available, *e.g.* tetrabromobisphenol A, compounds based on phosphate or phosphate and chlorine and compounds of zirconium and aluminium. However these cannot replace PBDEs for all applications so there remains the possibility that restrictions on the use of PBDEs would lead to an *increase* in the number of fires and hence human injury/death as a result. While it is recognized that brominated dioxins and furans may be emitted when treated textiles and fabrics are burned, the available evidence suggests that the quantities produced are low in comparison with total dioxin emission. There has been considerable discussion about the relative toxicity of dioxins and furans but, in general, governments tend to take a precautionary approach, reducing release and exposure wherever possible.

The need for, and nature of, any controls over PBDEs therefore require a careful balance of the risks posed by (a) the environmental hazards of PBDEs (particularly as a source of dioxins) and (b) the hazards presented by using less efficient flame retarding substances. In simple terms the lives saved by using PBDE fire retardants has to be weighed against the effects from exposure to PBDEs and/or dioxins. We currently do not have either the data (*e.g.* on toxicity or the hazards presented from alternative fire retardants) or indeed the techniques available to carry out such a risk assessment. Perhaps the most likely form of control will be voluntary agreements amongst manufacturers in introducing a range of risk reduction measures, *e.g.* during manufacture of the chemical and its products particularly in minimizing the production of unwanted by-products, and controls over disposal. In fact, the Organization for Economic Cooperation and Development (OECD) has recently accepted (June 1995) a voluntary agreement of this nature put forward by the main producers.

**1.6.2   Oestrogenic Chemicals** (see also Chapter 18)

Chemicals which possess oestrogenic activity (*i.e.* are able to mimic the action of, or inhibit, hormones such as oestrogen) form an extremely wide and diverse group. Chemicals in this category include chlorinated pesticides such as DDT, detergents such as alkylphenol ethoxylates and their breakdown products, plasticizers such as phthalates, and dioxins. These chemicals, in fact, display very weak oestrogenic activity compared to the hormones themselves; however, there is growing concern in the medical/scientific community that this activity is giving rise to adverse effects on the reproductive health of humans and wildlife.[19–22]

There is evidence of a decrease in the quality of human sperm over the last 40 years and there is good evidence for an increase in rates of testicular cancer and other male reproductive abnormalities.[19,20] A wide range of evidence is also available linking exposure from chemicals with oestrogenic activity to effects on the reproduction of wildlife. This ranges from studies on alligators in a lake in Florida exposed to the organochlorine insecticide dicofol, studies on gulls in California having high tissue levels of DDT, to the effects of tributyltin on a range of molluscs.[19] In the UK, recent research has shown that male fish exposed to sewage effluents can undergo changes in their biochemistry and produce an egg yolk protein, vitellogenin, normally found only in fertile females.[22]

A wide range of industrial chemicals have been implicated as giving rise to, or at least potentially contributing to, these effects. Many of these chemicals are already banned or restricted (*e.g.* DDT, tributyltin, PCBs) whilst others (*e.g.* nonyl phenol, various phthalates) are the subject of risk assessments *via* the EC Existing Chemicals Regulations. Particular attention has been paid recently to nonylphenol ethoxylates, largely because they are in widespread use as industrial cleaning agents or detergents. Annual UK consumption was estimated to be at about 18 000 tonnes in 1993.[23] Nonylphenol ethoxylates are the most common products from a group of similar chemicals (alkylphenol ethoxylates) manufactured from alkylphenols and ethylene oxide. They are invariably produced as a complex mixture of chemicals with a varying number of ethylene oxide units. This has led to difficulties with their analysis in environmental samples but methods are now available based on either HPLC or GLC coupled with various detectors including mass spectrometry. The ethoxylates degrade to the parent phenol during waste water treatment and in the environment, and both chemicals can be detected in the environment. For example they have been detected in both sewage effluents and receiving waters at relatively high concentrations (approx. 40–400 $\mu$g l$^{-1}$);[24] at present it is not clear whether such

[19]'Chemically Induced Alterations in Sexual and Functional Development: The Wildlife/Human Connection', eds. T. Colborn and C. Clement, Princeton Scientific Publishing Co., New Jersey, 1992.

[20]M. J. Wilkinson, I. Milne, R. Mascarenhas and J. Fawell, 3/94, WRc, 1994.

[21]Institute of Environmental Health, University of Leicester, 1995.

[22]S. Jobling and J. P. Sampter, *Aquat. Toxicol.*, 1993, **27**, 361.

[23]ENDS Report, 1993, **222**, 9.

[24]S. S. Talmage, 'Environmental and Human Safety of Major Surfactants', Lewis Publishers, Michigan, USA, 1994.

concentrations are within the oestrogenically active range. More research is required to provide evidence of a link between observed activity and measured water concentrations.

It is becoming increasingly likely that if the observed changes in human and animal reproduction are established to be caused primarily by environmental contamination, they are not due to a single chemical or exposure to a single source but to a range of chemicals and exposure routes. This clearly gives rise to problems for the control of chemicals having oestrogenic activity. What is certainly needed (and this is the subject of much research worldwide) is a 'test' for oestrogenic activity to assess the potential for damage to reproductive systems in both humans and animals. However, before a meaningful test can be developed and implemented more research needs to be carried out on the link between reproductive health and exposure to environmental oestrogens. In fact the recent 5th North Sea Conference (June 1995) called for internationally co-ordinated research on this subject.

The question arises as to what controls, if any, should be placed on chemicals with oestrogenic activity in the meantime. Environmental pressure groups argue that we cannot afford to wait for research and that chemicals shown to have oestrogenic activity should be phased out or banned now. This is already happening to a certain extent and most industrialized countries have, for example, endorsed an international agreement under PARCOM to phase out alkylphenol ethoxylates in domestic and industrial cleaning uses by the year 2000. The Soap and Detergent Industries Association has also announced that its members will phase out these chemicals by 1997.

There will almost certainly be increasing pressure on both governments and industry to phase out suspect chemicals, and more widespread product and other controls may need to be implemented in the future as and when chemicals with oestrogenic activity are identified.

### 1.6.3 Pesticides

Pesticides are designed to control living organisms, thus they can be expected to have an environmental impact and many examples of effects can be cited. However most of these were due to accidents or deliberate misuse and there is little evidence to suggest that when used properly pesticides cause any adverse effects in the aquatic environment. Recent studies in the UK and other countries have demonstrated, however, that low concentrations of a wide range of pesticides can be detected in the aquatic environment.

The EC Directive Relating to Water intended for Human Consumption (80/778/EEC) stipulated a maximum admissible concentration (MAC) of $0.1 \mu g \, l^{-1}$ for individual pesticides in drinking water, regardless of toxicity. Since some pesticides are not efficiently removed by conventional drinking water treatment, $0.1 \mu g \, l^{-1}$ is now one level of concern for pesticides in the aquatic environment in general. A further level of concern is that needed to protect the aquatic environment, although for most pesticides $0.1 \mu g \, l^{-1}$ tends to be the lower value.

Much more information is now available on pesticide concentrations in water than was the case a few years ago. Water companies regularly monitor drinking water for a range of pesticides (based on those used within the catchment concerned) and many companies also carry out routine monitoring of their raw water sources. In addition, the Environment Agency monitors many surface waters for a range of pesticides. The NRA (now the Environment Agency) recently issued a report on contaminant levels entering the seas around England and Wales, which includes data on pesticides in surface waters.[15] The Working Party on the Incidence of Pesticides in Water (WPIPW), set up in 1991 by the Department of the Environment to review the arrangements for the collection, evaluation and dissemination of information from the monitoring of pesticides in water in the UK, will be publishing its first report in late 1995. Thus a large amount of data are now emerging on pesticide levels in the environment. In addition to regular monitoring, the Environment Agency recently commissioned the development of a modelling system for the prediction of pesticide pollution in the environment (POPPIE). This model will predict ground and surface water quality concentrations, assess leaching/run-off potential for new pesticides and will be used to develop monitoring programmes and identify potential problem pesticides.

Under the Control of Pollution Act (COPA) 1974, Water Authorities, as pollution control authorities, had powers to set up areas in order to protect water resources and these powers could be applied to pesticide usage. Under the provisions of the Water Act 1989, these powers were transferred to the National Rivers Authority and now to the Environment Agency. A variety of actions are possible under these powers; for example pesticide users could be advised, wherever possible, to reduce pesticide usage in specified, sensitive areas. This action would probably have little effect on the agricultural use of pesticides but might reduce non-agricultural use, for example by Local Authorities, British Rail, golf clubs, *etc.* More formal restrictions on pesticide use could be achieved by setting up water protection zones (WPZs), for example to protect water sources used for drinking water supplies. However it might be difficult to identify such zones and the resource costs for policing and monitoring pesticide use in a WPZ would be high. Control of pesticide inputs to water bodies requires a co-ordinated approach. This has to involve liaison and co-operation between regulators and also educating pesticide users to adopt 'best practice' to minimize pesticide pollution of water.

It is impossible to assess, at the moment, how long it would take for any pollution control measures to reduce the concentration of pesticides currently in the aquatic environment, particularly in groundwaters. In the future, the registration requirements for new pesticides should prevent widespread contamination of the environment.

### 1.6.4 Mercury

Mercury is a metal, liquid at normal temperatures and pressures, which forms salts in two ionic states: mercury(I) and mercury(II). Mercuric salts [mercury (II)] are the more prevalent of the two. Mercury also forms a range of relatively stable organometallic compounds.

Natural degasing of the Earth's crust releases considerable quantities of mercury into the environment and this is probably the major source. Estimates of the load vary between 25 000–125 000 tonnes,[25] although a more recent estimate was much lower, and in the range 2700–6000 tonnes.[26] In the same study, anthropogenic release was estimated at 630–2000 tonnes. World production of mercury, from mining and smelting operations, was estimated at around 10 000 tonnes in 1973, with industries such as chloralkali, electrical and paint production the largest users. The use of mercury is now declining although it was widespread as a cathode in the production of chlorine and caustic soda, in electrical appliances, control instruments and dental amalgams. Current uses of mercury are predominantly in the production of chlorine and in the extraction of gold from ore, a use which has been estimated to release 50–70 tonnes per annum to the environment.[26] Mercury is still also used in dental amalgams – estimated at around 10–20 tonnes annually.[27] Organomercury compounds have been used as fungicides, antiseptics, preservatives, electrodes and reagents.

Previous industrial use has led to the release of significant amounts to the environment, *e.g.* from the chloralkali industry, wood pulping and burning fossil fuels. Elemental mercury is virtually insoluble in water, while the solubility of mercury compounds varies widely. It is known that inorganic mercury can be methylated in the environment by bacterial action under aerobic and anaerobic conditions.[25] Methyl mercury is released from the bacteria, rapidly becomes bound to proteins in aquatic biota and enters the food chain.

Some sediments close to areas of production are heavily contaminated and pollution of water from old mines is still a problem. General environmental levels thus vary considerably depending on the location. Atmospheric levels of mercury range from 2–10 ng m$^{-3}$, while the concentration in rainwater can vary from 5–100 ng l$^{-1}$. A large number of data are available on mercury levels in the aquatic environment; background levels of around 1–15 ng l$^{-1}$ are considered representative, but levels as high as 5 $\mu$g l$^{-1}$ have been reported in contaminated areas. Fewer data are available on concentrations of methyl mercury and other organomercury compounds. The analysis of inorganic mercury is straightforward either by flameless atomic absorption spectrometry or atomic fluorescence spectrometry. Alkyl mercury compounds are normally determined by GLC.[25]

The metal is toxic at low concentrations to a wide range of organisms, including humans. The organic form of mercury can be particularly toxic, and the methyl- and ethyl- forms have been the cause of several major epidemics of poisoning in humans resulting from the ingestion of contaminated food, *e.g.* fish. Two major epidemics in Japan were caused by the release of methyl and other mercury compounds from an industrial site, followed by accumulation of the chemicals in edible fish. These incidents gave rise to the effects of the poisoning becoming known as Minimata disease – following the name of the bay into which the discharges of mercury were released. The use of organomercury fungicides as

[25]'Environmental Health Criteria 86', International Programme on Chemical Safety, WHO, Geneva, 1993.
[26]W. C. Pfeiffer and L. Drude de Lacorda, *Environ. Health Lett*, 1988, **9**, 4, 325.
[27]M. Hutton and C. Symon, *Sci. Total Environ.*, 1986, **57**, 129.

seed dressing caused the death of many seed-eating birds in Europe, with many raptors also dying from eating the corpses.[25]

Mercury is heavily regulated, being covered by a wide range of legislation. In the EC it is designated as being of List I status under the Dangerous Substances Directive, with corresponding limit values set at a Community level, and the marketing and use of mercury and some of its compounds are covered under the 8th Amendment to the Marketing and Use Directives. In the UK, mercury and its compounds are on the Red List and classified as prescribed substances and are therefore subject to strict control with respect to industrial processes and discharge to sewer, *etc.* Nevertheless contamination of surface waters and other environmental compartments by mercury is still considered to be a problem, with attention being drawn recently to the accumulation of mercury in fish in acid upland waters. This highlights the great difficulty in controlling a chemical whose input to the environment is now largely from diffuse sources. By its very nature, as a relatively volatile metal, it can be released into the environment from many of man's activities not related directly to mercury use, *e.g.* burning fossil fuels, steel, cement and phosphate production and metal smelting. Atmospheric transport of elemental mercury (and dimethyl mercury) is thought to be a significant process and can give rise to environmental contamination far from its original source.

Future controls on mercury will therefore have to concentrate on the many relatively minor uses/sources of this chemical. For example, a considerable amount of mercury resides and is used in dental amalgams, control instruments (barometers, *etc.*) and laboratory apparatus (thermometers, *etc.*). An inventory of mercury-containing equipment followed by controlled withdrawal and disposal, and the use of mercury traps are measures not being adopted. In Germany and the Netherlands the installation of such traps by dentists is now required by law and The Paris Commission (PARCOM) have recently recommended that traps should be installed by the year 1997.

## 1.7  CONCLUSIONS

The difference between a 'priority' pollutant and a 'non-priority' pollutant, in terms of environmental impact, is not distinct. The distinction is also not without problems in relation to overall environmental improvement, because attention tends to be drawn to priority pollutants even when non-priority substances are causing the most damage. As indicated at the start of this chapter, any chemical can become a pollutant and a move to a risk-based approach to the assessment and control of chemicals seems an inevitable and desirable development. The relative risks between the beneficial effects of 'use' of a chemical and the potential for pollution are becoming less clear as the more obvious priority pollutants are brought under control. This is well illustrated by the brominated diphenyl ethers (Section 1.6.1) where the clear,

[28]U. Sellstrom, B. Jansson, A. Kierkegaard, C. de Witt, T. Odsjo and M. Olsson, *Chemosphere*, 1993, **26**, 1703.

undoubted benefits of fire retardancy have to be balanced against poorly characterized concerns about long-term toxicity. At the same time, our concerns about the potential effects of environmental contaminants are focusing on the longer term and, in particular, trying at an early stage to identify more subtle impacts, for example that represented by oestrogens (Section 1.6.2). Such effects are often difficult to detect but, nevertheless, are possibly more sinister than fish deaths or foaming rivers which provided the driving force for priority pollutant control a few decades ago.

CHAPTER 2

# Chemistry and Pollution of the Marine Environment

M. R. PRESTON and R. CHESTER

## 2.1 INTRODUCTION

The World Ocean acts as a giant 'chemical system' operating within the global biogeochemical cycles that control both the movement and fate of material on the planet. Within these cycles, the importance of the World Ocean is heightened because it acts as a large-scale 'dumping ground' for material which originates in other geospheres.

Material reaches the oceans following a series of source ↔ sink pathways which, according to Chester,[1] may be considered to consist of a number of individual, but inter-related, stages. These are: (i) Stage 1 (source), the initial release of material into the environment, (ii) Stage 2 (transport), the introduction of material to the ocean reservoir, (iii) Stage 3 (internal reactivity), the biogeochemical processes which operate within the ocean reservoir and (iv) Stage 4 (sink), the removal of material from the ocean reservoir.

Much of the material involved in this overall source ↔ sink pathway has a natural origin, but, in addition, material generated in association with man's activities (anthropogenic origin) can have an important effect on oceanic cycles. However, it is necessary to distinguish between two different types of anthropogenic material:

Type I – naturally occurring materials, for which the 'anthropogenic effect' arises only when man's activities result in its excessive release into the environment (*e.g.* by mining, smelting, waste incineration, sewage disposal, *etc.*).

Type II – non-naturally occurring material which has been created in the laboratory or industrial plant, and which is subsequently released into the environment.

[1] R. Chester, 'Marine Geochemistry', Chapman and Hall, London, 1990, 698 pp.

Both types of anthropogenic material can impinge on marine biogeochemical cycles, and once they have been released into the environment they follow the source ↔ sink pathway identified above. It follows, therefore, that a detailed understanding of the impact of anthropogenic material on marine systems can only be obtained from: (i) a knowledge of the natural processes which occur following the release of material into the environment and (ii) an awareness of the types of environmental stresses that are superimposed on the natural processes by the presence of the anthropogenic material itself.

The present chapter is intended to provide an overview of the most important features of the interactions between potential pollutants and the marine system, so that the relationships between natural processes and pollutant-driven environmental changes can be evaluated. In the treatment adopted, some general features of the source ↔ sink flux pathways in the World Ocean are identified. Following this, pollutants are divided into a number of categories, each of which is treated individually.

## 2.2  GENERAL FEATURES OF THE OCEANIC ENVIRONMENT

The World Ocean, including coastal and marginal seas, covers almost three quarters of the Earth's surface ($361110 \times 10^3$ km$^2$), and has an average depth of $\sim 3800$ m.

Four principal parameters must be considered when attempts are made to describe the distribution of material in the oceans. These are: (i) source terms, by which material is delivered to the ocean reservoir; (ii) circulation patterns, which govern the transport of material within the reservoir; (iii) biogeochemical processes, which govern the reactivity of material in the reservoir and (iv) sink terms, by which the material is removed from the reservoir. Some of the complexities of the system as they apply to coastal regions are shown in Figure 2.1.

### 2.2.1  Source Terms

The ocean reservoir is continuously subjected to series of material fluxes which are delivered *via* a number of transport pathways. With respect to the input of pollutant material, river run-off, atmospheric deposition and a number of anthropogenic pathways are the dominant natural input mechanisms.

(i)   River run-off delivers both particulate and dissolved material to the surface ocean at the land/sea margins with discharge usually being *via* estuaries; *i.e.* regions where fresh and saline waters mix. Estuaries are regions of intense physical, chemical and biological activity and the river flux undergoes considerable modification as it passes through them.

(ii)  Atmospheric deposition delivers both particulate material ('dry' deposition) and a combination of dissolved and particulate material ('wet' deposition, or precipitation scavenging) to the whole ocean surface; *i.e.* atmospheric inputs are not confined to the land/sea margins, and so need

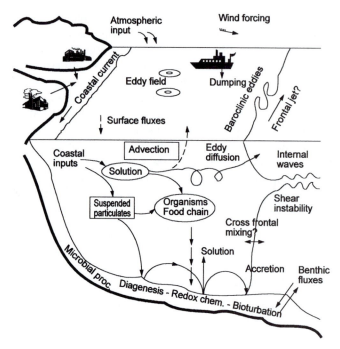

**Figure 2.1**   *Some of the processes determining pollutant behaviour in the coastal zone*[2]

not pass through the estuarine environment. However, the strength of the atmospheric signal is strongest in coastal regions closest to the continental material sources.

(iii) Anthropogenic pathways by which pollutants are delivered to the oceans include: (i) dumping (*e.g.* for the disposal of sewage sludge, radioactive waste, dredge spoil, military hardware, off-shore structures), (ii) deliberate coastal discharges (*e.g.* coastal pipeline discharges from power stations, industrial plants, oil refineries, radioactive reprocessing plants), (iii) off-shore operational discharges from tankers or other ships, and (iv) accidental discharges: *e.g.* from an oil-tanker wreck.

### 2.2.2   Circulation Patterns

Once material is delivered to the oceans it is subjected to transport *via* the marine circulation systems. On a global-scale, circulation in the surface ocean is mainly wind-driven, whereas in the deep-ocean it is gravity-driven. However, the strengths of both the river run-off and the atmospheric deposition signals bringing pollutant material to the oceans are strongest in coastal and marginal

[2]J. H. Simpson, in 'Understanding the North Sea System', eds. H. Charnock, K. R. Dyer, J. M. Huthnance, P. S. Liss, J. H. Simpson and P. B. Tett, The Royal Society, Chapman and Hall, London, 1994, pp. 1–4.

seas, and here the water circulation patterns are constrained by local influences.

### 2.2.3 Sea Water Reactivity; Biogeochemical Processes

Sea water is not simply a static reservoir in which the material supplied to it has accumulated over geological time. Thus, rather than being thought of as an 'accumulator', sea water should be regarded as a 'reactor' from which material is continually being removed on time scales (residence times) which vary considerably from one element to another.

The driving force behind the removal of elements from sea water to their marine sink, mainly sediments, is particulate ↔ dissolved reactivity. Vertical water column profiles of elements can provide data on the type of particulate ↔ dissolved reactivity that affects them in the oceans, and on this basis the elements in sea water can be divided into three principal types.

(i) Conservative-type unreactive major elements. These exhibit largely invariant concentration profiles down the water column, and have relatively large oceanic residence times (usually $> \sim 10^6$ yr). Conservative-type elements are mainly the salinity-contributing elements, such as sodium, potassium and chlorine.

(ii) Scavenging-type reactive trace elements. These are usually trace elements which are involved in passive particle scavenging throughout the water column and are not recycled, which results in a surface enrichment–subsurface depletion down-column concentration profile. Scavenging-type trace elements have oceanic residence times which can be as low as a few hundred years, and include Al, Mn and Pb.

(iii) Nutrient-type reactive trace elements. These are involved in active biological removal mechanisms. They exhibit surface depletion–subsurface enrichment, and as a result of their involvement with biota in the major oceanic biological cycles, they are involved in a major recycling stage similar to that exhibited by nutrients. Nutrient-type elements have residence times in the order of thousands of years.

Pollutants delivered to the oceans enter the circulation-driven transport and biogeochemically-driven removal processes, and, unless degraded, are finally deposited in the sediment sink. Thus, pollution can affect the water, the biota and the sediment compartments of the ocean reservoir and tends to have its greatest impact in coastal and marginal seas.

A wide range of pollutant, and potentially pollutant, substances can affect the marine environment, and the individual pollutants described in this chapter have been selected to cover examples of a variety of substances having different effects on the marine environment. The pollutants selected in this way are oil, sewage/nutrients, persistent organic compounds, trace metals and artificial radionuclides. Each type of pollutant is treated individually below.

## 2.3 SOURCES, MOVEMENTS AND BEHAVIOUR OF INDIVIDUAL POLLUTANTS OR CLASSES OF POLLUTANT

### 2.3.1 Oil

The world demand for oil is enormous, at a level of $> 3200 \times 10^6$ t a$^{-1}$, which represents around 39% of the world's commercial energy demands.[3] Production of this oil is concentrated in the Middle East, which accounts for 28.6% of world production and 66% of proven reserves. Much of this oil is moved from its point of production to its purchasers in tankers which are frequently of very considerable size. Over one-third of the present tanker fleet (between 400 and 460 vessels) has a capacity $> 250\,000$ t and over 60% has a capacity $> 100\,000$ t. These tankers are too large to follow the older trade routes (*e.g.* through the Suez Canal) and so, in the case of tankers heading towards western Europe, follow the African coast from the Middle East, around the Cape of Good Hope up through the Atlantic. In most cases the tankers follow a minimum depth contour (*e.g.* 200 m) which frequently takes them relatively close to coastlines. It is therefore not surprising that most tanker accidents result in widespread coastal pollution which subsequently receives a great deal of public attention.

*2.3.1.1 The Composition of Crude Oil.* Crude oil is a very complex mixture of many different chemicals. Consequently the effects of an oil spill on the marine environment depend on the exact nature and quantity of the oil spilled, as well as such other factors as the prevailing weather conditions and the ecological characteristics of the affected region.[4] An indication of the physical-chemical properties of the major components is shown in Table 2.1.

**Table 2.1**  *Typical physical-chemical properties for hydrocarbon groups*[5]

| Hydrocarbon | Molecular weight (approx.) | Solubility (g m$^{-3}$) | Vapour pressure (Pa) | Density (kg m$^{-3}$) | Oil–water partition coeff. | Solubility (g m$^{-3}$) |
|---|---|---|---|---|---|---|
| Lower alkanes (C$_3$–C$_7$) | 72 | 40 | 70 000 | 800 | 20 000 | 6 |
| Higher alkanes (> C$_8$) | 120 | 0.8 | 2000 | 800 | 1 000 000 | 0.08 |
| Benzenes | 100 | 200 | 1500 | 800 | 4000 | 20 |
| Naphthenes | 160 | 20 | 5 | 800 | 40 000 | 0.4 |
| Higher polycyclics | 200 | 0.1 | 0.003 | 800 | 8 000 000 | 0.002 |
| Residues | — | 0 | 0 | 800 | ∞ | 0 |

[3]Institute of Petroleum, Petroleum Statistics 1993, Institute of Petroleum, London, 1993, 2 pp.
[4]GESAMP, Rep. Stud., GESAMP No. 50, 1993, 180 pp.
[5]J. W. Doerffer, 'Oil Spill Response in the Marine Environment', Pergamon Press, Oxford, 1992, 391 pp.

*2.3.1.2 Fluxes of Oil to the Marine Environment.* The fact that there are many direct and indirect ways in which oil can reach the marine environment means that it is very difficult to construct reliable estimates of the total amounts of oil reaching the seas each year. Recent estimates range between 1.7 and 8.8 × $10^6$ t $a^{-1}$, with a best estimate of $\sim 3.25 \times 10^6$ t $a^{-1}$. However, improvements in tanker operations have reportedly reduced contributions from this source from around 1.47 × $10^6$ t $a^{-1}$ in 1981 to 0.0568 × $10^6$ t $a^{-1}$ in 1989.[6] The contributions from different sources are given in Table 2.2.

It is important to note that there are considerable natural sources of both crude oil and other hydrocarbons to the marine environment. Natural oil seeps are

**Table 2.2**  *Sources of oil to the marine environment*[7]

| Source | Estimate (m ta$^{-1}$) | Total (m ta$^{-1}$) |
|---|---|---|
| *Transportation* | | |
| Tanker operations | 0.7 | |
| Tanker accidents | 0.42 | |
| Bilge and fuel oils | 0.3 | |
| Dry docking | 0.03 | |
| Non-tanker accidents | 0.02 | |
| | Sub-total | 1.47 |
| *[Revised 1991 total]*[6] | | [0.568] |
| *Fixed installations* | | |
| Coastal refineries | 0.1 | |
| Offshore production | 0.05 | |
| Marine terminals | 0.02 | |
| | Sub-total | 0.17 |
| *Other sources* | | |
| Municipal wastes | 0.70 | |
| Industrial waste | 0.20 | |
| Urban run-off | 0.12 | |
| River run-off | 0.04 | |
| Atmospheric fallout of oil-derived compounds | 0.30 | |
| Ocean dumping | 0.02 | |
| | Sub-total | 1.38 |
| *Natural inputs* | 0.25 | 0.25 |
| Total | | 3.27 |
| | | [2.368] |
| *Natural sources/fluxes of organic carbon** | | |
| Primary production by marine phytoplankton | 36 000 | |
| Atmospheric fallout of naturally derived species | 220 | |

*from Chester (1990)[1]

[6]P. Ambrose, *Mar. Poll. Bull.*, **22**, 1991, 262.
[7]NAS, 'Oil in the Sea, Inputs, Fates and Effects', National Academy of Sciences, Washington, DC, 1985.

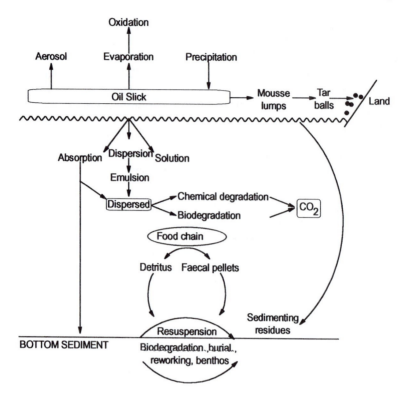

**Figure 2.2**   *The fate of a marine oil slick*

known to have existed for thousands of years, particularly in regions such as the
Persian Gulf or Gulf of Mexico. This has had the result that a variety of marine
organisms (particularly bacteria) have evolved which can feed on and degrade
crude oil.[8] Oil is therefore a Type I pollutant, unlike the chlorinated pesticides or
polychlorinated biphenyls (PCBs) which are both totally synthetic and recent in
origin and where degradation in the environment is not greatly assisted by pre-
adapted organisms (Type II).

*2.3.1.3   The Behaviour and Fate of Spilled Oil.*   Once crude oil is released onto the
sea surface a number of different processes immediately begin to act on it which
influence its composition and environmental toxicity. These include evaporation,
dissolution and advection, dispersion, photochemical oxidation, emulsification,
adsorption onto suspended particulate material, biodegradation and sedimentation (Figure 2.2) each of which has different effects on the oil (Table 2.3). In
addition, the action of surface waves and currents drive the oil slick away from its
point of release.

---

[8]M. R. Preston, 'Marine Pollution', in 'Chemical Oceanography Vol 9', ed. J. P. Riley, Academic
   Press, London, 1989, pp. 53–196.

**Table 2.3** *The effects of the weathering process on an oil slick*

| Process | Rate | Effects | Controlling factors |
|---|---|---|---|
| Evaporation | Rapid | Major influence in first 24–48 hr. removes volatiles (*e.g.* <C4 n-alkane) and reduces acute toxicity | Oil composition, wind speed and temperature, water temperature and roughness |
| Dissolution | Fairly rapid | Removes more polar components from slick producing high sub-slick concentrations | Oil composition, wind speed/water roughness, water temperature |
| Dispersion | Variable though more rapid in early stages | Formation of oil in water emulsions. Dispersion enhances degradation rates | Oil composition, wind speed/water roughness |
| Emulsification | More rapid in early stages | Formation of water in oil emulsions ('chocolate mousse'). Oil in this state is untreatable by chemical techniques | Oil composition, wind speed/water roughness |
| Photochemical oxidation | Dependent on light | Oxidation produces aliphatic/aromatic acids, alcohols, ethers, dialkyl peroxides which are more soluble (and possibly more toxic) than parent molecules | Oil composition, light intensity |
| Adsorption | Dependent on suspended particle loading | Removal of oil from surface slick to underlying sediment, potential impact on benthos | Dependent on suspended particle loading |
| Biodegradation | Initially fairly rapid then more slowly | Rapid removal of n-alkanes followed by other susceptible species | Oil composition, local bacterial/fungal populations, degree of oil dispersion, dissolved oxygen content and availability of nutrients |

**Table 2.4**   *The types of damage caused to marine organisms by oil spills*

| Affected organisms | Nature of damage |
| --- | --- |
| Plankton | Minor local damage, possible growth inhibition of phytoplankton by shading effect of slick |
| Seaweeds | Major damage to slick-affected inter-tidal species. Recovery rapid but removal of grazers may cause excessive growth in future years |
| Invertebrates | Large scale mortality in littoral communities through acute toxicity and smothering. Recovery of populations may take years |
| Fish | Normally only minor casualties but pollution of spawning grounds/ migration routes can cause greater damage |
| Seabirds | Diving birds badly affected by oil. Death through drowning, hypothermia or toxic effects of ingested oil |
| Marine mammals | Rarely affected but coastal populations (*e.g.* seal colonies) are vulnerable |

*2.3.1.4   The Environmental Impact of Marine Oil Spills.*   The most obvious problems of oil pollution are those associated with the aftermath of major events. These have included tanker accidents such as the Amoco Cadiz wreck off the Brittany coast in 1978 and, more recently, the Exxon Valdez wreck in Prince William Sound, Alaska 1989; the IXTOC I blowout in the Bay of Campeche in 1979, and the major releases in the Persian Gulf in the 1980s and early 1991 associated with the 8 year Iran–Iraq war and the subsequent war in Kuwait. These high profile events clearly demonstrate the major features of oil pollution damage. The general effects of oil spills on marine organisms are shown in Table 2.4.

The damage to coastlines is very much a function of the coastal ecosystem as well as the prevailing weather conditions. Various indices of coastal variability have been formulated, with high energy coastlines such as exposed rocky headlands, or eroding wave-cut platforms, being amongst the least vulnerable; with sheltered tidal flats, salt marshes, mangrove stands and coral reefs being most sensitive.[8,9]

Experience has shown that in some circumstances attempts to clean-up oil spills have caused more damage than if the oil had been left entirely alone; this was particularly the case in the Torrey Canyon incident.[10] Various guidelines have therefore been produced which indicate how to select the most appropriate treatment strategy from the various options available. Particularly clear examples of these include two field guides published by CONCAWE[9,11] on inland and coastal oil spill control and clean-up techniques.

[9]CONCAWE, 'A Field Guide to Coastal Oil Spill Control and Clean-up Techniques', Rep. No 9/81, CONCAWE, Den Haag, 1981, 112 pp.

[10]J. E. Smith, '"Torrey Canyon" Pollution and Marine Life', Cambridge University Press, Cambridge, 1970, 196 pp.

[11]CONCAWE, 'A Field Guide to Inland Oil Spill Clean-up Techniques', Rep. No 10/83, CONCAWE, Den Haag, 1983, 104 pp.

**Table 2.5** *A summary of oil spill treatment techniques*

| Treatment process | Advantages/disadvantages |
| --- | --- |
| *Chemical* | Modern treatments effective on appropriate oil types. Dispersion enhances degradation. Only applicable to fresh oil, requires good logistical support, may enhance local toxicity whilst protecting coastlines. |
| *Physical (containment)* | Effective for small spills in enclosed or calm waters. Poor efficiency in rough weather. |
| *Physical (recovery)* | Effective for small spills in enclosed or calm waters. Poor efficiency in rough weather. Significant disposal problems with recovered oil. |
| *Adsorption–sinking* | Removes oil to sediment. Trades short term protection for long term benthic contamination |
| *Burning* | May work on some fresh oils but not generally used. Leaves tarry residues |

*2.3.1.5 Control and Clean-up Techniques. Prevention.* Nearly all marine pollution incidents involving oil pollution are avoidable. Routine oil discharges from tank washing, and accidents involving tankers, have historically been amongst the most important sources of oil to the marine environment and have led to the development of a number of strategies designed to reduce the levels or risks of pollution. For the most part, pollution prevention has received the most attention with techniques such as the Load on Top (LOT), Crude Oil Washing (COW) and Segregated or Clean Ballast Tanks (SBT or CBT) being introduced. These techniques are now widely used, and when combined with the oil reception facilities required under international conventions, such as Annex 1 of the International Convention for the Prevention of Pollution from Ships (MARPOL 73/78),[10] have dramatically reduced routine oil discharges to the oceans. More recent requirements for double hull construction to be incorporated into all new tankers have been contentious both on grounds of cost and effectiveness, and it yet remains to be seen how great the benefits from this strategy will be. In principle the double hulls should allow for better protection against spillage in the event of either grounding or collision.

*Oil Spill Treatment Technologies.* Oil spill treatment technologies can be divided into a number of main types: chemical (including chemical enhancement of microbial processes), physical containment, and recovery systems, adsorption and burning (Table 2.5). Useful reviews of this subject have been provided by ITOPF[11] and Doerffer.[5]

## 2.3.2 Sewage

Domestic sewage, with or without the presence of industrial wastes, probably represents the commonest and most widespread contaminant of inshore and

[10]MARPOL 73/78 International Maritime Organization, 1992, 485 pp.
[11]ITOPF, 'Response to Marine Oil Spills', International Tanker Owners Pollution Federation Ltd, London, 1987.

nearshore waters (see Chapter 5 for details of the nature of sewage and sewage treatment techniques). Sewage poses aesthetic and health risks to human populations and also acts as a vector whereby a considerable variety of other contaminants reach the marine environment. Sewage may reach the seas in a number of different forms; ranging from untreated raw sewage discharges, through various degrees of treated discharge, to the dumping of associated sewage sludge at marine sites. The potential problems associated with sewage discharge are significant. Over $10^6$ tonnes of dry sewage sludge are produced in the UK alone each year,[12] and there is no doubt that some disposal strategies are working at or close to their practical limits. At present, a significant amount of both untreated and treated sewage and sewage sludge reaches coastal waters; for example, over $5 \times 10^6$ wet t a$^{-1}$ (equivalent to approximately $5 \times 10^5$ dry t a$^{-1}$) of sewage sludge was dumped in the North Sea by the UK in 1990 though this will cease by 1998.[13]

*2.3.2.1 Problems Associated with BOD* (see Chapter 4).   Whether a high bio-chemical oxygen demand (BOD) waste discharge to a natural water causes a serious environmental problem depends almost entirely on the characteristics of the receiving system. In essence, if the input of BOD is greater than the ability of the receiving water to supply new oxygen then there will be major problems of oxygen depletion and, in extreme cases, total anoxia. If the BOD input and the new oxygen supply are similar in magnitude then some oxygen depletion may be seen, and only if the renewal of oxygen is much greater than the supply of BOD will the discharge be innocuous (at least in this respect). Overall, therefore, the major problems arise from the discharge of untreated sewage to rivers, estuaries or enclosed coastal waters (see also Chapter 4).

One of the UK rivers most affected by high BOD discharges is the River Mersey[14] which has a legacy not only of high, direct BOD discharges to the estuarine system but also to the catchment area. This has led to significant oxygen depletion in the upper estuary, particularly on spring tides where the higher tidal energy causes sediment with a high BOD to be resuspended in the water column thus stripping dissolved oxygen from the overlying water (see Figure 2.3). Major effluent treatment schemes introduced to the region over the last thirty years have reduced the BOD coming into the estuary from about 42 mg l$^{-1}$ to about 4 mg l$^{-1}$. In addition, within the estuary itself new treatment schemes are presently under construction with a view to intercepting the large number of direct, raw sewage outfalls constructed in Victorian times, and directing the effluent to a treatment plant. The effects of severe oxygen depletion on organisms within the estuary are considerable. Many benthic animals, estuarine fish species, and those migratory species which pass from land to sea or *vice versa* as part of their natural life cycle, are effectively barred from transit through anoxic waters. Life in underlying sediments is also considerably restricted. Only when oxygen levels can

[12]DoE, 'Digest of Environmental Statistics, No 17', HMSO, London, 1995.
[13]NSQSR, North Sea Quality State Report 1993, Oslo and Paris Commissions, London, 1993, 132 pp.
[14]NRA, 'The Mersey Estuary, A Report on Environmental Quality', HMSO, London, 1995, 44 pp.

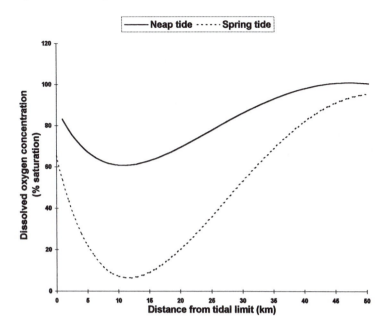

**Figure 2.3** *Dissolved oxygen in the River Mersey Estuary*

be maintained at a consistently high level over all tidal states can the full, natural range of organisms be sustained.

Problems of oxygen depletion associated with sewage discharge are quite common, and may extend beyond the river and estuarine environment into coastal waters. Oxygen depletion has been reported in areas of the New York Bight,[15] the German Bight, southern and eastern Kattegat and eastern Skagerrak as well as several Norwegian fjords.[13] Whilst the hydrographic conditions prevailing in these regions may sometimes lead to low oxygen levels as a result of natural processes, there can be little doubt that organic matter from sewage, agriculture, industry and sundry diffuse sources has also had a significant impact.

*2.3.2.2 Sewage and Nutrients.* Sewage effluents contain large quantities of micronutrient elements such as nitrogen and phosphorus. The greater the degree of treatment, the greater the mineralization of the nutrients so that, for example, a treated waste will contain primarily nitrate and an untreated one ammonia and organic nitrogen species. Nutrient inputs to marine systems have been a matter for considerable concern because of the potential eutrophication effects that they may cause. A simple view of eutrophication is that additional nutrients lead to the formation of excess biomass which, in turn, leads to an increase in BOD with subsequent oxygen depletion effects as the biomass decays.

[15]M. Takizawa, W. L. Straube, R. T. Hill and R. Colwell, *Appl. Environ. Microbiol.*, 1993, **59**, 3406–3410.

It is not only an increase in nutrient concentration that can cause deleterious effects. Anthropogenically induced alterations in nutrient ratios can also cause changes in prevailing phytoplankton species because of their differing physiological properties and requirements for nutrients. As a general rule marine phytoplankton assimilate nitrogen and phosphorus in a ratio of about 16:1 and under normal conditions in marine systems nitrogen is the growth limiting element. However, efforts to reduce nutrient inputs over recent years have primarily focused on phosphorus, with the result that nitrogen to phosphorus ratios have tended to increase. As a consequence, phosphorus has then become the growth limiting element in some regions (*e.g.* parts of the North Sea), so that phosphorus supplies are effectively exhausted during the spring/early summer phytoplankton bloom leaving a considerable excess of nitrogen unconsumed.[13] This excess nitrogen may be advected to other regions where it raises the total nitrogen levels, alters the prevailing N:P ratio, and has been implicated not only in changing dominant plankton species but also in stimulating toxin production in some species.

Examples of the problems associated with excessive nutrient enrichment have been seen in a number of regions in recent years. For example, a major bloom of the phytoplankton *Phaeocystis* spp. in the upper Adriatic in the summer of 1990 produced large quantities of scum-like organic material which washed up on beaches. There it had the appearance of sewage waste, had a very unpleasant smell as it decomposed, attracted flies and, not surprisingly, had a devastating effect on the local tourist industry. In 1988, similar problems with excessive plankton blooms caused major damage to coastal mariculture units on the Swedish and Norwegian coasts.[13]

*2.3.2.3  Marine Disposal of Sewage Sludge.*  Disposal of sewage sludge in an environmentally acceptable way has increasingly been recognized as a problem in recent years. In major societies the quantities involved are considerable and practical, and/or social, pressures have resulted in the marine disposal option being used by some countries, notably the UK which annually dumped over 5 million tonnes of sewage sludge in the North Sea alone (27% of current production). The declared intention for the UK is to cease all sewage sludge dumping by the end of 1998 (a decision taken at the 1990 North Sea Conference) at the latest, although there are some indications that this date may be deferred largely because of public resistance to many of the other disposal options.

Sewage sludge dumping takes place at designated sites, which are defined areas rather than a single point. The sludge particles gradually settle through the water column, where they may act as a food source to mid-water feeding fish species, finally settling on the sediment surface. The problems arising from this practice are essentially twofold. First, the sewage acts as a vector for the transfer of contaminants to the biota (notably trace metals and persistent organic chemicals). Second, the high organic input to the sediments cause changes in both species diversity (decreases) and biomass (increases) in the benthic community.

The extent of the changes induced depends to a considerable extent on the hydrodynamics of the receiving site. There is essentially a choice between a 'dispersive' site, where strong current systems disperse the sludge widely at low concentration, and an 'accumulating' site, which is hydrodynamically quiet and where there will be higher sludge concentrations over a limited area. With the exception of the Garroch Head site which received sludge from the Glasgow region, all UK sites are relatively dispersive in character.

*2.3.2.4 Marine Sewage Disposal and Public Health.* Sewage debris on beaches and in coastal waters is clearly an aesthetic problem, but it also poses a health risk. There is not only a direct risk of infection by sewage derived pathogenic organisms, but there are also considerable risks associated with the consumption of contaminated and improperly prepared sea food. Some examples of such infections are given in Table 2.6.

It should be noted that the lifetimes of most terrestrially derived bacteria in sea water are relatively short; typically of the order of 12–24 hours. This limited lifetime is due to the natural antibiotic properties of sea water which derive from a combination of the high salt concentrations, low concentrations of inorganic and organic chemicals with antibiotic properties and exposure to natural UV radiation in surface waters. However, not all bacteria or viruses necessarily die. Some bacteria may, for example, enter a dormant phase (non-platable bacteria)

**Table 2.6** *Sewage derived pathogenic organisms and their effects on humans*[16]

| Aetiological agent | Mode of transmission to humans* | Diseases/symptoms |
|---|---|---|
| *Salmonella typhi* | Fish or shellfish | Typhoid |
| *S. paratyphi* | - do - | Paratyphoid |
| *S. typhimurium* | - do - | Salmonellosis; gastroenteritis |
| *S. enteritidis* | - do - | - do - |
| *Vibrio parahaemolticus* | - do - | Diarrhoea, abdominal pain |
| *Clostridium botulinum* | - do - | Botulism (high fatality rate) |
| *Staphylococcus aureus* | - do - | Staphylococcal intoxication, nausea, vomiting, abdominal pain, prostration |
| *Clostridium perfringens* | - do - | Diarrhoea, abdominal pain |
| *Erysipelothrix insidiosa* | Skin lesions | Erysipeloid – severe wound inflammation |
| *Hepatitis virus* | Shellfish | Infectious hepatitis |
| *Heterophyes heterophyes* | Fish or shellfish | Heterophyiasis: abdominal pain, mucous diarrhoea (eggs may be carried to brain, heart *etc.*) |
| *Paragonismus westermani (P. ringeri)* | Crabs, crayfish or contaminated water | Flukes in lungs and other organs |
| *Anisakis matina* | Marine fish (notably herring) | Anisakiiasis; eosinophilic enteritis |
| *Angiostrongylus cantonensis* | Shrimps or crabs | Eosinophilic meningitis |

*Note that in most cases contamination by the organism is not transmitted to humans unless the food has been inadequately stored or cooked.

[16]GESAMP, Rep. Stud., GESAMP No 5, 1976, 23 pp.

which does not show up in routine test procedures, but which can become active again if ingested by bathers.

*2.3.2.5   The European Bathing Water and Seafood Directives.*   Within the European Union sea water quality standards as they affect human health are set by the European Bathing Water Directive (EEC/76/160). This directive requires certain microbial quality standards to be met for beaches designated as bathing beaches. The mandatory limit for faecal coliforms under this directive is 2000/100 ml.

The Seafood Directive (EEC/91/492), which became operative in January 1993, sets limits for the number of faecal coliforms, *Salmonella*, and toxins such as those which cause Paralytic Shellfish Poisoning (PSP) or Diarrhetic Shellfish Poisoning (DSP). These toxins are produced by dinoflagellates, such as *Alexandrium* and *Protogonyaulax*, and are accumulated in shellfish consumption of which is capable of causing human deaths. The relationship between blooms of these dinoflagellates and pollution is not clear because such events do not take place exclusively in waters recognized as being polluted; however, such a link cannot be entirely discounted.[13]

### 2.3.3   Persistent Organic Compounds

The European Commission listing of potential environmental pollutants is dominated by organic chemicals (see Chapter 1). The criteria for inclusion of a chemical in that, and other priority pollutant lists, are that they have relatively high environmental toxicity (to humans and other organisms), they are persistent and liable to undergo significant biomagnification and they are produced in sufficient quantities to represent a potential threat.

Of the huge number of chemicals that might be considered to be potential pollutants only comparatively few have been studied in any detail in marine systems and many of these are halogenated species.

*2.3.3.1   Halogenated Compounds* (see also Chapter 16).   *Chlorinated Pesticides.*   Dominant amongst the older formulations were the organochlorine pesticides such as DDT, dieldrin, aldrin and endrin (the 'drins'), lindane ($\gamma$-hexachlorocyclohexane), hexachlorobenzene (HCB) and toxaphene. Despite widespread bans on many persistent organochlorine pesticides in the industrialized world, usage on a global scale is still important because the cheapness and effectiveness of the chemicals makes them attractive options for developing countries.

Other persistent organic chemicals which have proved to be of some concern in marine systems include polychlorinated dibenzo-*p*-dioxins (PCDD) and polychlorinated dibenzofurans (PCDF); pesticides including 2,4-D (2,4-dichlorophenoxyacetic acid), 2,4,5-T (2,4,5-trichlorophenoxyacetic acid), and MCPA (2-methyl-4,6,-dichlorophenoxyacetic acid).[17] 2,4-D and 2,4,5-T became notorious through their use as defoliants in the Vietnam War (Agent Orange) and because

---

[17]D. Broman, 'Transport and Fate of Hydrophobic Organic Compounds in the Baltic Aquatic Environment', Ph.D. Thesis, University of Stockholm, 1990.

**Table 2.7**   *The distribution of PCBs between different environmental compartments*[18]

| Environment | PCB load (t) | Percentage of PCB load |
|---|---|---|
| *Terrestrial and coastal* | | |
| Air | 500 | 0.13 |
| River and lake water | 3500 | 0.94 |
| Sea water | 2400 | 0.64 |
| Soil | 2400 | 0.64 |
| Sediment | 130 000 | 34.73 |
| Biota | 4300 | 1.15 |
| Total (A) | 143 100 | 38.23 |
| | | |
| *Open Ocean* | | |
| Air | 790 | 0.21 |
| Sea water | 230 000 | 61.45 |
| Sediment | 110 | 0.03 |
| Biota | 270 | 0.07 |
| Total (B) | 231 170 | 61.77 |
| Total Load (A + B) | 374 270 | 100 |

of the contamination of the product with the dioxin 2,3,7,8-TCDD (2,3,7,8-tetrachlorodibenzo-*p*-dioxin). The 'dioxins' and the PCDF have become widespread contaminants of marine systems, although their links with any deleterious effects in the oceans remain slight.

*PCBs.*   PCBs are widely distributed amongst all marine systems. Their toxicity to marine organisms varies considerably, but the co-planar PCBs exhibit the greatest mammalian toxicity. However, such acute toxicity is unknown in marine organisms which are more likely to suffer sub-lethal effects from the biomagnifications of PCBs through the food web (see Table 2.7).

*The Origins of the Environmental Hazards.*   The environmental problems associated with the chlorinated pesticides and PCBs derive from similarities in their physical, chemical and biological properties. These may be summarized as persistence, widespread distribution amongst most environmental compartments, propensity to undergo biomagnification and high toxicity (including non-lethal effects; note however, that DDT has a low mammalian toxicity). In other words, nearly all of the properties that make a pollutant a high risk.

The organochlorines are relatively water insoluble and involatile. They also have high octanol-water partition coefficients (*e.g.* log $K_{ow}$ = 4.46–8.18 for different PCB congeners) and are therefore lipophilic. The low apparent volatility is, however, misleading. Under certain conditions (notably high temperatures and humidity) evaporative losses of applied pesticides can be high, and this accounts for their widespread occurrence through subsequent transport as vapour or condensates on atmospheric aerosol particles (Table 2.8).

[18]S. Tanabe, *Environ. Pollut.*, 1988, **50**, 5–28.

**Table 2.8**   *Estimated fluxes of PCBs to the Ocean Surface*[19]

| | Flux ng m$^{-2}$ a$^{-1}$ | | |
| --- | --- | --- | --- |
| | *Arochlor 1242* | *Arochlor 1254* | *Total* |
| *Particles* | | | |
| Dry | 16 | 41 | 57 |
| Wet | 250 | 650 | 900 |
| Gas Phase | 2209 | 1709.5 | 3919 |
| Total | 2475 | 2400.5 | 4876 |
| Total flux to Oceans ($\times 10^6$ g a$^{-1}$) | 8.6 | 8.3 | 16.9 |

The lipophilic nature of the organochlorines leads not only to their rapid uptake and storage in fatty tissues, but also to their slow elimination, because the fat reserves are only called upon at stages in the life cycle when energy demands are high or food supplies low. It is this feature which leads to their biomagnification. The extent of biomagnification is large. Concentrations of organochlorines in sea water are typically of the order of pg l$^{-1}$ to low ng l$^{-1}$, rising to 10s of ng g$^{-1}$ in marine invertebrates, low mg g$^{-1}$ values in mussels and up to 10–100s of mg g$^{-1}$ in fatty tissue of top predators such as seals, pelicans and terrestrial hawk species.

*The Effects of Marine Organochlorine Pollution.*   At the planktonic level primary production rates are reduced at DDT or PCB concentrations above ~1 mg l$^{-1}$. This is a low concentration but still very much higher than those likely to be found in sea water. Amongst marine invertebrates and fish, 96 hr LC$_{50}$ values generally fall within the range 1–100 mg l$^{-1}$. However, bivalve molluscs are very resistant with 96 hr LC$_{50}$ values $> 10\,000$ mg l$^{-1}$.

Amongst the top predators the main deleterious effects have been eggshell thinning in birds (DDE) and interference with the reproductive and immune system in mammals (PCBs). A considerable number of top predator birds showed major declines in abundance during the period when DDT usage was at its greatest and since the banning of these chemicals in the industrialized world, these populations have mostly demonstrated a major recovery. Where populations have not recovered, other factors, such as decline in suitable habitats, are probably responsible.

A number of seal populations, most notably in the Dutch Wadden Sea and the Baltic Sea, have exhibited reproductive abnormalities attributed to PCBs which have had a significant impact on populations. The main symptoms of PCB poisoning are changes in the uterus, implantation or abortion/premature pupping. There have also been suggestions that PCBs may cause carcinogenic, teratogenic and immunological effects. An outbreak of phocine distemper virus in the common seal populations in much of the North Sea in 1988 has sometimes

[19]E. Atlas, T. Bidleman and C. S. Giam, in 'PCBs in the Environment, Vol 1', ed. J. S. Waid, CRC Press, Boca Raton, FL, 1986.

been attributed to depression of immune systems by PCBs, but the evidential basis for this is weak and such epidemics have also been reported in seal populations in other, less contaminated, regions.

*2.3.3.2    Other Persistent Organic Compounds.*    Organochlorines are by no means the only organic compounds which are of concern as marine contaminants. There are many other chemicals which are sufficiently common, toxic and persistent to represent potential threats. These include, for example other pesticides, polycyclic aromatic compounds (PAH), plasticizers (*e.g.* phthalate esters), detergent residues, organic solvents *etc.* It is not possible within this chapter to review the behaviour and effects of all of these compounds. However, one aspect of organic pollution which has become particularly important recently is the relationship between the reported number of abnormalities in male sex development in wildlife and humans coinciding with the introduction of so called 'oestrogenic' chemicals.[20-22] Compounds apparently responsible for this effect include several members of the DDT family of compounds, chlordecone (Kepone), various sterols and possibly nonyl phenol. Of these, *p,p'*-DDE is the most potent. Research in this area is only just beginning, but the consequences of widespread disruption of fertility by organic contaminants is clearly a very serious one (see also Chapter 18).

### 2.3.4    Trace Metals

A large number of trace metals are transported to the oceans from natural sources. However, these natural sources are supplemented by releases from anthropogenic processes which, for some metals, can exceed natural inputs.

Trace metals are found in the water, biota and sediment compartments of the marine system, but potentially the most hazardous environmental effects to human health arise when they enter the food chain. The relationship between the total concentration of a trace metal in the environment and its ability to cause toxic effects in organisms is complex, and two important constraints must be considered; *i.e.* the speciation of the metals, and the condition of the organisms.

(i)    *Metal speciation.* All organisms have a requirement for certain trace metals which must be present in their diet (or growth medium) to sustain healthy development. Such 'essential' metals include iron, copper, vanadium, cobalt and zinc. Other, non-essential, metals exert neither beneficial nor deleterious effects if present at sufficiently low concentrations, but become increasingly harmful as the concentrations increase. However, it is the speciation of a trace metal, *i.e.* the way in which it is partitioned between host associations in the water and particulate phase, and not its total concentration, which constrains its effect on the environment. For

[20]E. Carlson, A. Giwercman, N. Keiding and N. Skakkebaek, *Br. Med. J.*, 1992, **305**, 609–612.
[21]R. M. Sharpe and N. E. Skakkebaek, *Lancet*, 1993, **341**, 1392–1395.
[22]W. R. Kelce, C. R. Stone, S. Laws, L. E. Gray, J. A. Kemppainen and E. M. Wilson, *Nature (London)*, 1995, **375**, 581–585.

example, not all forms of a trace metal are 'bioavailable', and numerous incidents have demonstrated that it is the organic form of metals which tend to be of the most damaging to marine organisms.

(ii) *Condition of the organisms.* A number of factors influence the effect that potentially toxic substances have on organisms. According to Preston[8] these include the following: (a) the stage of development, *e.g.* egg, larva, adult; (b) size and sex; (c) history, *e.g.* previous exposure to toxic substance; (d) location, which is very relevant for intertidal organisms; (e) diet; (f) general environmental conditions, such as temperature, salinity, pH, Eh, light intensity, dissolved oxygen, that might affect metal/organism interactions.

The hazardous effects of trace metals on the oceanic biogeochemical system are illustrated below with respect to mercury, lead and tin; all of which have been shown to be capable of generating stress in the marine environment.

*2.3.4.1 Mercury* (see also Chapter 1). Mercury is on the 'Black List' of chemicals, and probably represents the best known example of trace metal pollution in the marine environment through its role as the causative agent of Minimata disease. This was an extreme example, however, and mercury contamination of coastal waters affected by industrial effluents is not an uncommon problem. For a full discussion of mercury in the environment the reader is referred to Nriagu.[23]

Mercury has a very wide range of industrial uses and the global production of mercury for industrial purposes is $\sim 6000$ t a$^{-1}$, which has increased threefold since 1900.[24]

The marine cycle of mercury has been reviewed by Fitzgerald.[25] It is thought that there are three potentially important chemical forms of dissolved mercury in natural waters. These are elemental mercury (Hg), divalent mercury (Hg$^{2+}$), and methylmercury (CH$_3$Hg); although ethylmercury can also occur.

Four factors strongly influence the environmental effects of mercury. (i) Mercury has a particularly high affinity for organic species, which results in its accumulation in marine biota. (ii) Inorganic mercury can undergo bio-mediated transformation into 'alkylated' forms (methyl and dimethyl mercury) which are particularly toxic species of the element. In the environment methylmercury compounds can be synthesized by a variety of micro-organisms. This occurs almost entirely under aerobic conditions in the marine environment.[26] The principal form of mercury in fish is methylmercury. (iii) Mercury is non-conservative in estuaries where it tends to be accumulated in fine-grained near-shore sediments from which it can be remobilized.[27] (iv) In the particulate form,

[23]J. O. Nriagu, in 'The Bio-geochemistry of Mercury in the Environment', ed. J. O. Nriagu, Elsevier, Amsterdam, 1979, p. 23.

[24]A. Andren and J. O. Nriagu, in 'The Biogeochemistry of Mercury in the Environment', ed. J. O. Nriagu, Elsevier, Amsterdam, 1979, p. 7.

[25]W. F. Fitzgerald, in 'Chemical Oceanography, Vol 10', eds. J. P. Riley, R. Chester and R. A. Duce, Academic Press, London, 1989, p. 152.

[26]L. Landner, *Nature*, 1971, **230**, 452.

[27]J. A. Campbell, E. Y. Chan, J. P. Riley, P. C. Head and P. D. Jones, *Mar. Poll. Bull.*, 1986, **17**, 36.

mercury has a strong association with organic particles, *via* which it is readily transmitted to biota.

The residence time of mercury in the oceans has been estimated to be $\sim 350$ yr,[25] and indicates a high degree of biogeochemical activity and a rapid removal from the water column. Open-ocean concentrations of reactive mercury (which includes 'labile' organo-Hg associations, but not the more stable organo-Hg associations) are around 2–10 pmol $1^{-1}$,[28,25] although concentrations can rise to $\sim 100$ pmol $1^{-1}$ in polluted coastal waters.[29] The WHO maximum tolerable limits for mercury consumption are 0.3 mg per week, of which not more than 0.2 mg should be methylmercury. These levels are not likely to be exceeded except under 'extreme' circumstances where seafood from highly contaminated waters is consumed. Such 'extreme' circumstances occurred at Minimata in Japan, and in Minimata Bay concentrations as high as 10 nmol $1^{-1}$ were reported.[30]

In what Goldberg[31] described as a typical example of the 'surprise factor' in environmental problems, outbreaks of methylmercury poisoning at Minimata were first discovered during the 1950s. By the mid-1970s around a thousand cases had been diagnosed with many more suspected, and 41 deaths had occurred; there was also evidence of the long-term accumulation of mercury in the human brain from eating seafood caught in Minimata Bay.[32]

The Minimata incident provides a prime example of how a pollutant released into the environment can move along the food chain to affect humans. Mercury compounds were released into the water of Minimata Bay from a factory using mercury(II) chloride in the synthesis of vinyl chloride, and mercury(II) sulfate in the production of acetaldehyde.[31] According to Waldron[33] there is some disagreement over the form in which the mercury initially entered the environment. It was claimed that mercury was released by the factory in an inorganic form, and that this was subsequently methylated by bacteria in the sediments of Minimata Bay. However, the rate of conversion is probably too slow to have accounted for the high levels of methylmercury in the waters of the bay. The methylmercury was markedly accumulated by fish and shell fish and appeared in man from the consumption of the seafood.

*2.3.4.2 Lead.* The global cycle of lead has been strongly perturbed by anthropogenic effects, and the metal is an example of the input of a contaminant to the natural environment mainly through release into the atmosphere. For example, Nriagu[34] estimated that the global anthropogenic emission of Pb to the atmo-

[28]K. W. Bruland, in 'Chemical Oceanography, Vol. 8', eds. J. P. Riley and R. Chester, Academic Press, London, 1983, p. 157.
[29]C. W. Baker, *Nature*, 1977, **270**, 230.
[30]P. L. Bishop, 'Marine Pollution and its Control', McGraw-Hill, New York, 1983, 357 pp.
[31]E. D. Goldberg, in 'Chemical Oceanography, Vol. 3', eds. J. P. Riley and G. Skirrow, Academic Press, London, p. 39.
[32]S. Smith, in 'Understanding Our Environment: An Introduction to Environmental Chemistry and Pollution', ed. R. M. Harrison, Royal Society of Chemistry, Cambridge, 1992, p. 244.
[33]H. A. Waldron, in 'Pollution: Causes, Effects, and Control', ed. R. M. Harrison, Royal Society of Chemistry, Cambridge, 1990, p. 261.
[34]J. O. Nriagu, in 'Control and Fate of Atmospheric Trace Metals' eds. J. M. Pacyna and B. Ottar, Kluwer Academic Publishers, Dordrecht, 1989, p. 3.

sphere was $\sim 332 \times 10^6$ kg yr$^{-1}$, which exceeded the natural emission ($\sim 12 \times 10^6$ kg yr$^{-1}$) by a factor of $\sim 28$.

Lead is used in a number of industrial applications, but the main anthropogenic input to the environment is *via* the combustion of fossil fuels, especially from the use of lead alkyls (tetraethyl and tetramethyl lead) as an anti-knock additive in combustion engine fuel. However, the use of lead in fuels has decreased markedly over the past decade, or so, with the reduction of the anti-knock additive in Europe and North America; for example, the maximum lead concentrations in gasoline in the US fell from 2.5 g US gal$^{-1}$ in 1970 to 0.5 g US gal$^{-1}$ in 1979[35] and is now effectively zero.

Three factors strongly influence the distribution of lead in the marine environment. (i) Its global cycle is strongly perturbed by anthropogenic inputs. (ii) It is a 'scavenging-type' trace metal which is rapidly removed from surface waters, and which has an oceanic residence time of only a few hundred years. (iii) It has a large atmospheric signal to surface waters and because it is emitted into the atmosphere from anthropogenic sources as small particles[36] it undergoes long-range transport. As a result, both the open-ocean and the coastal cycles of lead have been strongly perturbed by man's activities. However, the major effects of Pb contamination are found in the coastal environment, and Pb can be used to illustrate the accumulation of a pollutant trace metal in the sediment reservoir.

In coastal regions accumulation rates are fast enough for sediments to record anthropogenic inputs. For example, with respect to Pb, Chow *et al.*[37] reported data on the distribution of the metal in dated sediment cores from a number of basins off the coast of southern California. Aluminium, which has a mainly natural origin, was used as a normalizing element and Pb:Al ratios were used to establish the background lead levels in the sediments. The data showed that in the sediments of some basins which received anthropogenic inputs there was an increase in both the concentrations of lead and the Pb:Al ratios in sediments deposited after $\sim 1940$. In addition, lead isotopic ratios indicated that the source of the 'excess' lead had been influenced by gasoline emissions. The study provides an example of how natural lead cycles in the marine environment can be perturbed by anthropogenic inputs. However, according to Bernhard[38] the marine environment represents only a negligible source of lead to humans. Nonetheless, situations can arise where the sudden release of lead into the marine environment is potentially harmful. There have also been examples of an increase in alkyl lead compounds in biota and sediments from areas where there has been no evidence of abnormal discharges of alkyl lead to the environment. For example, mortality among seabirds has been attributed to poisoning

[35] J. R. Ashby and P. J. Craig, in 'Pollution: Causes, Effects, and Control', ed. R. M. Harrison, Royal Society of Chemistry, Cambridge, 1990, p. 309.

[36] R. Arimoto and R. A. Duce, *J. Geophys. Res.*, 1986, **92**, 2787.

[37] T. J. Chow, K. W. Bruland, K. Bertine, A. Soutar, M. Koide and E. D. Goldberg, *Science*, 1973, **181**, 551.

[38] M. Bernhard, in 'Lead in the Marine Environment', eds. M. Branica and Z. Konrad, Pergamon Press, Oxford, 1980, p. 345.

from trialkyl lead, a stable decomposition product of tetra-alkyl lead, in the River Mersey estuary which receives a number of industrial effluent discharges.[39]

*2.3.4.3 Tin.* Tin is now recognized as being potentially a very serious marine pollutant and is an example of a substance introduced into the marine environment for a specific purpose, rather than one which enters the system either as a by-product of anthropogenic activity, or as a result of accidental release.

Tin has a number of industrial uses, but it is the organotin compounds (notably tributyl tin oxide – TBTO) which have given rise to the greatest concern. The principal uses of organotin compounds are as stabilizers for PVC and as biocides, especially anti-fouling paints used on ships hulls. Organotin compounds in the marine environment derive from two main sources; (i) *via* the bacterial methylation of inorganic tin, and (ii) by the sea water leaching of alkyl and aryl tin from some types of anti-fouling paints which may affect non-target species, especially in areas such as harbours, boat yards and marinas where there are high concentrations of ships.[8]

TBTO can produce major harmful responses in a number of marine organisms, such as commercial Pacific oysters (*Crassostrea gigas*) and dogwhelks (*Nucella lapillus*); In the latter, TBTO can cause 'imposex' – a phenomenon in which females develop male sexual characteristics leading to sterility and population collapse. Concern has also been expressed over TBTO in farmed salmon, where it has been used on breeding cages.[40,41] Other marine organisms affected by TBTO include phytoplankton and zooplankton, and evidence suggests that the TBTO can have harmful effects at concentrations as low as the ng $l^{-1}$ level. Waters in many regions associated with commercial or recreational shipping may have TBTO concentrations which exceed this level, and the recent North Sea Quality Status Report (1993) identified numerous areas around the North Sea where serious TBTO effects may occur; these include the Solent, the Wash and regions adjacent to the major continental European rivers such as the Elbe and the Rhine. Even remote areas, such as the Shetland islands, may be affected because of the use of TBTO on fishing boats. Although many countries have now banned the use of tin-based anti-fouling paints, either entirely or at least on small-sized vessels, there is a legacy of old paintwork that is likely to sustain elevated sea water concentrations of TBTO for some time.

### 2.3.5 Radioactivity (see also Chapter 17)

The release, or even the potential release, of radioactive substances to the environment is always a major issue, which is being clearly demonstrated at the time of writing by the decision of the French Government to resume weapons testing on Mururoa Atoll in the southern Pacific and the subsequent actions of the Greenpeace organization.

[39] J. P. Riley and J. V. Towner, *Mar. Poll. Bull.*, 1987, **15**, 153.
[40] I. M. Davies and J. C. McKie, *Mar. Poll. Bull.*, 1987, **18**, 405.
[41] I. M. Davies, S. K. Bailey and D. C. Moore, *Mar. Poll. Bull.*, 1987, **18**, 400.

The subject of marine radioactivity has been extensively reviewed by Burton[42] and Preston,[8] and only a brief description of some of the more important, or recent, issues is included in this section.

*2.3.5.1    The Natural Radioactivity of Sea Water.*    Sea water and marine sediments are naturally radioactive (Table 2.9). Sea water itself has a radioactivity of around 12.6 Bq $l^{-1}$, with that of marine sands and muds being around 200–400 Bq $kg^{-1}$ and 700–1000 Bq $kg^{-1}$, respectively. Most of the radiation comes from the isotope potassium-40, but there are numerous members of the uranium and thorium decay series present as well. In addition, the creation of lighter radio-isotopes through the interaction of cosmic rays and atmospheric gases, with the products subsequently transferred to the ocean surface, also makes a contribution (Table. 2.9). Chemically, radionuclides behave almost identically to their stable counterparts (where these exist).[44] They are, therefore, partitioned between water, sediments and biota according to their behavioural properties. Examples of this chemically determined behaviour include: (i) Cs-137, which is largely water soluble and behaves as a fairly conservative property of sea water moving with the prevailing currents, and (ii) plutonium-239/240 isotopes, which are highly non-conservative and form strong associations with fine-grained sediments.

One of the more important consequences of the natural radioactivity of marine systems is that marine organisms have evolved in a comparatively radioactive

**Table 2.9**    *The concentrations of some radio-nuclides in sea water*[43]

| Radionuclide | Concentration (Bq $kg^{-1}$) |
| --- | --- |
| Potassium-40 | 11.84 |
| Tritium | 0.022–0.11 |
| Rubidium-87 | 1.07 |
| Uranium-234 | 0.05 |
| Uranium-238 | 0.04 |
| Carbon-14 | 0.007 |
| Radium-228 | $(0.0037–0.37) \times 10^{-2}$ |
| Lead-210 | $(0.037–0.25) \times 10^{-2}$ |
| Uranium-235 | $0.18 \times 10^{-2}$ |
| Radium-226 | $(0.15–0.17) \times 10^{-2}$ |
| Polonium-210 | $(0.022–0.15) \times 10^{-2}$ |
| Radon-222 | $0.07 \times 10^{-2}$ |
| Thorium-228 | $(0.007–0.11) \times 10^{-3}$ |
| Thorium-230 | $(0.022–0.05) \times 10^{-4}$ |
| Thorium-232 | $(0.004–0.29) \times 10^{-4}$ |

[42] J. D. Burton, in 'Chemical Oceanography, Vol 3', eds. J. P. Riley and G. Skirrow, Academic Press, London, Vol. 3, pp. 91–191.
[43] R. B. Clarke, 'Marine Pollution', Oxford University Press, Oxford, 3rd edn, 172 pp.
[44] IAEA, International Atomic Energy Agency, Vienna, TECDOC, 329, 1985, 183 pp.

environment, so that the additional radiation introduced as a result of human activities is to a considerable extent a Type I rather than a Type II pollutant problem.

*2.3.5.2 Radiation Releases from Weapons Testing Programmes.* Between the first atomic weapon test explosion in 1945 and the present day, around 2000 tests have been carried out. Many of the early tests resulted in significant contamination of marine systems[45] although comparatively few of those conducted since 1963 have resulted in similar contamination. This is because from 1963, the countries which have ratified the Partial Nuclear Test Ban Treaty confined their tests to underground sites; however, non-signatory nations have continued to carry out 'open' tests.

*2.3.5.3 Routine Releases from Nuclear Power Plants.* Under normal operating conditions, power generating reactors release very small quantities of radioactivity to the environment. Such isotopes that are released include tritium, sulfur-35, zinc-65 and cobalt-60 which primarily derive from the neutron activation of cooling water or soluble species within it. In the UK, the station with the highest discharges is probably Heysham Station 1 which, in 1990, released 0.40 and 157 TBq of sulfur-35 and tritium respectively and 0.058 TBq of other activity.[46]

*2.3.5.4 Releases from Nuclear Fuel Reprocessing Plants to Marine Systems.* Over the years, the most important non-military source of routine radiation discharges to the marine environment has been the routine low-level discharges from nuclear fuel reprocessing plants such as those at Sellafield in the UK and Cap de la Hague in France. The main isotopes discharged in this manner include tritium, C-14, Co-60, Sr-90, Te-99, Ru-106, I-129, Cs-134 and 137, Ce-144, Pu-241 and Am-241.[47]

The consequences of the Sellafield discharges to radiation levels in the Irish Sea and human populations have recently been reviewed by Kershaw *et al.*[47] who report that between 1983 and 1988 collective radiation doses to the UK population have decreased from 70 to 40 man-Sievert with an increase in 1986 as a result of the Chernobyl accident (see below). The resultant individual doses to members of the local fishing community and 'typical members of the fish-eating public consuming fish landed at Whitehaven/Fleetwood' are 0.11–0.16 and 0.004 mSv respectively compared to a recommended principal dose limit of 1 mSv a$^{-1}$.

*2.3.5.5 Reactor Accidents on Land.* Reactor accidents on land have been rare, but the major incident at Chernobyl in the Ukraine (April, 1986) resulted in very widespread contamination of western Europe. In the UK contamination of both

[45]A. B. Joseph, P. F. Gustafson, I. R. Russell, E. A. Schuert, H. L. Volchok and A. Tamplin, in 'Radioactivity in the Marine Environment', National Academy of Sciences, Washington D.C., 1971, pp. 6–41.

[46]MAFF, 'Aquatic Environment Monitoring Report No 29, MAFF Directorate of Fisheries Research, Lowestoft, 1992, 68 pp.

[47]P. J. Kershaw, R. J. Pentreath, D. S. Woodhead and G. J. Hunt, 'Aquatic Environment Monitoring Report No. 32', MAFF Directorate of Fisheries Research, Lowestoft, 1992, 65 pp.

**Table 2.10**   *Amounts of intermediate and low level radioactive waste dumped in the North-east Atlantic*[43]

| Year | α emitters (TBq) | β/γ emitters (TBq) | Tritium (TBq) |
|------|------------------|--------------------|---------------|
| 1975 | 28.9 | 1130 | 1100 |
| 1976 | 32.6 | 1200 | 775 |
| 1980 | 70.3 | 3075 | 3630 |
| 1981 | 77.7 | 2930 | 2750 |
| 1982 | 51.8 | 1830 | 2860 |
| Total 1948–82 | 660 | 38 000 | 15 000 |

land and sea was considerable,[48] and whilst the aquatic burdens have now essentially reduced to previous levels the effects on upland pastures are still influencing the marketability of animals grazed on these areas.

*2.3.5.6   Dumping of Low Level Waste.*   Disposal of low-level, packaged radioactive waste in the oceans was used by a number of countries until 1983 when the authorizing body, the London Dumping Convention (now the London Convention), took the political decision to suspend it. This decision was confirmed in 1995 when the members of the Convention voted to cease all marine waste disposal. The total radiation inventory dumped in the north-east Atlantic is given in Table 2.10.

*2.3.5.7   Naval Sources of Radioactivity to the Oceans.*   Since 1945 at least 50 nuclear warheads and nine reactors have been introduced to the world's oceans, mainly as a result of accidents to submarines.[49] The total count is still rising as a result of the increased information flow from the former Soviet Union about disposal of radioactive wastes in the Russian Arctic. The incident of greatest concern appears to be the loss of a Soviet 'Mike Class' submarine containing two liquid metal nuclear reactors and two nuclear armed torpedoes, which sank some 270 miles north of Norway in April 1989 after a fire on board. The proximity of this wreck to land, and the damage sustained by the vessel prior to sinking, must make the risks of radiation releases in the relatively near future fairly high. The same factors also make any kind of recovery operation potentially very hazardous.

*2.3.5.8   The Effects of Artificial Radioactivity on the Marine Environment.*   Humans are the most sensitive of living organisms to the effects of radiation, and so all radiological population measures are designed to protect humans on the understanding that all other species will therefore automatically receive adequate protection.[50] Nevertheless, studies of the effects of anthropogenically derived

[48]W. C. Camplin, N. T. Mitchell, D. R. P. Leonard and D. F. Jeffries, MAFF Aquatic Environment Monitoring Report, 1986, **15**, 49 pp.
[49]M. Evans, *The Times*, 7th June 1989.
[50]ICRP, 'Recommendations of the International Commission on Radiological Protection', ICRP Publication 26, Pergamon Press, Oxford, 1977, 53 pp.

radiation on marine organisms have been conducted.[47] For example studies of the possible effects of Sellafield discharges on plaice eggs (the most sensitive developmental stage) indicate the $LD_{50}$ at metamorphosis to be 0.9 Gy, with local dose rates estimated at $\sim 1$ nGy hr$^{-1}$.[51] The general conclusions are that the lowest dose rates at which minor radiation-induced disturbances of physiology or metabolism might be detectable are around 0.4–8 mGy hr$^{-1}$,[52] with the highest (mid-1970s) Sellafield dose rates being about two orders of magnitude below those likely to cause observable effects at the population level. Since then discharges have declined by several orders of magnitude.

## 2.4 CONCLUSIONS

The influence of human activities on the marine environment can now be detected in even the most remote regions. The main impacts are, not surprisingly, in coastal waters which are both closest to the sources of pollutants and the most physically, chemically and biologically active zones. In such coastal areas, waste discharges combine with other pressures, such as coastal developments and fishing, to produce deleterious effects which may be environmentally significant. The situation in the more remote oceanic regions has not yet become as serious, and human activities are normally only detectable through the presence of artificially produced (Type II) chemicals and, with the exception of lead, all Type I pollutants normally fall within the range of natural variability.

Although the pressures on marine systems are now widely recognized, there is still a considerable gap between the scientific recognition of actual, or potential, problems and the social, and/or political, willingness to take appropriate action. The future health of the oceans therefore depends on integrated management strategies designed to provide the maximum possible levels of environmental protection.

[51]D. Woodhead, *Radiat. Res.*, 1970, **43**, 582–597.
[52]IAEA, Tec. Rep. International Atomic Energy Agency, Vienna, 172, 1976, 131 pp.

CHAPTER 3

# Drinking Water Quality and Health

R. F. PACKHAM

## 3.1 INTRODUCTION

The main consideration in ensuring the safety of public water supplies is the elimination of the agents of waterborne infectious disease. In the last century, major epidemics of typhoid and cholera in Britain and elsewhere in Europe were only eliminated when their bacteriological origin was recognized, the contamination of water supplies with sewage was eliminated and disinfection treatment was introduced. In the UK, the Croydon typhoid epidemic of 1937 led to chlorination of all public water supplies. From then until 1986, 34 outbreaks of waterborne disease were recorded in the UK.[1] Of these outbreaks, 21 were due to public water supplies and resulted from disinfection failure or contamination of treated water. Some new problems affecting the microbiological safety of water supplies have emerged in recent years,[2] but the basic approach of securing the adequacy of disinfection with several lines of defence, while maintaining an efficient monitoring strategy, has resulted in the achievement of an excellent record of safety.

Concern about the possibility of health effects associated with the chemical components of drinking water only really developed in the second half of the 20th century, although early drinking water standards included limits for lead and some other toxic elements. The main upsurge of interest in chemical aspects of water quality stemmed from the application of gas chromatography, later coupled with mass spectrometry, in the 1970s and 80s. Originally focused on pesticides, these and subsequent developments in analytical technology revealed for the first time the presence in water of a wide range of organic compounds at trace levels. The precise significance to health of these is still uncertain, but some would undoubtedly cause concern were they present at much higher concentrations. Acute health effects of the chemical constituents of drinking water are very unlikely, but effects due to exposure to low concentrations over a lifetime are feasible, although they are likely to be small and very difficult to measure.

[1] N. S. Galbraith, N. J. Barrett and R. Stanwell-Smith, *J. Inst. Water Environ. Managt.*, 1987, **1**, 7.
[2] Department of the Environment and Department of Health, 'Cryptosporidium in Water Supplies. Report of the Group of Experts', HMSO, London, 1990.

With a shortage of clear evidence of health effects on humans, precautionary limits for chemicals in drinking water, based on toxicological evaluations have been developed and applied by national and international organizations such as the World Health Organization, the US Environmental Protection Agency, Health and Welfare Canada, the European Union and others. These limits cover both inorganic and organic chemicals. In 1971,[3] the World Health Organization proposed a total of nine limits for chemicals in drinking water, eight inorganic and one organic. By 1993, in the WHO 'Guidelines for Drinking Water Quality',[4] this number had grown to a total of 94, of which 22 were for inorganics, 72 for organics. This period has therefore been associated with a remarkable increase in the complexity of drinking water standards.

Apart from any potential effects on health, the chemical composition of drinking water is of considerable importance in relation to its acceptability to the consumer in terms of appearance, taste, smell, hardness and corrosiveness.

## 3.2 DRINKING WATER STANDARDS

The present quality requirements for public water supplies in England and Wales are defined in regulations made under the Water Supply (Water Quality) Regulations 1989.[5] Essentially similar regulations have been established for Scotland and Northern Ireland. UK regulations must comply with a European Union Directive on the quality of water intended for human consumption.[6]

The Directive lays down limits for 44 contaminants that member states are required to regulate. Thus the Directive provides the basis of national legislation on minimum drinking water quality within the European Union. There is nothing to prevent more stringent limits or limits for a wider range of contaminants from being included in national regulations. For example, drinking water quality regulations for the UK include an additional 11 standards.

The World Health Organization (WHO) published 'Guidelines for Drinking Water Quality' in 1984 and in 1993.[4] These are up-to-date reviews and recommendations by international experts and are intended to provide guidance to national organizations worldwide with responsibilities for setting drinking water standards. The WHO guidelines have no legal status, but they are widely referred to as a source of reliable information on health aspects of drinking water quality.

The basis of the EU Directive on Drinking Water Quality, which came into force in 1985, has never been published, but it seemed to be based to some extent on WHO Drinking Water Standards published in 1970. The Directive has been widely criticised on scientific grounds and it is now under review with particular reference to the most recently published WHO Guidelines.

[3]World Health Organization, 'International Standards for Drinking Water', WHO, Geneva, 1971.
[4]World Health Organization, 'Guidelines for Drinking Water Quality', WHO, Geneva, 1993.
[5]Statutory Instrument No. 1147. 'The Water Supply (Water Quality) Regulations 1989'. HMSO, London, 1989.
[6]Council of the European Communities, 'Council Directive of 15 July 1980 Relating to the Quality of Water Intended for Human Consumption (80/778/EEC)', *Official Journal of the European Communities*, No. L229/11, August 30, 1980.

The WHO Guidelines are used as the basis of national drinking water standards worldwide, but there can be many variations to take account of local circumstances. WHO stress the need to consider guideline values in the context of local or national environmental, social, economic and cultural conditions.

## 3.3  LEAD

Lead levels in untreated natural waters are normally very low and lead in drinking water is almost entirely due to lead pipes in household plumbing and service connections, often but not invariably in combination with soft water. Where there is 'plumbosolvency' the concentration of lead increases with the period of contact with the lead pipe, maximum levels arising in morning 'first draw' samples. The use of lead is avoided in modern plumbing systems, but many older properties in the United Kingdom still have lead pipes, even in soft water areas. Apart from lead pipes, lead may be a constituent of solders, alloys used in water fittings and PVC pipes, all of which may be in contact with drinking water.

Lead is a general toxicant that accumulates in the skeleton and its presence in drinking water is highly undesirable. Blood lead levels are commonly used as an index of recent exposure to lead and several studies have indicated a correlation between blood lead and water lead extending down to concentrations below 50 $\mu g \, l^{-1}$ of lead in drinking water, the current EU limit. The fact that such low levels of lead in drinking water can produce a detectable elevation of blood lead in a population is not a harmful effect in itself although it gives rise to concern about the possibility of consequential problems. Of particular importance is the possibility of neuro-physiological effects influencing learning ability and general behaviour in children, who with pregnant women and infants are considered to be the most sensitive group of the population. On the basis of a thorough review of the evidence, a health based guideline value of $10 \, \mu g \, l^{-1}$ lead in drinking water was developed by WHO in 1993.

Of the possible remedies for plumbosolvency, water treatment is the most attractive, although it may present problems in isolated areas with a large number of small unmanned sources. Careful adjustment of the pH to between 8.0 and 8.5, sometimes with the addition of orthophosphate, has proved in several cases to be effective in reducing lead levels at the tap to below $50 \, \mu g \, l^{-1}$. There is no chance however, that a limit of $10 \, \mu g \, l^{-1}$ will be achievable by water treatment and the only remedy will be the complete replacement of lead with a more suitable material. This is an extremely expensive option and one that WHO has recognized will take time to achieve. Where plumbosolvency is known to be a problem, everything must be done to reduce the exposure of consumers by controlling the corrosiveness of the water and by advising them to avoid 'first draw' water.

The concentration of lead in water that has been through a lead pipe varies according to factors such as the contact time and flow and it is especially important to ensure that the conditions of sampling are adequately specified when water is monitored for compliance with a lead standard.

## 3.4 NITRATE

There is considerable evidence[7] that nitrate levels in water have increased in the United Kingdom since the 1960s and although the increased use of nitrogenous fertilizers has been blamed for this situation, it has been shown that additional factors need to be taken into account. These include changes in land use, in particular the conversion of pasture into arable land, and increased recycling of sewage effluent in lowland waters.

Limits for nitrate in drinking water are based on its effect on a blood disease, methaemoglobinaemia, in bottle fed infants. The 1970 WHO European Drinking Water Standards[8] included a desirable limit for nitrate of 50 mg $l^{-1}$ (as $NO_3$), levels between 50 and 100 mg $l^{-1}$ (as $NO_3$) being 'acceptable' provided that medical authorities are warned of the possible danger of infantile methaemoglobinaemia. The EU Drinking Water Directive included a Maximum Admissible Concentration (MAC) of 50 mg $l^{-1}$ (as $NO_3$) and this was set by WHO as a Guideline Value in 1993. Water supplies containing nitrate levels exceeding 50 mg $l^{-1}$ potentially affect approximately one million people in the UK, but there is no evidence of a problem with infantile methaemoglobinaemia.

Infantile methaemoglobinaemia is caused by the bacterial reduction of ingested nitrate to nitrite in the stomach. The nitrite combines with blood haemoglobin to produce methaemoglobin thereby reducing its oxygen carrying capacity. Many factors other than water nitrate level are now known to be important in relation to infantile methaemoglobinaemia and it is noteworthy that a very high proportion of cases reported in the world literature relate to rural wells where bacteriological contamination, which can influence methaemoglobinaemia, cannot be ruled out.

There has been considerable interest in the possibility that ingested nitrate can be reduced to nitrite in the adult stomach leading, in the presence of secondary amines, to the endogenous synthesis of *N*-nitroso compounds. The significance of *N*-nitroso compounds is that many are highly carcinogenic in laboratory animals and there has been concern that high nitrate levels in water could be associated with an increased incidence of cancer of the gastro-intestinal and urinary tracts. The overall picture in the UK does not, however, support this. Not only have gastric cancer rates been decreasing while nitrate levels have been increasing, but many of the areas showing the highest rates of gastric cancer have a low level of nitrate in their water supply.

A WHO Working Group[9] reviewed the available evidence for an association between nitrate levels in water and gastric cancer. It concluded: 'There is no convincing evidence of a relationship between gastric cancer and consumption of drinking water containing nitrate at or below the present Guideline Value: above this level the evidence is inconclusive'.

[7]Department of the Environment, Central Directorate of Environmental Pollution, 'Nitrate in Water', Pollution Paper No. 26, HMSO, London, 1986.
[8]World Health Organization, 'European Standards for Drinking Water', WHO, Geneva, 1970.
[9]World Health Organization, 'Health Hazards from Nitrates in Drinking Water', WHO, Copenhagen, 1985.

Although there is no evidence that current levels of nitrate in public water supplies present a health hazard, compliance with the EU Maximum Admissible Concentration of 50 mg $l^{-1}$ is a mandatory requirement which is necessitating remedial action in UK and other European countries. Changes in agricultural practice, including controls on the use of nitrogenous fertilizers and the specification of protection zones around wells and boreholes, are being introduced in an attempt to stem the upward trend in nitrate concentrations. Where possible, nitrate levels in drinking water are reduced by blending, but in some cases it has been necessary to install nitrate removal treatment. The two available options are ion exchange or biological denitrification, either of which add significantly to the cost and complexity of water treatment.

## 3.5 WATER HARDNESS AND CARDIOVASCULAR DISEASE

The hardness of water supplies in Britain tends to follow a north to south-east gradient with the softest supplies in Scotland, Northern England and Wales, and the hardest in East Anglia and Southern England. Mortality from cardiovascular disease (CVD) tends to follow the same pattern and several statistical studies have demonstrated a highly significant inverse relationship between cardiovascular disease mortality and water hardness even when other environmental and socio-economic factors are taken into account.

The British Regional Heart Study is a major epidemiological investigation of the role of environmental, socio-economic and personal risk factors in determining the regional variation of CVD mortality in Britain. The first phase of this research involved a statistical study[10] in which mortality from CVD in 253 large towns (population $> 50\,000$) in England, Scotland and Wales was examined in relation to a wide range of climatic and socio-economic factors, as well as a variety of drinking water quality parameters. This work confirmed that, despite considerable scatter in the data (see Figure 3.1), there is a highly significant inverse statistical relationship between the hardness of drinking water and mortality from CVD. This relationship persists when age, sex, socio-economic and climatic factors are taken into account (see Figure 3.2) and is not shown for mortality from non-cardiovascular diseases.

On average, very soft water towns have about 10% higher cardiovascular mortality than medium-hard or harder water towns after adjustment for socio-economic and climatic factors. Figures 3.1 and 3.2 show that most of the variation in mortality takes place at the soft end of the hardness range (below 150 mg $l^{-1}$ as $CaCO_3$).

Out of the large number of water variables examined, water hardness and certain associated water quality parameters, for example nitrate, calcium, conductivity, carbonate hardness and silica, give the strongest negative correlation with CVD mortality. The proportion of the water supply derived from upland sources gives an equally strong, but positive correlation. It could only be concluded that CVD is influenced by water softness or some factor closely

[10] S. J. Pocock, A. G. Shaper, D. G. Cook, R. F. Packham, R. F. Lacey, P. Powell and P. F. Russell, *Brit. Med. J.*, 1980, **280**, 1243.

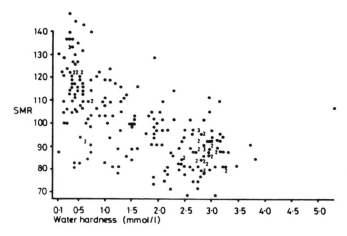

**Figure 3.1** *Water hardness plotted against SMR\* for all men and women aged 35–74 with cardiovascular disease for each town. (Water hardness: 1 mmol $l^{-1}$ = calcium carbonate equivalent 100 mg $l^{-1}$)*
\*SMR is Standardized Mortality Ratio (actual mortality divided by mortality estimated on basis of age/sex profile of populations).[10]

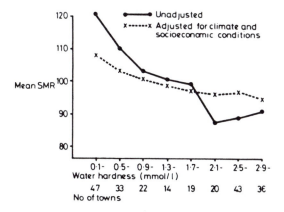

**Figure 3.2** *Geometric means of SMR\* (for all men and women aged 35–74 with cardiovascular diseases) for towns grouped according to hardness of water. (Water hardness: 1 mmol $l^{-1}$ = calcium carbonate equivalent 100 mg $l^{-1}$)[10]*
\*SMR Standardized Mortality Ratio.

associated with it. The relationship could be due either to a harmful factor associated with soft water or a protective factor associated with hard water.

The conclusions were strengthened by work[11] which showed that in situations where the hardness of water supplies changed, there was a corresponding change in the CVD mortality rate, at least for men. The size of the water factor is small in comparison with other CVD risk factors, *e.g.* heavy smoking doubles the risk of a CVD event while very soft water increases it by 10%.

[11] R. F. Lacey and A. G. Shaper, *Int. J. Epidemiol.*, 1984, **13**, 18.

Detailed examination of drinking water composition in hard and soft water towns showed that apart from obvious differences in the major mineral components, the levels of aluminium, iron, manganese and lead tended to be higher in the softer waters. Where the water was not only soft, but also acidic, levels of copper as well as lead could be very high. Apart from these extreme situations, copper levels tended to be higher in the hard water towns.

While there is strong statistical evidence that drinking water affects CVD, in the absence of a valid explanation, such drastic, unwelcome, costly and barely practicable actions as the hardening of soft water supplies would have no justification. Some restriction is placed on the level of central softening in the EU Drinking Water Directive and the installers of domestic water softening equipment are now offering a hard water tap to provide water for drinking and culinary purposes.

## 3.6 OTHER INORGANIC CONSTITUENTS

Drinking water standards and guidelines include limits for many inorganic ions in addition to lead and nitrate. Those set on the basis of toxicity include cadmium, mercury, chromium, arsenic, selenium and cyanide. None of these represent a general problem in the UK although problems may occur periodically in specific locations. There are a few inorganic water constituents that attract media attention from time to time and some notes on some of the more common of these are included here.

### 3.6.1 Aluminium (see also Chapter 18)

Aluminium sulfate and other aluminium compounds are widely used as coagulants in the treatment of water for public supply. On addition to water an insoluble 'floc' of hydrolysis products of aluminium is formed which engulfs colloidal and suspended material in the raw water. The floc is normally removed by sand filtration, usually preceded by sedimentation. Inefficient treatment can lead to the breakthrough of floc into the water distribution system where it can give rise to dirty water problems. Limits currently placed on the maximum level of aluminium in drinking water are all set with the objective of limiting such problems; they have no toxicological basis.

Concern about possible toxic effects of aluminium in drinking water stemmed from evidence of its role in dialysis encephalopathy (also called dialysis dementia). High aluminium levels in the brain tissue of patients dying of this disease and an association between its incidence and high aluminium concentrations in dialysate water have led to measures to minimize aluminium intake from this source. The very low limits for aluminium required (less than $10 \ \mu g \, l^{-1}$) can only be met by installing a special treatment unit in close proximity to the dialysis machine. Water companies provide information to renal dialysis centres on aluminium levels, but are not required to take responsibility for meeting such a low limit.

Higher than normal levels of aluminium are found in certain characteristic lesions of brain tissue in persons dying of a form of premature senile dementia – Alzheimer's disease. One of several mechanisms postulated for this disease ascribes an important role to ingested aluminium. Even in the worst cases it is unlikely that water contributes more than 5% of the total dietary intake of aluminium, although it has been argued that water aluminium may be more bioavailable than other dietary forms. Concern about water aluminium has been heightened by statistical studies indicating an association between water aluminium at low levels and the incidence of Alzheimer's disease.[12] WHO advises caution in the interpretation of these studies which need to be confirmed in analytical epidemiological studies. No health-based Guideline Value was set in 1993 because it was concluded that the balance of the epidemiological and physiological evidence did not support a causal role for aluminium.

### 3.6.2 Asbestos

Asbestos is a component of many materials used in contact with drinking water including asbestos-cement pipes, tanks, reservoir covers and certain glands and packing materials. Asbestos is a natural contaminant of water in some areas; its presence in water may also result from pollution from tips containing asbestos waste. It has been shown[13] that drinking water can contain of the order of one million chrysotile fibres per litre, usually as short fibres (less than 5 $\mu$m in length). Although there is clear evidence of harmful health effects associated with inhaled asbestos, the lack of any consistent evidence of a hazard associated with ingested asbestos led WHO to conclude in 1993 that there was no need to establish a health-based Guideline Value for asbestos in drinking water.

### 3.6.3 Fluoride

Fluoride in water has been the subject of an extended debate, but the practical and scientific evidence leaves no room to doubt the benefits to dental health arising from fluoridation to 1 mg l$^{-1}$ and the absence of harmful side effects. The moral issue of mass medication is still argued and fluoridation is not widely practised although it has official backing.

### 3.6.4 Sodium

Sodium is normally present in drinking water at concentrations below 50 mg l$^{-1}$ and the intake from this source will represent no more than 1–2% of the total dietary intake. The only group for whom water sodium is of any consequence is bottle-fed infants. The infant kidney is less effective than the adult at sodium elimination, so there is a danger of hypernatraemia if cow's milk, which has three times the sodium level of human milk, is fed to the infant. The sodium content of

[12]C. N. Martyn, D. J. P. Barker, C. Osmond, E. C. Harris, J. A. Edwardson and R. F. Lacey, *Lancet*, 1989, **1**, 59.
[13]D. M. Conway and R. F. Lacey, 'Asbestos in Drinking Water', Water Research Centre Technical Report TR202, WRc, Medmenham, Marlow, 1984.

the milk powders used in baby food is therefore reduced and it is important not to make the feed up in water with a high sodium concentration. Normally this will not arise except where a hard water has been softened in a domestic water softener by base exchange. The Department of Health[14] has advised against the use of such water in preparing baby food and it is now good practice for domestic water softener installations to be provided with a hard water tap for drinking and culinary use. For patients on a strictly reduced sodium diet, the use for drinking purposes of base exchange softened water would also be inadvisable.

## 3.7  ORGANIC MICROPOLLUTANTS  (see also Chapter 16)

The total organic content of drinking water as represented by the concentration of dissolved organic carbon is usually no greater than 5 mg $l^{-1}$. Following the application in water analysis of techniques such as GC-MS, information on the nature of the organic chemicals present in drinking water developed rapidly and it soon became apparent that thousands of different organic compounds could be present at minute concentrations, often well below 1 $\mu$g $l^{-1}$. Many surface derived waters contain a substantial fraction of material of natural origin derived from humic substances present in soil and peat, together with a wide range of contaminants derived from industry, agriculture and other human activities. Compounds formed during water treatment by the reaction of chlorine and other disinfectants principally with the 'natural' organic compounds present, have received a great deal of attention because it is believed that they represent a potential hazard to health. Drinking water derived from groundwater sources usually contains much lower levels of humic substances and their derivatives, but pesticide contamination may be apparent in agricultural areas. Volatile contaminants such as carbon tetrachloride and trichloroethylene are more persistent in groundwater than in surface waters.

The total land surface is the catchment for our drinking water sources and it is evident that any activity on that surface is potentially capable of affecting water quality. In this context, the diversity of the chemicals shown by highly sensitive analytical techniques to be present at trace levels in water must come as no surprise. A greater awareness of the nature of these organic micropollutants has led to the establishment of health-related limits for many of them and a major development programme aimed at improved pollution control and effective water treatment for their removal.

## 3.8  PESTICIDES

The term pesticides embraces a very diverse group of chemicals which individually can have very different functions in agriculture. Despite this, all substances covered by the term were treated in precisely the same way in the EU Drinking Water Directive and subsequently in the Water Supply (Water Quality) Regulations 1989. The statutory maximum admissible concentrations for pesti-

[14]Department of Health, 'Present Day Practice in Infant Feeding', Reports on Health and Social Subjects No. 20, HMSO, London, 1980.

cides in drinking water set by EU, were widely criticized as having no toxicological basis. The limit of 0.1 $\mu$g $l^{-1}$, which applies to any individual pesticide, is believed to have been based on a one time limit of detection for chlorinated insecticides. As such it had been regarded as a surrogate zero and was combined with a limit for total pesticides of 0.5 $\mu$g $l^{-1}$.

Limits set for pesticides in the USA and Canada were based on the toxicity of individual pesticides and an essentially similar approach was adopted by the World Health Organization in the Guidelines for Drinking Water Quality. It should be noted that because of the requirements of various schemes for the approval or registration of pesticides, there is a substantially greater body of toxicological data available for pesticides than for most other micropollutants of water.

For some compounds, an Acceptable (or Tolerable) Daily Intake (ADI or TDI) in mg/kg body weight, developed in a joint programme for food additives and pesticides by WHO and the Food and Agricultural Organization, is available and this has been used by WHO as the basis for drinking water guidelines. The ADI or TDI is obtained from an experimentally determined no-effect level by applying safety factors of up to 1000 to account for uncertainties involved in extrapolating toxicological data for animals to man and from high levels of exposure used in toxicological studies to the low levels experienced in drinking water. Safety factors are also incorporated to take account of any deficiencies in the data base. In setting limits, the ADI or TDI has to be allocated to the various possible routes of exposure, *i.e.* food, air and water. For pesticides, the major fraction is apportioned to food, only between 1 and 10% being allocated to drinking water. Thus taken together, very considerable safety factors are involved in these limits.

For carcinogenic substances, it cannot be assumed that below some threshold level of exposure there are no adverse effects on health, and for these the toxicological data is extrapolated to determine a concentration corresponding to an 'acceptable' level of risk. A number of extrapolation models are available for this purpose; the 'one hit multistage model' used by WHO was developed by the US Environmental Protection Agency. For the purpose of setting Guideline Values the acceptable level of risk was set at 1 in $10^5$ *i.e.* one additional cancer in 100 000 lifetimes of exposure. The uncertainties involved in these extrapolations are considerable, amounting in some cases to two orders of magnitude.

As shown in Table 3.1, many guideline values set by WHO for pesticides in 1993 exceeded the EU limit of 0.1 $\mu$g $l^{-1}$, but some were lower.

About 500 pesticides have been approved for use in the UK. The Department of the Environment in 1989 published 'Advisory Values', based on an approach similar to that adopted by WHO, for pesticides thought likely to be found in water. Advisory Values, like the WHO Guideline Values, assume lifetime consumption and incorporate safety factors of 1000 or more. A small or transient breach of an Advisory Value is not considered to represent a threat to public health.

Herbicides including atrazine, simazine, isoproturon and chlorotoluron have been detected most frequently in groundwater derived supplies. Not all of these

**Table 3.1**   *WHO   Guideline   Values   (1993)[4]   for*
                *certain pesticides*

| Compound or group of isomers | Guideline Value $\mu g\, l^{-1}$ |
|---|---|
| DDT | 2 |
| aldrin and dieldrin | 0.03 |
| chlordane | 0.2 |
| hexachlorobenzene | 1* |
| heptachlor and heptachlor epoxide | 0.03 |
| Lindane | 2 |
| methoxychlor | 20 |
| 2.4-D | 30 |

*derived by cancer risk extrapolation

are necessarily related to agricultural use; there has been a considerable use of atrazine and simazine for weed control on roadside verges and railway embankments. Supplies derived from surface waters may occasionally exceed the EU limit because of their vulnerability to contamination from run off, field drains and spray drift. There is evidence that the conventional treatment applied to surface water achieves some reduction in the concentration of certain pesticides, but this cannot be relied upon to achieve compliance. The treatment normally applied to groundwater has little effect on pesticide levels. 34 pesticides were detected at levels greater than the EU limit of 0.1 $\mu g\, l^{-1}$ during 1993 in water supplies in England and Wales.[15] With very few exceptions the levels found were below Advisory Levels of WHO Guideline Values.

Many water companies have given undertakings to provide information on pesticide levels to the National Rivers Authority on a regular basis so that pollution control action can be considered. They have also undertaken to evaluate and install water treatment plants by a specified date with the objective of achieving compliance with the EU limit. There is no doubt that the cost of the additional processes that will be required to do this will be very significant.

## 3.9   DISINFECTION BY-PRODUCTS

The recognition in the 1970s that the use of chlorine for the disinfection of water could lead to the formation of potentially hazardous trihalomethanes, had a significant impact on water treatment throughout the world. The limit of 100 $\mu g\, l^{-1}$ originally proposed by USEPA[16] for a total of four trihalomethanes (chloroform, bromodichloromethane, dibromochloromethane and bromoform), has been adopted in many countries including the UK. In waters found to contain

[15]Department of the Environment and the Welsh Office, 'Drinking Water 1993 – A Report by the Chief Inspector, Drinking Water Inspectorate', HMSO, London, 1994.
[16]US Environmental Protection Agency, 'National Interim Primary Drinking Water Regulations: Control of Trihalomethanes in Drinking Water; Final Rule', *Federal Register*, 1979, **44**, 68624.

more than this limit for 'Total THMs', three main techniques have been used to achieve compliance:

(i)  reducing the intensity of chlorination by reducing or eliminating pre-chlorination and by cutting the chlorine dose where this can be done without compromising disinfection,
(ii)  optimizing the removal of THM precursors during coagulation and
(iii)  considering the use of an alternative disinfectant.

In the UK, official advice[17] has been that any health hazards associated with disinfection by-products will be extremely small and that efficient disinfection should on no account be compromised by efforts to minimize them. There has nevertheless been a marked reduction in the level and use of prechlorination and, in the treatment of some lowland waters, ozone is being used prior to clarification, partly as a preliminary disinfectant and partly as an oxidant for trace organics.

There have been no marked changes in final disinfection on account of disinfection by-products and chlorine, and in some cases chloramine, continue to be used as before. In the USA and in some continental European countries, chloramination has become more widespread. There has also been considerable interest in chlorine dioxide, but questions relating to its acceptability on toxicological grounds, which have not yet been resolved, have limited its use. A generally cautious approach in some countries towards the adoption of disin-fectants other than chlorine, is due in part to the lack of information on the nature and toxicology of relevant by-products.

A large number of chlorination by-products in addition to trihalomethanes have now been identified and many of these are potentially hazardous com-pounds. In 1993, WHO presented Guideline Values for 15 disinfection by-products including the trihalomethanes, some chlorinated acetic acids and some halogenated acetonitriles. Provisional Guideline Values were included for bromate $(25\ \mu g\ l^{-1})$, a by-product of ozonization, and for chlorite $(200\ \mu g\ l^{-1})$, a by-product of the use of chlorine dioxide.

## 3.10  WASTEWATER REUSE AND HEALTH

Approximately one-third of water supplies in the UK are derived from lowland rivers, used also for conveying treated domestic and industrial wastes to the sea. These rivers are relatively short and often support large populations and considerable industrial activity. The proportion of the total river flow represented by wastewater can often be high, particularly during dry periods. The level of this indirect reuse of wastewater for potable purposes is likely to increase in future.

No precise information is available on the effect on drinking water quality of wastewater recycling. It may be assumed that there will be an increase in the concentration of those waste constituents that are not biochemically degraded

---

[17]Department of the Environment and the Welsh Office, 'Guidance on Safeguarding the Quality of Public Water Supplies', HMSO, London, 1990.

during sewage treatment, together with a wide range of metabolic products derived from degradable chemicals.

An investigation into the possible significance to health of wastewater reuse, funded by the Department of the Environment,[18] had as a main objective the possible effects of reuse on cancer incidence. The detailed analysis of different types of water supply by GC-MS provided no real evidence that drinking water derived from sources containing a high proportion of sewage effluent contained more potentially hazardous substances than those with a low proportion. The health statistics of populations receiving water supplies derived from source waters containing differing proportions of sewage effluent was examined, the most detailed study concentrating on the London boroughs.[19] Some of these have groundwater supplies, while others have supplies drawn from rivers containing at times a high proportion of sewage effluent. It was essential in this work to take account of socio-economic factors, which are known to have an influence on cancer incidence. When this was done, no strong association between cancer mortality and reuse was evident. A national study gave substantially similar results. It has to be recognized that because of the large number of factors that influence cancer incidence, epidemiological studies of this kind are rather insensitive. The feasibility of large scale case control studies has been considered, but on practical and financial grounds such work is unlikely to proceed in the near future.

## 3.11  CONTAMINATION OF DRINKING WATER IN DISTRIBUTION SYSTEMS

Potential sources of contamination of water during distribution include leachates and corrosion products from pipe materials and pipe linings and organic substances able to permeate through certain plastic pipe materials.

In the UK, about 60% of distribution pipes are made of ductile iron and until about 1977 the majority of these were coated internally with coal-tar pitch. Coal-tar pitch can contain up to 50% polycyclic aromatic hydrocarbons (PAH) and there is potential for the leaching of these substances, which include some known carcinogens, into the water supply. A survey of PAH levels in British waters[20] showed that low levels of these compounds can usually be detected in water that has passed through pipes lined with coal-tar pitch. Some iron pipes are now being lined with epoxy resin coatings. These are often applied *in situ* requiring strict adherence to a rigid code of practice to ensure that the coating is properly cured to minimize leaching of the coating chemicals.

Organic compounds added to the plastic formulations during pipe manufacture have been shown to leach into drinking water. Antioxidants in polyethylene (PE) pipes sometimes give rise to objectionable tastes. Low concentrations of UV

[18]R. F. Packham, S. A. Beresford and M. Fielding, *Sic. Total Environ.*, 1981, **18**, 167.
[19]S. A. Beresford, *Int. J. Epidemiol.*, 1981, **10**, 103.
[20]R. J. Crane, M. Fielding, T. M. Gibson and C. P. Steel, 'A Survey of Polycyclic Aromatic Hydrocarbon Levels in British Waters', Water Research Centre Technical Report TR 158, WRc, Medmenham, Marlow, 1981.

stabilizers and their degradation products can be detected in water distributed through PE pipes. New PVC pipes usually contain leachable levels of lead or tin stabilizers. The level of extraction of most of these compounds decreases with time.

The ability of organic compounds to permeate PE has been known for many years[21] and it has been shown in the laboratory that low molecular weight compounds such as phenol, benzene, toluene, trichloroethylene and chlorobenzene will readily permeate PE pipe. There have been several examples of service pipes coming into contact with chemicals and giving rise to water contamination in this way.

## 3.12 CONCLUSIONS

The remarkable increase over two decades in the complexity of assessments of chemical aspects of drinking water quality is due to developments in analytical technology and toxicology together with a growing concern about the possible chronic effects of long-term exposure to trace chemicals. This concern is not confined to water, but there is less opportunity with water than with food for the consumer to exercise choice. A development of particular significance in the UK has been the establishment of mandatory limits for water quality parameters which have provided for the first time an unambiguous basis for complying with the legal requirement placed on water companies to supply drinking water that is 'wholesome'.

There is no reason to believe that knowledge in this field has reached a plateau; indeed it is certain that there will be further increases in the number of limits for chemicals in water in the next decade. Although in the UK we should be reaping the benefit of a much tighter regime of pollution control, new chemicals will come into use and some will inevitably find their way into water supplies. There will be new pesticides and although it is to be hoped that these will be engineered to minimize environmental and water supply problems, the difficulty of controlling diffuse inputs to water remains. A major problem area which has by no means been worked out is that of disinfection by-products. This growing group of compounds includes many that are potentially hazardous and which present control problems because disinfection cannot be compromised.

An obvious conclusion is that water quality monitoring and water treatment will become even more complex. It is possible that, at some time, the situation will be simplified on the monitoring side by the identification of a number of satisfactory 'indicator parameters' and on the regulatory side by the specification of mandatory unit processes which, like disinfection, can be depended on to deal with particular groups of undesirable compounds. Such an approach would bring the chemical and the microbiological parameters of water quality into line.

---

[21]R. F. Packham and J. K. Fawell, Proceedings of Conference 'The Use of Plastics and Rubber in Water and Effluents,' Plastics and Rubber Institute, London, 1982.

CHAPTER 4

# Water Pollution Biology

C. F. MASON

## 4.1 INTRODUCTION

Water pollution can be defined as 'the introduction by man into the environment of substances or energy liable to cause hazards to human health, harm to living resources and ecological systems, damage to structure or amenity, or interference with legitimate uses of the environment'.[1] Pollutants are therefore chemical or physical in nature and can thus be measured more or less accurately in water. The measured quantities can then be compared with standards of allowable concentrations. Why then do we need to undertake quantitative studies of organisms and biological communities when we know that they can be defined with much less precision than chemical or physical parameters? There are a number of reasons why biological studies are important.

Firstly, the definition of pollution given above includes the adverse effects on living resources and ecological systems, so that impacts need quantifying. Man, of course, by drinking water, eating fish and using freshwaters as recreation areas, is also linked to the aquatic community. We can consider the effects of pollutants we record on the biological community as an early warning of potential effects on ourselves.

Secondly, animal and plant communities also respond to intermittent (episodic) pollution which may be missed in a chemical surveillance programme. A chemical survey will indicate to the water manager what is present in a sample at a particular moment in time. However, a plug of pollution could have passed down a river before the sample was taken or between sampling occasions, while a disreputable factory owner or farmer may well know which day the river inspector takes his sample so that he can discharge his toxic waste accordingly. The pollution will kill the most vulnerable members of the aquatic community, the most sensitive species acting as *indicators* of pollution. The amount of change in the community will be related to the severity of the incident. Because the community can only be restored to its former diversity by reproduction and

---

[1] C. F. Mason, 'Biology of Freshwater Pollution', Longman Scientific and Technical, Harlow, 3rd edn, 1996.

immigration, its recovery is likely to be slow. If the intermittent pollution occurs with some frequency, the community will remain impoverished. The biologist will be able to detect such damage and suggest a more detailed surveillance programme, both biological and chemical, to find the source of pollution.

Thirdly, biological communities may respond to unsuspected or new pollutants in the environment. There are well in excess of 1500 potential pollutants discharged to freshwaters, while only some 30 determinants will routinely be tested for. It is financially prohibitive to look for more compounds. If there is a change in the biological community, however, then a wider screening for pollutants can be initiated. For example, male rainbow trout (*Oncorhynchus mykiss*), caged in the River Lea, north of London, below the discharge of effluent from a sewage treatment works, exhibited marked increases in vitellogenin levels.[2] This compound is normally produced in the liver of female fish in response to the hormone oestradiol and is incorporated into the yolk of developing eggs. Clearly, some chemical in the effluent was behaving like a female hormone and the fish provided an early warning of a potential problem requiring urgent investigation. Increases in testicular cancer and falling sperm counts in the human male may be related to oestrogen-mimics released into the environment.[3] Nonyl-phenols (widely used as antioxidants) were suspected of stimulating vitellogenin production in the trout but many other substances, including DDT and some PCBs, exert similar effects.

Finally, some chemicals are accumulated in the bodies of certain organisms, concentrations within them reflecting environmental pollution levels over time. In any particular sample of water, the concentration of a pollutant may be too low to detect using routine methods, but nevertheless it may gradually accumulate within the ecosystem to levels of considerable concern in some species. Heavy metals, organochlorine pesticides and PCBs have caused particular problems in aquatic habitats and are potential threats to human health.

This review will describe the effects of major types of pollutants on aquatic life, though it must be remembered that any particular source of pollution may contain a range of compounds, such that the effect on ecosystem processes are likely to be complex. The review will conclude with a description of some of the methods of assessing the biological impacts of pollution.

## 4.2 ORGANIC POLLUTION

Organic pollution results when large quantities of organic matter are discharged into a watercourse to be broken down by microorganisms which utilize oxygen to the detriment of the stream biota. A simple measure of the potential of organic matter for deoxygenating water is given by the biochemical oxygen demand (BOD) which is determined in the laboratory by incubating a sample of water for five days at 20 °C and determining the oxygen used.

[2]C. E. Purdom, P. A. Hardiman, V. J. Bye, N. C. Eno, C. R. Tyler and J. P. Sumpter, *Chem. Ecol.*, 1994, **8**, 275.
[3]R. M. Sharpe and N. E. Skakkebaek, *Lancet*, 1993, **341**, 1392.

**Figure 4.1**  *A poor quality effluent discharge from a sewage treatment works*

One of the major sources of organic pollution is effluents from sewage treatment works. In the United Kingdom, such effluents are supposed, as a minimum requirement, to meet the Royal Commission Standard, allowing no more than 30 mg $l^{-1}$ of suspended solids and 20 mg $l^-$ BOD (a 30:20 effluent). A dilution with at least eight volumes of river water, having a BOD of no more than 2 mg $l^{-1}$, is required to achieve this standard. Unfortunately, the design capacities of many sewage treatment works are below the population they are now having to serve (Figure 4.1). This may cause chronic pollution of rivers or result in periodic flushes of poor quality water which damages the aquatic community. In the majority of poorer countries of the world there are few, or indeed often no, sewage treatment facilities and the faecal contamination of water results in many parasitic infections and waterborne diseases such as dysentery, cholera and poliomyelitis. Contaminated water supplies still cause more than two million deaths a year and countless more illnesses.

Other sources of organic pollution include industries such as breweries, dairies and food processing plants. Run-off from hard surfaces and roads in towns, especially during storm conditions, can be very polluting. Farm effluents have become a particular problem over the last few years, with the intensification of livestock rearing and the overwintering of animals in confined buildings, as well as the increased use of silage to feed them. Silage effluent can be 200 times as strong as settled sewage, measured in terms of BOD. In South-west England, a predominantly rural area, many rivers are failing to comply with their long-term quality objectives, agricultural pollution being the major reason for the decline.

Figure 4.2 outlines the general effects of an organic effluent on a receiving stream.[1] At the point of entry of the discharge there is a sharp decline in the

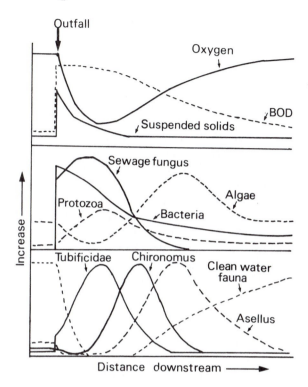

**Figure 4.2** *Changes in water quality and populations of organisms in a river below a discharge of an organic effluent*

concentration of oxygen in the water, known as the *oxygen sag curve*. At the same time there is a massive increase in BOD as the microorganisms added to the stream in the effluent, and those already present, utilize the oxygen as they break down the organic matter. As the organic matter is depleted, the microbial populations and BOD decline, while the oxygen concentration increases, a process known as *self-purification*, assisted by turbulence within the stream and by the photosynthesis of algae and higher plants (macrophytes).

The effluent will also contain large amounts of suspended solids which cut out the light immediately below the discharge, thus eliminating photosynthetic organisms. Suspended solids also settle on the stream bed altering the nature of the substratum and smothering many organisms living within it.

Under conditions of fairly heavy pollution, sewage fungus develops. This is an attached, macroscopic growth consisting of a whole community of microorganisms, dominated by *Sphaerotilus natans*, which consists of unbranched filaments of cells enclosed in sheaths of mucilage, and by zoogloeal bacteria. Sewage fungus may form a white or light brown slime over the surface of the substratum, or it may exist as a fluffy, fungoid growth with long streamers trailing into the water.

Protozoans are chiefly predators of bacteria and respond to changes in bacterial numbers. Attached algae are eliminated immediately below the outfall due to the diminished penetration of light, but they gradually reappear below the zone of gross pollution, *Stigeoclonium tenue* being the initial colonizer. With the decomposition of organic matter large quantities of nitrates and phosphates are released, stimulating algal growth and resulting in dense blankets of the filamentous green *Cladophora* smothering the stream bed. Similarly, macrophytes may respond to the increased nutrient concentration, though only *Potamogeton pectinatus* is very tolerant of organic pollution.[4]

Heavy organic pollution affects whole taxonomic groups of macroinvertebrates rather than individual sensitive species, and it is only in conditions of mild pollution that the tolerances of individual species within a group assume significance. The groups most affected are those which thrive in waters of high oxygen content and those which live on eroding substrata, the most sensitive being the stoneflies (Plecoptera) and mayflies (Ephemeroptera). The differences in tolerances of groups of macroinvertebrates form the basis of methods for monitoring, as will be described later in this chapter.

In the most severe pollution the tubificid worms, *Limnodrilus hoffmeisteri* and/or *Tubifex tubifex*, are the only macroinvertebrates to survive. The organic effluent provides an ideal medium for burrowing and feeding, while in the absence of predation and competition, the worms build up dense populations, often approaching one million individuals per square metre of stream bed. These worms contain the pigment haemoglobin, which is involved in oxygen transport, and they can survive anoxic conditions for up to four weeks. As conditions improve slightly downstream, the midge larva *Chironomus riparius*, which also contains haemoglobin, thrives in dense populations and, as the water self-purifies, other species of this large family of flies appear, the proportion of *Chironomus riparius* gradually declining. Below the chironomid zone, the isopod crustacean *Asellus aquaticus* becomes numerous, especially where large growths of *Cladophora* are present. At this point, molluscs, leeches and the predatory alder fly (*Sialis lutaria*) may also be present in some numbers. As self-purification progresses downstream the invertebrate community diversifies, though some stonefly and mayfly species, which are sensitive even to the mildest organic pollution, may not recolonize the stream.

Fish are the most mobile members of the aquatic community and they can swim to avoid some pollution incidents. In conditions of chronic organic pollution they are absent below the discharge, reappearing in the *Cladophora/Asellus* zone, the tolerant 3-spined stickleback (*Gasterosteus aculeatus*) being the first to take advantage of the abundant invertebrate food supply. Organic pollution is usually most severe in the downstream reaches of rivers and may prevent sensitive migratory species, such as salmon (*Salmo salar*) and sea trout (*S. trutta*), from reaching their pollution-free breeding grounds in the headwaters.

It must be remembered that many toxic compounds, such as ammonia, may occur in organic effluents, particularly those emanating from sewage treatment works. These make the prediction of the effects on the aquatic community of any particular discharge rather difficult.

[4] S. M. Haslam, 'River Plants', Cambridge University Press, Cambridge, 1978.

## 4.3 EUTROPHICATION

It has already been described how the release of nutrients during the breakdown of organic matter stimulates the growth of aquatic plants. The addition of nutrients to a waterbody is known as eutrophication. Other important sources of nutrients include phosphorus-containing detergents (much of this entering the river in sewage effluent), agricultural run-off and leaching of artificial fertilizers, the washing of manure from intensive farming units into water, the burning of fossil fuels which increases the nitrogen content of rain, the felling of forests which causes increasing erosion and run-off, and erosion resulting from recreational boating.

Nitrogen and phosphorus are the two nutrients most implicated in eutrophication and, because growth is normally limited by phosphorus rather than nitrogen, it is the increase in phosphorus which stimulates excessive plant production in freshwaters. Nitrogen is highly soluble and fertilizers form the main source of this element to rivers, accounting for some 70% of the annual mass flow of nitrogen in East Anglian rivers. Some 50% of the nitrogen applied to crops is lost to water. Phosphorus is largely insoluble, so that it enters water from land mainly by erosion. Some 90% of the phosphorus enters East Anglian rivers in the effluent from sewage works.

The concentrations of nitrate and phosphate in the water of Ardleigh Reservoir, a eutrophic waterbody in East Anglia, are shown in Figure 4.3.[5] Note that the concentration of nitrate increases during the late winter when fertilizer is applied to growing crops and is washed in large amounts into streams which feed the reservoir. By contrast, the concentration of phosphate peaks in late summer, when the low flow in the feeder streams consists largely of treated sewage effluent while, at this time of year, much phosphate is released from the reservoir sediments into the water.

Table 4.1 lists the guidelines for assessing the trophic status of a waterbody. Peak phosphorus concentrations in Ardleigh Reservoir were some 250 times the minimum concentration for assigning a waterbody as eutrophic, while peak concentrations of nitrogen were 10 times the minimum.

One of the major biological effects of eutrophication, resulting in considerable financial loss, is the stimulation of algal growth, especially in water supply reservoirs. As eutrophication progresses, there is a decline in the species diversity of the phytoplankton and a change in species dominance as overall populations and biomass increase. Figure 4.4 illustrates the seasonal changes in biomass of the dominant groups of algae in Ardleigh Reservoir.[5] Typical of temperate lakes is the early peak of diatoms (*Bacillariophyta*), followed by a late spring peak of green algae (*Chlorophyta*). Eutrophic lakes are characterized by enormous summer growths of cyanobacteria (*Cyanophyta* or blue-green algae), in the case of Ardleigh Reservoir mainly *Microcystis aeruginosa*, and these are the main cause of water treatment problems.

Large populations of algae, and in some cases the zooplankton which they support, may result in the blocking of filters in the treatment works. The reservoir

[5]M. M. Abdul-Hussein and C. F. Mason, *Hydrobiologia*, 1988, **169**, 265.

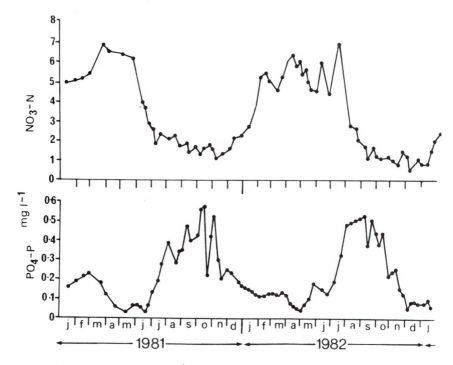

**Figure 4.3** *Concentrations* (mg l$^{-1}$) *of nitrate and phosphate in Ardleigh Reservoir, eastern England, over two years*

may have to be taken out of service, sometimes for several weeks, because the water becomes untreatable. If the reservoir is the only source of water this can have severe consequences. Some of the smaller algae may pass through the treatment process altogether, to decompose in the water main, giving drinking water an unpleasant taste and odour. The crash of an algal bloom and its subsequent decomposition may also result in a potable water so unpleasant to taste that it is almost undrinkable.

Drinking water which is high in nitrates presents potential health problems. In particular, babies under six months of age who are bottle fed may develop

**Table 4.1** *Eutrophication survey guidelines for lakes and reservoirs*

|  | Oligotrophic | Mesotrophic | Eutrophic |
|---|---|---|---|
| Total phosphorus ($\mu$g l$^{-1}$) | <10 | 10–20 | >20 |
| Total nitrogen ($\mu$g l$^{-1}$) | <200 | 200–500 | >500 |
| Secchi depth (m) | >3.7 | 3.7–2.0 | <2.0 |
| Hypolimnetic dissolved oxygen (% saturation) | >80 | 10–80 | <10 |
| Chlorophyll-a ($\mu$g l$^{-1}$) | <4 | 4–10 | >10 |
| Phytoplankton production (g C m$^{-2}$ d$^{-1}$) | 7–25 | 75–250 | 350–700 |

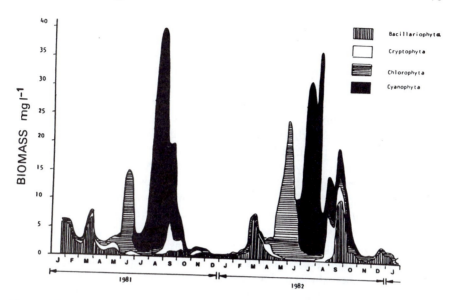

**Figure 4.4** *Seasonal variation in phytoplankton composition and total biomass (*mg l$^{-1}$ *wet weight)
in Ardleigh Reservoir, eastern England, over two years*

methaemoglobinaemia (blue baby syndrome), the nitrate in their feeds being
reduced to nitrite in their acid stomachs and then oxidizing ferrous ions in the
haemoglobin of their blood, so lowering its oxygen carrying capacity. There have
been several thousand cases world-wide, many of them fatal, but none in the
United Kingdom since 1972. The disease only occurs when bacteriologically
impure water, with nitrate levels approaching 100 mg l$^{-1}$, is supplied. Treated,
piped water presents no problems. There are also claims, so far unsubstantiated,
that nitrates may convert to carcinogenic nitrosamines in the adult stomach. The
European Health Standard for drinking water recommends that nitrates should
not exceed 50 mg $NO_3^-$ l$^{-1}$, but concentrations in some lowland areas of Britain
frequently exceed 100 mg $NO_3^-$ l$^{-1}$.

The algal blooms associated with excessive amounts of nutrients have other
consequences for the aquatic ecosystem. The macrophyte swards of many lakes
have been eliminated as the light is reduced on the lake bed, preventing
photosynthesis of germinating plants. Growths of epiphytic algae on the leaf
surface may also restrict light uptake by aquatic plants so that they become scarce
in the lake. Zooplankton use macrophyte canopies as a refuge from fish. Without
macrophytes, they are very vulnerable and this itself can accelerate the
eutrophication process as the grazing pressure on phytoplankton is reduced in
the absence of zooplankton, allowing denser algal blooms to develop.

The Norfolk Broads of eastern England are a series of shallow lakes, formed
during medieval times when peat workings became flooded, and famous for their
rich flora of *Charophytes* (stoneworts) and aquatic angiosperms, supporting a
diverse assemblage of invertebrates. During the 1960s a rapid deterioration set in
as nutrient loadings increased, and a survey of 28 of the main broads in 1972–73

revealed that 11 were devoid of macrophytes and only six had a well-developed aquatic flora. The invertebrate fauna was similarly impoverished. Since then the situation has deteriorated further with additional losses of macrophytes and the Norfolk Broads now have some of the highest total phosphorus concentrations of world freshwater lakes.[6]

It appears that shallow lakes have two alternative stable states over a range of nutrient concentrations.[7] They may either have clear water, dominated by aquatic vegetation, or turbid water with high algal biomass. The macrophyte community may be stabilized by luxury uptake of nutrients, making nutrients unavailable to plankton, by secreting chemicals to prevent plankton growth, and by sheltering large populations of grazers which eat algae. The phytoplankton community may be stabilized by an early growing season, shading the later germination of macrophytes, by increasing the vulnerability of herbivores to predation in the unstructured environment and by producing large, inedible algae. They also acquire carbon dioxide more easily, especially at high pH. In many lakes, including the Norfolk Broads, the stability of the macrophyte community has been broken. The causes of the switch may include the intro- duction of insecticides and herbicides widely into agriculture in the 1950s. Zooplankton are especially vulnerable to insecticides, while the cocktail of herbicides running off from land into watercourses could have weakened the growth of macrophytes. The loss of zooplankton would have allowed the build-up of large algal populations, switching the lakes into the alternative stable state.

Fish communities also change as oligotrophic lakes become eutrophicated, those cold water fish with high oxygen requirements, such as salmonids and whitefish, being replaced by less demanding cyprinids. The commercial value of the fishery declines in parallel. Some algae at high densities produce toxins which kill fish. In the Norfolk Broads *Prymnesium parvum* has caused several large fish kills over the last decade. The sudden collapse of an algal bloom may result in rapid deoxygenation of the water and massive kills of fish.

Cyanobacteria may also produce potent poisons which can induce rapid and fatal liver damage at low concentrations. Livestock and wildlife have been killed. Toxins do not always occur in blooms and can be highly variable with time, making them difficult to predict, detect and monitor.[8]

Problems of eutrophication are not restricted to standing waters. Nutrient-rich streams support dense growths of aquatic weeds which impede the flow of water, increasing the risk of severe flooding following summer storms. The weeds have to be cut, often twice a year, at great expense and the situation has been made worse by the clearance in the interests of 'river improvement' of bankside trees, which otherwise shade the stream and reduce weed growth.

It is possible to reverse eutrophication[9] though, in many cases, expensive remedial measures have proved much less successful than was hoped.[10] Methods

[6]B. Moss, *Biol. Rev.*, 1983, **58**, 521.
[7]M. Scheffer, S. H. Hosper, M.-L. Meijer, B. Moss and E. Jeppeson, *Trends Ecol. Evol.*, 1993, **8**, 275.
[8]L. A. Lawton and G. A. Codd, *J. IWEM*, 1991, **5**, 460.
[9]G. D. Cooke, 'Restoration and Management of Lakes and Reservoirs', Lewis, Boca Raton, 1993.
[10]P. Cullen and C. Forsberg, *Hydrobiologia*, 1988, **170**, 321.

include 'bottom-up' (nutrient control) and 'top-down' (biomanipulation) approaches. Bottom-up approaches involve controlling inputs of phosphorus because this element is normally limiting to plant growth and much of it in freshwaters is derived from point sources (*i.e.* sewage treatment works), whereas nitrogen enters the aquatic ecosystem diffusely, *via* land drainage. Phosphorus can be removed from the effluent at the treatment works by chemical flocculation (tertiary treatment) or before water is pumped into a reservoir. However the reduction in the inflow water may be counterbalanced by the regeneration of phosphorus from the sediments, so-called internal loading, which may necessitate either sediment removal or chemical flocculation and sealing, for example, with iron or aluminium sulfate.

Biomanipulation normally requires the management (or complete removal) of planktivorous fish, allowing the re-establishment of populations of the larger herbivorous zooplankton, which graze algae. Fish removal has led to increases in water clarity, encouraging the growth of macrophytes. Once the macrophyte stable state has developed it should be possible, indeed it is desirable, to introduce fish. Biomanipulation technology, however, is still rather new and the clear water phase may only be temporary, several lakes showing increased turbidity after the first few years of management. The control of phosphorus is in most cases, a necessary prerequisite to biomanipulation, which is likely to be most effective in shallow waters where macrophytes are a major component of the ecosystem.

## 4.4  ACIDIFICATION

Acid rain and acidification of freshwaters have received considerable attention over the past few years but the problem is not new, for there are observations of lakes losing their fish populations in Scandinavia as early as the 1920s. Studies of the diatom remains in cores of sediment from lochs in South-west Scotland have indicated that progressive acidification began around 1850, with an increase in *Tabellaria binalis* and *T. quadriseptata*, species characteristic of acid waters.[11,12] However, there has certainly been an acceleration of the process in the last three decades. For example, of 87 lakes in southern Norway surveyed in both the periods 1923–49 and 1970–80, 24% had a pH below 5.5 in the earlier period compared to 47% in the second period.[13] The major source of acid is undoubtedly the burning of fossil fuels, releasing oxides of sulfur and nitrogen to the atmosphere. Electricity generating stations are a large source of these pollutants but domestic and industrial sources are also significant, as are the exhausts of vehicles.

The acids either fall directly into waterbodies in precipitation or are washed in from vegetation and soils within the catchment. Three broad categories of water which differ in acidity can be recognized.[14]

[11]R. J. Flower, R. W. Battarbee and P. G. Appleby, *J. Ecol.*, 1987, **75**, 797.
[12]R. W. Battarbee, *Hydrobiologia*, 1994, **274**, 1.
[13]A Wellburn, 'Air Pollution and Acid Rain: The Biological Impact', Longman Scientific and Technical, Harlow, 1988.
[14]United Kingdom Acid Waters Review Group, 'Acidity in United Kingdom Fresh Waters', Department of the Environment, London, 1986.

(i)   Those which are permanently acid, with a pH less than 5.6, low electrical conductivity, and alkalinity close to zero. Such conditions occur in the headwaters of streams and in lakes, where the soils are strongly acid, or in the outflows of peat bogs.

(ii)  Those which are occasionally acid, where pH is normally above 5.6 but because they have low alkalinity (usually less than 5.0 mg l$^{-1}$ CaCO$_3$) the pH may drop below 5.6 periodically. These include streams and lakes in upstream areas of low conductivity on rocks unable to neutralize acid quickly. Such waters may show episodes of extreme acidity, for instance during snowmelt or following storms. These may be very damaging to aquatic life but the infrequency of acid events makes the problem difficult to detect.

(iii) Those which are never acid, the pH never dropping below 5.6 and the alkalinity always above 5 mg l$^{-1}$ CaCO$_3$.

Much of northern and western Britain has a solid geology consisting of granites and acid igneous rocks; there is little or no buffering capacity. The situation is exacerbated in those catchments which have been extensively afforested with conifers. The sulfate ion is very mobile and transfers acidity very efficiently from soils to surface waters. The nitrate ion behaves similarly but is normally taken up by plants first.

In Canada, one experimental lake was artificially acidified, from 1975, by adding sulfuric acid and the pH dropped from 6.8 to around 5.0 over an eight year period. Among the first animals to disappear from the lake were shrimps and minnows. Within a year, at pH 5.8 opussum shrimps (*Mysis relicta*), with an estimated population of almost seven million before the experiment, had all but disappeared. Fathead minnow (*Pimephales promelas*) failed to reproduce and died out a year after the shrimps. Young trout were failing to appear and many of their food items were killed at pH 5.8 and as the pH fell still further cannibalism was noted in trout. At pH 5.6 the exoskeleton of crayfish (*Oronectes virilis*) began to lose its calcium and the animals became infested with a microsporozoan parasite. Some species, such as chironomid midges, did well as they were released from the pressure of predation in the simplified ecosystem. Since no species of fish reproduced at values of pH below 5.4, it was predicted that the lake would become fishless within ten years, based on knowledge of the natural mortalities of long-lived species. Some observations were unexpected in this experiment. There was no decrease in primary production, in the rate of decomposition or in nutrient concentrations. The changes in the lake were caused solely by changes in hydrogen ion concentration and not by any secondary effects such as aluminium toxicity.[15].

Table 4.2 provides a generalized summary of the sensitivity of aquatic organisms to lowered pH based on studies in Scandinavian lakes. Changes in the community begin at pH 6.5 and most species have disappeared below pH 5.0

[15]D. W. Schindler, *Science*, 1988, **239**, 149.

**Table 4.2** *Sensitivities of aquatic organisms to lowered pH*

| | |
|---|---|
| pH 6.0 | Crustaceans, molluscs, *etc.*, disappear. |
| | White moss increases. |
| pH 5.8 | Salmon, char, trout and roach die. |
| | Sensitive insects, phytoplankton and zooplankton die. |
| pH 5.5 | Whitefish, grayling die. |
| pH 5.0 | Perch, pike die. |
| pH 4.5 | Eels, brook trout die. |

leaving just a few species of tolerant insects and some species of phyto- and zooplankton.

Considerable research has been directed towards the effects of acidification on fish because of their economic and recreational importance. Figure 4.5 shows the proportion of lakes in southern Norway which have lost their brown trout (*Salmo trutta*) over the 35 years since 1940. By 1975 over half a total of 2850 lakes were without trout. The effect of acidity on fish is mediated *via* the gills. The blood plasma of fish contains high levels of sodium and chloride ions and those ions which are lost in the urine or from the gills must be replaced by active transport, against a large concentration gradient, across the gills. When calcium is present in the water it reduces the egress of sodium and chloride ions and the ingress of hydrogen ions. The main cause of mortality in acid waters is the excessive loss of ions such as sodium which cannot be replaced quickly enough by active transport. When the concentrations of sodium and chloride ions in the blood

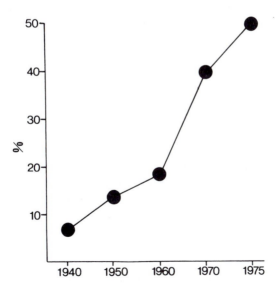

**Figure 4.5** *Percentage of lakes* (n = 2850) *in Scandinavia which have lost their population of brown trout. The number of fishless lakes in southern and southwestern Norway in 1986 had doubled since the survey a decade earlier ( and which forms part of the data in this figure)*

plasma fall by about a third, the body cells swell and extracellular fluids become more concentrated. To compensate for these changes, potassium may be lost from the cells but, if this is not eliminated quickly from the body, depolarization of nerve and muscle cells occurs, resulting in uncontrolled twitching of the fish prior to death.

Aluminium has been shown to be toxic to fish in the pH range 5.0–5.5 and during episodes of acidity aluminium ions are frequently present in high concentrations, especially where water is draining from conifer plantations. Aluminium ions apparently interfere with the regulation by calcium of gill permeability so enhancing the loss of sodium in the critical pH range. They also cause clogging of the gills with mucus and interfere with respiration. The developmental stages of fish are particularly sensitive to acidification[16] and it is thought that aluminium may interfere with the calcification of the skeleton of fish fry resulting in a failure of normal growth. This failure in recruitment results in a gradual decline of the fish population to extinction.

The simplification of the aquatic ecosystem due to acidification may also affect higher levels of the food chain. The dipper (*Cinclus cinclus*) lives along swiftly flowing streams where it feeds mainly on aquatic invertebrates which it collects by 'flying' underwater or foraging on the stream bottom. Dippers in Wales are scarcest along streams with low mean pH (less than 5.5–6.0) and elevated concentrations of aluminium (greater than $0.80–1.0 \text{ g m}^{-3}$). A decline in pH of 1.7 units on one river over the period 1960–1984 resulted in a 70–80% decline in the dipper population. On acidic streams nesting started later and clutch size was smaller than in those dippers breeding elsewhere.[17] If fish populations are eliminated from headwaters of streams by acidification, then the piscivorous otter (*Lutra lutra*) may not use them. In general, however, acidification results in a reduction in the carrying capacity for otters rather than a decrease in distribution.[18]

The process of acidification has been reversed by liming, enabling normal plant and animal communities to be maintained. However, liming is very expensive and, because of the vast number of fishless lakes, is hardly realistic on a large scale. In West Germany alone it has been estimated that the liming of soils and waters to halt the process of acidification would cost £15 000 million.

Re-acidification may occur, resulting in high mortality of stocked trout.[19] For long-term effects liming of catchments rather than water might prove more effective. Clearly the control of emissions at power stations and from vehicle exhausts is a more sensible alternative. Nevertheless it has been predicted that a 30% reduction in sulfur emissions in Europe would reduce the number of acid lakes in southern Norway by only 20%, while a 50% reduction would reduce the number by 35%.[20] In South-west Scotland mathematical modelling has

[16]M. Appelberg, E. Degerman and L. Norrgren, *Finnish Fish. Res.*, 1992, **13**, 77.
[17]S. J. Tyler and S. J. Ormerod, *Environ. Pollut.*, 1992, **78**, 49.
[18]C. F. Mason and S. M. Macdonald, *Water, Air, Soil, Pollut.*, 1989, **43**, 365.
[19]B. T. Barlaup, Å. Åtland and E. Kleiven, *Water, Air, Soil Pollut.*, 1994, **72**, 317.
[20]A. Henriksen, L. Liem, B. O. Rosseland, T. S. Traaen and I. S. Sevaldrud, *Ambio*, 1989, **18**, 314.

indicated that a 50% reduction in acid emissions would be necessary to prevent further increases in stream acidity on moorlands, more for afforested catchments.[21]

It is necessary briefly to mention the problems of acid mine drainage which, for example, affects some 19 300 km of streams and 73 000 ha of lakes in the USA alone.[22] Coal mines are the most important source of pollution. Chemosynthetic bacteria oxidize the mineral iron pyrite with the formation of sulfuric acid, the water flowing from the mine having a pH often below 3.5, resulting in the extermination of much of the stream biota. The problem becomes much worse when mines are abandoned and pumping ceases, as is currently being observed in many areas of Great Britain following the demise of the coal-mining industry. The acidic conditions bring metals, from ores exposed by the mining activity, into solution so that the effluent from the mine is also highly toxic. Ferric hydroxide precipitates out on to the stream bottom covering the substratum with a brown slime and smothering benthic organisms and macrophytes. Many mines are situated in rural localities so aquatic ecosystems in otherwise pristine environments may be destroyed.

## 4.5 TOXIC CHEMICALS

Some aspects of toxic pollution have already been mentioned but it is now appropriate to describe the modes of action of toxic chemicals. The major types of toxic compounds are:

(i) metals, such as zinc, copper, mercury, cadmium;
(ii) organic compounds, such as pesticides, herbicides, polychlorinated biphenyls (PCBs), phenols;
(iii) gases, such as chlorine, ammonia;
(iv) anions, such as cyanide, sulfide, sulfite;
(v) acids and alkalis.

There are a number of terms in regular use in the study of toxic effects:

*acute* – causing an effect (usually death) within a short period;
*chronic* – causing an effect (lethal or sublethal) over a prolonged period of time;
*lethal* – causing death by direct poisoning;
*sub-lethal* – below the level which causes death but which may affect growth, reproduction or behaviour so that the population may eventually be reduced;
*cumulative* – the effect is increased by successive doses.

A typical example of a toxicity curve is given in Figure 4.6. Note that both axes are on a logarithmic scale. The median periods for survival are plotted against a

[21]P. G. Whitehead, C. Neal and R. Neale in 'Reversibility of Acidification', ed. H. Barth, Elsevier, London, 1987, p. 126.
[22]R. L. P. Kleinmann and R. Hedin, in 'Tailings and Effluent Management', eds. M. E. Chalkly, B. R. Conrad, V. I. Lakshmanan and K. G. Wheeland, Pergamon, New York, 1990, p. 140.

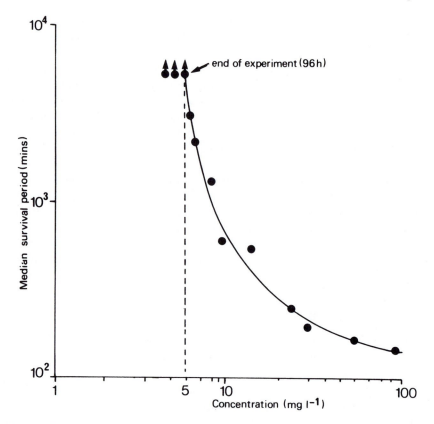

**Figure 4.6**   *A typical toxicity curve*

range of concentrations. In many cases the relationship between survival time and concentration is curvilinear and there is a concentration below which the organism is likely to survive for long periods (5 mg $l^{-1}$ in Figure 4.6), a concentration known as the incipient $LC_{50}$. The lethal concentration (LC) is used where death is the criterion of toxicity. The number indicates the percentage of animals killed at that concentration and it is also usual to indicate the time of exposure. Thus 96 hour $LC_{50}$ is the concentration of toxic material which kills fifty per cent of the test organism in ninety six hours. The incipient level is usually taken as the concentration at which fifty per cent of the population can live for an indefinite period of time. Where effects other than death are being sought, for example respiratory stress or behavioural changes, the term used is the effective concentration (EC) which is expressed in a similar way, *e.g.*, 96 hour $EC_{50}$.

Effluents are frequently complex mixtures of poisons. If two or more poisons are present together they may exert a combined effect on the organism which is additive, for example mixtures of zinc and copper or ammonia and zinc. In other cases, there may be antagonism, the overall toxicity being less than when compounds are present alone; calcium, for example, antagonizes the toxic effect of lead and aluminium. In some cases the overall effect of mixtures of poisons on an organism may be more than additive (synergistic), for example, mixtures of nickel and chromium.

There has been a large amount of data collected on the acute toxicity of chemicals, especially to fish and invertebrates,[23,24] and this has undoubtedly been of great value in elucidating the mechanisms of toxicity. The value of these data to the river manager is more questionable. Incidents resulting in large mortalities of fish and other organisms are usually accidents over which the river manager has no control. He can merely assess the damage and perhaps restock when conditions improve. In addition, detailed information on toxicity of a range of compounds is available for only a few test organisms, such as rainbow trout or *Daphnia*, and it is well known that even closely related species may show very different responses to particular pollutants. It is the sub-lethal effects of pollutants which are of particular concern in many field situations, for low levels of pollutants may result in the gradual loss of populations, without any overt signs of a problem.

Experiments on sub-lethal effects are more difficult to carry out because they invariably take longer and individuals under test may respond very differently to low levels of pollution. Furthermore, the response to pollutants may vary over the lifetime of an organism, developmental stages often being more susceptible. It is necessary to study an organism under experimental conditions over its lifetime to find the weak link in its response to pollution and such long-term experiments, possibly over several generations, are essential to discover any carcinogenic, teratogenic or mutagenic effects of pollutants. Sub-lethal effects may be manifested at the biochemical, physiological, behavioural and life cycle level.[1] Although it is possible to show small effects, for example, in biochemistry or on growth, at very low levels of pollution, it is essential that they reduce the fitness of an organism in its environment and are not merely within the organism's range of adaptation. Nevertheless these sub-lethal effects may be quite subtle and can be measured long before any outward toxic effects are manifested. They can be used as *biomarkers*, to show that an organism has been exposed to contaminants at levels which exceed the normal detoxification and repair capabilities.[25,26] Biomarkers can be used to predict what concentrations of a pollutant are likely to cause damage, rather than merely to measure concentrations when damage has been recorded.

Porphyrins, involved in the synthesis of haem, have been used as a biomarker. Concentrations of porphyrins in the livers of herring gulls (*Larus argentatus*) from various sites in the Great Lakes of North America were compared with samples from the Atlantic coast. Samples from Hamilton Bay, Lake Ontario and Green Bay, Michigan, had porphyrin levels respectively 38 times and 28 times those in

[23]J. M. Hellawell, 'Biological Indicators of Freshwater Pollution and Environmental Management', Elsevier Applied Science, London, 1986.
[24]G. Mance, 'Pollution Threats of Heavy Metals in Aquatic Environments', Elsevier Applied Science, London, 1987.
[25]D. Peakall, 'Animal Biomarkers as Pollution Indicators', Chapman and Hall, London, 1992.
[26]G. A. Fox, *J. Great Lakes Res.*, 1993, **19**, 722.

livers from the Atlantic coast, though it was not possible to determine which compound was responsible for the increase; gull livers contain many xenobiotics.[26]

Organisms which are regularly subjected to toxic pollutants may develop tolerance to them. This may be achieved either by functioning normally at high loadings of pollutants or by metabolizing and detoxifying pollutants. Algae living in streams receiving mine drainage are highly tolerant to metals and this adaptation has been shown to be genetically determined. Populations of the isopods *Asellus aquaticus* and *A. meridianus* collected from sediments contaminated with copper and lead were found to be less susceptible to these metals than animals from uncontaminated sediments in acute toxicity tests. The metals are stored in the hepato-pancreas. Copper and lead appear to compete for binding sites, lead being more readily bound.[27]

Of particular concern to environmental toxicologists are those compounds which accumulate in tissues, especially the heavy metals and organochlorines (pesticides and PCBs). From often undetectable concentrations in water, organisms may accumulate levels of biological significance. Furthermore, these compounds are passed along the food chain so that top carnivores feeding on contaminated prey may accumulate enormous concentrations of pollutants. For example, it has been estimated that the concentration factor of PCBs from water to carnivorous mammals may be as high as ten million times.[28]

Figure 4.7 illustrates the frequency distribution of concentration of total mercury in the muscle of 85 eels (*Anguilla anguilla*) and 79 roach (*Rutilus rutilus*) collected from waters in five regions of Great Britain.[29] It is likely that in the vast majority or rivers and lakes from which the fish originated mercury concentrations in water would have been below the limit of detection. Eels are probably more contaminated than roach because they spend much of the winter buried within sediments which also accumulate heavy metals. The EC directive for mercury levels in fish flesh taken for consumption sets a limit of 300 $\mu$g kg$^{-1}$. None of the roach exceeded this standard but 25% of the eels, a species which is commercially exploited, especially for the export market, contained more than 300 $\mu$g kg$^{-1}$ Hg. Most of the fish contained lead and cadmium also, a good proportion of them above recommended standards for consumption. Eels contained more cadmium than roach but roach were more heavily contaminated with lead. It has been cogently argued that we have little knowledge of the long term, low level exposure of human populations to toxic metals and over ten billion individuals are being unwittingly exposed to elevated levels of toxic metals and metalloids in the environment.[30]

Much of the evidence of the effects of PCBs on wild populations of vertebrates has come from the Great Lakes region of North America. In Green Bay, Lake Michigan, Forster's terns (*Sterna forsteri*) exhibit impaired reproduction. The

[27]B. E. Brown, *Nature (London)*, 1978, **276**, 388.
[28]S. Tanabe, *Environ. Pollut.*, 1988, **50**, 5.
[29]C. F. Mason, *Chemosphere*, 1987, **16**, 901.
[30]J. O. Nriagu, *Environ. Pollut.*, 1988, **50**, 139.

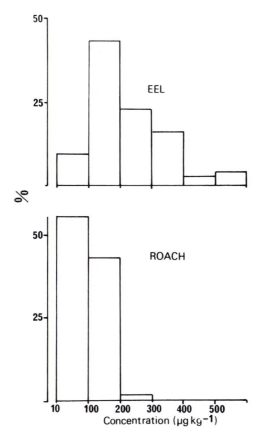

**Figure 4.7** *Concentration* $(\mu g\ kg^{-1})$ *of total mercury in eels and roach from freshwaters in Great Britain*

incubation is extended, few eggs hatch, the chicks have lower body weight and their livers are larger than normal. They show a high incidence of congenital deformities. The parents are inattentive at nesting and this further reduces their reproductive success.

It was concluded, following a detailed toxicological analysis, that those PCB congeners which induce the enzyme aryl hydrocarbon hydroxylase (AHH) were the only contaminants present in sufficient amounts to cause the observed effects on eggs and chicks. More than 90% of the effect could be explained by two pentachlorobiphenyls. The behavioural abnormalities in adults were caused by total PCBs.[31] In addition to the Forster's tern, six other species of fish-eating bird from the Great Lakes have exhibited growth deformities and physiological effects.[32] These symptoms have only been observed in the past two decades and

[31]T. J. Kubiak, H. J. Harris, L. M. Smith, D. L. Starling, T. R. Schwartz, J. A. Trick, L. Sileo, D. E. Docherty and T. C. Erdman, *Arch. Environ. Contam. Toxicol.*, 1989, **18**, 706.
[32]J. P. Giesy, J. P. Ludwig and D. E. Tillitt, *Environ. Sci. Technol.*, 1994, **28**, 128A.

it is considered that, in earlier years, symptoms were masked by the effects of DDE, which thinned eggs to such an extent that they did not survive long enough for deformities to be expressed.

Recent studies on humans also give cause for concern. Mothers giving birth in hospitals close to Lake Michigan were questioned about their history of eating fish caught in the lake. Blood was taken from the umbilical cords of their newborn babies and analysed for PCBs. Babies of those mothers who had eaten fish, equivalent to only two–three salmon or trout meals per month, weighed significantly less than controls (whose mothers had eaten no fish) at birth, while their head circumferences were disproportionately small for their weight. Weight was still lower than average at age four years, while babies showed a number of behavioural defects.[33]

There has been considerable, detailed research on PCBs in the Great Lakes but the problem, of course, is not confined to that region. There is, for example, a considerable body of information which suggests that PCBs have been largely responsible for the decline of the otter over large areas of Western Europe, in some places to extinction. There is a strong relationship between the mean amount of PCBs in tissues of otters and the extent of the population decline, relationships which do not hold for other contaminants.[34]

Although there have been restrictions on the use of PCBs, it is considered that, because of their persistence and the transfer across the generations, concentrations in top carnivores and humans are unlikely to decrease significantly for some years to come.[35] This should be a matter of great concern in view of the apparent potential consequences for human health at very low levels of long-term exposure.

## 4.6 THERMAL POLLUTION

Cooling water discharges from electricity generating stations are the main sources of pollution by heat. Such effluents also contain a range of chemical contaminants which, though small in relation to the volume of cooling water, may in fact have a greater impact on the ecology of the receiving stream.[36] An increase in temperature alters the physical environment, in terms of a reduction in both the density of the water and its oxygen concentration, while the metabolism of organisms increases. Cold water species, especially of fish, are very sensitive to changes in temperature and they will disappear if heated effluents are discharged to the headwaters of streams. As the temperature increases, the oxygen consumption and heart rate of a fish will increase to obtain oxygen for increased metabolic processes but, at the same time, the oxygen concentration of the water is

[33] J. L. Jacobson and S. W. Jacobson, *J. Great Lakes Res.*, 1993, **19**, 776.
[34] S. M. Macdonald and C. F. Mason, 'Status and Conservation Needs of the Otter (*Lutra lutra*) in the Western Palearctic', Council of Europe, Nature and Environment 67, 1994.
[35] S. J. Harrad, A. P. Stewart, R. Alcock, R. Boumphrey, V. Burnett, R. Duarte-Davidson, C. Halsall, G. Sandars, K. Waterhouse, S. R. Wild and K. C. Jones, *Environ. Pollut.*, 1994, **85**, 131.
[36] T. E. Langford, 'Electricity Generation and the Ecology of Natural Waters', Liverpool University Press, Liverpool, 1983.

decreased. For example, at 1°C a carp (*Cyprinus carpio*) can survive in an oxygen concentration as low as 0.5 mg l$^{-1}$, whereas at 35°C the water must contain 1.5 mg l$^{-1}$. The swimming speeds of some species declines at higher temperatures, *e.g.* trout at 19°C, making them less efficient predators. Resistance to disease may also change. The bacterium *Chondrococcus columnaris* is innocuous to fish below 10°C but it invades wounds between 10°C and 21°C while it can invade healthy tissues above 21°C.

However, it must be remembered that temperature changes are a feature of natural ecosystems so that organisms have the ability to adapt to the altered conditions provided by thermal effluents and, although much research has been carried out into thermal pollution, it is now considered to be of little importance compared to other sources of pollution.[36] Indeed, there are some benefits of heated effluents, for growth and productivity may be enhanced.

Global warming could have a major impact in future on freshwaters. One problem is the increased frequency of unpredictable events, such as storms, leading to widespread flooding and financial loss. Summer drought, by contrast, may lead to the drying out of river headwaters with severe consequences for the biota. Surface water temperatures may rise so that the effects described above for thermal discharges may become general. Effluents already make up a substantial proportion of the low summer flow of many rivers and, should these become more concentrated, only the most resilient aquatic species may survive. Higher concentrations of bioaccumulating compounds (*e.g.* metals, PCBs) could have ramifications for the food-chain far beyond the boundaries of the river. Finally, with predicted rises in sea-level, many coastal freshwaters may become brackish, while salt intrusions may push far inland up rivers.

## 4.7 OIL

Compared to the marine situation comparatively little work has been done on the effects of oil in freshwater ecosystems. Nevertheless the chronic pollution of freshwaters with hydrocarbons is widespread. Much of it derives from petrol and oil washed from roads together with the illegal discharge of engine oil. Other sources include boats and irrigation pumps, while accidents involving transporters and spillages from storage tanks are also significant. Over the period 1989–92 in England and Wales, 23% of all reported pollution incidents affecting rivers were caused by oils.[37]

The water soluble components of crude oils and refined products may prove toxic to freshwater animals though the prediction of toxic effects is rather difficult owing to the complex chemical nature of discharges. Eggs and young stages of animals are especially vulnerable.

In general terms, the aliphatic compounds of oils are relatively innocuous while the monohydric aromatic compounds are generally toxic, the degree of toxicity increasing with increasing unsaturation. Some contaminants of oils, such as PCBs and lead, will accumulate in tissues. Emulsifiers and dispersants, used to

---

[37]J. A. Cole, R. L. Norton and H. A. C. Montgomery, *Water Sci. Technol.*, 1994, **29**, 203.

clean up spillages, are themselves often highly toxic. The surface active agents which they contain make membranes more permeable and increase the penetration of toxic compounds into animals. In this way mixtures of oils and dispersants are often more toxic than either applied separately. There are also marked species differences in the susceptibility to particular compounds of oils, for instance the polynuclear aromatic hydrocarbons, further adding to the difficulties of making predictions about toxicity.

The physical properties of floating oil are a special threat to higher vertebrates, especially aquatic birds because contamination reduces buoyancy and insulation, while the ingestion of oil, frequently the result of attempts to clean the plumage, may prove toxic. A further problem is the tainting of flesh, especially of fish, which is detectable to the human palate at very low levels of contamination and renders fish inedible. The major sources of taint are light oils and the middle boiling range of crude oil distillates but there are a number of other sources, such as exhaust from outboard motors, waste from petrochemical factories, refinery wastes and all crude oils.

## 4.8  RADIOACTIVITY (see also Chapter 17)

Chemically, radionuclides behave in the same way as their non-radioactive isotopes but, if they accumulate up the food chain the radioactive isotopes have much greater significance. Radionuclides come mainly from fall-out from weapons testing and the effluent from nuclear power stations. Because ionizing radiation is highly persistent in the environment, causing cancer and genetic disorders in humans, it has always attracted special concern and the release of radionuclides is strictly monitored and controlled.

Until recently, it was thought that the natural environment was little affected by radionuclides, which are discharged very locally. At only one site in Britain, Trawsfynydd in North Wales, was cooling water from a nuclear power-station discharged into freshwater though this station is now being de-commissioned. All other nuclear power stations discharge into estuaries or at sea. However, the explosion at the nuclear power station at Chernobyl in the Ukraine in April, 1986 and the subsequent spread and deposition of radionuclides over large areas of western Europe, emphasized how potentially damaging such pollution can be.

In Sweden, levels of radioactivity in freshwaters rose quickly but fish species varied in the amount and rate of accumulation of radionuclides. In one lake, bream (*Abramis brama*), which feed on bottom-living invertebrates had accumulated 1000 Bq kg$^{-1}$ of caesium-127 within two weeks. By July accumulation had increased to 3800 Bq kg$^{-1}$. Perch (*Perca fluviatilis*) which feed on zooplankton, accumulated cesium more slowly but to higher levels, 8000 Bq kg$^{-1}$ by July. Concentrations in the fish-eating pike (*Esox lucius*) were little higher than those in bream in July.[38] Between 4000 and 7000 Swedish lakes had fish with a Cs-137

[38] R. C. Petersen, L. Landner and H. Blanck, *Ambio*, 1986, **15**, 327.

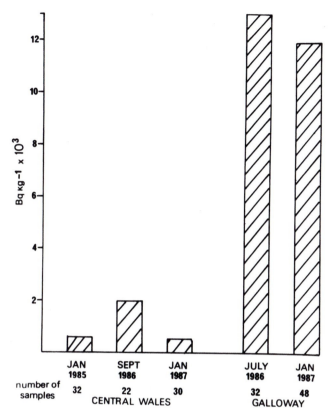

**Figure 4.8** *Mean concentrations* $(Bq\ kg^{-1})$ *of radioactivity in otter spraints*

load more than the consumption guideline of 1500 Bq kg$^{-1}$ and activity levels in pike could exceed this value well into the 21st century.[39]

Because wild caught freshwater fish feature little in the diet of the vast majority of the human population it was considered that the Chernobyl accident presented no cause for concern to human health away from the vicinity of the reactor. However, other species within the aquatic ecosystem are more exclusively piscivorous. Figure 4.8 shows the radioactivity in otter faeces (spraints) collected from river banks in central Wales and Galloway. Central Wales was outwith the main deposition area for Chernobyl fallout. In the September following the accident, radioactivity in otter spraints was more than double that in a sample collected some fifteen months prior to the accident and stored in a deep freeze. By the following January levels had returned to normal. By contrast, Galloway, which is an area of soft waters, received large quantities of Chernobyl fall-out and average radioactivity was more than six times that in Wales (13 000 Bq kg$^{-1}$, with a maximum of 79 500 Bq kg$^{-1}$ in one sample).[40]

[39]L. Lundgren, *Ambio*, 1993, **22**, 369.
[40]C. F. Mason and S. M. Macdonald, *Water, Air, Soil Pollut.*, 1988, **37**, 131.

Unfortunately, there was no pre-Chernobyl sample but levels were still high in the following January, as has been found for other biological materials from this area. Whether or not such high levels of radioactivity could adversely affect otters is unknown and it would be impossible to disentangle any mortality caused by radiation from other mortality factors, a major problem with many ecotoxicological studies.

## 4.9  BIOLOGICAL MONITORING OF POLLUTION IN FRESHWATERS

The biological assessment of pollutants includes both laboratory and field techniques.[1,23] Two widely used laboratory methods are biostimulation tests and toxicity testing. Biostimulation tests are used mostly for evaluating the nutrient status of water bodies and are normally carried out with algae, the most frequently used species including *Selenastrum capricornutum*, *Asterionella formosa* and *Microcystis aeruginosa*. The Standard Bottle Test measures either the maximum specific growth rate or the maximum standing crop. The specific growth rate ($\mu$) is:

$$\mu = \frac{\ln(X_2/X_1)}{t_2 - t_1}$$

where $X_1$ is the initial biomass concentration at time $t_1$ and $X_2$ is the final biomass concentration at time $t_2$. The maximum standing crop is the maximum algal biomass achieved during incubation and may be obtained by direct determination of dry weight, by counting cells, by absorbance, by chlorophyll measurement or by total cell carbon. The algae initially present in the test water must be removed by filtration or autoclaving, which may alter the water quality, while care must be taken to avoid problems of nutrient depletion and the build up of metabolic wastes.

The chief uses of toxicity tests are for a preliminary screening of chemicals, for monitoring effluents to determine the extent of risk to aquatic organisms and, for those effluents which are toxic, to determine which component is causing death so that it can receive especial treatment. The simplest type of test is the static test in which an organism is placed in a standard tank in the water under investigation for 48–96 hr. There are normally a series of tanks with test water of different dilutions. More sophisticated techniques involve the periodic replacement of test water or indeed continuous flow systems. Fish have traditionally been used as test organisms. In the United States, the main test species have been fathead minnows and bluegill sunfish (*Lepomis macrochirus*), though the tendency now is to use a range of test species. Much toxicity work in Britain has been with rainbow trout but the tropical harlequin (*Rasbora heteromorpha*) has become increasingly popular because it is smaller and has a similar sensitivity to pollutants. Fish require large volumes of clean water for maintenance and, because tests need replicating, much space is needed. Furthermore there are growing ethical objections to using vertebrates for routine toxicological assessments. There is therefore growing interest in developing other test organisms. The planktonic crustacean *Daphnia* is widely used for it is easily cultured, has a high reproductive rate and is sensitive to a range of pollutants.[41] Bacterial tests are also being developed, of which one, the Microtox test, is commercially available. It utilizes the bioluminescence of the marine *Photobacterium phosphoreum*, the reduction in light output being the measure of toxicity. The test is sensitive, precise and reproducible. It has been used, for example, to demonstrate that the quality of water in the River Meuse is, over most of its length, below environmentally acceptable standards for a range of organic compounds.[42]

Fishes show distinct physiological and behavioural responses to low levels of pollutants and recent studies have attempted to use these to devise automatic alarm systems. Automatic fish monitors should provide rapid indications of a deteriorating water quality and have potential uses for monitoring river waters and raw waters which are abstracted for potable supply and for monitoring effluents from treatment plants. Fish alarm systems have monitored movement and respiration. Movements are monitored with photosensitive cells and respiration by changes of potential between two electrodes. Multiple species monitors, using invertebrates as well as fish, are also being developed; two or more species are more likely to respond to a wider range of pollutants than a single species.[43]

It has already been shown that pollutant levels in water are frequently below the limits of detection but that some organisms can accumulate large quantities of some contaminants which reflect the pollution level over periods of time. Because the degree to which organisms concentrate pollutants varies, for large-scale surveys a single, widespread species is needed. Care must be taken in the interpretation of results because many factors may influence the total pollutant content and concentration. These include the age of the organism, its sex, size and weight as well as the time of year, sampling position and the relative levels of other pollutants in the tissues. In addition to sampling organisms from the field to measure pollutant levels, they can also be placed in the field in cages so that uptake rates can be measured over defined periods of time.

Fish have frequently been used as biomonitors because they can be considered as a *critical material*, forming part of the diet of man. Eels occur in most freshwater habitats and are hence particularly useful for large-scale surveys. Molluscs and macrophytes have also been used locally while mosses have proved especially valuable for monitoring pollution by metals.[44]

The biological assessment of water quality in the field may involve a number of levels of effort: survey, surveillance, monitoring or research.[1] It is usually impossible to investigate the entire fauna and flora in a pollution study because of constraints of time and the wide range of sampling methods required for different groups of organisms. A survey or monitoring programme should therefore be based on those organisms which are most likely to provide the right

[41]R. Baudo, *Mem. Ist. Ital. Idrabiol.*, 1987, **45**, 461.
[42]H. J. G. Polman and D. de Zwart, *Water Sci. Technol.*, 1994, **29**, 253.
[43]D. Gruber, C. H. Frago and W. J. Rasnake, *J. Aquat. Ecosystem Hlth.*, 1994, **3**, 87.
[44]J. Lopez, M. D. Vasquez and A. Carballeira, *Freshwater Biol.*, 1994, **32**, 185.

information to answer the particular question being posed. To be suitable, the chosen group must meet a number of criteria.[23] For example, the presence or absence of organisms must be a function of water quality rather than other ecological factors and the group must be relatively easy to sample and identify. In Great Britain the most favoured group for routine biological surveillance are the macroinvertebrates living in the stream bed.

To sample sites for macroinvertebrates a pond net (mesh size 900 $\mu$m) is used. The net is placed on the stream bed, facing downstream, and the feet are shuffled on the substratum, animals being dislodged and floating into the net. As much of the stream bed as possible is sampled and additional habitats, such as aquatic macrophytes or marginal vegetation, are also sampled by sweeping the net through them. The sampling period is three minutes and the objective is to obtain a comprehensive list of taxa with the minimum of sampling effort.[45] These samples are normally collected from eroding substrata (rocks, gravels, *etc.*). If there are long stretches of stream consisting of depositing substrata (silts and muds) or which are too deep to wade then sampling may have to be restricted to the margins. Alternatively artificial substrata, which are colonized by invertebrates, can be left *in situ* for later collection and analysis.[1,23]

Once the invertebrates have been identified and counted, there are two commonly used methods for data presentation and interpretation, diversity indices and biotic indices. Diversity indices take into account the number of species within the collection (species richness) and the relative abundance of species within the collection (evenness). It is argued that a community from an unstressed, *i.e.* pollution-free, environment will contain a large number of species (high richness), many at fairly low densities (high evenness), so that the calculated diversity index will be high. As pollution stress increases, species will gradually decline in number and disappear (low richness), while a few tolerant species will build up big populations in the absence of predation and competition (low evenness) resulting in a low diversity index. Diversity indices take no account of the tolerances of individual species to pollution. Many diversity indices have been devised.[46]

Biotic indices take account of the sensitivities of different species to pollution, that is, species are used as indicators of pollution. Those species which are sensitive to pollution (such as stoneflies) are given a high score; tolerant species (such as tubificid worms) are given a low score. In Britain, the biotic index most in favour is the Biological Monitoring Working Party (BMWP) Score (Table 4.3). Identification is necessary only to the family level. Score values for individual families reflect their pollution tolerance based on current knowledge of distribution and abundance. Each family is given a score depending on its position in Table 4.3 and a site score is calculated by summing the individual family scores. This total score can then be divided by the number of families recorded in the sample to derive the average score per taxon (ASPT), which is less sensitive to sample size and sampling effort, and hence gives more informa-

[45]M. T. Furse, J. F. Wright, P. D. Armitage and D. Moss, *Water Res.*, 1981, **15**, 679.
[46]J. L. Metcalfe, *Environ. Pollut.*, 1989, **60**., 101.

**Table 4.3**   *The BMWP score system*

| Families | Score |
|---|---|
| Siphlonuridae, Heptageniidae, Leptophlebidae, Ephemerellidae, Potamanthidae, Ephemeridae, Taeniopterygidae, Leuctridae, Capniidae, Perlodidae, Perlidae, Chloroperlidae, Aphelocheiridae, Phryganeidae, Molannidae, Beraeidae, Odontoceridae, Leptoceridae, Goeridae, Lepidostomatidae, Brachycentridae, Sericostomatidae | 10 |
| Astacidae, Lestidae, Agriidae, Gomphidae, Cordulegasteridae, Aeshnidae, Corduliidae, Libellulidae, Psychomyiidae, Philopotamidae | 8 |
| Caenidae, Nemouridae, Rhyacophildae, Polycentropidae, Limnephilidae | 7 |
| Neritidae, Viviparidae, Ancylidae, Hydroptilidae, Unionidae, Corophiidae, Gammaridae, Platycnemididae, Coenagriidae | 6 |
| Mesoveliidae, Hydrometridae, Gerridae, Nepidae, Naucoridae, Notonectidae, Pleidae, Corixidae, Haliplidae, Hygrobiidae, Dytiscidae, Gyrinidae, Hydrophilidae, Clambidae, Helodidae, Dryopidae, Elminthidae, Chrysomelidae, Curculionidae, Hydropsychidae, Tipulidae, Simuliidae, Planariidae, Dendrocoelidae | 5 |
| Baetidae, Sialidae, Piscicolidae | 4 |
| Valvatidae, Hydrobiidae, Lymnaeidae, Physidae, Planorbidae, Sphaeriidae, Glossiphoniidae, Hirudidae, Erpobdellidae, Asellidae | 3 |
| Chironomidae | 2 |
| Oligochaeta (whole class) | 1 |

tion for less effort. High ASPT scores characterize clean, upland sites which contain large numbers of high scoring taxa such as stoneflies, mayflies and caddisflies. Lower ASPT values are obtained from slow-flowing, lowland sites which are dominated by molluscs, chironomids and tubificid worms.

Multivariate techniques (TWINSPAN and DECORANA) have been used to classify the running waters of Great Britain on the basis of their macroinvertebrate fauna. Substratum type and alkalinity explain much of the variation between invertebrate groups, long-term average flow and distance from source also being important.[47] From this study it has become possible to predict the probability with which a given species or family will be captured at a site, using environmental data, and a software package, RIVPACS (River Invertebrate Prediction and Classification System) has been developed.[48] The predicted target assemblage of macroinvertebrates can be used to generate expected BMWP or ASPT scores against which the results of field surveys can be assessed. This procedure was used in the 1990 River Quality Survey of the United Kingdom. A system of banding was used to place the biological class into four categories, derived from the ratio of observed to expected ASPT, number of taxa and BMWP score. The overall classification of biological quality was the median of

[47]J. F. Wright, P. D. Armitage, M. T. Furse and D. Moss, *Regul. Rivers*, 1989, **4**, 147.
[48]J. F. Wright, M. T. Furse and P. D. Armitage, *Eur. Water Pollut. Control*, 1993, **3(4)**, 15.

the three individual classes, unless the individual class for ASPT was lower, when this was taken as the classification (more of the variation in the ASPT score than in the BMWP and total taxa is explained by the environmental variables in RIVPACS).[48]

## 4.10 CONCLUSIONS

Statutory water quality objectives (SWQOs) are currently being developed for individual stretches of river in England and Wales. These WQOs are classifications of water defined by quantitative standards of environmental quality related to the use, or potential use, of a watercourse. The five uses are River Ecosystem (to protect aquatic ecosystems of high conservation value), Abstraction for Potable Supply, Agricultural/Industrial Abstractions and Watersports. It is clear that the biologist will have an important role in defining the standards for some of these uses.

The River Ecosystem classification is the most developed and is based on the concentrations of seven common determinates which have effects on fish populations: dissolved oxygen, BOD, total ammonia, unionized ammonia, pH, copper and zinc. The quality targets will consist of two parts, a target class and a target date by which compliance will be achieved.[49]

Alongside WQOs, catchment management plans are being developed which include a consideration of the uses required for the SWQOs and a range of non-statutory objectives related to activities which take place within the catchment. A multifunctional and multi-use appraisal of the catchment is undertaken, which results in an agreed strategy for achieving the environmental potential of a catchment. For the catchment management plan to be successful, detailed scientific investigations at the local level will be needed to determine the precise environmental parameters of each aspect requiring action.[50]

Despite the vast amount of scientific information gathered on pollution over the past four decades, many of our freshwaters still suffer from poor water quality, with severe consequences for the flora and fauna they support. Scientific know-how is subverted by economic and political constraints, and in many instances by careless accidents within the catchment. A continuing programme of surveillance and monitoring is therefore essential. Newly synthesized materials are also constantly being added to our waterways as traces or in effluents and the long-term effects of these are largely unknown. Constant vigilance is required to protect our water resources and the biologist should have a central role in the management team.

[49]M. Everard, *Freshwater Forum*, 1994, **4**, 179.
[50]A. S. Gee and F. H. Jones, in 'The Ecological Basis for River Management', eds. D. M. Harper and A. J. D. Ferguson, Wiley, Chichester, 1995, p. 475.

CHAPTER 5

# Sewage and Sewage Sludge Treatment

J. N. LESTER

## 5.1 INTRODUCTION

It is estimated that the volume of water used daily in England and Wales (exclusive of water abstracted for cooling purposes) amounts to $5000 \times 10^6$ gal ($23 \times 10^6$ m$^3$) or approximately 95 gal (430 l) per capita per day. Domestic use accounts for nearly $1800 \times 10^6$ gal ($8 \times 10^6$ m$^3$) of this average daily total. Nearly all of the water used domestically and approximately $1500 \times 10^6$ gal ($6.8 \times 10^6$ m$^3$) of the water used by industry each day is discharged to the sewers, yielding a total sewage flow of $3100 \times 10^6$ gal ($14.1 \times 10^6$ m$^3$) or about 60 gal (275 l) per capita per day.

The sewage from approximately 44 million people in England and Wales is treated by conventional wastewater treatment processes, that from about a further six million people is discharged without treatment to the sea and some one to two million people are not connected to the sewerage system. To achieve this degree of wastewater treatment requires some 5000 sewage treatment works serving populations in excess of 10 000; these are distributed throughout the ten Water Authorities, which have been privatized, in England and Wales. The sewerage systems which carry the sewage to the site of treatment, or point of discharge, are of two types. Foul sewers carry only domestic and industrial effluent. In areas serviced in this way there are entirely separate systems for the collection of stormwater which is discharged directly to natural water courses. However, in older towns and cities considerable use has been made of combined foul and stormwater systems. The use of combined sewerage systems leads to very significant changes in the flow of sewage during storms. However, even in foul sewers significant changes in the flow occur due to variations in the pattern of domestic and industrial water usage which is essentially diurnal, and at its greatest during the day. Infiltration will also influence the flow in the sewage system. Although a properly laid sewer is watertight when constructed, ground movement and aging may allow water to enter the sewer if it is below the water table. The combined total of average daily flows to a sewage treatment works is called the dry weather flow (*DWF*). The *DWF* is an important value in the design and operation of the sewage treatment works and other flows are expressed in terms of it. *DWF* is defined as the daily rate of flow of sewage (including both domestic and trade waste), together with infiltration, if any, in a sewer in dry weather. This may be measured after a period of 7 consecutive days during which the rainfall has not exceeded 0.25 mm.

The *DWF* may be calculated from the following formula:

$$DWF = PQ + I + E$$

where,

$P$ = population served
$Q$ = average domestic water consumption ($1d^{-1}$)
$I$ = rate of infiltration ($1d^{-1}$)
$E$ = volume (in litres) of industrial effluent discharged to sewers in 24 hours

### 5.1.1  Objectives of Sewage Treatment

Water pollution in the United Kingdom was already a serious problem by 1850. It is probable that the early endeavours to control water pollution were considerably stimulated by the state of the lower reaches of the River Thames which, at the point where it passed the Houses of Parliament, was grossly polluted. An early solution to these problems was sought through the construction of interceptor sewers. These collected all the sewage draining to the River Thames and carried it several miles down the river before discharging it to the estuary on the ebb tide. From this it moved towards the sea and in so doing received greater dilution. Despite these measures, and the passing in 1876 of the first Act of Parliament to control water pollution the situation continued to deteriorate. The requirement for, and the objectives of, sewage treatment were first outlined by the Royal Commission on Sewage Disposal (1898–1915). The objectives of sewage treatment have developed significantly since this report; however, the standards described then are still applicable in many areas and this report provided the framework around which the United Kingdom wastewater industry has developed.

Originally the objective of sewage treatment was to avoid pestilence and nuisance (disease and odour) and to protect the sources of potable supply.

During sewage treatment disease-causing organisms may be destroyed or concentrated in the sludges produced; similarly, offensive materials may be concentrated in the sludges or biodegraded. As a consequence the quantities of these agents present in the sewage effluent is much less than in the untreated sewage and their dilution in the receiving water far greater. The benefits of sewage treatment are not limited to greater dilution, however, since each receiving water has a certain capacity for 'self-purification'. Providing sewage treatment reduces the burden of polluting material to a value less than this capacity then the ecosystem of the receiving water will complete the treatment of the residual materials present in the sewage effluent. Thus sewage treatment in conjunction with the selection of appropriate points for sewage effluent discharge has resulted in the elimination of waterborne disease in the UK and many other advanced countries. However, as the population has expanded and become urbanized with a concomitant development of water-consuming industries an additional requirement has been placed upon sewage treatment.

It is now the objective of sewage treatment in many parts of the UK to produce a sewage effluent which after varying degrees of dilution and self-purification is suitable for abstraction for treatment to produce a potable supply. This indirect re-use affects some 30% of all water supplies in the UK.

### 5.1.2 The Importance of Water Re-use

That the United Kingdom practises indirect re-use to a greater extent than most other countries may appear surprising given the annual rainfall. Indeed, that re-use should be important in global terms given the abundance of water on the earth's surface may also be considered improbable in all but the most arid regions. However, two important factors readily explain this situation; firstly a vast amount of the available water is too saline to be used as a potable supply (the salinity is too costly to remove in all but the most extreme cases) and secondly the non-uniform distribution of the population and the available water supply. The available water supply is determined by the rainfall, the ability of the environment to store water (essentially the size of lakes and rivers which are small in the United Kingdom) and their location, *e.g.* Wales has an abundance of suitable water supplies, but limited population, whilst South East England has a large population with limited water resources.

It has been estimated that of the water falling on the United Kingdom 50% is not available for use as a result of run-off to the sea. Of the remainder approximately 17% is utilized. Current predictions suggest that by 2000 AD the amount of water used will have doubled. Thus the potential reserves are very limited. However, because demand and supply are not geographically proximate re-use is already essential. As a consequence the traditional concept of water supply employing single-purpose reservoirs impounding unused river water has been abandoned in favour of multi-purpose schemes designed to permit repeated use of the water before it reaches the sea. In these schemes sewage treatment plays a vital role in addition to being an integral part of the hydrological cycle.

### 5.1.3 Criteria for Sewage Treatment

Sewage is a complex mixture of suspended and dissolved materials; both categories constitute organic pollution. The strength of sewage and the quality of sewage effluent are described in terms of their suspended solids (SS) and biochemical oxygen demand (BOD); these two measures were either proposed or devised by the Royal Commission (1898–1915).

The SS are determined by weighing after the filtration of a known volume of sample through a standard glassfibre filter paper, the results being expressed in $mg\,l^{-1}$.

Dissolved pollutants are determined by the BOD they exert when incubated for 5 days at 20°C. Samples require appropriate dilution with oxygen saturated water and suitable replication. The oxygen consumed is determined and the results again expressed in $mg\,l^{-1}$.

The two standards for sewage effluent quality proposed by the Royal Commission were for no more than 30 mg $l^{-1}$ of suspended solids and 20 mg $l^{-1}$ for BOD, the so called 30 : 20 standard. The Royal Commission envisaged that the effluent of this standard would be diluted 8 : 1 with clean river water having BOD of 2 mg $l^{-1}$ or less. This standard was considered to be the normal minimum requirement and was not enforced by statute because the character and use of rivers varied so greatly. It was intended that standards would be introduced locally as required. For example, a river to be used for abstraction of potable supplies would require a higher standard such as the 10 : 10 standard imposed by the Thames Conservancy. Whilst other countries which are members of the European Union have adopted 'uniform emission standards', that is, the same quality of effluent regardless of the state or use of the river, the United Kingdom has continued with its pragmatic approach whereby effluent standards are set depending on the 'water quality objectives' of the river, which in turn is determined by its function or use. In the 1970s with the reorganization, the water industry's reliance solely on the 30 : 20 standard was abandoned, although this standard is probably still the most commonly applied. Sewage treatment now attempts to consistently produce an effluent with a quality superior to its 'Legal Consent' and attempts to achieve an 'Operating Target', frequently half the Legal Consent. In addition considerable importance has been placed upon the concentration of ammonia in the effluent. In the case of a works attempting to nitrify the effluent (see Section 5.2.3.8) the ammonia concentration is frequently limiting. Typical Legal Consent and Operating Targets are outlined in Table 5.1. It is evident that the Operating Targets included in Table 5.1 are the same as the Royal Commission 30 : 20 standard.

**Table 5.1**  *Legal Consent and Operating Target values for a conventional two stage sewage treatment works*

| Parameter | Legal consent value $(mg\,l^{-1})$ | Operating target value $(mg\,l^{-1})$ |
|-----------|------------------------------------|---------------------------------------|
| SS        | 50                                 | 30                                    |
| BOD       | 35                                 | 20                                    |
| Ammonia   | 25                                 | 12                                    |

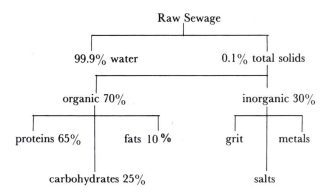

**Figure 5.1** *Composition of a typical raw sewage*

### 5.1.4 Composition of Sewage

Domestic sewage contains approximately 1000 mg $l^{-1}$ of impurities of which about two-thirds are organic. Thus sewage is 99.9% water and 0.1% total solids upon evaporation (see Figure 5.1). When present in sewage approximately 50% of this material is dissolved and 50% suspended. The main components are: nitrogenous compounds – proteins and urea; carbohydrates – sugars, starches and cellulose; fats – soap, cooking oil and greases. Inorganic components include chloride, metallic salts and road grit where combined sewerage is used. Thus sewage is a dilute, heterogeneous medium which tends to be rich in nitrogen.

## 5.2 SEWAGE TREATMENT PROCESSES

Conventional sewage treatment is a three stage process including preliminary treatment, primary sedimentation and secondary (biological) treatment; these are presented schematically in Figure 5.2. In addition some form of sludge treatment facility is frequently employed, typically anaerobic digestion (see Section 5.3.1).

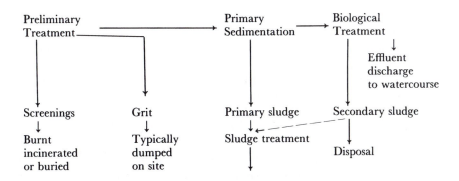

**Figure 5.2** *Flow diagram of a conventional sewage treatment works*

### 5.2.1  Preliminary Treatment

These treatment processes are intended to remove the larger floating and suspended materials. They do not make a significant contribution to reducing the polluting load, but render the sewage more amenable to treatment by removing large objects which could form blockages or damage equipment.

Floating or very large suspended objects are frequently removed by bar screens. These consist of parallel rods with spaces between them which vary from 40 to 80 mm, through which the influent raw sewage must pass. Material which accumulates on the screen may be removed manually with a rake at small works, but on larger works some form of automatic raking would be used. The material removed from the screens contains a significant amount of putrescible organic matter which is objectionable in nature and may pose a disposal problem. Typically the material is buried or incinerated and less frequently burnt.

If screens have been used to remove the largest suspended and virtually all the floating objects, then it only remains to remove the small stones and grit, which may otherwise damage pumps and valves, to complete the preliminary treatment. This is most frequently achieved by the use of constant velocity grit channels. The channels utilize differential settlement to remove only the heavier grit particles whilst leaving the lighter organic matter in suspension. A velocity of 0.3 m s$^{-1}$ is sufficient to allow the grit to settle whilst maintaining the organic solids in suspension. If the grit channels are to function efficiently the velocity must remain constant regardless of variation of the flow to the works (typically between 0.4 and 9 *DWF*). This is achieved by using channels with a parabolic cross section controlled by venturi flumes. The grit is removed from the bottom of the channel by a bucket scraper or suction, and organic matter adhering to the grit is removed by washing, with the wash water being returned to the sewage. Small sedimentation tanks from which the sewage overflows at such a rate that only grit will settle out may also be used. These are compact and by the introduction of air on one side a rotary motion can be induced in the sewage which washes the grit *in situ*. However, these tanks do not cope with the variation in hydraulic load in such an elegant and effective manner as the grit channel.

To avoid the problems associated with the disposal of screenings, comminutors are frequently employed in place of screens. Unlike screens, which precede grit removal, the comminutors are placed downstream of the grit removal process. The comminutors shred the large solids in the flow without removing them. As a result they are reduced to a suitable size for removal during sedimentation. Comminutors consist of a slotted drum through which the sewage must pass. The drum slowly rotates carrying material which is too large to pass through the drum towards a cutting bar upon which it is shredded before it passes through the drum.

The total flow reaching the sewage treatment works is subjected to both these preliminary treatment processes. However, the works is only able to give full treatment up to a maximum flow of 3 *DWF*. When the flow to the works exceeds this value the excess flows over the weir to the storm tanks which are normally empty. If the storm is short, no discharge occurs and the contents of the tanks are pumped back into the works when the flow falls below 3 *DWF*. If the storm is prolonged then these tanks will begin to discharge to a nearby watercourse, inevitably causing some pollution. However, this excess flow has been subjected to sedimentation which removes some of the polluting material. Moreover, as a consequence of the storm, flow in the watercourse will be high giving greater dilution.

## 5.2.2  Primary Sedimentation

The raw sewage (containing approximately 400 mg $l^{-1}$ SS and 300 mg $l^{-1}$ BOD) at a flowrate of 3 *DWF* or less and with increased homogeneity as a result of the preliminary treatment process enters the first stage of treatment which reduces its pollutant load, primary sedimentation, or mechanical treatment. Circular (radial flow) or rectangular (horizontal flow) tanks equipped with mechanical sludge scraping devices are normally used (see Figure 5.3). However, on small works hopper bottom tanks (vertical flow) are preferred; although more expensive to construct these costs are more than offset by savings made as a result of eliminating the requirement for scrapers (see Figure 5.3).

Removal of particles during sedimentation is controlled by the settling characteristics of the particles (their density, size, and ability to flocculate), the retention time in the tank (h), the surface loading ($m^3$ $m^{-2}$ $d^{-1}$) and to a very limited degree the weir overflow rate ($m^3$ $m^{-1}$ $d^{-1}$). Retention times are generally between 2 and 6 hr. However, the most important design criterion is the surface loading; typical values would be in the range 30 to 45 $m^3$ $m^{-2}$ $d^{-1}$. The surface loading rate is obtained by dividing the volume of sewage entering the tank each day ($m^3$ $d^{-1}$) by the surface area of the tank ($m^2$). The retention time may be fixed independently of the surface loading by selection of the tank depth, typically 2 to 4 m, which increases the volume without influencing the surface area. Because they strongly influence the value for surface loading selected, the nature of the particles in the sewage is one of the most important factors in determining the design and efficiency of the sedimentation tank. Of the three factors mentioned before, flocculation is perhaps the most significant.

Four different types of settling can occur:

*Class 1 Settling*: settlement of discrete particles in accordance with theory (Stokes' Law).
*Class 2 Settling*: settlement of flocculant particles exhibiting increased velocity during the process.
*Zone Settling (Hindered Settlement)*: at certain concentrations of flocculant particles, the particles are close enough together for the interparticulate forces to hold the particles fixed relative to one another so that the suspension settles as a unit.

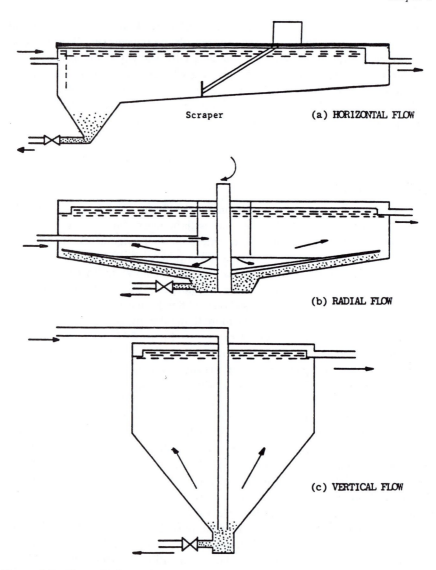

Scraper   (a) HORIZONTAL FLOW

(b) RADIAL FLOW

(c) VERTICAL FLOW

**Figure 5.3**   *Types of sedimentation tank*

*Compressive Settling*: at high solids concentrations the particles are in contact and the weight of the particles is in part supported by the lower layer of solids.

During primary sedimentation settlement is of the Class 1 or 2 types. However, in secondary sedimentation (see Section 5.2.4) zone or hindered settlement may occur. Compressive settlement only occurs in special sludge thickening tanks.

Primary sedimentation removes approximately 55% of the suspended solids and because some of these solids are biodegradable the BOD is typically reduced by 35%. The floating scum is also removed and combined with the sludge. As a result the effluent from the primary has a SS of approximately 150 mg $l^{-1}$ and a BOD of approximately 200 mg $l^{-1}$. This may be acceptable for discharge to the sea or some estuaries without further treatment. The solids are concentrated into the primary sludge which is typically removed once a day under the influence of hydrostatic pressure.

### 5.2.3 Secondary (Biological) Treatment

There are two principal types of biological sewage treatment:

(i)  The percolating filter (also referred to as a trickling or biological filter).
(ii) Activated sludge treatment.

Both types of treatment utilize two vessels, a reactor containing the microorganisms which oxidize the BOD, and a secondary sedimentation tank, which resembles the circular radial flow primary sedimentation tank, in which the microorganisms are separated from the final effluent.

The early development of biological sewage treatment is not well documented. However, it is established that the percolating filter was developed to overcome the problems associated with the treatment of sewage by land at 'sewage farms', where large areas of land were required for each unit volume of sewage treated. It was discovered that approximately 10 times the volume of sewage could be treated in a given area per unit by passing the sewage through a granular medium supported on underdrains designed to allow the access of air to the microbial film coating the granular bed.

The origins of the percolating filter are present in land treatment and its development was an example of evolution. The second and probably predominant form of biological sewage treatment, the activated sludge process, arose spontaneously and represents an entirely original approach. This process involves the aeration of freely suspended flocculant bacteria, 'the activated sludge floc' in conjunction with settled sewage which together constitute the 'mixed liquor'. Activated sludge treatment continues the trend established by the change from land treatment to the percolating filter in that, at the expense of higher operating costs, it is possible to treat very much larger volumes of sewage in a smaller area.

The activated sludge process is probably the earliest example of a continuous bacterial (microbial) culture deliberately employed by man, and certainly the largest used to date. Development of the activated sludge process was announced by its originators Fowler, Ardern and Lockett in 1913, based upon their research at the Davyhulme Sewage Treatment Works, Manchester. These scientists very generously did not patent the process to facilitate its rapid and widescale application.

Development of these two forms of biological sewage treatment has been largely empirical and undertaken without the benefit of information about the fundamental principles of continuous bacterial growth, which began to be developed from the late 1940s when Monod published his work on continuous bacterial growth, although the relevance was not perceived until approximately

10 years later. This lack of microbiological knowledge is highlighted by the fact that the role of microorganisms in the activated sludge process was not fully accepted until after 1931; prior to this it was accepted by several workers that coagulation of the sewage colloids was the principal mechanism in the activated sludge process, although in the USA the role of bacteria in percolating filters was first recognized in 1889.

*5.2.3.1   Percolating Filter.*   These units consist of circular or rectangular beds of broken rock, gravel, clinker or slag with a typical size in the range of

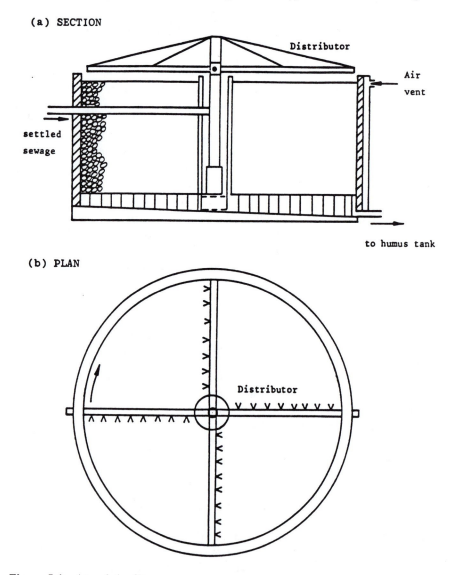

**Figure 5.4**   *A percolating filter*

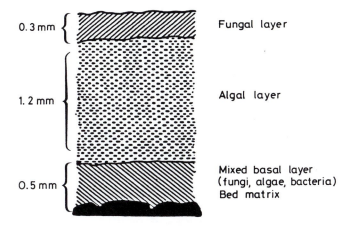

**Figure 5.5** *Cross-section of the surface layers of a percolating filter*

50–100 mm. The beds are between 1.5 and 2.0 m deep and of very variable diameter or size depending on the population to be served. The proportion of voids (empty spaces) in the assembled bed is normally in the range 45 to 55% (see Figure 5.4). The settled sewage trickles through interstices of the medium which constitutes a very large surface area on which a microbial film can develop. It is in this gelatinous film containing bacteria, fungi, protozoa and on the upper surface algae that the oxidation of the BOD in the settled sewage takes place. The percolating filter is in fact a continuous mixed microbial film reactor. Settled sewage is fed onto the surface of the filter by some form of distributor mechanism. On circular filters a rotation system of radial sparge pipes is used which are usually reaction-jet propelled although on larger beds they may be electrically driven. With rectangular beds electrically powered rope hauled arms are used.

The microorganisms which constitute the gelatinous film appear to be organized, at least near the surface of the filter where algae are present, into three layers (see Figure 5.5). The upper fungal layer is very thin (0.33 mm), beneath it the main algal layer is approximately 1.2 mm and both are anchored by a basal layer containing algae, fungi and bacteria of approximately 0.5 mm. However, algae do occur to some extent in all three layers. Beneath the surface, where sunlight is excluded and as a consequence the algae are absent, this structure is significantly modified, probably into a form of organization with only two layers. It has been calculated that photosynthesis by algae could provide only 5% or less of the oxygen requirements of the microorganisms in the filter. Furthermore, photosynthesis would only be an intermittent source of oxygen since it would not occur in the dark and algae are often present only in the summer months. Carbon dioxide generated by other organisms in the filter might however increase the rate of photosynthesis. It has been proposed that algae derive nitrogen and minerals from the sewage and that some may be facultative heterotrophs. The nitrogen fixing so-called 'blue–green algae', really bacteria, are frequently present in filters.

Whilst fungi are efficient in the oxidation of the BOD present in the settled sewage they are not desirable as dominant members of the microbial community. They generate more biomass than bacteria, per unit of BOD consumed, thus increasing the sludge disposal problem. Moreover, an accumulation of predominantly fungal film quickly causes blockages of the interstices of the filter bed material, impeding both drainage and aeration. The latter may result in a reduction in the efficiency of treatment which is dependent upon the metabolic activity of aerobic microorganisms.

Protozoa and certain metazoans (macrofauna) play an important role in the successful performance of the biological filter, although the precise nature of this role is dependent on the extrapolation of observations made in the activated sludge process, which is more amenable to study. However, the similarity in the distribution of organisms within the two processes suggests strongly that their roles are the same in both. The protozoa in particular remove free-swimming bacteria thus preventing turbid effluents, since freely suspended bacteria are not settleable. Certain metazoans may also ingest free-swimming bacteria, but their most important function is to assist in breaking the microbial film which would otherwise block the filter. This film is 'sloughed off' with the treated settled sewage. Protozoa (principally ciliates and flagellates) tend to dominate in the upper layers of the filter, whilst the macrofauna (nematodes, rotifers, annelids and insect larvae) dominate the lower layers.

If film is not removed satisfactorily, frequently as the result of excessive fungal growth, the condition known as 'ponding' develops. In this condition the surface of the filter is covered in settled sewage, air flow ceases, treatment stops and the bed becomes anaerobic. Ponding may also be caused by the growth of a sheet or felt of large filamentous algae principally *Phormidian* sp. on the face of the filter.

To minimize film production recirculation of treated effluent is often employed. This reduces film growth by dilution of the settled sewage, improves the flushing action for the removal of loose film, and promotes more uniform distribution of the film with depth.

Treated sewage is subject to secondary sedimentation which is similar to primary sedimentation as a result of which the suspended sloughed off film is consolidated into humus sludge and the final effluent discharged to the receiving water.

*5.2.3.2   Activated Sludge.*   In the activated sludge process the majority of biological solids removed in the secondary sedimentation tank are recycled (returned sludge) to the aerator. The feedback of most of the cell yield from the sedimentation tank encourages rapid adsorption of the pollutants in the incoming settled sewage and also serves to stabilize the operation over a wide range of dilution rates and substrate concentrations imposed by the diurnal and other fluctuations in the flow and strength of the sewage. Stability is also provided by the continuous inoculation of the reactor with microorganisms in the sewage and airflows, which are ultimately derived from human and animal excreta, soil run-off, water and dust. The reactor of the activated sludge plant is usually in the form of long deep channels. Before entering these channels the returned sludge and settled sewage are mixed thereby forming the 'mixed liquor'. The retention time of the 'mixed liquor' in the aerator is typically three to six hours; during this period it moves down the length of the channel before passing over a weir, prior to secondary sedimentation. The sludge which is not returned to the aerator unit is known as surplus activated sludge and has to be disposed of. In practice the conditions in the aeration unit diverge from the completely mixed conditions commonly used for industrial fermentations and it may be best described as a continuous mixed microbial deep reactor with feed-back.

The design of the concrete tanks which form the reactor is strongly influenced by the type of aeration to be employed. Two types are available, compressed (diffused air) (see Figure 5.6) and mechanical (surface aeration) (see Figure 5.7).

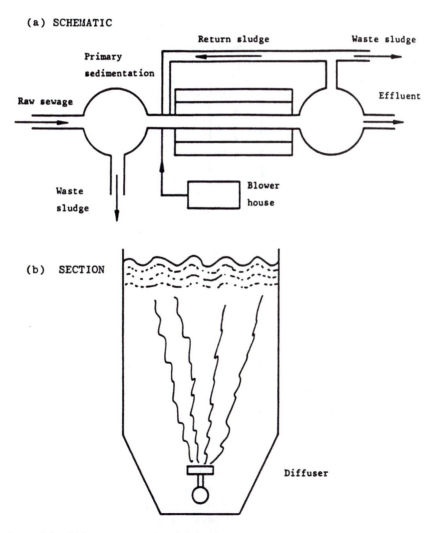

**Figure 5.6** *Diffused aeration activated sludge plant*

**(a)   SCHEMATIC**

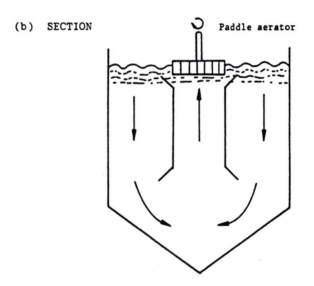

**(b)   SECTION**

**Figure 5.7**   *Mechanically aerated activated sludge plant*

In the diffused air system much of the air supplied is required to create turbulence, to avoid sedimentation of the bacteria responsible for oxidation. Surface aeration systems introduce the turbulence mechanically and only provide sufficient air for bacterial oxidation. Both types of system aim to maintain a dissolved oxygen concentration of between 1 and 2 mg l$^{-1}$.

In the diffused air system the air is released through a porous sinter at the base of the tank and this system is characterized by long undivided channels which may be quite narrow (see Figure 5.6). Mechanical aeration utilizes rotating paddles to agitate the surface thereby incorporating air and creating a rotating current which maintains the bacterial flocs in suspension. Each paddle is located in its own cell which has a hopper shaped bottom, this gives the plant the appearance of a square lattice (see Figure 5.7). However, beneath the face of the mixed liquor all the cells are connected forming a channel. In both systems the channels are 2–3 m deep and 40–100 m long.

The success of the activated sludge process is dependent on the ability of the microorganisms to form aggregates (flocs) which are able to settle. It is generally accepted that flocculation can be explained by colloidal phenomena and that bacterial extracellular polymers play an important role, but the precise mechanism is not known. The significance of flocculation to the success of the process is not the only characteristic to distinguish it from other industrial continuous cultures. There are four additional and very significant differences: it utilizes a heterogeneous microbial population, growing in a very dilute multi-substrate medium, many of the bacterial cells are not viable and finally the objectives of the process, which are the complete mineralization of the substrates (principally carbon dioxide, water, ammonia and/or nitrate) with minimal production of both biomass and metabolites are also unique.

The heterogeneous population present in activated sludge includes bacteria, protozoa, rotifers, nematodes and fungi. The bacteria alone are responsible for the removal of the dissolved organic material, whilst the protozoa and rotifers 'graze', removing any 'free-swimming' and hence non-settleable bacteria, the protozoans and rotifers being large enough to settle during secondary sedimentation. The role of protozoa in activated sludge has been extensively studied; there are three groups involved: the ciliates, flagellates and amoebae. It is probably the ciliates (*Ciliophora*), which constitute the greatest number of species with the greatest number present in each species, play the major role in the clarification process. The effect on effluent quality as a consequence of grazing by protozoa is summarized in Table 5.2. Not only do the protozoa remove free-swimming activated sludge bacteria but they play an important role in the reduction of pathogenic bacteria, including those which cause diphtheria, cholera, typhus and streptococcal infections. In the absence of protozoa approximately 50% of these types of organisms are removed while in their presence removals rise to 95%. Nematodes have no significant role in the process, whilst the effects of the fungi are generally deleterious and contribute to or cause non-settleable sludge known as 'bulking'. Members of the following bacterial genera have been regularly isolated from activated sludge, *Pseudomonas, Acinetobacter, Comamonas, Lophomonas, Nitrosomonas, Zoogloea, Sphaerotilus, Azotobacter, Chromobacterium, Achromobacter, Flavobacterium, Alcaligenes, Micrococcus* and *Bacillus*. Attributing the appropriate importance to each genus is a problem which confounds bacteriologists.

**Table 5.2** *Importance of ciliated protozoa in determining effluent quality*

| Effluent property | Ciliates absent | Ciliates present |
|---|---|---|
| Chemical Oxygen Demand (mg $l^{-1}$) | 198–254 | 124–142 |
| Organic Nitrogen (mg $l^{-1}$) | 14–20 | 7–10 |
| Suspended Solids (mg $l^{-1}$) | 86–118 | 26–34 |
| Viable bacteria ($10^7$ ml$^{-1}$) | 29–42 | 9–12 |

Of the principal groups of substrates listed in Section 5.1.4, only one single substrate (cellulose) was included. Each of the groups includes many substrates; for example the 'sugars' identified in sewage include glucose, galactose, mannose, lactose, sucrose, maltose and arabinose, whilst the nitrogenous compounds include proteins, polypeptides, peptides, amino acids, urea, creatine and amino-sugars. Since bacteria normally only utilize a single carbon substrate or at the most two, this diversity of substrates in part explains the numerous genera of bacteria isolated from activated sludge, because each substrate under most conditions will sustain one species of bacterium. Moreover as a consequence of the large number of substrates present in the settled sewage the concentration of individual substrates is far less than the 200 mg l$^{-1}$ of BOD present, perhaps 20–40 mg l$^{-1}$ for the most abundant and less than 10 mg l$^{-1}$ for the less common ones. The concentration of each substrate is further reduced in the aeration tank by dilution with the returned activated sludge which is typically mixed 1:1 with settled sewage resulting in a 50% reduction in substrate concentration.

The low substrate concentration means that the bacteria are in a starved condition. As a consequence many of them are 'senescent', *i.e.* in that phase between death, as expressed by the loss of viability, and breakdown of the osmotic regulatory system (the moribund state); thus the bacterium is a functioning biological entity incapable of multiplication. That bacteria could exist in this condition was established at an early stage in a series of inspired experiments by Wooldridge and Standfast who published their results in 1933. They determined the dissolved oxygen concentrations and bacterial numbers (by viable counts) in a series of biochemical oxygen demand bottles containing diluted raw sewage on a daily basis. The viable count reached a maximum on the second day and thereafter fell rapidly. However the consumption of oxygen increased by equal amounts until the fourth day and fell to a negligible value on the fifth day. There was no obvious relationship between viability and oxygen consumption. They tested experimentally the hypothesis that non-viable bacteria were apparently capable of oxygen uptake by destroying the capacity for division without significantly diminishing enzyme activity. Treatment of *Pseudomonas fluorescens* with a 0.5% formaldehyde solution prevented division but these bacteria exhibited vigorous oxygen uptake in both sewage and other media. Subsequently they were able to determine the presence of active oxidase and dehydrogenase enzymes in these non-viable bacteria. The effects of low substrate concentration on the viability of the bacteria are compounded by their specific growth rate. It is intended that biological wastewater treatment should result in the production of a final effluent containing negligible BOD. The biochemically oxidizable material in the effluent is composed of compounds originally present in the settled sewage, which have not been completely biodegraded, and bacterial products. Moreover this is to be achieved with the minimal production of biomass. These twin objectives are concomitant with the utilization of a bacterial population with a very low specific growth rate.

Unlike the percolating filter, bacterial growth in the activated sludge process is amenable to the type of description used by bacteriologists for conventional continuous cultures. However, although it is amenable to this type of treatment it inevitably appears to be very different from all other continuous cultures. The dilution rates (rate of inflow of settled sewage/aeration tank volume) used are invariably low by the standards of industrial fermentations, typically $0.25 \text{ hr}^{-1}$, *i.e.* one quarter of the aeration tank volume is displaced every hour, therefore the *hydraulic retention time* is four hours. Although in the conventional single pass reactor the dilution rate and the specific growth rate (time required for a doubling of the population) are identical; that is the state in which the rate of production of cells through growth equals the rate of loss of cells through the overflow. In the activated sludge process because of the recycling of the biomass the specific growth rate is very much lower than the dilution rate, typically in the range $0.002–0.007 \text{ hr}^{-1}$. Since, under steady-state conditions, the bacteria are only able to grow at the same rate as they are lost from the system, recycling them dramatically lowers their specific growth rate and allows it to be controlled independently of the dilution rate. Under steady-state conditions the specific growth rate is equivalent to the specific rate of sludge wastage (mass of suspended solids lost by sludge wastage and discharged in the effluent in unit time as a proportion of the total mass in the plant) which is the reciprocal of the 'sludge age' or mean cell retention time which is typically 4–9 days. Thus, whilst the retention of the aqueous phase in the system is only four hours, the retention of the bacterial cells or sludge age is several days. The sludge age $(\theta c)$ is a value which describes a great deal about the type of activated sludge plant; its purpose, quality of effluent and the bacteriological and biochemical states are all summarized by this item.

The activated sludge process may have up to four phases:

(i) clarification, by flocculation of suspended and colloidal matter;
(ii) oxidation of carbonaceous matter;
(iii) oxidation of nitrogenous matter (see Section 5.2.3.8);
(iv) auto-digestion of the activated sludge.

The occurrence of these four phases is directly dependent on increasing sludge age. Those processes which operate at low sludge ages give rapid removal of BOD per unit time, but the effluent is of poor quality. Plants which have high sludge ages give good quality effluents but only a slow rate of removal. Low sludge ages result in actively growing bacteria, and consequently high sludge production, whilst bacteria grown at high sludge ages behave conversely. Figure 5.8 illustrates the relationship between the growth curve of the bacterial culture and the type of activated sludge plant. By operating continuously the activated sludge process functions only over a small region of the batch growth curve; this region is determined by the specific sludge wastage rate. The region selected determines the type of plant and its performance. These are summarized in Figure 5.8.

*5.2.3.3 Dispersed Aeration.* This type of process is rarely used and is not applicable to the treatment of municipal sewage but may be of use in the preliminary treatment of some industrial wastes. The bacteria are growing rapidly (exponential phase), thus the process has the ability to remove a large

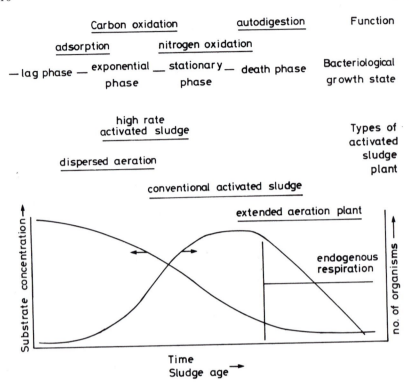

**Figure 5.8**   *Relationship between batch culture and type of activated sludge plant*

quantity of BOD per unit of biomass and as a consequence a small reactor may be used which is cheap to construct. However, because of their high rate of growth, the bacteria convert much of the BOD into biomass, causing a sludge disposal problem, flocculation is limited so additional treatment is essential to remove solids. Furthermore, although BOD removal per unit biomass is high, the effluent BOD is also high.

*5.2.3.4   High-rate Activated Sludge.*   This shares many features of the previous process; however, flocculation proceeds satisfactorily and secondary sedimentation will remove the solids effectively. The growth rate of the bacteria is still high, but only carbonaceous material will be oxidized. However, some 60 to 70% of the influent BOD will be removed with a hydraulic retention time of approximately two hours. This type of process is probably most frequently used for industrial wastes prior to discharge to the sewers, although it is also used for domestic sewage treatment, perhaps most appropriately where effluents are to be discharged to estuarine waters where standards are less stringent.

*5.2.3.5   Conventional Activated Sludge.*   The two previous processes utilize actively growing bacteria in the exponential phase of growth. They achieve the oxidation of carbon compounds utilizing an exclusively heterotrophic bacterial population.

Conventional activated sludge plants operate in the stationary or declining growth phases utilizing senescent bacteria. This very slow growth results in very low residual substrate concentrations and hence low values for effluent BOD. In addition, plants operating at sludge ages towards the upper end of this range contain autotrophic nitrifying bacteria. These organisms convert ammonia to nitrite and nitrate. This further improves the quality of the effluent since ammonia can exert an oxygen demand but nitrate cannot. In addition to maximizing effluent quality, conventional activated sludge plants limit the production of new cells. Bacteria which are growing slowly use much of the organic matter available in the maintenance of their cells rather than in the production of new cells. These features have made conventional activated sludge the most widely adopted biological sewage treatment process for medium and large communities. The rate of oxidation is highest at the inlet of the tank and it can be difficult to maintain aerobic conditions. Two solutions to this problem have been adopted. With *tapered aeration* rather than supplying air uniformly along the length of the tank the air is concentrated at the beginning of the tank and progressively reduced along its length. The volume of air supplied remains unchanged but it is distributed according to demand. Alternatively *stepped loading* may be utilized. This aims to make the requirement for air uniform by adding the settled sewage at intervals along the tank, thus distributing the demand.

*5.2.3.6 Extended Aeration.* This process operates at very high sludge ages exclusively in the declining phase of growth. The retention time in the aeration tank is between 24 hr and 24 days. As a consequence, the available substrate concentration is low and the bacteria undertake endogenous respiration (see Figure 5.8), that is respiration after the consumption of all available extracellular substrate. The result of utilizing endogenous materials is the breakdown of the sludge, sometimes referred to as auto-digestion. By this means sludge production is minimized and the small amount of material that must be disposed of is highly mineralized and inoffensive. This type of treatment has been extensively used for small communities; whilst capital costs of such plants are high, operating and sludge disposal costs are very low.

*5.2.3.7 Contact Stabilization.* The contact stabilization process is a variation of conventional activated sludge used for treating wastes with a high content of biodegradable colloidal and suspended matter. The process utilizes the adsorptive properties of the sludge to remove the polluting material very rapidly (0.5–1 hr) in a small aeration tank. The mixed liquor is then settled and passed into a second aeration tank and aerated for a further 5 to 6 hr, during which period the adsorbed material is oxidized. After this the sludge with its adsorptive capacity restored is returned to the contact basin. Although this process requires two aeration tanks, the two are very much smaller than the equivalent single tank, since the mixed liquor suspended solids in the contact basin are typically 2000 mg l$^{-1}$ and in the second tank (digestion unit) they are about 20 000 mg l$^{-1}$.

*5.2.3.8 Nitrification.* The production of a final effluent with the minimum BOD value is dependent upon the complete nitrification of the effluent, which involves the conversion of the ammonia present to nitrate. This is a two-stage process undertaken by autotrophic bacteria principally from the genera *Nitrosomonas* and *Nitrobacter*. Nitrification occurs in percolating filters and activated sludge plants operated in a suitable manner. The first stage, sometimes referred to as 'nitrosification' involves the oxidation of ammonium ions to nitrite and follows the general formula:

$$NH_4^+ + 1.5O_2 \xrightarrow{\textit{Nitrosomonas}} NO_2^- + 2H^+ + H_2O$$

In the second stage nitrite is oxidized to nitrate:

$$NO_2^- + 0.5O_2 \xrightarrow{\textit{Nitrobacter}} NO_3^-$$

The overall nitrification process is described by the formula:

$$NH_4^+ + 2O_2 \longrightarrow NO_3^- + 2H^+ + H_2O$$

Two important points are evident from this last formula. Firstly, nitrification requires a considerable quantity of oxygen. Secondly, hydrogen ions are formed and hence the pH of the wastewater will fall slightly during nitrification.

The settled sewage is effectively self buffering but a fall of 0.2 of a pH unit is frequently observed at the onset of nitrification. In this autotrophic nitrification process, ammonia or nitrite provide the energy source, oxygen the electron acceptor, ammonia the nitrogen source and carbon dioxide the carbon source. The carbon dioxide is provided by the heterotrophic oxidation of carbonaceous nutrient, by reaction of the acid produced during nitrification with carbonate or bicarbonate present in the wastewater, or by carbon dioxide in the air. Whereas for carbonaceous removal the oxygen requirement is roughly weight for weight with the nutrients oxidized, in the case of ammonia removal by nitrification requires approximately seven times as much oxygen as is required to achieve the removal of the same quantity of nutrient.

Nitrification significantly increases the cost of sewage treatment since more air is required. Furthermore, because these autotrophic organisms grow only slowly, longer retention periods are also required resulting in higher capital costs. Nor does nitrification result in the production of an entirely acceptable sewage effluent. In areas where water re-use is practised the concentration of nitrate in river waters causes concern. There exists a limit on the concentration of nitrate in drinking water to avoid the occurrence of methaemoglobinaemia (so called 'blue baby' syndrome). As a consequence denitrification is now practised after nitrification in some activated sludge treatment plants. In this anoxic hetero-trophic bacterial process, nitrite and nitrate replace oxygen in the respiratory mechanism and gaseous nitrogen compounds are formed (nitrogen gas, nitrous and nitric oxides). However, this procedure is not part of conventional sewage treatment practice at present.

### 5.2.4 Secondary Sedimentation

Both types of biological treatment require sedimentation to remove suspended matter from the oxidized effluent. Tanks similar to those normally employed for primary sedimentation are generally employed, although at a higher loading of approximately 40 $m^3$ $m^{-2}$ $d^{-1}$, at 3 *DWF*. Because of the lighter and more homogenous nature of secondary sludge, simple sludge scrapers are possible and scum removal is not necessary. The association of primary sedimentation tanks and a biological process for secondary treatment, results in a sewage treatment works, as opposed to sewage farms where only land treatment was (is) employed. As an awareness of environmental pollution, in addition to public health, developed in the fifties and sixties, the term water pollution control works was introduced to describe sewage treatment works, although this change of terminology was merely cosmetic. With the recognition of the importance of water re-use the term water reclamation works has found favour in some areas. Such works frequently apply additional tertiary treatment processes.

Sewage treatment results in the production of a final effluent suitable for discharge in the selected receiving water, and one or more sludges which may require treatment prior to disposal.

## 5.3 SLUDGE TREATMENT AND DISPOSAL

Sludge treatment and disposal is a facet of wastewater treatment which is often given insufficient attention. Sludge treatment and disposal may account for 40% of the operating costs of a wastewater treatment facility. Prior to treatment the sludges contain between 1 and 7% solids (they are therefore nearly all water) which are usually highly putrescible and offensive. A total of 40 million tonnes of sludge, equivalent to 1.3 million tonnes of dry solids, is produced in the UK every year. The sludges are the product of primary sedimentation of raw sewage and the by-product of secondary aerobic treatment of settled sewage. Primary sludge is particularly offensive, with a pronounced faecal odour, and is liable to become putrescent thus causing a nuisance. Secondary sludge consists very largely of bacterial solids. It is less offensive than primary sludge but may still become putrescent. These sludges are sometimes combined during sewage treatment as a consequence of co-settlement of waste activated sludge in the primary sedimentation tanks. The main aims of sludge treatment are to make it easier and cheaper to dispose of the sludge consistent with minimizing any nuisance or adverse effects on the environment generally. A wide range of treatment processes and disposal options has been used, although, recently, the cost of energy has reduced the numbers currently employed because of economic considerations.

The most convenient and economical method of disposal at any given site depends on a number of factors. Treatment of sludge is frequently influenced by the final disposal option selected. If sludge is to be disposed to sea from a works where the sludge may be pumped directly to the disposal vessel then little treatment is required. Should the treatment works be close enough to the sea to make that type of disposal feasible, but not close enough to allow direct pumping to the disposal vessel, then economies in transport costs may be achieved by utilizing some type of treatment process to thicken the sludge and reduce its water content prior to transport to the disposal vessel. If sludge is to be disposed of to land it is desirable to reduce transport costs, since the sludge will have to be spread over a wide area, and, in the case of treatment works in urban locations, transported a significant distance to reach suitable land.

At present 67% of the sludge produced in the UK is disposed to land, 29% to sea, and 4% is incinerated. Of the sludge disposed to land approximately two-thirds is applied to agricultural and horticultural land and the remainder is used for land reclamation and landfill.

The processes available for sludge treatment include: thickening by stirring or flotation; digestion, aerobically or anaerobically; heat treatment; composting with domestic refuse; chemical conditioning with either organic or inorganic materials; dewatering on drying beds, in filter presses, by vacuum filtration or centrifugation; heat drying; incineration in multiple hearth or fluidized bed furnaces and wet air oxidation. It is not feasible within this presentation to deal with all these processes in depth and the following is confined to the predominant sludge treatment process, anaerobic digestion, mechanical dewatering, and the most frequently utilized disposal option, that to agricultural land.

### 5.3.1 Anaerobic Digestion

During anaerobic digestion the organic matter present in the sewage sludge is biologically converted to a gas typically containing 70% methane and 30% carbon dioxide. The process is undertaken in an airtight reactor usually equipped with a floating gas collector. Sludge may be introduced continuously, but more frequently is added intermittently, and the digester operates on a 'fill and draw' process. The methane produced is generally utilized for maintaining the process temperature, heating and power production by combustion in dual fuel engines which use oil in the absence of methane.

Methane production is only significant at elevated temperatures, when 1 $m^3$ of methane at STP is produced for every 3 kg of BOD degraded. Digesters are characterized by the temperature at which they operate, those in which gas production is optimum at $35°C$ are described as 'mesophilic' whilst those in which gas production is optimum at $55°C$ are 'thermophilic', these terms describing the temperature preferences of the bacteria undertaking the process.

Heat exchangers are used to transfer heat from the treated sludge to the influent sludge. The additional heat is provided by the combustion of methane. Digesters in the UK operate in the mesophilic range, since heat loss from thermophilic digesters would be unacceptable. To minimize heat loss, digesters are frequently surrounded by earth banks to provide insulation. For efficient operation the digester requires a mixing system which may be mechanical or utilize the gas produced in the process to provide turbulence. A conventional anaerobic digester is illustrated in Figure 5.9. The result of anaerobic digestion is to reduce the volatile solids present in the original sludge by 50% and the total

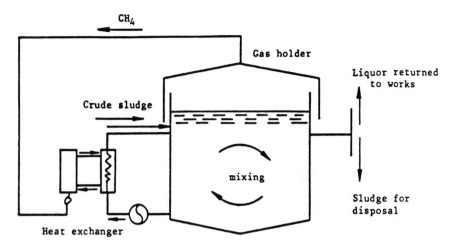

**Figure 5.9** *Schematic diagram of an anaerobic digester*

solids by 30%. In addition the unpleasant odour associated with the raw sludge is drastically reduced. During the 20 to 40 days required for digestion the sludge is stabilized and emerges with a slightly tarry odour.

Traditionally, anaerobic digestion has been considered a two-stage process, a non-methanogenic stage followed by a methanogenic stage. The non-methanogenic stage has also been referred to as the acid forming stage since volatile fatty acids are the principal products. However it is now recognized that the first stage may include as many as three steps. The first, involving the hydrolysis of the fats, proteins and polysaccharides present in the sludge, produces long chain fatty acids, glycerol, short chain peptides, amino acids, monosaccharides and disaccharides. The second step (acid formation) involves the formation of a range of relatively low molecular weight materials including hydrogen, formic and acetic acids, other fatty acids, ketones and alcohol. It is now recognized that only hydrogen, formic acid and acetic acid can be utilized as substrates by the methanogenic bacteria. Thus in the third step compounds other than hydrogen, formic acid and acetic acid are converted by the obligatory *hydrogen producing* acetogenic (OHPA) bacteria. Some bacteria are able to undertake both steps one and two and produce hydrogen, formic acid and acetic acid which therefore do not require step three. These stages are summarized in Figure 5.10. Once in operation, with reasonable retention times and volatile solids loadings, routine operation of digesters must include careful monitoring of certain parameters which are used to indicate whether the process is about to fail. The main parameters are volatile acids and hydrogen ion concentration (pH).

Anaerobic digestion is quite sensitive to fairly low concentrations of toxic pollutants, such as heavy metals and chlorinated organics, and to variations in loading rates and other operational aspects. If the balance of the process is upset it is most likely that the methanogenic organisms become inhibited first. This

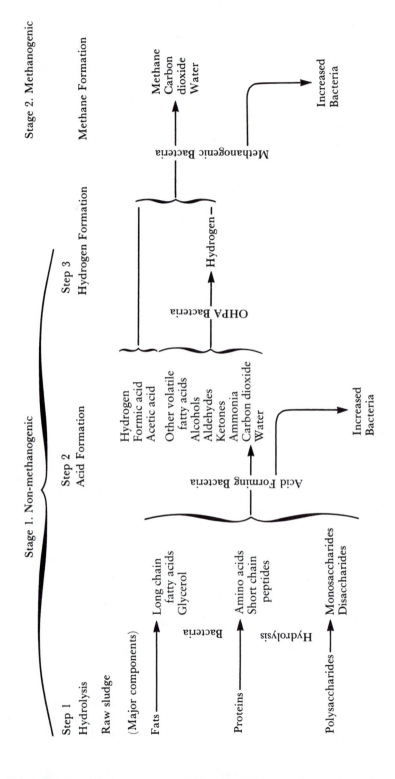

**Figure 5.10** *Biochemical transformations involved in anaerobic digestion*

results in a build-up of the intermediate compounds at the stage immediately prior to methane formation. These intermediates collectively are called volatile fatty acids. They include formic, acetic and butyric acids and can be monitored to determine the state of the process. The volatile acids are important because of their acidic nature. Normally digesters operate in the pH range of neutrality (6.5–7.5). They also have some resistance to pH change. High concentrations of the volatile acids can cause a reduction in pH sufficient to inhibit bacterial activity to the extent where irreversible failure of the process occurs. Because of their capacity to resist changes in pH, volatile acid concentrations can build up to significant levels before pH change occurs. Therefore they can act as an early warning indicator of impending process failure. Normal levels of volatile acids are 250–1000 mg $l^{-1}$. If they exceed 2000 mg $l^{-1}$ this could lead very quickly to failure; if they exceed 5000 mg $l^{-1}$ failure is almost inevitable. The adverse effects of volatile acid build-up can be rectified by the addition of lime to restore the balance between acidity and alkalinity.

Anaerobically digested sludge is frequently further dewatered in lagoons prior to disposal. Supernatant liquors are pumped from the surface of the lagoons to the head of the works for treatment.

## 5.3.2 Dewatering of Sludge

One of the major objectives of sludge treatment is to reduce the water content. The advantages of this are two-fold. First it reduces the volume of sludge to be handled, which can very often lead to savings in transport costs, and secondly it can improve the physical properties of the sludge making it easier to handle. Sludge can be dewatered in two main ways. Either it can be allowed to dry out naturally or it can be dewatered by forcing the water out mechanically, typically by either pressure filtration or vacuum filtration.

If sludge is to be dried naturally it is usually spread in layers up to about 2–3 cm thick in special drying beds. These very thin layers permit water loss both by evaporation from the surface and drainage. The sludge lies on a layer of fine ash, over a layer of coarse ash, under which are laid underdrains. The liquor which drains off the sludge goes to a central sump. From there it is pumped back to the main treatment works to undergo aerobic treatment. After a period of about two months the solids content increases to about 25% and the sludge can be dug up. This can be done manually although mechanical scrapers which transfer the sludge onto a moving conveyor are sometimes used. Because the sludge is spread in such thin layers, a large land area is required for drying beds. In the UK a drying area of about 0.3 $m^2$ per head of population is normally required.

Sludges are usually difficult to dewater and their dewaterability can usually be improved by the use of conditioners and this is normally the practice if mechanical dewatering is to be employed. Frequently aluminium and iron compounds are used, such as aluminium chlorohydrate and ferrous sulfate, or alternatively organic polymers called polyelectrolytes may be used. It is not certain how these work but it is probable that they react with the surfaces of the small sludge particles which would appear to be the cause of poor dewatering.

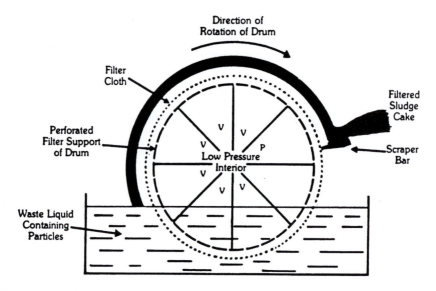

**Figure 5.11** *Vacuum filter*

Following conditioning, sludge may be dewatered by pressure filtration. The sludge is pumped at a pressure of 700 kPa into a cloth-lined chamber; at this pressure it requires between 2 and 18 hours to form a cake of 25–50% solids. The process can be operated either in batch or continuous manner.

Alternatively, vacuum filtration may be used; this is invariably a continuous process. A drum, containing several internal segments, revolves on its horizontal axis partially submerged in sludge. A partial vacuum inside the drum causes the sludge to adhere to it in a thin layer and as it rotates out of the wet sludge the water is sucked out. A scraper separates the dried sludge from the outside of the drum. At the point where the scraper is positioned, the vacuum in the corresponding segment inside the drum is released, aiding the release of the sludge from the outside. The pressure difference between inside and outside is 70 kPa, which is only about 10% of the pressure attained in pressure filtration. Hence the dried sludge typically only contains 15–20% solids. A diagram of a vacuum filter is shown in Figure 5.11.

### 5.3.3 Disposal of Sewage Sludge to Land

The practice of disposing sewage sludge to land has several potential benefits. In 1981 a Standing Technical Committee Report on the Disposal of Sewage Sludge to Land had this to say about the objectives of good practice:

'Good practice in sewage sludge disposal, whether to land, to sea, or by incineration, involves striking a balance between economic constraints and the avoidance as far as possible of adverse effects on man, animals and the environment. Low cost is not always compatible with limited adverse environmental effects and sometimes compromises have to be reached. The best option of treatment within a disposal method may change with time due to changing costs, improved knowledge of treatment processes and environmental effects, experience and research.

Disposal of sewage sludge involving application to agricultural land has the benefit of resource recovery and the value of the nutrients utilized should be taken into account in assessing the minimum cost to the nation. Disposal of sludge to land can have greater environmental impact than other options and in assessing likely benefits and potential adverse effects consideration should also be given to amenity, formal and informal recreation and wildlife.'

This report concludes:

'Where ever it is economically justified and environmentally acceptable sewage sludge should be utilized on agricultural land in accordance with ... recommendations ...'

In coming to this conclusion, the Committee had to weigh up the advantages and disadvantages prior to setting out their recommendations, which included guidelines designed to regulate the quantities of sludge applied to land.

Most of the sludge used in agriculture (about 97%) receives some form of treatment. Nearly half of it is anaerobically digested. Slightly more goes to general arable land than to grazing land. Grass grown for hay or silage is particularly suitable to receive sludge because it reduces the risks of transmission of disease. Fields used for forage crops and cereals are also suitable to receive sludge.

Both dried and liquid sludges are applied to land, the latter from tankers by spraying. Arable land can be ploughed, following the application, which speeds up incorporation of the sludge and can reduce any odour problems. However, when it is applied, even spreading is important to prevent localized 'hotspots'. Liquid sludges may also be injected directly into the soil giving uniform application and almost complete elimination of odour problems.

Ideally, disposal sites should be well away from housing, but have good access. The risk of leachate or run-off contaminating groundwater or surface streams must be carefully considered. If the sludge is to be spread by spraying, care should be taken not to allow drift, especially in windy conditions.

Agricultural land is an important outlet for sewage sludge disposal, with about 40% of the sludge produced at inland wastewater treatment works being disposed of in this way. Nevertheless only a very small fraction of the agricultural land (less than 2%) in the UK receives sludge.

The application of sludge to land may help to slow down the decline in organic matter in soils under modern farming methods, leading to improvements in water holding capacity, porosity and aggregate stability. The main value of sludge as a fertilizer lies in its nitrogen and phosphorus contents. However, much of the nutrient content may be in organic forms, and thus be unavailable to plants until mineralization occurs. Although from an economic point of view it may be desirable to apply dried sludge to land, significant quantities of available forms of the nutrients may be lost during drying. Liquid digested sludge may contain up to 10% (w/w) of nitrogen but only a fraction of this may be in available forms. For the purpose of calculating the available nitrogen in sludge, it is assumed that 85% of the total nitrogen in liquid digested sludge and 33% of that in dried sludge is available to crops during the growing season.

Since most agricultural soils are deficient in nutrients, fertilizers nearly always have to be added. However, sewage sludge is deficient in potassium, and therefore cannot fulfil complete fertilizer requirements. Furthermore, if all of the sewage sludge produced in the UK were to be applied to agricultural land, it would only provide 4.5% of the country's fertilizer requirement.

Liquid sludges consist mainly of water. Some farmers value sludge solely for its water content which can often help to overcome irrigation problems during dry weather.

Sewage sludge generally represents the non-degradable residue left after the treatment of domestic or mixed industrial/domestic wastewater. This means that it contains many of the materials originally present in the wastewater which could be classified as pollutants.

The potential hazards from the application of sludge to land are protozoal, viral, bacterial and other pathogens, which are present to the greatest extent in untreated sludges, persistent toxic organic compounds and toxic heavy metals.

Due to these hazards, salad or other crops which may be eaten raw should not be sown until one year after the application of treated sludge. If treated sludge is applied to pasture, animals should not be grazed until three weeks after the application. In the case of dairy cattle whose milk is not to be pasteurized, the period of delay should be five weeks.

Heavy metals are of particular concern because they may be detrimental to crop growth or mobilized through the food chain. In many countries there exist guidelines designed to maintain the addition of sludge to land within safe limits. In the UK the permissible concentrations of heavy metals in sludge to be applied to land and the quantities of such sludge which may be spread on the land are based on what is known about normal concentrations of heavy metals in soil and the levels at which adverse effects are likely to occur. Considerable research has been done and continues to be done on this subject. The guidelines in use now represent the present state of knowledge, and will inevitably be revised in the future as more is discovered about the behaviour of heavy metals (and other pollutants) in soil.

There are limits placed on:

(a) the cumulative quantities of metals which may be added to the soil over a period of 30 years. These quantities are expressed in terms of kilograms of metal per hectare of land.
(b) the total concentrations of metals which may be present in the soil. These concentrations are expressed in terms of milligrams of metal per kilogram of soil dry matter. This unit is the same as micrograms of metal per gram of soil dry matter and is sometimes referred to as parts per million.

**Table 5.3**  *Limits on metal additions to arable soils through sewage sludge disposal*

| Metal | Cumulative limit kg ha$^{-1}$ | Annual limit kg ha$^{-1}$ | Soil concentration limit mg kg$^{-1}$ |
|---|---|---|---|
| Arsenic | 10 | 0.33 | 10 |
| Cadmium | 5 | 0.17 | 3.5 |
| Chromium | 1000 | 33 | 600 |
| Copper | 280 | 9.3 | 140 |
| Mercury | 2 | 0.067 | 1 |
| Molybdenum | 4 | 0.133 | 4 |
| Nickel | 70 | 2.3 | 35 |
| Lead | 1000 | 33 | 550 |
| Selenium | 5 | 0.167 | 3 |
| Zinc | 560 | 18.6 | 280 |
| Boron | — | 3.5 | 3.25 |
| Fluorine* | 600 | — | 500 |

The cumulative period referred to in the Table is 30 years.
The values shown are for arable land.
The values for pasture are the same except for copper, nickel and zinc which are double the values shown.
Where the soil pH is >7 the values for copper, nickel and zinc can be doubled.
*Fluorine is not actually a metal but is included because it is desirable to limit its concentration in soil.

Table 5.3 shows the limits placed on each of these quantities. In addition, the annual average quantities of metal which can be added have also been calculated by dividing the cumulative totals by 30, and these are included in the table. The guidelines also specify that the maximum quantity which can be applied in any one year is six times the annual average.

The additions of Zn, Cu and Ni are governed by the Zinc Equivalent Concept which appears to be the limiting factor for up to 75% of the sludges produced in the UK. The Zinc Equivalent Concept assumes that the phytotoxic effects of Zn, Cu and Ni are additive in the ratio 1:2:8. The recommended limit of addition of zinc equivalent is 560 kg ha$^{-1}$. The following is an example of the calculation of the zinc equivalent:

assuming a sludge contains 1000 mg kg$^{-1}$ Zn, 500 mg kg$^{-1}$ Cu, and 60 mg kg$^{-1}$ Ni, then the zinc equivalent is given by:

$$\text{zinc equivalent} = (1 \times 1000) + (2 \times 500) + (8 \times 60)$$
$$= 2480 \text{ mg kg}^{-1}$$

The recommended limit of application of the sludge (in tonnes of dry solids per hectare) over 30 years is given by:

$$\frac{\text{limit of addition of a particular metal (kg ha}^{-1}) \times 1000}{\text{concentration of metal in sludge (mg kg}^{-1})}$$

which in the case of the above example is:

$$\frac{560 \times 1000}{2480} = 225 \text{ t ha}^{-1} \text{ over 30 years}$$

the 560 kg ha$^{-1}$ being the limit of addition for zinc (zinc equivalent).

This calculation is repeated, substituting values for the limits of addition of the other metals and their concentrations in the sludge, and the lowest value obtained is the maximum quantity of the sludge that may be applied.

The behaviour, fate and significance of organic micropollutants during sludge treatment and disposal is not well understood and, as a consequence, no guidelines for the disposal of sludges contaminated with these materials exist. In the United States the Environmental Protection Agency has indicated 114 compounds of particular concern. Many of these materials fall into the following groups, polynuclear aromatic hydrocarbons, halogenated aliphatic and aromatic hydrocarbons, organochlorine pesticides, polychlorinated biphenyls and phthalate esters. Most of these materials are hydrophobic and not readily amenable to chemical or biological degradation. During sewage treatment they are intimately associated with the solids and therefore strongly concentrated into the sludges produced. They are therefore in the sludges to be disposed of but knowledge of their significance is very limited. It appears that some chlorinated organics present in sewage sludge have been ingested (with soil) by grazing cows and that these materials have been passed to the milk associated with its lipid content.

## 5.4 BIBLIOGRAPHY

C. R. Curds and H. A. Hawkes (ed.), 'Ecological Aspects of Used-Water Treatment', Volume 1 – The Organisms and their Ecology, Academic Press, London, 1975.

Government of Great Britain, 'Water Pollution Control Engineering', HMSO, London, 1970.

Government of Great Britain, Department of the Environment, 'Taken for Granted, Report of the Working Party on Sewage Disposal', HMSO, London, 1970.

Government of Great Britain, Department of the Environment/National Water Council, 'Report of the Working Party on the Disposal of Sewage Sludge to Land', Standing Technical Committee Report No. 5, HMSO, London, 1977.

R. M. Sterritt and J. N. Lester, 'Microbiology for Environmental and Public Health Engineers', E. and F. N. Spon, London, 1988.

Metcalf and Eddy Inc., 'Wastewater Engineering: Treatment, Disposal, Re-use', McGraw-Hill Inc., New York, 1979.

The Open University, Environmental Control and Public Health, 'Units 7–8 Water: Distribution, Drainage, Discharge and Disposal', Administrative Control, The Open University Press, Milton Keynes, 1975.

T. H. Y. Tebbutt, 'Water Science and Technology', John Murray, London, 1973.

M. Winkler, 'Biological Treatment of Wastewater', Ellis Horwood Ltd., Chichester, 1981.

J. N. Lester, (ed.), 'Heavy Metals in Wastewater and Sludge Treatment Processes, Volume II: Treatment and Disposal', CRC Press, Boca Raton, 1987.

J. N. Lester and R. M. Sterritt, 'Water Pollution Control, Module 3, Unit 2: The Basic Principles of Biological Wastewater Treatment', Manpower Services Commission, Sheffield, 1988.

R. M. Sterritt and J. N. Lester, 'Water Pollution Control, Module 3, Unit 4: The Treatment and Disposal of Sludge', Manpower Services Commission, Sheffield, 1988.

CHAPTER 6

# The Treatment of Toxic Wastes

A. JAMES and F. R. McDOUGALL

## 6.1 INTRODUCTION

The presence of toxic substances in wastewater has always been a matter for concern. This concern has become much more pressing with the intentional or unintentional release of an ever larger variety of substances into the environment.

A whole spectrum of difficulties has arisen in attempting to control the toxicity problem. Three major issues may be summarized as follows:

(a) Assessment of environmentally safe concentrations – most toxicity testing measures lethal concentrations but subtle sub-lethal effects may impair ecological success at concentrations well below those causing death. Chronic effects due to prolonged exposure or bio-accumulation, toxic interactions and geo-chemical cycles involving toxins all add further complications.
(b) Assessment of biodegradability – persistence of toxins in the environment is clearly undesirable but there are still difficulties in designing suitable tests to assess the biodegradability of newly synthesized compounds.
(c) Clean technology – the treatment of toxic wastes may be technically difficult and expensive or in cases like metals consist of mere segregation. The current emphasis is therefore on developing clean technologies to avoid or reduce the production of toxic wastes.

Because of these doubts and uncertainties the treatment and disposal of toxic wastewaters has remained on an empirical level. This is reflected in the following notes which cover:

(a) Sources of Toxic Wastewaters
(b) Toxicity Problems in the Collection System
(c) Pretreatment
(d) Primary and Secondary Treatment
(e) Sludge Treatment and Disposal

123

(f)  Direct Disposal
(g)  Case Studies.

## 6.2  SOURCES AND TYPES OF TOXIC WASTES

Toxic substances are primarily associated with industrial wastes but may be
found in all types of wastewaters as shown below:

(a)  Domestic Wastewaters – these wastes contain ammoniacal nitrogen in
     concentrations up to 50 mg $l^{-1}$ and when septic may contain sulfide at
     levels up to 50 mg $l^{-1}$. Both of these can cause damage to aquatic fauna
     unless diluted and dispersed. The former is particularly damaging to
     freshwater fish ($TL_m$ for Rainbow Trout is around 2 mg $l^{-1}$) and the latter
     acts as an enzyme inhibitor in a wide variety of aquatic organisms at levels
     of a few mg per litre.
(b)  Stormwater – the composition of stormwater is much more varied than
     domestic wastes and is influenced by the nature of the drainage area and
     the frequency of storms. But the toxic potential is mainly associated with
     heavy metals like zinc and lead and is invariably found in the first flush of
     run-off from a storm.
(c)  Agricultural Wastes – these contain a wide variety of materials that are
     used for fertilizers and pest control. Fertilizers containing oxidized nitro-
     gen can cause human toxicological problems. Pesticides and herbicides are
     the most potent aquatic toxins. Because of their toxicity and persistence in
     the environment some of these substances like eldrin have had to be
     prohibited. Wastes from animal husbandry can also be toxic especially
     from silage.
(d)  Industrial Wastes – the range of toxic substances present in industrial
     wastes is too wide to catalogue but Table 6.1 gives an indication of the
     main types of toxic industrial waste and the toxins they contain.
(e)  Leachates – a wide range of toxic substances can leach out of sites used for
     dumping solid wastes particularly where surface water or ground water
     has access. This particularly applies to the dumping of wastes from
     industry.

## 6.3  TOXICITY PROBLEMS IN THE COLLECTION SYSTEM

The cost of constructing a wastewater collection system in an urban area is
extremely high, often accounting for 70% of the total cost for treatment and
disposal. Damage to the fabric of the sewer is therefore to be avoided and strict
controls are usually imposed on substances that may be discharged to the sewer.
As shown in Table 6.2 these controls are also intended to control the discharge of
substances like cyanides or sulfide that may give rise to poisonous gases which
could damage the health of sewer workers. Levels of HCN and $H_2S$ of 0.03% in
the atmosphere are toxic and with $H_2S$ there is an additional problem of
anaesthesia which makes detection difficult.

**Table 6.1**   *Sources of some common toxins*

| Toxin | Sources |
| --- | --- |
| Acids – mainly inorganic but some organic causing pH <6 | Acid Manufacture<br>Battery Manufacture<br>Chemical Industry<br>Steel Industry |
| Alkalis – causing pH >9 | Brewery Wastes<br>Food Industry<br>Chemical Industry<br>Textile Manufacture |
| Antibiotics | Pharmaceutical Industry |
| Ammoniacal nitrogen | Coke Production<br>Fertilizer Manufacture<br>Rubber Industry |
| Chromium – mainly hexavalent but also less toxic trivalent form | Metal Processing<br>Tanneries |
| Cyanide | Coke Production<br>Metal Plating |
| Detergents – mainly anionic but some cationic | Detergent Manufacture<br>Textile Manufacture<br>Laundries<br>Food Industry |
| Herbicides and Pesticides – mostly chlorinated hydrocarbons | Chemical Industry |
| Metals – mainly copper, cadmium, cobalt, lead, nickel and zinc | Metal Processing and Plating<br>Chemical Industry |
| Phenols | Coke Production<br>Oil<br>Refining<br>Wood Preserving |
| Solvents – mostly benzene, acetone, carbon tetrachloride and alcohols | Chemical Industry<br>Pharmaceuticals |

Some organic solvents may cause similar difficulties. They tend to be immiscible with water, volatile and intoxicating, and may also form explosive mixtures.

## 6.4   PRE-TREATMENT OF TOXIC WASTES

In general, industrial wastewaters are most readily and most economically treated in admixture with domestic wastewaters rather than in isolation. Many benefits of scale, balancing nutrient supplementation as well as skilled operation, may be obtained by discharging the industrial wastewaters to a sewer. But there are a number of occasions when this is not possible or not desirable:

**Table 6.2** *Typical consent conditions for discharge to sewers*

| Parameter | Consent condition |
|-----------|-------------------|
| Maximum temperature | 40–45°C |
| pH | 6–10 |
| Substances producing inflammable vapours | Nil |
| Cyanide concentration | 5–10 mg l$^{-1}$ |
| Sulfide concentration | 1 mg l$^{-1}$ |
| Soluble sulfates | 1250 mg l$^{-1}$ |
| Synthetic detergents | 30 mg l$^{-1}$ |
| Free chlorine | 100 mg l$^{-1}$ |
| Mercury | 0.1 mg l$^{-1}$ |
| Cadmium | 2 mg l$^{-1}$ |
| Chromium | 5 mg l$^{-1}$ |
| Lead | 5 mg l$^{-1}$ |
| Zinc | 10 mg l$^{-1}$ |
| Copper | 5 mg l$^{-1}$ |
| Zinc equivalent (Zn + Cd + 2Cu + 8Ni) | 35 mg l$^{-1}$ |
| Total non-ferrous metal | 30 mg l$^{-1}$ |
| Total soluble non-ferrous metal | 10 mg l$^{-1}$ |

Note: There are also a large number of specific toxic substances whose discharge to sewers is controlled.

(a) Rural areas without sewerage
(b) By-product recovery is economically and technically feasible
(c) Domestic effluent is used ultimately in irrigation
(d) Industrial wastewater does not meet consent conditions for discharge to a sewer.

Under these circumstances some form of pretreatment is needed to render the wastewater suitable for discharge, further treatment or disposal. The main advantages of treatment on site are the possibilities of recovering specific substances in an uncontaminated condition and economies which result from treatment at higher temperatures or concentrations. There may be an important additional advantage; that is the avoidance of contamination of a much larger wastewater stream which would cause difficulties in disposal. Where toxic materials are organic in nature there is often a problem in treatment due to inhibition of bacterial growth. It is often easier and cheaper to develop the necessary bacterial flora in an on-site treatment plant.

This to some extent depends upon the concentration and toxicity of the substances concerned. In some cases dilution of the wastes by admixture with sewage reduces the toxic inhibition making it preferable to treat the industrial waste and sewage together. Also many industrial wastes are deficient in some nutrient such as nitrogen or phosphorus. The desirable ratio of BOD:N:P is 100:5:1 and the ratio in domestic waste is commonly 100:18:2.5 so that deficiencies in industrial wastes can be balanced.

There are other considerations in deciding for or against pre-treatment such as:

(a) Availability of space – the site may be too restricted or land may be too valuable to be used for a treatment plant.
(b) Availability of expertise – the company may not wish to get involved in effluent treatment.
(c) Sludge and/or odour production may create a nuisance.
(d) Possibility for the introduction of clean technologies.

Even where it is decided to carry out pre-treatment of the toxic waste by chemical or biological methods it is often useful to install devices to improve the effluent quality by simple physical means. These include some form of screening, coarse or fine, to reduce solids. Also some form of balancing to reduce variations in concentration, flow, pH, *etc.*, some traps to prevent the escape of oil and grease and some grit arrestors.

Every attempt should be made to minimize the quantity of material discharged through good housekeeping. This can take the form of any or all of the following techniques:

(a) Extending the life of process solutions by filtration, topping up, adsorption, *etc.*
(b) Altering the production process to use less toxic compounds, *e.g.* substituting copper pyrophosphate for copper cyanide in electroplating solutions.
(c) Dry cleaning prior to wash-down, which can remove a large proportion of the pollutant in solid form.
(d) Evaporation of strong organic liquors, which can often produce a burnable product.
(e) Minimizing and segregating any flows which contain toxic materials. In some cases it is necessary to separate wastes for safety reasons, *e.g.* cyanides or sulfides and acid wastes, trichlorethylene and alkaline wastes. In other cases it may be desirable to segregate for treatment reasons. However segregation can be very expensive.

In Table 6.3 the first 5 items are common methods of physical treatment employed by a wide range of industrial processes while the others are more costly and complex, require more skilled operation and are therefore less common forms of pre-treatment.

Having minimized so far as possible the types, quantities and concentration of any toxic wastes, it may still be necessary to treat them prior to discharge either to a sewer or a water course. The processes which are used may be classified as physical, chemical and biological. The physical processes are summarized in Table 6.3. Where the toxic wastes contain or are composed of organic materials, it may also be necessary to provide some biological treatment especially if the effluent is to be discharged directly into a watercourse. Many different types of process are used but the following are the most popular:

**Table 6.3** *Physical methods of pre-treatment*

| Process | Aim | Examples |
|---------|-----|----------|
| Screening | Removal of coarse solids | Vegetable canneries, paper mills |
| Centrifuging | Concentration of solids | Sludge dewatering in chemical industry |
| Filtration | Concentration of fine solids | Final polishing and sludge dewatering in chemical and metal processing |
| Sedimentation | Removal of settleable solids | Separation of inorganic solids in ore extraction, coal and clay production |
| Flotation | Removal of low specific gravity solids and liquids | Separation of oil, grease and solids in chemical and food industry |
| Freezing | Concentration of liquids and sludges | Recovery of pickle liquor and non-ferrous metals |
| Solvent extraction | Recovery of valuable materials | Coal carbonizing, plastics manufacture |
| Ion Exchange | Separation and concentration | Metal processing |
| Reverse osmosis | Separation of dissolved solids | Desalination of process and wash water |
| Adsorption | Concentration and removal | Pesticide manufacture, dyestuffs removal |

(a) High-rate filtration using plastic media and very high rates of recirculation.

(b) Activated sludge using contact stabilization.

Like all biological processes these can suffer from toxicity problems especially where the concentration of toxin is not constant. In general terms it is easier for bacteria and other microorganisms to adapt to toxic substances than for organisms like worms, fly larvae, *etc*. For this reason conventional percolating filters have not proved successful – the lack of grazing fauna has led to persistent ponding.

Biological processes may be either aerobic or anaerobic. The latter have proved especially popular for treating high strength industrial wastes. As aerobic and anaerobic utilize different groups of bacteria, a waste that is toxic to aerobic organisms may not have the same effect on anaerobes.

Due to a combination of high organic strength and inhibition from toxic substances it is unusual to obtain complete treatment of toxic industrial wastes by conventional primary and secondary treatment. The effluent from high-rate

**Table 6.4**   *Chemicals used in industrial waste treatment*

| Chemical | Purpose |
|---|---|
| Calcium hydroxide | pH adjustment, precipitation of metals and assisting sedimentation |
| Sodium hydroxide | Used mainly for pH adjustment in place of lime |
| Sodium carbonate | pH adjustment and precipitation of metals with soluble hydroxide |
| Carbon dioxide | pH adjustment |
| Aluminium sulfate | Solids separation |
| Ferrous sulfate | Solids separation |
| Chlorine | Oxidation |
| Anionic polyelectrolytes | Enhance coagulation and flocculation |

filters often has a BOD and COD* similar to settled sewage and is suitable either for discharge to a sewer or for further biological treatment on site.

Chemical treatment of industrial wastes may be used in addition to and to some extent in place of biological treatment. The aims are somewhat different since biological treatment is mainly a way of oxidizing organic matter or a way of converting it into a settleable form. Chemical treatment is used only for oxidizing particular compounds, like cyanide, since it is expensive and liable to lead to the production of undesirable chlorinated organics. It is mainly used for pH correction and improving the removal of solids. The commonest chemicals in use are shown in Table 6.4.

## 6.5   PRIMARY AND SECONDARY TREATMENT

Provided that the pre-treatment of toxic industrial wastes is successful then no difficulties should be encountered in subsequent treatment. However no pre-treatment system is perfect and malfunction will occasionally occur mostly due to variations in the manufacturing process. As a result toxic material together with possible overload of organics and solids may be passed on to the subsequent treatment stages.

Wastewater treatment is conventionally divided into preliminary, primary and secondary treatment (plus tertiary treatment if necessary for some special purpose). The following notes omit any consideration of preliminary treatment as this usually consists of screening and grit removal which have little effect on toxic materials.

The effect of toxic materials on primary sedimentation is insignificant since this is a purely physical process of sedimentation and flocculation. However the effect of primary sedimentation on toxic wastes can be very important. Toxic materials in suspension such as particulate metals are effectively removed. Also the

---

* Chemical Oxygen Demand (COD), which always exceeds BOD, is the oxygen consumed by a sample when reacted with acidic dichromate solution. It results from oxidation of both organic and inorganic constitutents.

**Table 6.5**  *Amounts of heavy metal ions removed from sewage by sludges*

| Heavy metal ion | Primary sedimentation | | Percolating filter treatment | | Activated-sludge process | |
|---|---|---|---|---|---|---|
| | Metal concentration in crude sewage (mg l$^{-1}$) | Proportion removed by treatment (per cent) | Metal concentration in settled sewage (mg l$^{-1}$) | Proportion removed by treatment (per cent) | Metal concentration in settled sewage (mg l$^{-1}$) | Proportion removed by treatment (per cent) |
| Copper | Up to 0.8 | 45 | | | 0.4 | 54 |
| Copper | | | Up to 0.44 | 20 | Up to 0.44 | 80 |
| Copper | Up to 5 | 12 | | | 0.4–25 | 50–79 |
| Copper | | | | | 28 | 90–93 |
| Copper | | | | | | |
| Dichromate | (as Cr) Up to 1.2 | 28 | (as Cr) Up to 0.86 | 32 | (as Cr) Up to 0.86 | 67–70 |
| Dichromate | | | | | 4.0 | 6.3 |
| Dichromate | | | | | 0.5–2 | ca. 100 |
| Dichromate | | | | | 5 | 50 |
| Dichromate | | | | | 50 | 10 |
| Iron (Ferric) | 3–9 | 40 | 1.8–5.4 | Nil | 1.8–5.4 | 80 |
| Lead | 0.3–0.9 | 40 | 0.18–0.54 | 30 | 0.18–0.54 | 90 |
| Nickel | 0.1–0.3 | 20 | 0.08–10 | 40 | 0.08–0.24 | 30 |
| Nickel | | | | | 2.0 | 31 |
| Nickel | | | | | 2.5–10 | 30 |
| Zinc | 0.7–1.6 | 40 | 0.4–1.0 | 30 | 0.4–1.0 | 60 |
| Zinc | | | | | 2.5 | 90 |
| Zinc | | | | | 2.5 | 95 |
| Zinc | | | | | 7.5 | 100 |
| Zinc | Up to 5 | 12 | | | 15 | 78 |
| Zinc | | | | | 20 | 74 |

flocculant material has a great capacity for adsorption and removes the majority of dissolved metals, pesticides and other toxic organics. In one respect this is beneficial since it renders the waste material less inhibitory for biological treatment but it selectively concentrates the toxins in the sludge and may give rise to problems in digestion and in sludge disposal. Some indication of the removal of metals during primary treatment is given in Table 6.5.

Chemicals may be used to enhance the effectiveness of primary sedimentation, in some cases, removing additional material by precipitation at the same time. Chemical addition can be expensive, often requires pH correction, and may produce large quantities of sludge with a disposal problem. For these reasons chemical enhancement of primary sedimentation is rarely practised at plants treating domestic wastes. Nevertheless for many toxic industrial wastewaters this is an attractive treatment option since it enables industry to avoid secondary biological treatment and enables the waste to be discharged to a sewer, estuary or the sea.

Where chemically enhanced sedimentation is used the main aim is generally to increase removal of solids, but since many toxins such as metals and chlorinated organics adsorb strongly, their removal is also increased to levels similar to combined primary and secondary treatment. The material employed for enhancement is lime or less frequently aluminium salts sometimes supplemented by polyelectrolytes. Some indications of metal removal achievable by sedimentation, with and without lime, are given in Table 6.6.

Whatever the form of primary treatment employed, further treatment is generally brought about by biological processes, either aerobic or anaerobic. The key to successful secondary treatment of wastewaters containing toxins is the adaption of the microorganisms to the presence of the toxin. Bacteria, and to a lesser extent protozoa, show a remarkable ability to acclimatize to the presence of toxic substances and a great adaptability in degrading new synthetic organic compounds. Metazoa are less adaptable so forms of treatment that rely on metazoa are best avoided in dealing with toxic wastes.

It is important in biological treatment of toxic waste that a microbial population is developed which is acclimatized to the presence of the toxin and, in the case of degradable toxins, it is essential that it contains sufficient numbers of

**Table 6.6** *Metal removed by sedimentation*

| Metal | Concentration in wastewater $(\text{mg l}^{-1})$ | % Removal by sedimentation | % Removal (with lime) by sedimentation |
|---|---|---|---|
| Iron | 6.3 | 48 | 80 |
| Copper | 0.6 | 28 | 60 |
| Chromium | 0.34 | 40 | 58 |
| Lead | 0.12 | 33 | 55 |
| Mercury | 0.028 | 15 | 50 |
| Nickel | 0.08 | 15 | 15 |
| Zinc | 0.7 | 38 | 70 |

**Table 6.7**   *Biological processes for treating toxic wastewaters*

| Aerobic/ anaerobic | Reactor type | Advantages and disadvantages | Examples of process |
|---|---|---|---|
| Aerobic | Dispersed growth | Tend to be completely mixed therefore dilutes toxin but affects whole biomass. Liable to cause settling problems as well as interfere with oxidation | Activated sludge and modifications |
| Aerobic | Fixed film | Tend to be plug flow so no dilution unless recirculation is used. Biomass more robust for shock loads but metazoa more sensitive | High rate filters (good) Standard rate filters (less suitable) |
| Anaerobic | Dispersed growth | Tend to be completely mixed and suffer from washout of methanogens. The latter are also more sensitive than acidogens to toxic effects and have a low growth rate | UASB and contact digester |
| Anaerobic | Fixed film | Tend to be plug flow but level of attachment not as good as aerobic filters. Need recirculation to dilute toxins | Anaerobic filters and fluidized bed |

organisms which can metabolize the toxins. These twin aspects of acclimatization require great care in the start-up operation and may need a period of several months before successful operation is achieved. Even after start-up is complete, particular processes involving sensitive bacterial species like nitrification or methane production can be easily disrupted by shock loads.

Biological processes for treating wastewaters may be divided into aerobic and anaerobic and each division may be subdivided into dispersed growth and fixed film. The resulting four categories of treatment process have advantages and

**Table 6.8**   *Toxic levels in aerobic biological treatment*

| Toxin | Significant level |
|---|---|
| Hydrogen ions | pH $<6$ or $>9$ |
| Phenols | 50–100 mg l$^{-1}$ |
| Ammoniacal-N | 500–1000 mg l$^{-1}$ |
| Zinc | 10–50 mg l$^{-1}$ |
| Chromium | 5–20 mg l$^{-1}$ |
| Lead | 5–30 mg l$^{-1}$ |
| Alkyl Benzene Sulfonates | 3–20 mg l$^{-1}$ |
| Sulfide | 5–50 mg l$^{-1}$ |

Note: The wide ranges given in the Table are in part a reflection of the ability of bacteria to acclimatize to toxins but are also caused by physical/chemical interactions which adsorb or precipitate or otherwise remove toxins from solution.

**Table 6.9** *Toxic effects in anaerobic treatment*

| Toxin | Inhibitory concentration | |
|---|---|---|
| | In sewage | In sludge |
| Chromium | — | 2 |
| Cadmium | 2 | 2 |
| Copper | 1.5 | — |
| Iron | 10 | — |
| Lead | 100 | — |
| Nickel | 80 | — |
| Zinc | 50 | — |
| Detergent | — | 2% of Suspended Solids |
| Benzene | — | 50–200 |
| Carbon tetrachloride | — | 10 |
| Chloroform | — | 0.1 |
| Dichlorophen | 1 | — |
| $\gamma$-BHC | — | 48 |
| Toluene | — | 430–860 |

All concentrations in mg $l^{-1}$

disadvantages for the treatment of toxic wastewaters, as shown in Table 6.7. The tolerance levels of these processes are difficult to define precisely but some indication is given in Tables 6.8 and 6.9.

## 6.6 SLUDGE TREATMENT AND DISPOSAL

This topic is discussed in Chapter 5 so the remarks below are limited to the treatment of toxic sludges. These are generated almost exclusively by industrial sources although sludges from silage and other agricultural practises may cause occasional problems. Industrial sludges are generally formed during primary sedimentation and tend to contain high levels of toxins like metals and chlorinated organics which are adsorbed onto the solids. Where domestic wastes are treated along with industrial discharges a similar problem may arise.

The treatment and disposal of toxic wastes may be broken into a series of steps:

(a) Dewatering
(b) Biological treatment
(c) Disposal
(d) Detoxification

with two possible strategies, dispersion or concentration.

Where the industrial wastewater contains substances that are not permitted in the environment, the overall strategy is to segregate the waste into a small volume of hazardous material for containment in a long term storage site (at sea or inland). Where the toxic substances in the wastewater are less dangerous, or in dealing with the larger volume of less hazardous waste, the aim is to dispose of these to the environment by ensuring adequate dilution.

Examples of these processes are as follows:

(a) Containment of radioisotopes by acid extraction and precipitation, followed by vitrification for long term storage.
(b) Dispersion of metals by spraying the sludge on to farm land.

## 6.7 DISPOSAL OF TOXIC WASTES

Ultimately toxic wastes have to be disposed of and this disposal must be in a manner which does not present a short term or long term hazard to man or the environment. Some hazardous wastes may be rendered innocuous by treatment prior to disposal. Incineration is frequently used for decomposition of organic toxins. However care is required particularly in dealing with halogenated materials since irritant corrosive gases may be produced. There is also a danger that the treatment plant may become too complex. Examples of the successful use of incineration may be found in the treatment of toxic sulfide liquor from the Kraft process for making paper pulp and in the treatment of steel pickle liquor. Both of these have the added advantage of regenerating useful compounds.

The toxic wastewaters that give rise to the most serious problem are metallic wastes and radioactive wastes. These types of waste have common characteristics in that they contain hazardous elements which cannot be broken down (or not for many decades) and there appear to be less toxic substances that could be used to replace them.

The two major alternatives for disposal of toxic waste may be described briefly as follows:

(a) To land – where wastewaters contain human toxins great care is required to avoid contamination of groundwater. Where aquifers are at great depth and not directly connected to streams or underground supplies, *e.g.* some Middle Eastern countries, then land disposal may be preferable to using shallow semi-landlocked seas.

   The usual disposal arrangement for less hazardous wastes is a lagoon, which may have a connection with a watercourse, but which also permits infiltration (plus some evaporation and possibly some degradation). In the long term, swelling and blinding of the soil may reduce the infiltration capacity.

   For more hazardous wastewaters the land disposal policy is segregation followed by long term containment of the hazardous material in impervious disposal sites.

(b) To sea – the seas has an almost unlimited capacity for dilution but the corollary is that it has an almost infinite retention time. Disposal to sea is therefore capable of diluting acute toxins below their toxic threshold but problems may arise with substances that accumulate due to geochemical or biochemical mechanisms.

A further complication with marine disposal is the international aspect. Waste material discharged into the ocean may be transported around the world.

Two means of disposal are commonly practised:

(i)   Discharge by pipeline to inshore waters. Dispersion in the buoyant jet can give adequate initial dilution although inshore areas are particularly sensitive to pollution, being used as shellfisheries and recreational zones.

(ii)  Deep sea disposal of toxic wastes has in recent years been the subject of several international agreements as the result of which the volume of hazardous disposal has declined and the nature of the waste has changed. Substances like organohalogens, carcinogenic substances, mercury and cadmium compounds and plastics are all banned. Deep sea disposal of less hazardous material still takes place but in packaged form and only in deep sea.

Solidification can be used to render a variety of hazardous wastes suitable for disposal including oily wastes, sludges contaminated with PCBs and fly ash contaminated with heavy metals.

The technique relies upon reducing the mobility of the hazardous constituents by binding them into a solid matrix which has low permeability and is therefore resistant to leaching.

The mechanism of binding depends upon the agent employed which may be cement based, pozzolanic or silicate based, thermoplastic based or organic polymer based. The cheaper agents like cement, asphalt and pozzolanic-based have been most widely used. Solidification has given promising results in short term tests but the longer term is less certain except for vitrification. This technique is only financially possible for nuclear waste – and apart from cost has the additional disadvantage of being virtually irreversible.

## 6.8   INDUSTRIAL WASTE TREATMENT – CASE STUDIES

The foregoing notes have dealt in general with the problem of treating toxic wastes but it is not possible to explain the complexities in so general a discussion. The following notes therefore describe in more detail two examples of industries that produce toxic wastes and the methods used in treating them.

### 6.8.1   Tannery Wastes

Tanning is the process by which hides are converted into leather. The hides, after removal of flesh and fur, are treated with chemicals which cross-link the microscopic collagen fibres to form a stable and durable material. After tanning the hides will usually be further processed according to their intended end use. This will consist of trimming, drying, buffing and surface coating.

Wastes will arise from surplus, spent or washed-out chemicals used in the process; some chemical constituents may be toxic, while others are powerful pollutants in water and soil. The release of volatile sulfides gives rise to toxic and obnoxious odours. Certain solvent vapours can have adverse health effects after prolonged exposure. There will also be residues from operations such as cleaning, scraping, splitting and trimming. Each of these generates waste products which must be disposed of or reused. Solid waste products of animal origin are powerful pollutants in water, and are also highly odorous when they decompose in their solid forms.

A schematic diagram of the tanning process is presented in Figure 6.1, and this indicates the type of waste stream generated.

The composition of a combined tannery effluent that has not been treated is characterized by a high oxygen demand and a high salt content, and is strongly alkaline. It also contains a high level of suspended solids and possibly a persistent high load of chrome. Levels experienced in an actual tannery will vary from these values depending for example on water use *etc.* The generally accepted range of water use is 25–80 m$^3$ t$^{-1}$ of hide processing. Table 6.10 presents the chemical characteristics of a typical untreated combined tannery effluent.

Effluents from tanneries may be disposed of in one of four ways:

(a)  Discharge to sewer
(b)  Discharge to river or estuary
(c)  Discharge to sea
(d)  Discharge to underground.

**Table 6.10**   *Chemical characteristics of a typical untreated combined tannery effluent*

| Parameter (mg l$^{-1}$) | Chrome tannage | Vegetable tannage |
|---|---|---|
| pH (units) | 9 | 9 |
| Total solids | 10 000 | 10 000 |
| Total ash | 6000 | 6000 |
| Suspended solids | 2500 | 1500 |
| Ash in suspended solids | 1000 | 500 |
| Settled solids (2 hr) | 100 | 50 |
| BOD$_5$ | 900 | 1700 |
| KMnO$_4$ value | 1000 | 2500 |
| COD | 2500 | 3000 |
| Sulfide | 160 | 160 |
| Total nitrogen | 120 | 120 |
| Ammonia nitrogen | 70 | 70 |
| Chrome | 70 | — |
| Chloride | 2500 | 2500 |
| Sulfide | 2000 | 2000 |
| Phosphorus | 1 | 1 |
| Ether extractable | 200 | 200 |

(Source – UNIDO, 1990)

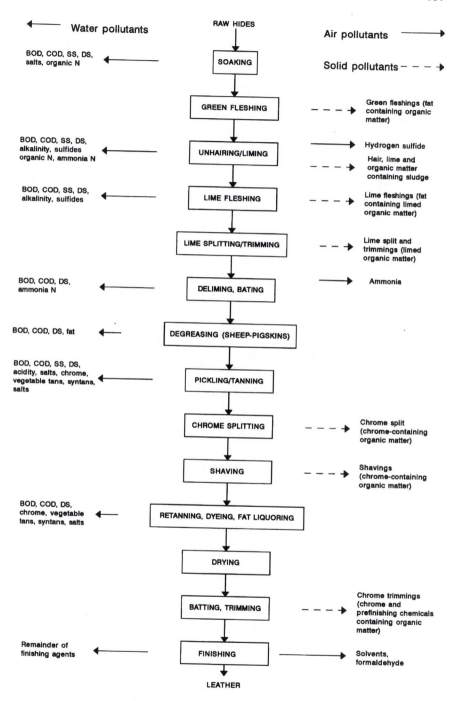

**Figure 6.1**   *Schematic diagram of tanning process*

The discharge of tannery waste to a river or estuary is usually subject to severe constraints on solids, BOD, toxins, pH *etc.* Few tanneries have access to the sea and although underground disposal is popular in the USA it is not favourably regarded in the UK.

Whether effluents are being discharged to sewers or to rivers the form of pre-treatment is often similar. In cases where space is limited a mechanically brushed perforated screen is all that is installed but wherever possible a balancing tank is added. This will reduce the need for pH correction, reduce the load of BOD and SS and enable any subsequent treatment units to operate continuously. Mechanical rotating screens which are self-cleaning have worked well within the industry; up to 30–40% of total suspended solids in the raw waste stream can be removed by a properly designed and operated screen. As a significant part of the COD load of the raw wastewater is due to organic solids, a preliminary settling operation can remove up to 30% of this COD, saving flocculating chemicals and reducing sludge volume in later treatment.

If the effluent is to be discharged to sewers some reduction in sulfide and metals will often be necessary. As $H_2S$ gas is liberated when sulfide liquors become neutralized, these should be separated and treated. Three methods have been used: aeration in the presence of manganese catalyst; direct precipitation using ferrous sulfide or ferric chloride; and oxidation by chlorine. The relative economics of these processes tend to favour the aeration technique. The precipitation of chrome is relatively simple. The success of the operation depends on the ability to collect the major chrome-bearing liquors for treatment. Precipitation is achieved by raising the pH above 8.0 by the addition of lime, followed by the addition of aluminium salts (200 ppm) and anionic polyelectrolytes (5 ppm) to give a fast-settling flocculant.

Where tannery effluents have been discharged to sewers there have been some reports of difficulties at the municipal treatment plants. These difficulties seem to have arisen mainly because of the increase in organic concentration rather than any effect due to toxins. As undiluted tannery wastes may have BOD values around 2000 mg $l^{-1}$ they require a minimum of 3–4 times dilution to avoid problems of oxygen transfer (care is needed in the use of COD data since the ratio of COD:BOD is often twice that found in raw sewage). The sulfide present in tannery waste may have an adverse effect on biological treatment in admixture with sewage, this is unlikely to be serious at concentration of $<10$ mg $l^{-1}$, concentrations around 25 mg $l^{-1}$ can be tolerated by the activated sludge process and even loads of up to 50 mg $l^{-1}$ can be withstood for short periods of time. Percolating filters are very resistant to sulfide and concentrations of up to 100 mg $l^{-1}$ can be treated successfully. Metal toxicity from tannery wastes has not been found to be a problem as 80% of the chrome is removed during primary sedimentation. The resulting sludges do not give rise to difficulties in digestion provided that the retention time exceeds 21 days.

Experience of treating tannery waste on its own has confirmed the general rule that it is better treated along with domestic waste. Anaerobic treatment has not been too successful, with moderate BOD removals and some difficulty in treating the resultant liquor.

Activated sludge appears to have good potential for treating tannery waste although this has not been exploited. Loading rates of 0.5–10 kg BOD kg$^{-1}$ MLSS day$^{-1}$ can be applied with BOD removals >90%. The oxidation ditch, a low-load activated sludge system which is relatively cheap to install, with low maintenance and a long retention time, is well able to withstand the variable character of effluent and the shock loads experienced in the tanning industry.

Biological filters are often discounted as it is now generally recognized that a tannery waste treatment system must be robust enough to withstand the occasional shock load and any other operational irregularities.

## 6.8.2 Metal Processing Wastes

Discharge of metals to the aquatic environment has been a major cause of concern and the treatment of these wastes has consequently attracted considerable attention. Wastes containing metals may arise from a variety of industrial and agricultural operations including tanneries, paint manufacture, battery manufacture, pig wastes, *etc.*, but the main source is from metal processing. The wastes from metal processing may be classified as follows:

(a) Mining – ore production and washing – also contains inert SS
(b) Ore processing – smelting, refining, quenching, gas, scrubbing, *etc.* – also contains sulfides, ammonia and organics
(c) Machining – metal particles from machining usually mixed with lubricants
(d) Degreasing – metals mostly in solution with cyanides, alkalis and solvents
(e) Pickling – acids with metals and metallic oxides in solution
(f) Dipping – alkalis with sodium carbonate, dichromate, *etc.*, plus metals
(g) Polishing – particles of metals and abrasives together
(h) Electrochemical or chemical brightening and smoothing – acids, mainly sulfuric, phosphoric, chromic and nitric with metals in solution
(i) Cleaning – hot alkalis with detergents, cyanides and dilute acids plus metals in solution
(j) Plating – acids, cyanides, chromium salts, pyrophosphates, sulfamates and fluoroborates plus metals in solution
(k) Anodizing – chromium, cobalt, nickel and manganese in solution.

The sources of wastes in metal processing are numerous and also extremely variable both in quantity and quality. Metals in the wastes occur in forms ranging from large particles of pure metal in suspension to metallic ions and complexes in solution. The most appropriate method of treatment depends upon the form of the metal, its concentration, pH, other constituents of the waste and the desired effluent standard. The technique most commonly employed in treating metal processing wastes is precipitation using pH adjustment. The optimum pH for precipitation varies depending on the particular metal and where several metals are involved a compromise pH is used. A typical value is in the range 8.0–9.0. With amphoteric metals, notably zinc, care must be taken to avoid too high a pH to prevent the formation of zincates. It should also be appreciated that other constituents of the waste, *e.g.* ammonia, can significantly affect the solubility of the metal hydroxides and it is therefore not possible to predict accurately the level of residual metal in the treated effluent.

Whilst the hydroxide precipitation method is satisfactory for most metals encountered in effluents both hexavalent chromium and lead are not precipitated in this way. Hexavalent chromium is present in wastes from metal plating and must first be reduced to the trivalent form before treatment with lime or caustic soda. The reducing agents commonly used are sodium bisulfate, sulfur dioxide and occasionally ferrous sulfate. The reduction is carried out under acid conditions and subsequent addition of alkali precipitates trivalent chromium hydroxide.

In the case of lead, the hydrated oxide formed when lime or caustic soda is added to the lead waste has an appreciable solubility and the resulting effluent after removal of solids would normally be unsatisfactory for discharge to sewer or watercourse. However basic lead carbonate has a very low solubility and therefore sodium carbonate can be used in place of lime as the precipitating agent. Like zinc, lead is amphoteric and redissolves as plumbate at high pH, so careful pH control must be exercised.

A particular type of precipitation system used in the metal plating industry is known as the Integrated Method of Treatment. The principal feature of this system is that the rinsing stage immediately after the metal plating stage is a chemical rinse which precipitates the metal from the liquid around the article being plated. A further water rinse is then required to wash off the treatment chemical. In the case of nickel plating the chemical rinse would contain sodium carbonate to precipitate nickel carbonate, whilst with chromium, a prior stage to effect reduction from hexavalent to trivalent form would be required. The integrated system has the advantages that water reuse can be readily practised and that the metals are not precipitated in a mixture and so can also be recovered. However, it is sometimes difficult to adapt the system to existing plating lines, since it necessitates the placement of an extra tank in the line.

Once the metals have been precipitated from solution, liquid and solid phases must be separated. The traditional method for this stage of treatment is settlement in either a circulate or rectangular tank. In small installation where the effluent flow is less than say 25 $m^3$ $day^{-1}$, it is convenient to carry out the effluent treatment on a batch basis and to allow settlement to take place in the same tanks as that used for reaction. For larger installations a continuous flow system is required. The size of tanks depends on both the maximum effluent flow rate and on the configuration adopted for the tank. The most common type of settlement vessel is of the vertical upward flow pattern having a central feed well, a peripheral collection launder, and a sludge cone at the bottom. Clarification of the effluent can be enhanced by the use of flocculating agents. Obviously the size and mode of operation of the precipitation system significantly affects quality of the effluent but typical figures for a well-designed, efficiently operated, settlement system for metal hydroxide precipitates would be in the range 10–30 mg $l^{-1}$ suspended solids.

Where space is at a premium, a compact settling system utilizing parallel tilted plates or tubes can be used to perform the separation stage.

There are two factors which make this system efficient in terms of ground area used. These are:

(a) the distance through which a settling particle has to fall to become 'settled' is considerably reduced;
(b) the configuration produces laminar flow conditions which enhance the settling rate and overall efficiency.

Tilted plates can also be used to uprate settling tanks.

Flotation may be used as an alternative to settlement. This process, which is gaining in popularity, consists in the carrying of metal hydroxides and other particles in suspension to the surface of liquid in the flotation vessel by increasing particle buoyancy using gas bubbles which adhere to the particles. The scum containing the gas bubbles and separated solids is skimmed off. Variations in the process lie mainly in the method of producing the carrier gas bubbles. This may be done by injecting a super-saturated solution of air in water under pressure into the tank – dissolved air flotation – or by injecting air through a diffuser – dispersed air flotation – or by the electrolysis of water to yield fine bubbles of hydrogen and oxygen – electrolytic flotation. The gas bubbles produced in these processes are extremely small, normally in the range 70–150 $\mu$m.

The use of direct filtration appears to be a very attractive process for the phase separation but unfortunately is seldom appropriate, mainly because of the tendency for the filter media to blind (*i.e.* clog) rapidly. This tendency is largely due to the gelatinous nature of the metal hydroxide precipitates. Occasionally, where a more granular precipitate is obtained, direct filtration can be satisfactory and a high quality effluent can be obtained.

Whilst filtration has only limited application as the main means of solids removal it is frequently used to polish the effluent from a settlement or flotation system to produce a higher quality effluent.

Where the metal is substantially in solution there are various techniques for separation or concentration of the metal so that a high quality treated effluent may be obtained.

*6.8.2.1  Ion Exchange.*  Ion exchange is a chemical treatment process used to remove dissolved ionic species from contaminated aqueous streams. Treatment for both anionic and cationic contaminants can be effected by ion-exchange processes. Ion exchangers are insoluble high-molecular weight polyelectrolytes that have fixed ionic groups attached to a solid matrix. Both natural and synthetic ion exchangers are available. However, due to their greater stability, higher exchange capacity and greater homogeneity of their exchange properties, synthetic ion exchange materials are predominantly used today. Synthetic ion exchangers are generally polymeric materials (resins) that have been chemically treated to render them insoluble, and to exhibit ion exchange capacity. Often the exchanger is in the form of spherical resin beads, although ion-exchange membranes are also available. Various different polymers have been used as ion exchangers but copolymers of styrene and divinylbenzene (DVB) are the most common synthetic ion-exchange materials.

*6.8.2.2  Evaporation.*  Evaporation is one of the most common methods used in industry for the concentration of aqueous solutions. Nevertheless, use of this process as a means of effluent treatment is rare and occurs only under special circumstances where the effluent contains a high concentration of a valuable material. The only application of note here is on the concentration of static rinses (drag out) from electroplating operations, especially chromium plating. In this application the rinse liquor is evaporated to a metal concentration which makes the concentrate suitable for direct reuse in the plating bath.

*6.8.2.3  Molecular Filtration.*  Molecular filtration is divided into two categories: ultrafiltration (UF) and reverse osmosis (RO) processes. The differentiating characteristic between RO and UF is the molecular weight cutoff of the membrane, and corresponding pressure differentials required to achieve a given membrane flux. RO membranes operate with molecular cutoffs that are much smaller than UF membranes (100–200 daltons compared with 2000–1 000 000 daltons), therefore the RO membrane will retain most organic materials, as well as many of the inorganic solutes. RO membranes operate with trans-membrane pressures of up to 500 psi, whereas UF membranes generally operate with pressure differentials only as high as 50 psi since most of the inorganics will pass through the UF membrane, minimizing the osmotic pressure resistance. This pressure difference has a significant economic implication. Often when reverse osmosis is used, upstream UF is provided as a pre-treatment for RO. The main operational problems associated with membrane processes are chemical and biological fouling of the membrane, and particularly with RO, membrane deterioration. The RO process has been used on effluents from electroplating in the electronic components industry. The continuous development of the process and improved mechanical strength of the membranes will almost certainly increase the range of its applications.

*6.8.2.4  Solvent Extraction.*  In general, the solvents used in extraction operations are too expensive to be used just once, and furthermore the contaminants are highly concentrated in the extract. Therefore, the spent solvent from a liquid-liquid extraction operation needs to be further treated to reclaim the solvent for reuse and to reduce further the volume in which the contaminant is contained. Some solvent repurification sequences include the use of distillation or adsorption.

*6.8.2.5  Electrodialysis.*  Dissolved inorganics, the mineral content of a waste-water, can be removed by electrodialysis. When an inorganic salt is dissolved in a water solution it ionizes to produce positively charged cations and negatively charged anions. When an electrical potential is then passed through the solution, the cations migrate to the negative electrode and the anions to the positive electrode. Semi-permeable membranes are commercially available that allow the passage of ions of only one charge: cation-exchange membranes are permeable only to positive ions and anion-exchange membranes are permeable only to negative ions. When a series of these membranes is placed alternatively in a solution, and a voltage applied, the solution between one pair of electrodes becomes clarified as the ions concentrate in the solution in the adjacent compartments.

## 6.9  BIBLIOGRAPHY

R. E. Train, 'Quality Criteria for Water', Castle House Publishers Ltd., 1979.

J. S. Alabaster and R. Lloyd, 'Water Quality for Freshwater Fish', Butterworths, London, 1980.

R. C. Curds and H. A. Hawkes, 'Ecological Aspects of Used Water Treatment Processes and their Ecology', Vol. 2 and 3, Academic Press, London, 1983.

A. Porteous, 'Hazardous Waste Management Handbook', 1985.

S. J. Arcievala, 'Wastewater Treatment and Disposal Engineering and Ecology in Pollution Control', Dekker, 1981.

M. J. Hammer, 'Water and Wastewater Technology', Wiley, New York, 2nd edn, 1986.

UNIDO, 'Technical Report Series No. 7: Audit and Reduction Manual for Industrial Emissions and Wastes', UNIDO, 1990, pp. 60–76.

CHAPTER 7

# Air Pollution: Sources, Concentrations and Measurement

R. M. HARRISON

## 7.1 INTRODUCTION

Before commencing the description of individual air pollutants, it is useful to start with a consideration of the terminology. Air pollutants may exist in *gaseous* or *particulate* form. The former includes substances such as sulfur dioxide and ozone. Concentrations are commonly expressed either in mass per unit volume ($\mu$g m$^{-3}$ of air) or as a volume mixing ratio (1 ppm $= 10^{-6}$ v/v; 1 ppb $= 10^{-9}$ v/v). Particulate air pollutants are highly diverse in chemical composition and size. They include both solid particles and liquid droplets and range in size from a few nanometres to hundreds of micrometres in diameter. Concentrations are expressed in $\mu$g m$^{-3}$. Gaseous and particulate air pollutants may be separated operationally by use of a filter.

Air pollutants emitted directly into the atmosphere from a source are termed *primary*. Thus, carbonaceous particles from diesel engine exhaust and sulfur dioxide from power stations are examples of primary pollutants. In contrast, *secondary* pollutants are not emitted as such, but are formed within the atmosphere itself. Thus, sulfuric acid and nitric acid, formed respectively from sulfur dioxide and nitrogen dioxide oxidation, are examples of secondary pollutants for which the atmospheric formation route far exceeds any primary emissions. The most commonly considered secondary pollutant is ozone, formed as a result of photolysis of molecular oxygen in the stratosphere, and of nitrogen dioxide in the troposphere (lower atmosphere).

This chapter will address the sources, concentrations and measurement methods for those pollutants considered most important in terms of human health effects and damage to crops and materials. Issues relating to the regional and global atmosphere will be considered in the following chapter.

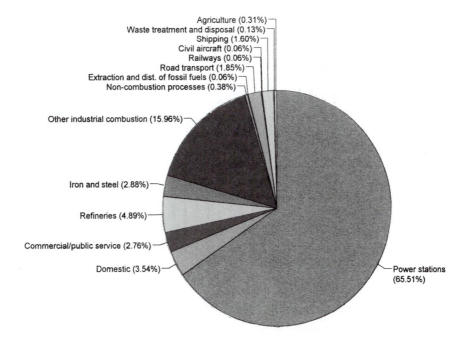

**Figure 7.1**   *Estimated UK emissions of sulfur dioxide by source category in 1993*[1]

## 7.2   SPECIFIC AIR POLLUTANTS

### 7.2.1   Sulfur Dioxide

The major source of sulfur dioxide is the combustion of fossil fuels containing sulfur. These are predominantly coal and fuel oil since natural gas, petrol and diesel fuels have a relatively low sulfur content. Until recently, emissions of sulfur dioxide from diesel engines led to a small but perceptible increment in sulfur dioxide alongside busy roads, but regulations coming into force in 1996 require a substantial reduction in the sulfur content of diesel fuel. Figure 7.1 shows in diagrammatic form the sources of sulfur dioxide emissions by source category for the United Kingdom.[1] Combustion of coal in power stations is far the most major single source of $SO_2$ emissions.

Figure 7.2 shows the trend in total UK emissions of sulfur dioxide[1] from 1970 to 1993. It may be seen that a substantial improvement has been achieved and, as a result of further international agreements, $SO_2$ emissions are set to fall significantly further in the coming years. Figure 7.2 also shows the 98 percentile concentration of sulfur dioxide at the Stepney (5) measurement site which may be seen to have fallen far faster than the fall in emissions. This results from a massive decline in the emission of sulfur dioxide from low level sources, mostly domestic coal fires within urban areas. Up to the 1960s these were the major source of

[1]Department of Environment, 'Digest of Environmental Statistics', No. 17, HMSO, London, 1995.

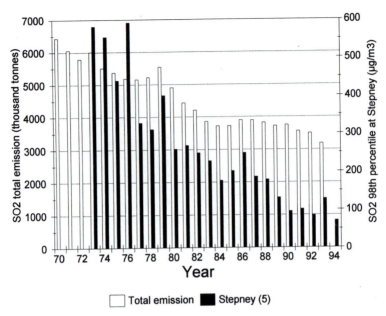

**Figure 7.2**  *UK national emissions of sulfur dioxide and 98 percentile ambient concentration at Stepney, London, 1970–1994*[1]

urban sulfur dioxide, but have now disappeared almost completely, to the point where urban and rural concentrations of sulfur dioxide are virtually indistinguishable. The short-term temporal pattern in sulfur dioxide concentrations has also changed, consistent with the main source of sulfur dioxide being emission from power station chimneys. The sulfur dioxide is present in relatively narrow plumes of pollution which cause short-term excursions in concentration to higher levels on an infrequent basis. This is superimposed on a very low background from diffuse sources.

Atmospheric sulfur dioxide is oxidized forming sulfuric acid. The chemical mechanisms are elaborated in Chapter 8. Sulfuric acid may be incorporated into rain or dry-deposited as fine droplets and, in doing so, causes acidification of soils and surface waters. Such acidification processes have been associated with the well publicized problems which go under the all-embracing term of 'acid rain'. Within Europe, such problems have been most acute in Scandinavia, which itself emits little sulfur dioxide, but which is subject to deposition of acid arising from emissions in many other countries, such as those of eastern Europe and the UK. Some of the problems associated with acidification are described in Chapter 4. In order to control sulfur dioxide, flue gas desulfurization plants are being installed on major UK power stations. The process involves use of a slurry of limestone to scrub the sulfur dioxide, but leads to production of large quantities of gypsum which require disposal.

A number of air quality standards and guidelines for sulfur dioxide have been set in order to protect human health. Those standards set a few years ago were

**Table 7.1** *United States federal primary and secondary ambient air quality standards*

| Pollutant | Type of standard | Averaging time | Frequency parameter | Concentration $\mu g\ m^{-3}$ | ppm |
|---|---|---|---|---|---|
| Sulfur oxides (as sulfur dioxide) | Primary | 24 hr | Annual maximum* | 365 | 0.14 |
| | | 1 year | Arithmetic mean | 80 | 0.03 |
| | Secondary | 3 hr | Annual maximum* | 1300 | 0.5 |
| Particulate Matter ($PM_{10}$) | Primary and Secondary | 24 hr | Annual maximum* | 150 | – |
| | | 24 hr | Annual arithmetic mean | 50 | – |
| Carbon monoxide | Primary and Secondary | 1 hr | Annual maximum* | 40 000 | 35 |
| | | 8 hr | Annual maximum* | 10 000 | 9 |
| Ozone | Primary and Secondary | 1 hr | Annual maximum* | 235 | 0.12 |
| Nitrogen dioxide | Primary and Secondary | 1 year | Arithmetic mean | 100 | 0.05 |
| Lead | Primary and Secondary | 3 months | Arithmetic mean | 1.5 | – |

*Not to be exceeded more than once a year

National primary ambient air quality standards define levels of air quality designed with an adequate margin of safety to protect the public health.

National secondary ambient air quality standards define levels of air quality judged necessary to protect the public welfare from any known or anticipated adverse effects of a pollutant.

based upon the effects of combined exposure to smoke and sulfur dioxide[2] and were determined by use of data collected during smogs in the 1950s and 1960s. The current perception is that the effects of sulfur dioxide and of particulate matter (smoke) are probably largely independent of one another and that joint standards are not warranted. Thus, standards for sulfur dioxide are now being set more in relation to the effect of short-term exposures upon asthmatics in whom the gas causes a rapid narrowing of the upper airways with a consequent diminution in respiratory function. The World Health Organization Guideline of 188 ppb as a 10 minute average is designed to protect against this effect (see Chapter 11), as is the UK standard of 100 ppb/15 minute average. The USEPA air quality standards are listed in Table 7.1, and Table 7.2 gives the European Union air quality standard for sulfur dioxide and smoke, which is a combined standard in which the limit values for sulfur dioxide are dependent upon the airborne concentration of smoke. At the time of writing, this and a number of other such air quality standards are under revision.

*7.2.1.1 Measurement of Sulfur Dioxide.* In the UK two techniques are commonly used for the determination of sulfur dioxide in the atmosphere. One is a very simple method developed originally for use in the 'National Survey of Smoke and

---

[2]World Health Organization, 'Air Quality Guidelines for Europe', European Series No. 23, WHO Regional Publications, Copenhagen, 1987.

**Table 7.2**   *EEC air quality standards for smoke and sulfur dioxide*

*(a) Smoke*

| | | |
|---|---|---|
| Yearly | Median of daily values throughout the year | $80 \ \mu g \ m^{-3}$ $(68 \ \mu g \ m^{-3})$ |
| Winter | Median of daily values from 1 October to 31 March | $130 \ \mu g \ m^{-3}$ $(111 \ \mu g \ m^{-3})$ |
| Peak | 98 percentile of daily values throughout the year | $250 \ \mu g \ m^{-3}$ $(213 \ \mu g \ m^{-3})$ |

*(b) Sulfur dioxide*

| | | | |
|---|---|---|---|
| | Smoke concentration | $\leqslant 40 \ \mu g \ m^{-3}$ $(34 \ \mu g \ m^{-3})$ | $120 \ \mu g \ m^{-3}$ |
| Yearly | | | |
| | Smoke concentration | $> 40 \ \mu g \ m^{-3}$ $(34 \ \mu g \ m^{-3})$ | $80 \ \mu g \ m^{-3}$ |
| | Smoke concentration | $\leqslant 60 \ \mu g \ m^{-3}$ $(51 \ \mu g \ m^{-3})$ | $180 \ \mu g \ m^{-3}$ |
| Winter | | | |
| | Smoke concentration | $> 60 \ \mu g \ m^{-3}$ $(51 \ \mu g \ m^{-3})$ | $130 \ \mu g \ m^{-3}$ |
| | Smoke concentration | $\leqslant 150 \ \mu g \ m^{-3}$ $(128 \ \mu g \ m^{-3})$ | $350 \ \mu g \ m^{-3}$ |
| Peak | | | |
| | Smoke concentration | $> 150 \ \mu g \ m^{-3}$ $(128 \ \mu g \ m^{-3})$ | $250 \ \mu g \ m^{-3}$ |

Figures in parentheses relate to the UK method of smoke measurement

Sulfur Dioxide' which has now been discontinued, although many of the measurement sites still exist within the EC Directive monitoring network for sulfur dioxide, and the UK basic urban network. The method involves absorption of sulfur dioxide in hydrogen peroxide solution to form sulfuric acid. The resultant acid has traditionally been determined by acid-base titration which is subject to interference by other gaseous, acidic or basic compounds such as nitric acid or ammonia respectively. As sulfur dioxide levels have fallen, so the reliability of measurements made by titration has been reduced and many measurements are now made by determination of sulfate by ion chromatography, which yields a result specific to sulfur dioxide.

The most commonly used instrumental technique for measurement of sulfur dioxide is based upon measurement of fluorescence excited by radiation in the region of 214 nm. Commercial instruments are available, capable of measurement of sulfur dioxide to less than 0.1 ppb, as well as source instruments with ranges into the thousands of ppm. The method is potentially subject to interferences from water vapour, which quenches the $SO_2$ fluorescence, and hydrocarbons capable of fluorescence at the same wavelength as $SO_2$. Commercial instruments are generally equipped with diffusion dryers and hydrocarbon scrubbers to overcome these problems. The commonly used techniques for

**Table 7.3** *Summary of commonly employed methods for measurement of air pollutants*

| Pollutant | Measurement technique | Sample collection period | Response time (continuous technique)[a] | Typical minimum detectable concentration |
|---|---|---|---|---|
| Sulfur Dioxide | Absorption in $H_2O_2$ and titration or sulfate analysis | 24 hr | | 1 ppb |
| | Gas phase fluorescence | | 2 min | 0.1 ppb |
| Oxides of Nitrogen | Chemiluminescent reaction with ozone | | 1 s | 0.1 ppb |
| Total Hydrocarbons | Flame ionization analyser | | 0.5 s | 10 ppbC |
| Specific Hydrocarbons | Gas chromatography/ flame ionization detector | [b] | | < 1 ppb |
| Carbon Monoxide | Electrochemical cell | | 25 s | 1 ppm |
| | Non-dispersive infrared | | 5 s | 0.5 ppm |
| | Gas filter correlation | | 90 s | 0.1 ppm |
| Ozone | UV absorption | | 30 s | 1 ppb |
| Peroxyacetyl Nitrate | Gas chromatography/ electron capture detection | [c] | | 1 ppb |
| Particulate Matter | High volume sampler | 24 hr | | 5 $\mu$g m$^{-3}$ |
| | TEOM | 1 hr | | 4 $\mu$g m$^{-3}$ |

[a] Time taken for a 90 percent response to an instantaneous concentration change
[b] Samples of air concentrated prior to analysis
[c] Instantaneous concentrations measured on a cyclic basis by flushing the contents of a sample loop into the instrument

analysis of $SO_2$ and other pollutants are summarized in Table 7.3 and are described in more detail elsewhere.[3,4]

## 7.2.2 Suspended Particulate Matter

Airborne particles are very diverse in character, including both organic and inorganic substances with diameters ranging from less than 10 nm to greater than 100 $\mu$m. Since very fine particles grow rapidly by coagulation and large particles sediment rapidly under gravitational influence, the major part (by mass) generally exists in the 0.1–10 $\mu$m range. A schematic representation of the typical size distribution for atmospheric particles appears in Figure 7.3. There are three peaks, or modes, in the distribution. The smallest one relates to the transient nuclei, which are very tiny particles formed by condensation of hot vapours, or gas to particle conversion processes. Thus, primary particles from

[3] R. M. Harrison and R. Perry (eds), 'Handbook of Air Pollution Analysis', Chapman & Hall, London, 2nd edn, 1986.
[4] J. P. Lodge (ed), 'Methods of Air Sampling and Analysis', Lewis Publishers, Chelsea, Michigan, 3rd edn, 1989.

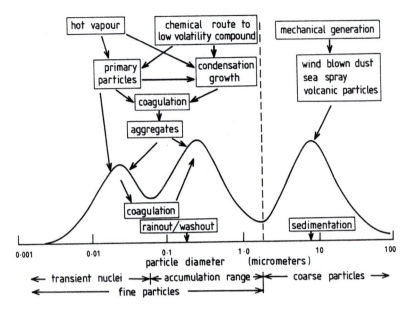

**Figure 7.3**  *Schematic diagram of the size distribution and formation mechanisms for atmospheric aerosols* (adapted from Ref 5)

motor vehicle exhaust and sulfuric acid formed from $SO_2$ oxidation are initially in the transient nuclei mode. Such particles, when emitted, are present in very high numbers and are subject to rather rapid coagulation both with other fine particles and also with coarser particles already in the atmosphere. Through this mechanism they enter the accumulation range of particles typically with diameters between about 100 nm and 2 $\mu$m. Such particles are also capable of growth through the condensation of low volatility materials. Removal of the accumulation range particles by rainwater scavenging or dry deposition to surfaces is inefficient and such particles have a typical lifetime in the atmosphere of around one to two weeks. This renders them capable of very long range transport. The third mode, termed coarse particles, are mostly greater than 2 $\mu$m with sizes extending up to about 100 $\mu$m. This is comprised in the main of mechanically-generated particles such as wind-blown dust, sea spray and primary volcanic particles. These are formed by attrition of bulk materials and tend to be appreciably larger than transient nuclei or accumulation range particles.

The suspended particles in the atmosphere are typically referred to as the *atmospheric aerosol* and a number of other terms are used, which in the main are related to the method of collection or analysis.

In the UK, the largest set of measurements are of *black smoke* which (see later) is a measurement related to the blackness or soiling capacity of the particles. Other

[5]K. T. Whitby, *Atmos. Environ.*, 1978, **12**, 135.

measures of suspended particulate matter depend upon a gravimetric determination of particle mass. The term total suspended particulate matter (TSP) has been used to describe the fraction of particles collected on a filter using the US high volume sampler. Since large particles, by virtue of their inertia, are not readily able to enter the inlet of such samplers, the measurement is dependent both on the orientation of the sampler with respect to the wind and the strength of wind, both of which influence the efficiency of particle aspiration.[3] To overcome this problem, the USEPA moved to use of a high volume sampler with a size selective inlet. This inlet has a 50% efficiency at 10 micrometres aerodynamic diameter and hence in simple terms may be considered to sample only those particles less than 10 micrometres diameter, irrespective of orientation or wind speed. Measurements made using inlets meeting the EPA criterion are referred to as $PM_{10}$. A number of devices, other than the high volume sampler, are now available for $PM_{10}$ measurement (see later).

*7.2.2.1 Black Smoke.* Because of the tradition of measuring atmospheric concentrations of particulate matter as black smoke, the United Kingdom has traditionally compiled emission inventories also in terms of black smoke. Inventories are generated by measuring the mass emission rates of particles from the various sources, and then multiplying by a blackness index which describes the relative darkness of the different smokes. Coal smoke is conventionally assigned an index of 1 and, for example, diesel smoke has a darkness index of 3 as it is three times blacker per unit mass than coal smoke.[6] The inventories show that road transport, and particularly the diesel fuel component, and coal burning in domestic premises are now the main sources of black smoke emissions. Within urban areas there is a prohibition on burning bituminous coal in domestic premises and thus road transport is far the major source of black smoke emissions. For example, in London 94% of black smoke emissions arise from road traffic.[7]

Figure 7.4 shows the trends in emission of black smoke in the UK together with the 98 percentile concentration measured at the Stepney (5) monitoring site. Urban concentrations have fallen very much faster than the national emissions of black smoke, although concentrations have stabilized in recent years as a result of the predominant influence of road traffic. During the London smog of December 1952, airborne concentrations of black smoke exceeded 1500 $\mu$g m$^{-3}$ (see Figure 7.5) accompanied by massive concentrations of sulfur dioxide. As shown by Figure 7.5, the daily death rate also increased greatly during the week of the pollution episode and this led to legislation to control urban air pollution by smoke and $SO_2$. The European air quality standards for smoke and sulfur dioxide appear in Table 7.2.

Measurement of black smoke is still widely performed in the UK and many other counties around the world. Air is drawn through a cellulose filter upon which the particles are collected. At the end of the sampling period the ability of

[6]Quality of Urban Air Review Group, 'Urban Air Quality in the United Kingdom', QUARG, London, 1993.
[7]Quality of Urban Air Review Group, 'Diesel Vehicle Emissions and Urban Air Quality', QUARG, Birmingham, 1993.

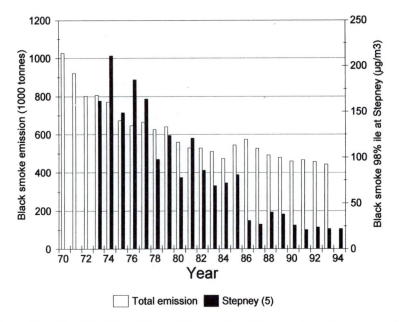

**Figure 7.4** *Trends in UK emissions of black smoke and 98 percentile ambient concentrations at Stepney, London, 1970–1994* [1]

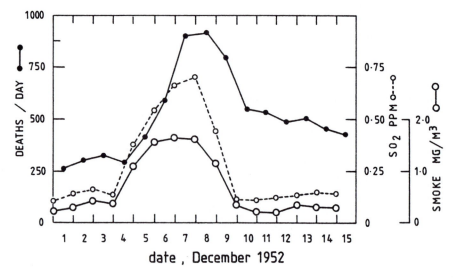

**Figure 7.5** *Daily concentrations of smoke and sulfur dioxide, together with mortality rates in the London smog of December 1952*

the filter surface to reflect light is determined quantitatively and the reflectance is related to a concentration of 'standard smoke' using a calibration graph constructed many years ago when urban particles were dominated by coal smoke. Nowadays, there are many other important sources of particles and the intrinsic darkness of urban particulate matter is much lower. Research has shown that the method measures elemental carbon, and thus the dramatic reduction in the inefficient combustion of bituminous coal in domestic grates over recent years has led to a large drop in the emissions of particles containing black elemental carbon.

*7.2.2.2 Gravimetrically Determined Particulate Matter: PM₁₀.* $PM_{10}$ particulate matter is determined gravimetrically and therefore includes both primary and secondary particles. In UK urban areas two sources are dominant; one is road traffic emissions and the other is secondary particulate matter, mostly ammonium sulfate and ammonium nitrate particles. Whilst the former are black in colour, the latter are white and hence are not determined by the black smoke method. Other sources such as sea spray and wind-blown dust also contribute to urban $PM_{10}$, although their precise impact on airborne concentrations is more difficult to determine.[6] There are many other minor sources such as particles from building work and demolition, and biological particles such as pollens, spores and bacteria. The health effects of $PM_{10}$ outlined in Chapter 11 appear to be largely independent of chemical composition.

In addition to adverse health effects on human health, airborne particles are responsible for the soiling of buildings and loss of visibility. Airborne particles both scatter and absorb light and thus cause deterioration in the quality of image transmission through the atmosphere which manifests itself as a loss of visibility. Although the efficiency of light scattering and absorption per unit mass of particles is dependent upon the size distribution and chemical composition of the particles, the aerosol mass loading is a fairly good predictor of visibility impairment.[8]

Typical UK urban concentrations of $PM_{10}$ are shown in Table 7.4. These concentrations and those of black smoke are orders of magnitude below those associated with the major smog episodes of the 1950s and 60s. The smogs were caused primarily by low level emissions from coal combustion during periods of meteorology unsuitable for effective pollutant dispersal (low windspeeds and shallow mixing depth). The combination with fog (smog = smoke + fog) led to dramatic losses in visibility. The smog of December 1952 is believed to have caused some 4000 premature deaths. For many years after, UK pollution control policy focused on smoke and sulfur dioxide, and it is in only recent years that the major focus has transferred to motor traffic as the major source of urban air pollution. In view of the health risks posed by $PM_{10}$ the UK Expert Panel on Air Quality Standards has recommended a maximum of 50 $\mu$g m$^{-3}$ as a rolling 24 hour average.

[8]R. J. Charlson, *J. Air Pollut. Control Assoc.*, 1968, **18**, 652.

**Table 7.4**   Summary statistics for $PM_{10}$ at U.K. urban sites in 1994

| Site | Annual mean ($\mu g\ m^{-3}$) | 98th percent ($\mu g\ m^{-3}$) | Max. hour ($\mu g\ m^{-3}$) | Peak/mean |
|------|------|------|------|------|
| Belfast | 26 | 66 | 490 | 18.8 |
| Bristol | 24 | 59 | 612 | 25.5 |
| Birmingham (Central) | 23 | 55 | 311 | 13.5 |
| Birmingham (East) | 21 | 50 | 319 | 13.9 |
| Cardiff | 34 | 76 | 564 | 16.6 |
| Edinburgh | 20 | 41 | 307 | 15.4 |
| Kingston-upon-Hull | 26 | 56 | 264 | 10.2 |
| Leeds | 26 | 64 | 310 | 11.9 |
| Leicester | 21 | 50 | 203 | 9.7 |
| Liverpool | 25 | 68 | 257 | 10.3 |
| London (Bexley) | 25 | 52 | 140 | 5.6 |
| London (Bloomsbury) | 27 | 56 | 307 | 11.4 |
| Newcastle | 26 | 60 | 297 | 11.4 |
| Southampton | 23 | 48 | 291 | 12.7 |
| Swansea | 22 | 32 | 73 | 3.3 |

*Measurement of $PM_{10}$.*   The simplest method for determination of $PM_{10}$ involves use of a high volume air sampler capable of drawing air through a filter at a rate of about 1 $m^3\ min^{-1}$ through a 10 $\mu m$ size selective inlet. The sampler is run for 24 hours and the gain in mass of the filter, together with the volume of air passed, is used to calculate the airborne concentration of $PM_{10}$ over the 24 hour sampling period.

Measurements in pseudo-real-time may be made using the Tapered Element Oscillating Microbalance (TEOM). In this device, particles are collected on a filter which is attached to a vibrating element whose vibrational frequency changes with the accumulation of particles on its tip. Measurement of the vibrational frequency is used to estimate the mass of particles collected. An example of averaged diurnal data for $PM_{10}$ and carbon monoxide collected in Birmingham is shown in Figure 7.6 which clearly illustrates the influence of road traffic emissions on both $PM_{10}$ and carbon monoxide concentrations.

*Specific Components of Suspended Particulate Matter.*   The components of airborne particles which give rise to most interest in the context of air pollution tend to be trace metals and trace organic compounds. A summary of concentrations of trace metals measured in Central London appears in Table 7.5. Also indicated are trends in concentration which in general have been clearly downwards for several years. Interest in trace metals relates primarily to their potential toxicity. Arsenic, chromium(VI) and nickel are believed to be genotoxic carcinogens and the World Health Organization has estimated unit risk factors for them.[2] Cadmium and mercury have elicited a great deal of interest because of their high toxicity and there have been moves to control these elements throughout the developed world.

The trace metal which has stimulated far the greatest interest in relation to public health has been lead. The main source of airborne lead is its combustion in leaded petrol (gasoline) to which it is added as an octane improver in the form of

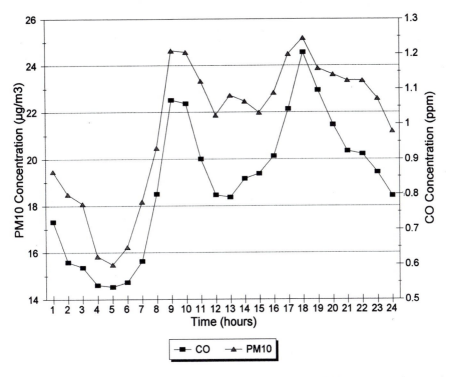

**Figure 7.6** *Averaged diurnal variation of carbon monoxide and $PM_{10}$ measured in central Birmingham, Winter 1993*

**Table 7.5** *Trace metals in air in Central London in 1992/93*

| Metal | Concentration in 1992/3 (ng m$^{-3}$) | % reduction since 1984/5 |
|---|---|---|
| Cadmium | 0.4 | 86 |
| Chromium | 5.4 | 55 |
| Copper | 13 | 52 |
| Iron | 891 | 9 |
| Manganese | 21 | −10* |
| Nickel | 5.2 | 69 |
| Zinc | 59 | 41 |
| Vanadium | 12 | 63 |
| Cobalt | 0.3 | 86† |

*Manganese increased by 10%
†Decline since 1988/89

tetraalkyllead compounds, notably tetramethyllead $(CH_3)_4Pb$ and tetraethyl-lead $(C_2H_5)_4Pb$. Upon combustion in the engine, these are emitted predominantly as an aerosol of fine particles of inorganic lead. Developed countries have without exception introduced regulations to limit the lead content of gasoline and the United States has been essentially lead-free for many years. In Europe, low-lead gasolines were introduced in response to concerns over the health effects of lead, and more recently, there have been substantial fiscal incentives to the purchase of unleaded gasoline because of its requirement for use in vehicles fitted with catalytic converters, and in order to further reduce exposure of the general population to the metal. The maximum permitted level of lead in UK gasoline stood in 1972 at 0.84 g $1^{-1}$, although such high concentrations were not generally used. By 1981, this had been reduced to 0.40 g $1^{-1}$, and in January 1986 fell sharply to 0.15 g $1^{-1}$ in line with most other western European countries. The decline in the use of lead in petrol between 1980 and 1994 is documented in Figure 7.7, which also shows the decline of lead in air concentrations in central London. In general, it may be seen that the two have proceeded very much in parallel with one another.

The abrupt fall in the lead emissions from road traffic at the begining of 1986 was exploited as the basis for an experiment to evaluate its impact upon population blood leads. In the event, blood leads fell slightly between 1985 and 1986, but little more, if at all, than between 1984 and 1985, and between 1986 and 1987 (when petrol lead remained almost constant) and only slightly more for groups heavily exposed to airborne lead than for less exposed groups. As blood

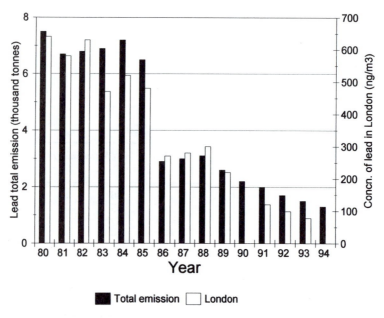

**Figure 7.7**  *Trends in emissions of lead from motor traffic and airborne concentrations in central London, 1980–1994*[1]

lead is a good indicator of recent lead exposure, the data points strongly to the importance of other non-atmospheric routes of exposure to lead such as diet, drinking water and other beverages. Ambient air quality standards for lead currently in force in the United States and Europe are 1.5 $\mu g$ m$^{-3}$ (quarterly average) and 2 $\mu g$ m$^{-3}$ (annual average) respectively, but concern over lead remains at a high level and these standards are likely to be reviewed downwards. Various authors have questioned whether the projected benefits of further reductions in lead exposure of the general population are cost effective in terms of the public health benefits gained.[9,10]

Concentrations of lead and most other metals in air are determined by use of air filtration to collect a sample. Non-destructive analytical methods are available, and these include $X$-ray fluorescence (both wavelength dispersive and energy dispersive) and instrumental neutron activation analysis. More commonly, destructive methods are used which involve dissolving the sample in oxidizing acids and analytical procedures such as atomic absorption spectrometry, inductively coupled plasma (ICP) emission spectrometry, ICP-mass spectrometry or anodic stripping voltammetry.

The organic components of atmospheric particles of most interest as air pollutants are the semi-volatile groups of compounds including the polynuclear aromatic hydrocarbons (PAH), polychlorinated biphenyls (PCB) and polychlorinated dibenzo-dioxins and dibenzo-furans (PCDD and PCDF). The sampling, measurement and environmental behaviour of such compounds is discussed in Chapter 16.

### 7.2.3 Oxides of Nitrogen

The most abundant nitrogen oxide in the atmosphere is nitrous oxide, $N_2O$. This is chemically rather unreactive and is formed by natural microbiological processes in the soil. It is not normally considered as a pollutant, although it does have an effect upon stratospheric ozone concentrations and there is much evidence that use of nitrogenous fertilizers is increasing atmospheric levels of nitrous oxide.

The pollutant nitrogen oxides of concern are nitric oxide, NO and nitrogen dioxide, $NO_2$. By far the major proportion of emitted $NO_x$ (as the sum of the two compounds is known) is in the form of NO, although most of the atmospheric burden is usually in the form of $NO_2$. The major conversion mechanism is the very rapid reaction of NO with ambient ozone, the alternative third order reaction with molecular oxygen being relatively slow at ambient air concentrations, although it can be of importance when NO concentrations reach about 1 ppm.[11]

The major source of $NO_x$ is the high temperature combination of atmospheric nitrogen and oxygen, in combustion processes, there being also a lesser contribu-

[9]R. M. Harrison, *J. Roy. Soc. Health*, 1993, **113**, 142–148.
[10]S. J. Pocock, M. Smith and P. Baghurst, *Br. Med. J.*, 1994, **309**, 11–89.
[11]R. M. Harrison, *Chemistry in Britain*, 1994, **30**, 987–1000.

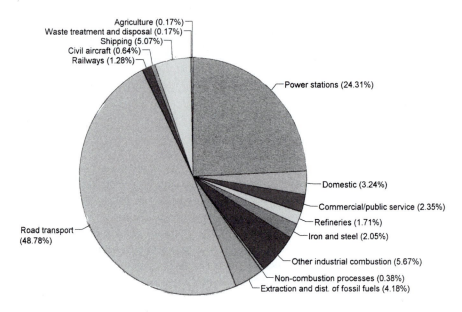

**Figure 7.8**   *Estimated UK emissions of $NO_x$ by source category in 1993*[1]

tion from combustion of nitrogen contained in the fuel. An emission inventory for the UK appears in Figure 7.8.

Typical hourly average ambient air concentrations of $NO_x$ are normally within the range 5–100 ppb in urban areas and $<20$ ppb at rural sites. A plot of the trend in annual mean and 98 percentile $NO_2$ concentrations in central London appears in Figure 7.9. The US Environmental Protection Agency (USEPA) ambient air quality standard for $NO_2$ is 50 ppb (annual average). The EEC Directive limit is a 98 percentile value of 105 ppb hourly average, not to be exceeded in a year, and guide values are 71 ppb and 26 ppb as 98 and 50 percentile hourly averages respectively, not to be exceeded in a year. The direct effects of exposure to oxides of nitrogen include human respiratory tract irritation and damage to plants. Indirect effects arise from the central role of $NO_2$ in photochemical smog reactions, and its oxidation to nitric acid contributing to acid rain problems (see Chapter 8).

*7.2.3.1   Measurement of Oxides of Nitrogen.*   Instrumental analysers have been available for a good number of years. The currently favoured technique for determination of oxides of nitrogen is based upon the chemiluminescent reaction of nitrogen oxide and ozone to give an electronically excited nitrogen dioxide which emits light in the 600–3000 nm region with a maximum intensity near 1200 nm:

$$NO + O_3 \rightarrow NO_2^* + O_2$$
$$NO_2^* \rightarrow NO_2 + h\nu$$

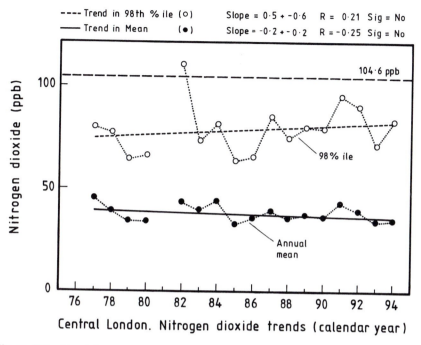

**Figure 7.9**   *Trends in annual mean and 98 percentile concentrations of nitrogen dioxide measured in central London, 1976–1994*[1]

In the presence of excess ozone generated within the instrument, the light emission varies linearly with the concentration of nitrogen oxides from 1 ppb to $10^4$ ppm. The apparatus is shown schematically in Figure 7.10.

The method is believed free of interference for measurement of NO and may be used to measure $NO_x$ by prior conversion of $NO_2$ to NO in a heated stainless steel or molybdenum converter. Dependent upon which converter is used, some interference in the $NO_x$ mode is likely from compounds such as peroxyacetyl nitrate and nitric acid. Some instruments incorporate two reaction chambers, one running permanently in the NO mode, the other analysing $NO_x$ after $NO_2$ to NO conversion. Thus, pseudo-real-time $NO_2$ concentrations may be measured.

A widely used inexpensive technique for measuring nitrogen dioxide is based upon the use of diffusion tubes. These are straight, hollow tubes of length about 7 cm and diameter 1 cm, sealed at one end which is placed upwards, and open at the other end. At the sealed end a metal grid is coated with triethanolamine which acts as a perfect sink for nitrogen dioxide. Access of nitrogen dioxide from ambient air to the triethanolamine is by molecular diffusion along the tube, and analysis of nitrite collected by the triethanolamine reagent and application of Fick's Law, allows calculation of airborne concentrations of nitrogen dioxide. The tubes have a tendency to over-estimate nitrogen dioxide concentrations because wind-induced turbulence in the entry to the tube reduces the effective diffusion length. Nonetheless, diffusion tubes have been used very widely, and

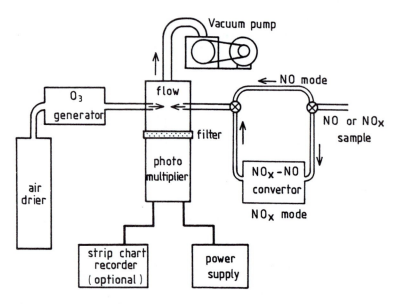

**Figure 7.10**   *Schematic diagram of chemiluminescent analyser for oxides of nitrogen*

particularly in dense networks, to evaluate the spatial distribution of pollutants within an urban area.[6,12] Nitrogen dioxide has also been mapped across the entire United Kingdom on the basis of a network of diffusion tube samplers.

### 7.2.4   Carbon Monoxide

As exemplified by the inventories,[1] carbon monoxide is a pollutant very much associated with emissions from petrol vehicles. Within urban areas where concentrations tend to be highest, motor traffic is responsible for about 98% of emissions of carbon monoxide.

The major sink process is conversion to $CO_2$ by reaction with the hydroxyl radical (see Chapter 8). This process is, however, rather slow and the reduction in CO level away from the source areas is almost entirely a function of atmospheric dilution processes. Carbon monoxide can be a problem in heavily trafficked areas especially in confined 'street canyons' where concentrations may reach 50 ppm or more. The adverse effect of CO is due to reaction with haemoglobin to form carboxyhaemoglobin which is relatively stable, causing a reduction in the oxygen-carrying capacity of the blood. The UK ambient air quality standard for CO is 10 ppm measured as an eight-hour running average.[13]

*7.2.4.1   Measurement of Carbon Monoxide.*   Non-dispersive infrared may be used to measure carbon monoxide in street air where levels encountered normally lie

[12]R. G. Clark and J. R. Wright, 'London Environmental Supplement' No. 14, 1986.
[13]Expert Panel on Air Quality Standards, 'Carbon Monoxide', HMSO, London, 1994.

within the range 1–50 ppm. Using a long-path cell, the IR absorbance of polluted air at the wavelength corresponding to the C–O stretching vibration is continuously determined relative to that of reference air containing no CO. This is achieved without dispersion of the IR radiation by using cells containing CO at a reduced pressure as detectors for two beams which are chopped at a frequency of about 10 Hz and passed respectively through sample and reference cells. Absorption of radiation by the CO causes a differential pressure between the two detector cells which is sensed by a flexible diaphragm between them and used to generate an electrical signal. Because of partial overlap of absorption bands, carbon dioxide and water vapour interfere. The latter may be removed by passing the air sample through a drying agent, and the former interference by interposing a cell of carbon dioxide between the sample and reference cells and the detectors.

The other most common type of instrumental analyser is based upon gas filter correlation (see Figure 7.11). Infrared broad band radiation passes sequentially through gas cells containing carbon monoxide and molecular nitrogen contained within a spinning wheel, prior to passage of the radiation through a multi-pass optical cell through which ambient air is drawn. There are thus two beams separated in time but not space, one of which is absorbed by carbon monoxide in the ambient air passing through the sample cell; the other of which is already depleted in the wavelengths absorbed by carbon monoxide and is hence affected only by absorption by components other than carbon monoxide in the sample air. The difference in signal between the two beams is thus the result of absorption by carbon monoxide within the sample cell. Interferences from water vapour and carbon dioxide are claimed to be negligible. The instrument has a minimum detection limit of about 0.1 ppm, and reads up to 50 ppm.

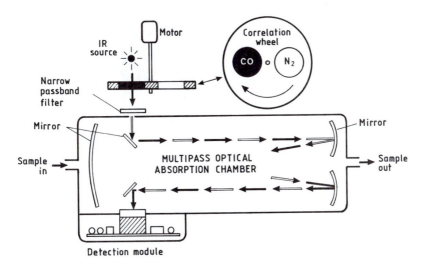

**Figure 7.11**  *Schematic diagram of gas filter correlation analyser for carbon monoxide*

An analyser for continuous determination of carbon monoxide at levels down to 1 ppm uses an electrochemical cell. Gas diffuses through a semi-permeable membrane into the cell and at an electrode CO is oxidized to $CO_2$ at a rate proportional to the concentration of CO in the air. The response time is fairly short and interferences from other air pollutants at normally encountered levels are minimal. Analysers based upon electrochemical cells are also available for measurement of sulfur dioxide and oxides of nitrogen.

### 7.2.5  Hydrocarbons

The major sources of volatile organic compounds (which are mainly but not exclusively hydrocarbons) in the UK atmosphere are shown in Figure 7.12. It may be seen that these are rather more diverse than for many of the pollutants and include natural sources such as release from forest trees. In urban areas, road transport is probably the major contributor, although use of solvents, for example in paints and adhesives, can be a very significant source. Emissions from road transport include both the evaporation of fuels and the emission of unburned and partially combusted hydrocarbons and their oxidation products from the vehicle exhaust.

Many sources emit a range of individual compounds and careful analytical work has shown measurable levels of in excess of 200 hydrocarbons in some ambient air samples. In the UK the Hydrocarbon Network, which makes

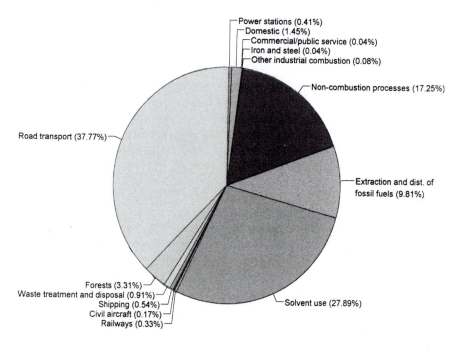

**Figure 7.12**   *Estimated UK emission of volatile organic compounds by source category for 1993*[1]

**Table 7.6** *Hydrocarbon concentrations in London*[14]*, December 1994*

|  | Average (ppb) | Hour max (ppb) |
|---|---|---|
| Ethane | 1.93 | 13.7 |
| Ethene | 2.30 | 15.3 |
| Propane | 3.14 | 23.6 |
| Propene | 2.72 | 22.6 |
| Ethyne | 8.83 | 71.8 |
| i-Butane | 3.03 | 44.4 |
| n-Butane | 6.66 | 66.8 |
| trans-2-Butene | 0.46 | 5.4 |
| 1-Butene | 0.65 | 6.9 |
| cis-2-Butene | 0.22 | 2.9 |
| i-Pentane | 3.50 | 44.4 |
| n-Pentane | 0.94 | 44.6 |
| 1,3-Butadiene | 0.53 | 40.1 |
| trans-2-Pentene | 0.32 | 3.1 |
| cis-2-Pentene | 0.18 | 2.0 |
| 2-Methyl-pentene | 1.15 | 14.5 |
| 3-Methyl-pentene | 0.76 | 10.1 |
| Isoprene | 0.48 | 5.7 |
| n-Hexane | 0.35 | 7.3 |
| n-Heptane | 0.20 | 3.0 |
| Benzene | 1.92 | 22.5 |
| Toluene | 2.47 | 52.9 |
| Ethylbenzene | 2.52 | 38.3 |
| m,p-Xylene | 2.41 | 28.4 |
| o-Xylene | 1.10 | 11.4 |

automated hourly measurements of volatile organic compounds, reports data on some 26 individual hydrocarbons. A data summary for 1993 appears in Table 7.6. Methane, which is not often measured, far exceeds the other hydrocarbons in concentration. The Northern Hemisphere background of this compound is approximately 1.8 ppm and elevated levels occur in urban areas as a result particularly of leakage of natural gas from the distribution system.

There are two major reasons for interest in the concentrations of hydrocarbons in the polluted atmosphere. The first is the direct toxicity of some compounds, particularly benzene and 1,3-butadiene, both of which are chemical carcinogens. The UK Expert Panel on Air Quality Standards has recommended for benzene an ambient air quality standard of 5 ppb measured as a rolling annual average, with a longer term target of 1 ppb rolling annual average.[15] The recommended standard for 1,3-butadiene is 1 ppb rolling annual average.[16] The second cause of concern regarding hydrocarbons is due to their

[14]Data Courtesy of National Environmental Technology Centre.
[15]Expert Panel on Air Quality Standards, 'Benzene', HMSO, London, 1994.
[16]Expert Panel on Air Quality Standards, '1,3-Butadiene', HMSO, London, 1994.

role as precursors of photochemical ozone (see Chapter 8). Compounds differ greatly in their potential to promote the production of ozone which has led to a system of classifying hydrocarbons according to their photochemical ozone creation potential (see also Chapter 8).

*7.2.5.1 Measurement of Hydrocarbons.* Determination of specific hydrocarbons in ambient air normally requires a pre-concentration stage in which air is drawn through an adsorbant such as a porous polymer or activated carbon, or a tube where freeze-out of the compounds by reduced temperature occurs, followed by injection into a gas chromatograph. Excellent separations of many compounds have been achieved, the best results coming from use of capillary columns. Detection may be by flame ionization or mass spectrometer (GC-MS), the latter technique allowing a more positive identification of individual compounds. Recent years have seen the development of automated systems for gas chromatographic measurements of hydrocarbon on a cyclic basis of about one hour to complete sampling and analysis. The UK Hydrocarbon Network was the first national network to adopt such instrumentation on a routine basis. It is now operated at some 13 sites. It is also possible to measure 'total hydrocarbons' by passage of a full air sample to a flame ionization detector. Results are reported as ppb C (parts per billion carbon) since the reponse of the FID is related closely to the rate of introduction of organic carbon atoms into the flame. Non-methane hydrocarbons may be determined with such instruments by alternate selective removal of hydrocarbons other than methane from the air stream prior to analysis and determination by difference.

### 7.2.6 Secondary Pollutants: Ozone and Peroxyacetyl Nitrate

*7.2.6.1 Ozone.* Atmospheric reactions involving oxides of nitrogen and hydrocarbons cause the formation of a wide range of secondary products. The most important of these is ozone. In severe photochemical smogs, such as occur in southern California, levels of ozone may exceed 400 ppb.

In Europe the classic Los Angeles type of urban smog is not experienced. Nonetheless, the same chemical processes give rise to elevated concentrations of ground-level ozone, often in a regional phenomenon extending over hundreds of miles simultaneously. Thus hydrocarbon and $NO_x$ emissions over wide areas of Europe react in the presence of sunlight causing large scale pollution, which is further extended by atmospheric transport of the ozone.[17] The phenomenon is crucially dependent upon meteorological conditions and hence, in Britain, is observed on only perhaps 10–30 days in each year on average. Concentrations of ozone measured at ground-level commonly exceed 100 ppb during such 'episodes' and have on one severe occasion been observed to exceed 250 ppb in southern England. These levels may be compared with a background of ozone at ground-level arising from downward diffusion of stratospheric ozone and general

---

[17]Photochemical Oxidants Review Group, 'Ozone in the United Kingdom 1993', PORG, London, 1993.

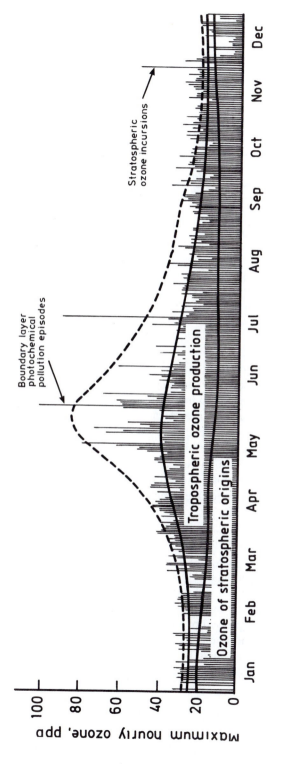

**Figure 7.13** *Contribution from the major ozone sources to the daily maximum hourly mean observed at ground-level at a rural site*[18]

[18]R. G. Derwent and P. J. A. Kaye, *Environ. Pollut.*, 1988, **55**, 191.

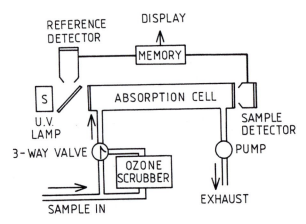

**Figure 7.14** *Schematic diagram of a U.V. photometric analyser for ozone*

tropospheric production of 20–50 ppb. This is seen in Figure 7.13 showing measurement data from southern Scotland.

There is considerable evidence of a detrimental effect of ozone exposure upon human lung function, although this appears to be reversible when exposure ceases. The USEPA ambient air quality standard is 120 ppb, not to be exceeded as an hourly average more than once per year. This is very much less stringent than the recommendation of the UK Expert Panel on Air Quality Standards of 50 ppb measured as a rolling eight-hour average.[19] Both standards are currently exceeded several times each year in most parts of the United Kingdom; the greatest number of exceedences occurring in rural areas of the south-eastern UK. The least frequent exceedences occur in the urban sites where high levels of nitric oxide emissions from road traffic suppress ozone (see Chapter 8). For the rural sites, there is a general gradient in ozone with highest levels in the south and east of Britain and lowest in the north and west. Damage to crop plants can occur at ozone concentrations as low as 40 ppb; this is elaborated on in Chapter 12.

*Measurement of Ozone.* The UV absorption of ozone at 254 nm may be used for its determination at levels down to 1 ppb. Interferences from other UV-absorbing air pollutants such as mercury and hydrocarbons may be minimized by taking two readings. The first reading is of the absorbance of an air sample after catalytic conversion of ozone to oxygen and the second is of an unchanged air sample, the difference in absorbance being due to the ozone content of the air. Available instruments perform this procedure automatically and give a read-out in digital form. Although truly continuous measurement of ozone levels is not possible, response is fast and readings may be taken at intervals of less than one minute. The instrument is shown diagrammatically in Figure 7.14.

*7.2.6.2 Peroxyacetyl Nitrate (PAN)* PAN is a product of atmospheric photo-chemical reactions and is a characteristic of photochemical smog (see Chapter 8).

[19]Expert Panel on Air Quality Standards, 'Ozone', HMSO, London, 1994.

$$CH_3-\underset{\underset{O}{\|}}{C}-O-O-NO_2 \quad PAN$$

Levels in southern California lie typically within the range 5–50 ppb on smoggy days. In Europe, the formation is far less favoured[17] and concentrations are more usually well below 10 ppb.

*Chemical Analysis of Peroxyacetyl Nitrate.* PAN may be determined by long-path IR measurements, the greatest sensitivity being achieved when Fourier transform methods are used. The most sensitive and specific routine technique involves gas chromatographic separation and detection of specific peroxyacyl nitrates by electron capture. The detection limit of below 1 ppb permits it to be used as a direct atmospheric monitor under circumstances of high pollution. Lower concentrations may be measured by pre-concentration of PAN from the air.

## 7.3 INDOOR AIR QUALITY

Until rather recently, the emphasis in air quality evaluation has centred upon the outdoor environment. Recently, however, it has become clear that exposures to pollutants indoors may be very important (see also Section 18.4.1).

Pollutants with indoor sources may build up to appreciable levels because of the slowness of air exchange. An example is oxides of nitrogen from gas cookers and flueless gas and kerosene heaters which can readily exceed outdoor concentrations. Kerosene heaters can also be an important source of carbon monoxide and sulfur dioxide. Building materials and furnishings can also release a wide range of pollutants, such as formaldehyde from chipboard and hydrocarbons from paints, cleaners, adhesives, timber and furnishings. The tendency towards lower ventilation levels (energy efficient houses) has tended to exacerbate this problem.

Pollutants with a predominantly outdoor source may be reduced to rather low levels indoors due to the high surface area/volume ratios indoors leading to extremely efficient dry deposition of pollutants such as ozone and sulfur dioxide.

## 7.4 APPENDIX

### 7.4.1 Air Pollutant Concentration Units

Probably the most logical unit of air pollutant concentration is mass per unit mass, *i.e.* $\mu g\ kg^{-1}$ or $mg\ kg^{-1}$. This is, however, very rarely used. The commonest units are mass per unit volume (usually $\mu g\ m^{-3}$) or volume per unit volume, otherwise known as a volume mixing ratio (ppm or ppb). For particulate pollutants the volume mixing ratio is inapplicable.

Much confusion arises in the interconversion of $\mu g\ m^{-3}$ and ppm. Whilst the volume mixing ratio is independent of temperature and pressure for an ideal gas (and air pollutant behaviour is close to ideal), the mass per unit volume unit is dependent on $T$ and $P$ conditions, and hence these will be taken into account.

*7.4.1.1   Example 1.*   Convert 0.1 ppm nitrogen dioxide to $\mu$g m$^{-3}$ at 20°C and 750 torr.

46 g NO$_2$ occupy 22.41 l at STP

46 g NO$_2$ occupy 22.41 $\times \dfrac{293}{273} \times \dfrac{760}{750}$ l

   = 24.37 l at 20°C and 750 torr

0.1 ppm NO$_2$ is $10^{-7}$ l NO$_2$ in 1 l, or

..... $10^{-4}$ l NO$_2$ in 1 m$^3$

$10^{-4}$ l NO$_2$ at 20°C and 750 torr contain 46 $\times \dfrac{10^{-4}}{24.37}$ g

   = 189 $\mu$g NO$_2$

∴ NO$_2$ concentration = 189 $\mu$g m$^{-3}$

*7.4.1.2   Example 2.*   Convert 100 $\mu$g m$^{-3}$ ozone at 25°C and 765 torr to ppb

48 g ozone occupy 22.41 l at STP

....... occupy 22.41 $\times \dfrac{298}{273} \times \dfrac{760}{765}$ l

   = 24.30 l at 25°C and 765 torr

100 $\mu$g ozone occupy 24.30 $\times \dfrac{100 \times 10^{-6}}{48}$ l

   = 50.6 $\times$ 10$^{-6}$ l at 25°C and 765 torr

∴ Volume mixing ratio = 50.6 $\times$ 10$^{-6}$ (l) $\div$ 1000 (l)

                     = 50.6 $\times$ 10$^{-9}$

                     = 51 ppb

CHAPTER 8

# Chemistry of the Troposphere

R. M. HARRISON

## 8.1 INTRODUCTION

The atmosphere may conveniently be divided into a number of bands reflective of its temperature structure. These are illustrated by Figure 9.1 in Chapter 9. The lowest part, typically about 12 km in depth, is termed the troposphere and is characterized by a general diminution of temperature with height. The rate of temperature decrease, termed the lapse rate, is typically around 9.8 K km$^{-1}$ close to ground level but may vary appreciably on a short-term basis. The troposphere may be considered in two smaller components: the part in contact with the earth's surface is termed the boundary layer; above it is the free troposphere. The boundary layer is normally bounded at its upper extreme by a temperature inversion (a horizontal band in which temperature increases with height) through which little exchange of air can occur with the free troposphere above. The depth of the boundary layer is typically around 100 m at night and 1000 m during the day, although these figures can vary greatly. Pollutant emissions are generally into the boundary layer and are mostly constrained within it. Free tropospheric air contains the longer-lived atmospheric components, together with contributions from pollutants which have escaped the boundary layer, and from some downward mixing stratospheric air.

The average composition of the unpolluted atmosphere is given in Table 8.1. Some of the concentrations are very uncertain since:

(i)   analytical procedures for some components have only recently reached the stage where good data can be obtained.
(ii)  some components such as $CH_4$ and $N_2O$ are known to be increasing in concentration at an appreciable rate.
(iii) it is questionable whether any parts of the atmosphere can be considered entirely free of pollutants.

Table 8.1 also includes estimates of the lifetime of the various components. Those with lifetimes of a few days or less are cycled mainly within the boundary layer.

**Table 8.1** *Average composition of the dry unpolluted atmosphere (based upon Seinfeld[1] and Brimblecombe[2])*

| Gas | Average concentration (ppm) | Approx. residence time |
|---|---|---|
| $N_2$ | 780 820 | $10^6$ years |
| $O_2$ | 209 450 | 5000 years |
| Ar | 9340 | |
| Ne | 18 | not |
| Kr | 1.1 | cycled |
| Xe | 0.09 | |
| $CO_2$ | 360 | 100 years |
| CO | 0.12 | 65 days |
| $CH_4$ | 1.8 | 15 years |
| $H_2$ | 0.58 | 10 years |
| $N_2O$ | 0.31 | 120 years |
| $O_3$ | 0.01–0.1 | 100 days |
| $NO/NO_2$ | $10^{-6}$–$10^{-2}$ | 1 day |
| $NH_3$ | $10^{-4}$–$10^{-3}$ | 5 days |
| $SO_2$ | $10^{-3}$–$10^{-2}$ | 10 days |
| $HNO_3$ | $10^{-5}$–$10^{-3}$ | 1 day |

Components with longer lifetimes mix into the free troposphere more substantially and those with lifetimes of a year or more will penetrate the stratosphere to a significant degree. An indication of polluted air concentrations of some of these components is given in Chapter 7.

### 8.1.1 Pollutant Cycles

Pollutants are emitted from *sources* and are removed from the atmosphere by *sinks*. A typical cycle appears in Figure 8.1. Most pollutants have both natural and man-made sources; although the natural source is often of sizeable magnitude in global terms, on a local scale in populated areas pollutant sources are usually predominant.

Sink processes include both dry and wet mechanisms. Dry deposition involves the transfer and removal of gases and particles at land and sea surfaces without the intervention of rain or snow. For gases removed at the surface, dry deposition is driven by a concentration gradient caused by surface depletion; for particles this mechanism operates in parallel with gravitational settling of the large particles. The efficiency of dry deposition is described by the *deposition velocity*, $V_g$, defined as:

$$V_g (m\ s^{-1}) = \frac{\text{Flux to surface } (\mu g\ m^{-2}\ s^{-1})}{\text{Atmospheric concentration } (\mu g\ m^{-3})}$$

[1] J H. Seinfeld, 'Atmospheric Chemistry and Physics of Air Pollution', Wiley, New York, 1986.
[2] P. Brimblecombe, 'Air Composition and Chemistry', Second Edition, Cambridge University Press, Cambridge, 1996.

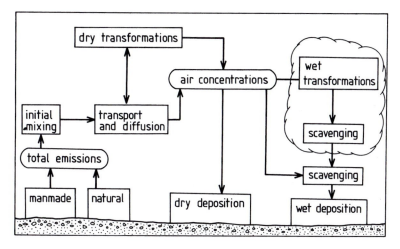

**Figure 8.1** *Typical atmospheric cycle of a pollutant* (Reprinted with permission from *Environ. Sci. Technol.*, 1988, **22**, 241, by permission of the American Chemical Society)

Some typical values of deposition velocity are given in Table 8.2. For gases, such as sulfur dioxide, which have a fairly high $V_g$, dry deposition has little influence upon near-source concentrations, but may appreciably influence ambient levels at large downwind distances.

Wet deposition describes scavenging by precipitation (rain, snow, hail, *etc.*) and is made up of two components, *rainout* which describes incorporation within the cloud layer, and *washout* describing scavenging by falling raindrops. The overall efficiency is described by the *scavenging ratio*, $W$, often rather misleadingly referred to as the Washout Factor.

$$W = \frac{\text{Concentration in rainwater (mg kg}^{-1})}{\text{Concentration in air (mg kg}^{-1})}$$

**Table 8.2** *Some typical values of deposition velocity*

| Pollutant | Surface | Deposition Velocity (cm s$^{-1}$) |
|-----------|---------|-----------------------------------|
| SO$_2$ | grass | 1.0 |
| SO$_2$ | ocean | 0.5 |
| SO$_2$ | soil | 0.7 |
| SO$_2$ | forest | 2.0 |
| O$_3$ | dry grass | 0.5 |
| O$_3$ | wet grass | 0.2 |
| O$_3$ | snow | 0.1 |
| HNO$_3$ | grass | 2.0 |
| CO | soil | 0.05 |
| Aerosol ($<2.5\ \mu$m) | grass | 0.15 |

**Table 8.3**   *Typical scavenging ratios*

| Species | W |
|---------|------|
| $Cl^-$ | 600 |
| $SO_4{}^{2-}$ | 700 |
| Na | 560 |
| K | 620 |
| Mg | 850 |
| Ca | 1890 |
| Cd | 390 |
| Pb | 320 |
| Zn | 870 |

Typical values of scavenging ratio are given in Table 8.3. A large value implies efficient scavenging, perhaps resulting from extensive vertical mixing into the cloud layer, where scavenging is most efficient. A related deposition process termed 'occult' deposition occurs when pollutants are deposited by fogwater deposition on surfaces. Pollutant concentrations in fogwater are typically much greater than in rainwater, hence the process may be significant despite the modest volumes of water deposited.

Another sink process involves chemical conversion of one pollutant to another (termed dry transformations in Figure 8.1). Thus atmospheric oxidation to sulfuric acid is a sink for sulfur dioxide. For many pollutants, a major sink is atmospheric reaction with the hydroxyl radical (OH). Such reactions are described later in this Chapter. Since pollutants are continually emitted into and removed from the atmosphere, they have an associated atmospheric lifetime or residence time defined below.

Many are chemically reactive and are transformed to other chemical species within the atmosphere. In some instances the products of such reactions, termed secondary pollutants, are more harmful than the primary pollutants from which they are formed. Thus an appreciation of atmospheric chemical processes is fundamental to any attempt to limit the adverse effects of air pollutant emissions.

Table 8.1 includes lifetimes (also termed residence times) of the atmospheric gases. In this context, lifetime, $\tau$, is defined as:

$$\tau = \frac{A}{F}$$

where $A$ = global atmospheric burden (Tg) (1 Tg = $10^{12}$ g)
    $F$ = global flux into and out of the atmosphere (Tg year$^{-1}$)

This treatment assumes a steady-state between the input and removal fluxes.

Sources of trace gases are not normally evenly spaced over the surface of the globe. Thus spatial variability in airborne concentrations occurs. For gases with an atmospheric lifetime comparable with the timescale of mixing of the entire troposphere (a year, or more), there is little spatial variation in concentration, as mixing processes outweigh the local variability in source strengths. For gases with short lifetimes, atmospheric mixing cannot prevent a substantially variable concentration. High spatial variability will also be associated with high temporal variability at one point as differing air mass sources and different mixing conditions will advect different concentrations of trace gas to the fixed receptor. An example of a rather well mixed gas of long lifetime is carbon dioxide which shows little variation in concentration over the globe and only small fluctuations at a given site (except very close to major combustion sources). At the other extreme, ammonia, which is chemically reactive and subject to efficient dry and wet deposition processes, is highly variable on both spatial and temporal scales. The spatial variability may be described by the coefficient of variation (equal to the standard deviation of the mean concentration divided by the mean) which relates to residence time as shown in Figure 8.2.

Early books on tropospheric chemistry tended to consider the cycle of one substance in isolation from those of others. It is now well recognized that many of the important atmospheric chemical cycles are closely inter-linked and that a more integrated approach to study is appropriate. In this context, the hydroxyl radical, a species not recognized as significant until recent years, has been recognized as having an immensely important role. It is responsible for the breakdown of many atmospheric pollutants, whilst its formation is dependent upon others. It is appropriate to commence a description of tropospheric chemistry with this short-lived free radical species.

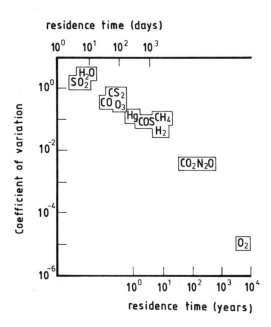

**Figure 8.2** *The relationship between coefficient of variation in spatially averaged concentration and residence time for atmospheric gases*

## 8.2   ATMOSPHERIC CHEMICAL TRANSFORMATIONS

### 8.2.1   The Importance of the Hydroxyl Radical (OH)

The primary source of OH in the background troposphere is photolysis of ozone by light of short wavelengths ($\lambda < 315$ nm) to form singlet atomic oxygen $O(^1D)$, which may either relax to the triplet state, $O(^3P)$ or may react with water vapour to form OH:

$$
\begin{array}{lll}
O_3 + h\nu & \rightarrow O(^1D) + O_2 \quad \lambda < 315 \text{ nm} & (1) \\
O(^1D) + M & \rightarrow O(^3P) + M & (2) \\
O(^1D) + H_2O & \rightarrow 2OH & (3)
\end{array}
$$

(M is an unreactive third molecule such as $N_2$)
Minor sources are available also through reactions of $O(^1D)$ with $CH_4$ and $H_2$:

$$
\begin{array}{lll}
CH_4 + O(^1D) & \rightarrow CH_3 + OH & (4) \\
H_2 + O(^1D) & \rightarrow H + OH & (5)
\end{array}
$$

Photolysis of both HONO and $H_2O_2$ produces OH directly:

$$
\begin{array}{lll}
HONO + h\nu & \rightarrow OH + NO \quad \lambda < 400 \text{ nm} & (6) \\
H_2O_2 + h\nu & \rightarrow 2OH \qquad\quad \lambda < 360 \text{ nm} & (7)
\end{array}
$$

Formation from nitrous acid may be of significance in polluted air. The route from hydrogen peroxide is not likely to represent a net source of OH since the main source of $H_2O_2$ is from the $HO_2$ radical:

$$
HO_2 + HO_2 \qquad \rightarrow H_2O_2 + O_2 \qquad\qquad (8)
$$

In polluted atmospheres, however, $HO_2$ is able to give rise to OH formation by a more direct route:

$$
HO_2 + NO \qquad \rightarrow NO_2 + OH \qquad\qquad (9)
$$

A review of tropospheric concentrations of the hydroxyl radical[3] found considerable variations in concentrations estimated by direct spectroscopic measurement, indirect measurement and modelling. There are also genuine variations with latitude, season and the presence of atmospheric pollutants. The overall consensus was of tropospheric concentrations within the ranges $(0.5–5) \times 10^6$ cm$^{-3}$ daytime mean and $(0.3–3) \times 10^6$ cm$^{-3}$ 24 hr mean. A seasonal variation of about threefold is suggested by model studies.[3]

As the following sections will demonstrate, the hydroxyl radical plays a central role, *via* the peroxy radicals with which it is intimately related, in the production of ozone and hydrogen peroxide. It also itself contributes directly to formation of

[3]C. N. Hewitt and R. M. Harrison, *Atmos. Environ.*, 1985, **19**, 545.

sulfuric and nitric acids in the atmosphere, as well as indirectly contributing *via* ozone and hydrogen peroxide. Thus atmospheric processes leading to ozone formation will also tend to favour production of other secondary pollutants, including the strong acids $HNO_3$ and $H_2SO_4$.

## 8.3 ATMOSPHERIC OXIDANTS

### 8.3.1 Formation of Ozone

Mid-latitude northern hemisphere sites show background ozone concentrations typically within the range 20–50 ppb. These concentrations were for many years attributed solely to downward transport of stratospheric ozone and the seasonal fluctuation, with a pronounced spring maximum and broad winter minimum,[4,5] relating to adjustments in the altitude of the tropopause.

In the late 1970s, Fishman and Crutzen showed that ozone could be formed from oxidation of methane and carbon monoxide in the troposphere in processes involving the hydroxyl radical:

$$CH_4 + OH \rightarrow CH_3 + H_2O \tag{10}$$
$$CH_3 + O_2 + M \rightarrow CH_3O_2 + M \tag{11}$$
$$CO + OH \rightarrow CO_2 + H \tag{12}$$
$$H + O_2 + M \rightarrow HO_2 + M \tag{13}$$

In the presence of NO:

$$CH_3O_2 + NO \rightarrow CH_3O + NO_2 \tag{14}$$
$$CH_3O + O_2 \rightarrow CH_2O + HO_2 \tag{15}$$
$$HO_2 + NO \rightarrow OH + NO_2 \tag{9}$$

Thus *via* reactions of $HO_2$ and $RO_2$ (R = alkyl), NO is converted to $NO_2$. Then:

$$NO_2 + h\nu \rightarrow NO + O(^3P) \quad \lambda < 435\ nm \tag{16}$$
$$O(^3P) + O_2 + M \rightarrow O_3 + M \tag{17}$$

The magnitude of this source of ozone is presently uncertain. However, recent evidence suggests that background northern hemisphere tropospheric ozone concentrations have approximately doubled since the turn of the century.[6] If this is indeed the case, the cause is almost certainly enhanced formation from the oxidation cycles of methane (whose concentration is known to be increasing), other less reactive hydrocarbons and carbon monoxide involving anthropogenic nitrogen oxides. Such a source is also consistent with a spring maximum in ozone caused by reaction of hydrocarbons accumulated through the less reactive winter months. Figure 8.3 shows near-surface ozone concentrations

[4] J. Fishman and P. J. Crutzen, *J. Geophys. Res.*, 1977, **82**, 5897.
[5] J. Fishman and P. J. Crutzen, *Nature (London)*, 1978, **274**, 855.
[6] D. Anfossi, S. Sandroni and S. Viarengo, *J. Geophys. Res.*, 1991, **96D**, 17349.

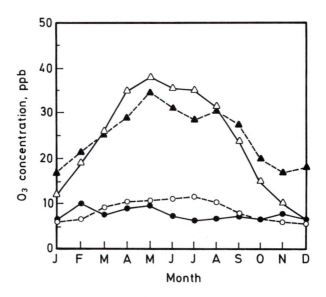

**Figure 8.3** *Average monthly concentrations of ozone at Moncalieri near Turin (Italy), open circles 1868–1893, and at Montsouris (France), solid circles 1876–1886, and at Arkona (Germany), solid triangles 1983, and Ispra (Italy), open triangles 1986–1989* (from Ref 6)

measured in France and Italy in the nineteenth century compared with recent data from Germany and Italy. These measurements provide strong evidence of an increase in ground-level ozone by a factor of approximately two over the past hundred years.

In polluted air there is abundant $NO_2$ whose photolysis leads to ozone formation. However, fresh emissions of NO lead to ozone removal and urban concentrations of ozone are usually lower than those in surrounding rural areas; the full cycle of reactions is:

$$NO_2 + h\nu \quad \overset{\mathcal{J}_1}{\rightarrow} NO + O(^3P) \quad \lambda < 435\,nm \quad (16)$$
$$O(^3P) + O_2 + M \rightarrow O_3 + M \quad (17)$$
$$NO + O_3 \quad \overset{k_3}{\rightarrow} NO_2 + O_2 \quad (18)$$

$\mathcal{J}_1$ and $k_3$ are the rate constants for the $NO_2$ photolysis and $O_3$ removal reactions respectively. All three reactions are rapid and an equilibrium is reached when the rate of ozone formation (equal to the rate of $NO_2$ photolysis if all $O(^3P)$ leads to $O_3$ formation) equals the rate of $O_3$ removal. Then:

$$\mathcal{J}_1[NO_2] = k_3[O_3][NO_2] \quad (19)$$
$$\text{and} \qquad [O_3] = \frac{\mathcal{J}_1[NO_2]}{k_3[NO]} \quad (20)$$

This is termed the *photostationary state*, and thus the ozone concentration is determined by the value of $J_1$, highest at peak sunlight intensity, and the ratio of $NO_2/NO$. Hydrocarbons, and to a lesser extent CO, play a crucial role in producing $HO_2$ and $RO_2$ radicals which convert NO to $NO_2$ *without* consumption of $O_3$. For example, propene is attacked by hydroxyl:

$$CH_3CH=CH_2 + OH \quad \rightarrow \quad CH_3CH-CH_2OH \tag{21}$$

$$CH_3CH-CH_2OH + O_2 \quad \rightarrow \quad \underset{\underset{OO}{|}}{CH_3CH-CH_2OH} \tag{22}$$

$$\underset{\underset{OO}{|}}{CH_3CH-CH_2OH} + NO \quad \rightarrow \quad \underset{\underset{O}{|}}{CH_3CH-CH_2OH} + NO_2 \tag{23}$$

$$\underset{\underset{O}{|}}{CH_3CH-CH_2OH} \quad \rightarrow \quad CH_2OH + CH_3CHO \tag{24}$$

$$CH_2OH + O_2 \quad \rightarrow \quad HCHO + HO_2 \tag{25}$$

In this case atmospheric photochemistry acts as a source of $O_3$ and aldehydes. It also leads to formation of $HO_2$, a source of hydrogen peroxide. None of the above processes is an effective free radical sink and hence, during hours of daylight, reactive free radicals such as OH and $HO_2$ are constantly recycled.

### 8.3.2 Formation of PAN

Peroxyacetyl nitrate (PAN) is of interest as a characteristic product of atmospheric photochemistry, as a probable reservoir of reactive nitrogen in remote atmospheres and because of its adverse effects upon plants. The formation route is *via* acetyl radicals ($CH_3CO$) formed from a number of routes, most notably acetaldehyde oxidation.

$$CH_3CHO + OH \quad \rightarrow \quad CH_3CO + H_2O \tag{26}$$

$$CH_3CO + O_2 \quad \rightarrow \quad CH_3C(O)OO \tag{27}$$

$$CH_3C(O)OO + NO_2 \quad \rightarrow \quad \underset{\underset{O}{\parallel}}{CH_3C-OONO_2} \tag{28}$$

$$\text{(PAN)}$$

In heavily polluted urban areas, subject to photochemical air pollution (*e.g.* Los Angeles), the concentrations of NO, $NO_2$, $O_3$, and other secondary pollutants tend to follow characteristic patterns, illustrated in Figure 8.4. Primary pollutants NO and hydrocarbons tend to peak with heavy traffic around 7 a.m. This is followed by a peak in $NO_2$ some time later when the atmosphere has developed sufficient oxidizing capability to oxidize the NO emissions of motor vehicles. The $NO_2/NO$ ratio is now high, favouring ozone production, and ozone peaks with peak sunlight intensity around noon, or a little

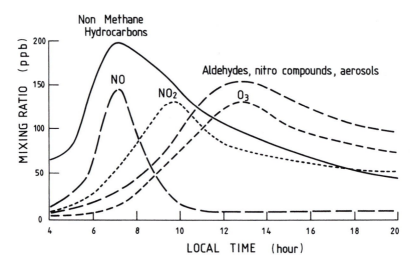

**Figure 8.4**   *Typical mixing ratio profiles, as a function of time of day, in the photochemical smog cycle*

later. In the United Kingdom, such diurnal profiles have been observed in urban areas during summer anticyclonic conditions, but peak ozone concentrations are generally rather later, around 3 p.m. The situation is more complex, however, as in the UK much of the ozone is generated during long-range transport of air pollution, often from continental European sources.

Perhaps surprisingly, UK rural sites exhibit diurnal variations in ozone which are remarkably similar to those at urban sites. Figure 8.5 shows average

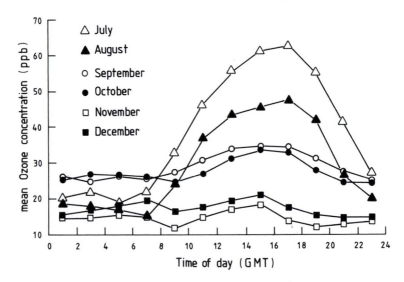

**Figure 8.5**   *Mean diurnal variations of ground-level ozone, July–December 1983 at Stodday, near Lancaster*[7]

[7]I. Colbeck and R. M. Harrison, *Atmos. Environ.*, 1985, **19**, 1577.

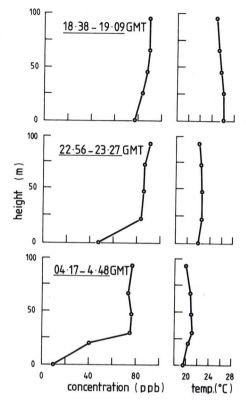

**Figure 8.6** *Vertical profiles of ozone and temperature on 11–12 July 1983*[8]

variations at Stodday, near Lancaster, for various months in 1983. This diurnal change is not due to the same causes as that in urban areas, where fresh NO emissions destroy ozone in the night-time. Measurement of vertical profiles of ozone and temperature (Figure 8.6) show that nocturnal depletion of ozone is a surface phenomenon due to dry deposition with little change in airborne concentrations at 50 m to 100 m altitude. The temperature profiles show the depletion occurs from a non-turbulent stable layer at the surface indicated by temperature rising with height (termed a temperature inversion). During the winter months, little diurnal change in ozone is seen (see Figure 8.5) since the atmospheric temperature structure shows a less marked change between night and day.

## 8.4 ATMOSPHERIC ACIDS

### 8.4.1 Weak Acids

Well-known weak acids in the atmosphere which may contribute to corrosion processes and influence the pH of precipitation are carbon dioxide and sulfur

[8] I. Colbeck and R. M. Harrison, *Atmos. Environ.*, 1985, **19**, 1807.

dioxide. Atmospheric $CO_2$ at a concentration of 340 ppm leads to an equilibrium pH of 5.6 at 15°C in otherwise unpolluted rainwater. This is normally taken as the boundary pH below which rain is considered acid. Sulfur dioxide is a stronger acid than $CO_2$ and at a concentration of only 5 ppb in air will, at equilibrium, cause a rainwater pH of 4.6 at 15°C.[2] In many instances, this pH is not attained due to severe kinetic constraints upon achievement of equilibrium, as is the case with many atmospheric trace gases. Dissolved $SO_2$ after oxidation may contribute appreciably to total sulfate and acidity in urban rainwater.

Organic acids such as formic acid may be formed in the atmosphere from formaldehyde:

$$HO_2 + HCHO \leftrightarrows HOOCH_2O \leftrightarrows OOCH_2OH \tag{29}$$
$$OOCH_2OH + NO \rightarrow OCH_2OH + NO_2 \tag{30}$$
$$OCH_2OH + O_2 \rightarrow HCOOH + HO_2 \tag{31}$$

Carboxylic acids are believed to contribute significantly to rainwater acidity in remote areas, although their contribution is not at all well quantified. In polluted regions they are unlikely to be of much importance.

### 8.4.2 Strong Acids

Strong acids of major importance in the atmosphere are as follows:

$H_2SO_4$
$HNO_3$
$HCl$
$CH_3SO_3H$ (methanesulfonic acid)

### 8.4.3 Sulfuric Acid

Atmospheric oxidation of $SO_2$ proceeds *via* a range of mechanisms. Consequently, dependent upon the concentrations of the responsible oxidants, the oxidation rate is extremely variable with space and time but typically lies around $1\% \, h^{-1}$.

Several gas phase mechanisms have been investigated including photo-oxidation, reaction with hydroxyl radical, Criegee biradical, ground state atomic oxygen, $O(^3P)$ and peroxy radicals. Table 8.4, based upon the treatment of Finlayson-Pitts and Pitts,[9] summarizes the rate data and clearly indicates the overwhelming importance of the hydroxyl radical reaction in the gas phase.

The mechanism of the $SO_2$–OH reaction is as follows:

$$SO_2 + OH \rightarrow HOSO_2 \tag{32}$$
$$HOSO_2 + O_2 \rightarrow HO_2 + SO_3 \tag{33}$$
$$SO_3 + H_2O \rightarrow H_2SO_4 \tag{34}$$

**Table 8.4**   *Homogeneous oxidation mechanisms for $SO_2$ (based on Ref 9)*

| Oxidizing species | Concentration* $(cm^{-3})$ | Rate constant $(cm^3\ molec^{-1}\ s^{-1})$ | Loss of $SO_2$ $(\%\ hr^{-1})$ |
|---|---|---|---|
| OH | $5 \times 10^6$ | $9 \times 10^{-13}$ | $1.6$ |
| $O_3$ | $2.5 \times 10^{12}$ | $<8 \times 10^{-24}$ | $<7 \times 10^{-6}$ |
| Criegee biradical | $1 \times 10^6$ | $7 \times 10^{-14}$ | $3 \times 10^{-2}$ |
| $O(^3P)$ | $8 \times 10^4$ | $6 \times 10^{-14}$ | $2 \times 10^{-3}$ |
| $HO_2$ | $1 \times 10^9$ | $<1 \times 10^{-18}$ | $<4 \times 10^{-4}$ |
| $RO_2$ | $3 \times 10^9$ | $<1 \times 10^{-18}$ | $<1 \times 10^{-3}$ |

*Assuming a moderately polluted atmosphere

In the presence of water droplets, which can take the form of fogs, clouds, rain or hygroscopic aerosols, sulfur dioxide will dissolve opening the possibility of aqueous phase oxidation. Upon dissolution, the following equilibria operate.

$$SO_{2(g)} + H_2O \leftrightarrows SO_2.H_2O \qquad (35)$$
$$SO_2.H_2O \leftrightarrows HSO_3^- + H^+ \qquad (36)$$
$$HSO_3^- \leftrightarrows SO_3^{2-} + H^+ \qquad (37)$$

These equilibria are sensitive to pH, and $HSO_3^-$ is the predominant species over the range pH 2–7. The other consequence of these equilibria is that the more acidic the droplet, the greater the degree to which the equilibria move towards gaseous $SO_2$ and limit the concentrations of dissolved S(IV) species. Some rate constants are pH-dependent in addition.

The major proposed oxidation mechanisms for $SO_2$ in liquid droplets include the following:

(i)   uncatalysed oxidation by $O_2$
(ii)  transition metal-catalysed oxidation by $O_2$
(iii) oxidation by dissolved oxides of nitrogen
(iv)  oxidation by ozone
(v)   oxidation by $H_2O_2$ and organic peroxides.

Relative rates for typical specified concentrations of reactive species are indicated in Table 8.5.

At high pH, all mechanisms in Table 8.5 are capable of oxidizing $SO_2$ at appreciable rates, generally far in excess of those observed in the atmosphere over a significant averaging period. Slower rates are observed, because of mass transfer limitations to the rate of introduction of $SO_2$ and oxidant to the water droplet and the previously noted effect of pH reduction (due to $H_2SO_4$ formation) on the solubility of $SO_2$. At pH 3, only the reaction with hydrogen peroxide is very fast, due to an increasing rate constant at reduced pH compensating for the lower

[9]B. J. Finlayson-Pitts and J. N. Pitts Jr., 'Atmospheric Chemistry', Wiley, New York, 1986.

**Table 8.5**  *Aqueous phase oxidation mechanisms for $SO_2$ (adapted from Ref 9)*

| Oxidant | Concentration* | Oxidation rate (% hr$^{-1}$) pH 3 | pH 6 |
|---------|----------------|------------------------------------|------|
| $O_3$ | 50 ppb (g) | $3 \times 10^{-2}$ | $5 \times 10^3$ |
| $H_2O_2$ | 1 ppb (g) | $8 \times 10^2$ | $5 \times 10^2$ |
| Fe-catalysed | $3 \times 10^{-7}$ M (1) | $2 \times 10^{-2}$ | $5 \times 10^1$ |
| Mn-catalysed | $3 \times 10^{-8}$ M (1) | $3 \times 10^{-2}$ | 7 |
| $HNO_2$ | 1 ppb (g) | $5 \times 10^{-4}$ | 3 |

*(g) and (1) denote gas and liquid phase concentrations respectively. Conditions are 5 ppb gas phase $SO_2$ at 25°C with no mass transfer limitations.

solubility of $SO_2$. Experimental studies indicate that this reaction can be very important in the atmosphere, but is limited by the availability of $H_2O_2$ which rapidly becomes depleted. The rate of formation of $H_2SO_4$ is therefore a function more of atmospheric mixing processes than of chemical kinetics in this case.

Several studies have emphasized the importance of $SO_2$ oxidation upon the surface of carbonaceous aerosols.[10-11]. Our own work,[12] including experiments at high relative humidity and in aqueous suspension, indicates that this reaction pathway is likely to make a negligible contribution to atmospheric sulfate formation under most conditions.

### 8.4.4  Nitric Acid

The main daytime route of nitric acid formation is from the reaction:

$$NO_2 + OH \rightarrow HNO_3 \tag{38}$$

The rate constant is $1.1 \times 10^{-11}$ cm$^{-3}$ molec$^{-1}$ s$^{-1}$ at 25°C[9] implying a rate of $NO_2$ oxidation of 19.8% hr$^{-1}$ at an OH concentration of $5 \times 10^6$ cm$^{-3}$ (*cf.* Table 8.4). This is thus a much faster process than gas phase oxidation of $SO_2$. This process is not operative during hours of darkness due to near zero OH radical concentrations.

At night-time, reactions of the $NO_3$ radical become important which are not operative during daylight hours due to photolytic breakdown of $NO_3$. The radical itself is formed as follows:

$$NO_2 + O_3 \rightarrow NO_3 + O_2 \tag{39}$$

[10]P. Middleton, C. S. Kiang and V. A. Mohnen, *Atmos. Environ.*, 1980, **14**, 463.
[11]S. G. Chiang, R. Toosi and T. Novakov, *Atmos. Environ.*, 1981, **15**, 1287.
[12]R. M. Harrison and C. A. Pio, *Atmos. Environ.*, 1983, **17**, 1261.

It is converted to $HNO_3$ by two routes. The first is hydrogen abstraction from hydrocarbons or aldehydes:

$$NO_3 + RH \rightarrow HNO_3 + R \tag{40}$$

Typical reaction rates imply formation of $HNO_3$ at about 0.3 ppb $hr^{-1}$ in a polluted urban atmosphere by this route.[9] This is modest compared to daytime formation from $NO_2$ and OH.

The other night-time mechanism of $HNO_3$ formation is *via* the reaction sequence:

$$NO_3 + NO_2 \overset{M}{\rightleftarrows} N_2O_5 \tag{41}$$
$$N_2O_5 + H_2O \rightleftarrows 2HNO_3 \tag{42}$$

The reaction involving water is rate determining and may be fairly slow at low relative humidity, contributing $HNO_3$ at $\sim 0.3$ ppb $hr^{-1}$.[9] As humidity increases, and especially in the presence of liquid water, more rapid reactions may be observed. This is presently supported only by indirect evidence and requires further experimental investigation.

Aqueous phase oxidation of $NO_2$ is of little importance due primarily to low aqueous solubility of $NO_2$.

### 8.4.5 Hydrochloric Acid

Hydrochloric acid differs from sulfuric and nitric acids in that it is emitted into the atmosphere as a primary pollutant and is not dependent upon atmospheric chemistry for its formation.

Some perspective of its significance as an acidic pollutant may be gained from Table 8.6 which shows the estimated emissions of $SO_2$, $NO_x$ and HCl in the UK and the potential hydrogen ion equivalent. At only 5.3%, the HCl contribution appears small until it is realized that this is immediately available acidity. Using diurnally averaged concentrations of reactive species, oxidations of $SO_2$ and $NO_2$ may be proceeding at only $\sim 1\%$ and 10% $hr^{-1}$ respectively, or less, and thus

**Table 8.6**  *Total UK acid emission*[13]

| Species | Emission (kt $a^{-1}$) | Potential $H^+$ equivalent (kt $a^{-1}$) | % total potential acidity |
|---------|---------|---------|---------|
| $SO_2$ | 3188 | 99.6 | 62.6 |
| $NO_x$ | 2347 | 51.0 | 32.1 |
| HCl | 310 | 8.5 | 5.3 |

[13]P. J. Lightowlers and J. N. Cape, *Atmos. Environ.*, 1988, **22**, 7.

over appreciable distances from a major source of all three pollutants (*e.g.* a power plant), HCl may be the predominant strong acid.

Another source, not considered in Table 8.6, may also contribute to HCl in the atmosphere. HCl is a more volatile acid than either $H_2SO_4$ or $HNO_3$ and thus may be displaced from aerosol chlorides, such as sea salt:

$$2NaCl + H_2SO_4 \rightarrow Na_2SO_4 + 2HCl \tag{43}$$

$$NaCl + HNO_3 \rightarrow NaNO_3 + HCl \tag{44}$$

This can be an appreciable source of HCl in areas influenced by maritime air masses.

### 8.4.6   Methanesulfonic Acid (MSA)

This strong acid is unlikely to be of importance in polluted regions but can contribute appreciably to acidity in remote areas. It is a major product of the oxidation of dimethylsulfide, $(CH_3)_2S$. A likely mechanism[9] is:

$$CH_3SCH_3 + OH \xrightarrow{M} \underset{\underset{OH}{|}}{CH_3SCH_3} \tag{45}$$

$$\underset{\underset{OH}{|}}{CH_3SCH_3} \rightarrow CH_3SOH + CH_3 \tag{46}$$

$$CH_3SOH + O_2 \xrightarrow{M} CH_3SO_3H \tag{47}$$

At night-time, the main breakdown mechanism of dimethylsulfide is reaction with $NO_3$ radicals. Dimethylsulfide is a product of biomethylation of sulfur in seawater and coastal marshes and its production shows a strong seasonal pattern, with peak emissions in spring and early summer.

Oxidation of dimethylsulfide can also lead to formation of sulfuric acid, both with and without the involvement of sulfur dioxide as an intermediate. Whether DMS oxidation gives MSA or sulfuric acid, the product is particulate and contributes to the number of cloud condensation nuclei. It has been postulated that biogenic dimethylsulfide may act as a climate regulator through this mechanism.[14]

## 8.5   ATMOSPHERIC BASES

Carbonate rock such as calcite or chalk, $CaCO_3$, or dolomite, $CaCO_3 . MgCO_3$, exist in small concentrations in atmospheric particles and provide a small capacity for neutralizing atmospheric acidity. In western Europe, however, the major atmospheric base is ammonia. Although not emitted to any major extent

[14]R. J. Charlson, J. E. Lovelock, M. O. Andreae and S. G. Warren, *Nature (London)*, 1987, **326**, 655.

by industry or motor vehicles, ammonia is arguably a man-made pollutant as it arises primarily from the decomposition of animal wastes; atmospheric concentrations relate closely to the density of farm animals in the locality. Release from chemical fertilizers can also be significant.[1]

In areas with a moderate or high ammonia source strength, ground-level atmospheric acidity is generally low. Sulfuric acid is present as highly neutralized $(NH_4)_2SO_4$ and $HNO_3$ and $HCl$ predominantly as $NH_4NO_3$ and $NH_4Cl$. The latter two salts are appreciably volatile and may release their parent acids under conditions of low atmospheric ammonia, high temperature or reduced humidity.

$$NH_4NO_3 \leftrightarrows HNO_3 + NH_3 \qquad (48)$$
$$NH_4Cl \leftrightarrows HCl + NH_3 \qquad (49)$$

The relative neutrality of ground-level air at such locations may, however, not be reflective of far greater acidity at greater heights above the ground, as is shown from simultaneous sampling of ground-level air and rainwater (see later). Additionally, once deposited in soils, ammonium salts are slowly oxidized to release strong acid in a process which may be represented as:

$$(NH_4)_2SO_4 + 4O_2 \rightarrow H_2SO_4 + 2HNO_3 + 2H_2O \qquad (50)$$

Thus the neutralization process has only a temporary influence and ultimately causes additional acidification.

Deposition of ammonia and ammonium may also contribute directly to damage to vegetation, which has proved to be a particular problem in the Netherlands. Since the reaction of ammonia with the OH radical is slow, the main sinks lie in wet and dry deposition processes.

## 8.6 ATMOSPHERIC AEROSOLS

Much of the atmospheric aerosol over populated areas is termed secondary as it is formed in the atmosphere from chemical reactions affecting primary pollutants. In the UK water-soluble materials account typically for about 60% of the aerosol by mass[15] and comprise nine, or ten, major ionic components:

Anions: $SO_4^{2-}$; $NO_3^-$; $Cl^-$; $(CO_3^{2-})$
Cations: $Na^+$; $K^+$; $Mg^{2+}$; $Ca^{2+}$; $NH_4^+$; $H^+$

If concentrations are expressed in gram equivalents per cubic metre of air, some interesting relationships appear:

$$\text{Concentration (g equiv m}^{-3}) = \text{Concentration (g m}^{-3}) \times \frac{Z}{M}$$

where $Z$ = Ionic charge and $M$ = Molecular weight.

[15] R. M. Harrison and C. A. Pio, *Environ. Sci. Technol.*, 1983, **17**, 169.

This is an expression of charge equivalents and hence if all major components are accounted for:

$$\Sigma \text{ anions} = \Sigma \text{ cations} \tag{51}$$

Typically for UK aerosol:

$$Na^+ \simeq 4.35 \; Mg^{2+} \tag{52}$$

and

$$Na^+ + Mg^{2+} \simeq Cl^- \tag{53}$$

when expressed in gram equivalents, which are relationships very similar to those pertaining in seawater, suggesting the latter as the major source of these components. In some instances land-derived $Mg^{2+}$ may be significant and upsets this relationship.

Another relationship commonly observed is:

$$SO_4{}^{2-} + NO_3{}^- \simeq NH_4{}^+ + H^+ \tag{54}$$

In UK air, the ratio $H^+/NH_4{}^+$ is normally very low, although in other parts of the world it can be much higher. This indicates the existence of $NH_4NO_3$ and $(NH_4)_2SO_4$ arising from ammonia neutralization of nitric and sulfuric acids respectively. Close examination of a large data set[14] revealed that the $NH_4{}^+$, which is not accounted for by the above relationship, and $Cl^-$, not accounted for by association with $Na^+$ and $Mg^{2+}$ in seawater, are approximately equal indicating the presence of $NH_4Cl$.

$K^+$ and $Ca^{2+}$ are mainly soil-derived, although some seawater contribution is likely, whilst $CO_3{}^{2-}$, not always observed, arises from carbonate minerals such as $CaCO_3$ in rocks and soils.

The water-insoluble fraction of atmospheric aerosols represents the minor part in the UK but may be far more significant in other countries, especially those with a dry climate. It comprises such materials as soil (*e.g.* clay materials), rock fragments (*e.g.* α-quartz) and elemental carbon from combustion processes.

Another approach to identifying the chemical compounds within the atmospheric aerosol is to use *X*-ray powder diffraction analysis. This technique is capable only of identifying crystalline components representing a significant proportion of the total mass; thus minor crystalline components and amorphous materials are not quantified. Table 8.7 shows the result of application of this method to seven winter samples and six spring/summer samples collected in Toronto, Canada.[16] A rather similar composition has been observed at other mid-latitude northern hemisphere sites, including some in the UK although the mineral composition is influenced by local geology. In the winter samples, a total

[16]W. T. Sturges, R. M. Harrison and L. A. Barrie, *Atmos. Environ.*, 1989, **23**, 1083.

**Table 8.7** *Abundance of major crystalline compounds identified in the atmosphere of Toronto, Canada* (from Ref 15)

| Compound | Concentration ($\mu$g m$^{-3}$) Summer | Winter |
|---|---|---|
| Total suspended particulate | $26.2 \pm 14.0$ | $19.4 \pm 8.1$ |
| $\alpha$–$SiO_2$ (quartz) | $1.37 \pm 0.67$ | $0.51 \pm 0.39$ |
| $CaCO_3$ (calcite) | $2.43 \pm 1.42$ | $1.19 \pm 0.86$ |
| $CaMg(CO_3)_2$ (dolomite) | $1.41 \pm 0.89$ | $0.71 \pm 0.64$ |
| $CaSO_4 . 2H_2O$ (gypsum) | $0.13 \pm 0.21$ | $0.32 \pm 0.45$ |
| NaCl (halite) | <d.l. | $1.40 \pm 1.00$ |
| $Al_{1-2} Si_{2-3}O_8$ (K,Na,Ca) (feldspar) | $1.07 \pm 0.07$ | $0.21 \pm 0.26$ |
| $(NH_4)_2SO_4$ | $3.36 \pm 2.90$ | $2.70 \pm 1.49$ |
| $3(NH_4)_2SO_4 . NH_4HSO_4$ | $0.38 \pm 0.60$ | $0.07 \pm 0.18$ |
| $CaSO_4 . (NH_2)_2SO_4 . H_2O$ | $0.58 \pm 0.51$ | $0.43 \pm 0.50$ |
| $PbSO_4 . (NH_4)_2SO_4$ | $0.14 \pm 0.13$ | $0.07 \pm 0.12$ |
| $(NH_4)_2SO_4 . 2NH_4NO_3$ | $0.25 \pm 0.22$ | $0.56 \pm 0.55$ |
| $(NH_4)_2SO_4 . 3NH_4NO_3$ | <d.l. | $0.86 \pm 1.67$ |
| $NH_4Cl$ | $0.01 \pm 0.02$ | $0.12 \pm 0.12$ |

of $52 \pm 4\%$ of the total suspended particulate matter was accounted for by the compounds listed; in the spring/summer samples it was $46 \pm 10\%$.

As Table 8.7 shows, $X$-ray powder diffraction is capable of identifying mixed salts (*e.g.* $(NH_4)_2SO_4 . 2NH_4NO_3$), which could not be distinguished from a mixture of $(NH_4)_2SO_4$ and $NH_4NO_3$ in the ionic balance work mentioned above. Lead, in the form of ammonium lead sulfate, $PbSO_4 . (NH_4)_2SO_4$ is observed in most localities where leaded gasoline is utilized. This is formed by a liquid phase reaction process occurring after coagulation of vehicle-emitted lead bromochloride, PbBrCl, with ambient ammonium sulfate, present at higher relative humidities as solution droplets:

$$2PbBrCl + 2(NH_4)_2SO_4 \rightarrow PbSO_4 . (NH_4)_2SO_4 + PbBrCl . (NH_4)_2BrCl \tag{55}$$

The other lead compound produced in this reaction, $PbBrCl.(NH_4)_2BrCl$ has also been identified in roadside air.[17]

The mixed salts of ammonium sulfate with ammonium nitrate and calcium sulfate are assumed to arise from coagulation mechanisms similar to that proposed above. Halite, NaCl, arises from marine sources at sites within 100 km or so of the sea. In Toronto, its presence in the winter, but not the summer, in the aerosol is suggestive of a source in road deicing salt. Quartz, calcite, dolomite, gypsum and feldspar are very commonly observed in atmospheric aerosols, arising from local rocks and soils and in some instances also from building materials and roads.

[17]P. D. E. Biggins and R. M. Harrison, *Environ. Sci. Technol.*, 1979, **13**, 558.

Recent years have seen a realization that aerosol particles can play a substantial active role in atmospheric chemistry, rather than being simply an inert product of atmospheric processes. They can act as a surface for heterogeneous catalysis or, probably more importantly, can provide a liquid phase reaction medium. Soluble particles deliquesce at humidities well below saturation and many compounds, such as the abundant $(NH_4)_2SO_4$, probably exist predominantly in solution droplet form in humid climates such as that of the UK. As indicated above these droplets can form a medium for reaction of solutes. They can also provide a medium for reaction with gaseous components, for example the dissolution of nitric acid vapour in sodium chloride aerosol, with release of hydrochloric acid vapour:

$$HNO_3 + NaCl \rightarrow NaNO_3 + HCl \qquad (44)$$

One fascinating aspect of aerosol chemistry is the fact that at typical tropospheric temperatures both ammonium nitrate and ammonium chloride aerosols are close to dynamic equilibrium with their gaseous precursors:

$$NH_4NO_3(\text{aerosol}) \rightleftharpoons NH_3(g) + HNO_3(g) \qquad (48)$$
$$NH_4Cl(\text{aerosol}) \rightleftharpoons NH_3(g) + HCl(g) \qquad (49)$$

The position of these equilibria may be predicted from chemical thermodynamics, both for crystalline particles, and at higher humidities for solution droplets.[18] Deviations from equilibrium probably arise from kinetic constraints upon the achievement of equilibrium. The implication for the UK, where ammonia levels are fairly high, is that the ammonia concentration at a given atmospheric temperature and relative humidity controls the nitric acid concentration *via* the ammonium nitrate dissociation equilibrium. It appears also that very similar considerations apply to the concentration of HCl vapour in the air which is subject to control by the ammonium chloride dissociation equilibrium.

Atmospheric aerosols are responsible for the major part of the visibility reduction associated with polluted air. In the UK and other heavily industrialized regions, visibility reduction tends to correlate closely with the concentration of aerosol sulfate, a component tending to be present in sizes optimum for scattering of visible light. In very dry climates, visibility reduction is primarily associated with dust storms and is thus correlated with airborne soil.

## 8.7 THE CHEMISTRY OF RAINWATER

Substances become incorporated in rainwater by a number of mechanisms. The major ones are as follows:

[18]A. G. Allen, R. M. Harrison and J. W. Erisman, *Atmos. Environ.*, 1989, **23**, 1591.

(a) particles act as cloud condensation nuclei, *i.e.* they act as centres for water condensation in clouds and fogs when the relative humidity reaches saturation
(b) particles are scavenged by cloudwater droplets or falling raindrops as a result of their relative motion
(c) gases may dissolve in water droplets either within or below the cloud.


As mentioned earlier, in-cloud scavenging is referred to as rainout, whilst below-cloud processes are termed washout. Incorporation in rain is a very efficient means of cleansing the atmosphere as evidenced by the substantial improvements in visibility occasioned by passage of a front.

The dissolved components of rainwater are the same nine or, at high rainwater pH, ten major ions listed earlier for atmospheric aerosol. There are also insoluble materials, again similar to the insoluble components of the atmospheric aerosol. When the composition of rainwater collected in the Lancaster, UK, area is compared with that of aerosol collected over the same periods close to ground-level, the average composition as percentages of total anion or cation load is very similar for many components, but is markedly different for some ions in aerosol and rainwater.[19] The most obvious difference is in the $H^+/NH_4^+$ ratio: rainwater has a far higher ratio and is thus much more acidic. This arises for two main reasons:


(a) the neutralizing agent, ammonia, has a ground-level source and is thus more abundant at ground-level where the aerosol is sampled than at cloud level where the major pollutant load is incorporated into the rain
(b) in-cloud oxidation processes for $SO_2$ (see above) may lead to appreciable acidification of the cloudwater.


The regional distribution of rainwater pH over the UK is shown by Figure 8.7. The trend of increasing acidity from west to east arises from the prevailing south-westerly circulation bringing relatively clean air into the west of the country. Primary pollutants emitted over the country and the secondary pollutants formed from them lead to acidification which increases progressively as the air moves eastwards. The quantitative deposition of acidity in rain (as $g\ H^+\ m^{-2}$ per year) shows a rather different pattern which arises from the fact that the annual deposition field is derived from the product of the volume-weighted mean concentration and the annual rainfall. Thus areas of high rainfall (*e.g.* Scottish Highlands) exhibit high hydrogen ion inputs despite a relatively high pH. One feature of acidic inputs is the substantial influence of episodicity: a very small number of rainfall events in any year combining high acidity with a large rainfall amount can contribute a large proportion of the total annual hydrogen ion input.

[19] R. M. Harrison and C. A. Pio, *Atmos. Environ.*, 1983, **17**, 2539.

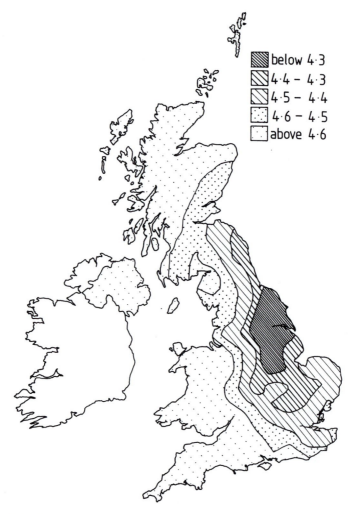

**Figure 8.7**  *Precipitation-weighted mean hydrogen ion concentration expressed as pH for 1986*[20]

## 8.8  ATMOSPHERIC COMPOSITION AND CLIMATE CHANGE

Some atmospheric gases, even at very low concentrations, can have important influences on the radiation budget of the atmosphere by absorbing infrared radiation emitted from the surface of the earth. As a consequence, more heat is retained in the lower atmosphere and thus there is a tendency towards gradual heating of the atmosphere, termed global warming. The gases with this property are termed *greenhouse gases* and the properties of the major contributors are listed in Table 8.8.

---

[20]Warren Spring Laboratory, 'United Kingdom Acid Rain Monitoring', Stevenage, 1988.

**Table 8.8** *A summary of key greenhouse gases affected by human activities*[21]

| | $CO_2$ | $CH_4$ | $N_2O$ | CFC-12 | HCFC-22 (a CFC substitute) | CF$_4$ (a perfluorocarbon) |
|---|---|---|---|---|---|---|
| Pre-industrial concentration | 280 ppmv | 700 ppbv | 275 ppbv | zero | zero | zero |
| Concentration in 1992 | 355 ppmv | 1714 ppbv | 311 ppbv | 503 pptv | 105 pptv | 70 pptv |
| Recent rate of concentration change per year (over 1980s) | 1.5 ppmv/yr | 13 ppbv/yr | 0.75 ppbv/yr | 18–20 pptv/yr | 7–8 pptv/yr | 1.1–1.3 pptv/yr |
| | 0.4%/yr | 0.8%/yr | 0.25%/yr | 4%/yr | 7%/yr | 2%/yr |
| Atmospheric lifetime (years) | (50–200)* | (12–17)† | 120 | 102 | 13.3 | 50 000 |

*No single lifetime for $CO_2$ can be defined because of the different rates of uptake by different sink processes

†This has been defined as an adjustment time which takes into account the indirect effect of methane on its own lifetime.

1 pptv = 1 part per trillion (million million) by volume

[21] Inter-governmental Panel on Climate Change, 'Radiative Forcing of Climate Change - Summary for Policy Makers', WMO, UNEP, 1995.

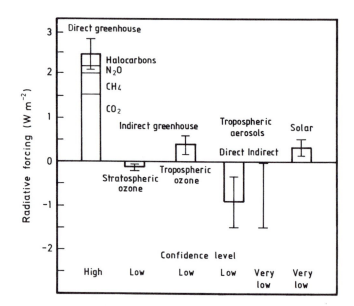

**Figure 8.8**  *Estimates of the globally averaged radiative forcing due to changes in greenhouse gases and aerosols from pre-industrial times to the present day and changes in solar variability from 1850 to the present day*[21]

The characteristic which the gases have in common is that they are of rather long atmospheric lifetime, and all are currently increasing in concentration. However, as a result of the Montreal Protocol the chlorofluorocarbons should soon peak and begin to decrease. The relative impacts of the gases per unit mass are expressed in terms of Global Warming Potentials. These are estimates of the radiative forcing per kilogram of substance relative to the effect of 1 kilogram of carbon dioxide. Thus, gases, such as CFC-12, although present in low concentration, have a relatively large Global Warming Potential (GWP) and thus can impact considerably upon climate. Different time frames of effect applying to GWPs are 20, 100 and 500 years to account for the differences in atmospheric longevity of the various substances and the fact that their relative impacts vary according to time into the future for which the prediction is made. 'Indirect effects' upon climate occur due to, for example, depletion of stratospheric ozone due to CFC emissions and enhancement of tropospheric ozone. Such effects, and those due to atmospheric aerosol are particularly hard to quantify.

Figure 8.8 shows the contribution globally averaged of the various greenhouse gases from pre-industrial times to the present day to radiative forcing. The main effect is the direct infrared absorbing property of the greenhouse gases. The estimated impact through stratospheric ozone is opposite in sign, but relatively small, and the indirect greenhouse effect of tropospheric ozone is smaller in magnitude than the direct effect of the greenhouse gases. There is a negative influence on radiative forcing due to aerosol particles in the lower atmosphere. These exert a direct effect by back scattering solar radiation to space, and hence act to cool the lower atmosphere. There is also a very poorly quantified indirect effect which arises from the ability of water soluble particles to act as cloud condensation nuclei, and in doing so to influence the albedo or reflectivity of clouds which can lead to reflection of solar radiation back to space. Currently, this effect is very poorly quantified indeed, and influences on cloud cover are one of the least well understood facets of climate change predicton. The final column in Figure 8.8 relates to changes in solar intensity between 1850 and the present day.

It is very difficult to discern from temperature records whether climate is actually changing. It is clear, however, that the last few decades have been notably warmer than the preceding 100 years and this is considered as evidence for global warming, although in the absence of a complete understanding of long, medium and short-term cycles in global climate, the interpretation cannot be certain.

# CHAPTER 9

# Chemistry and Pollution of the Stratosphere

I. COLBECK and A. R. MACKENZIE

## 9.1 INTRODUCTION

In the early 1970s the possibility of polluting the stratosphere and, in particular, depleting the ozone layer, was first raised. Concern centred on the potential adverse effects on human health and on the aquatic and terrestrial ecosystems, of allowing biologically harmful ultraviolet radiation to reach the ground. Among the chemical agents considered harmful were nitrogen oxides, produced by the detonation of nuclear weapons or from the exhaust of supersonic aircraft, and chlorofluorocarbons (CFCs) which were used mainly as refrigerants and aerosol propellants. Ten years later, concern that man's activities would affect significantly stratospheric ozone had diminished. Observations and models agreed; observed and predicted changes were less than 1 percent per decade. However, concern over ozone resurfaced abruptly in 1985 with the report by Farman *et al.*[1] that the total ozone column over Antarctica in September and October had decreased by up to 40% over the previous decade. Such a change had not been predicted by any model of future stratospheric composition. Scientific experiments – involving satellites, aircraft, balloons and ground stations – produced the first firm evidence that there had been an ozone decrease in the stratosphere over the heavily populated northern mid-latitudes. The annual average ozone concentration over mid-latitudes in the Northern Hemisphere has declined by more than 7 percent over the last 15 years, with the largest percentage loss in winter/spring.[2] Chemical compounds resulting from the breakdown of chlorine and bromine containing compounds are implicated.

Recent results from SESAME (Second European Stratospheric Arctic and Mid-latitude Experiment) provide clear evidence that ozone destruction has occurred within the lower stratosphere over the Arctic. The consequences of an

[1] J. C. Farman, B. G. Gardiner and J. D. Shanklin, 'Large Losses of Total Ozone Reveal Seasonal $ClO_x/NO_x$ Interactions', *Nature*, 1985, **315**, 207–210.
[2] WMO (World Meteorological Organization), 'Scientific Assessment of Ozone, Depletion: 1994', Global Ozone Research and Monitoring Program Report 37, WMO, Geneva, Switzerland, 1994.

ozone decline over the Arctic are potentially greater than that over the Antarctic. An Arctic ozone hole could endanger some of the most populated parts of the globe, including northern Europe, North America and the former Soviet Union.

Intensive theoretical and experimental investigations of stratospheric ozone have been underway for many years. Laboratory studies have improved the accuracy of reaction rates and have identified significant species and reactions which had been omitted from numerical models. Many trace constituents of the atmosphere are now measured from balloon-borne platforms and satellites, providing much needed data on the temporal and spatial variability of chemical species. If we want to protect our environment, it is essential that we know more about the various atmospheric trace species and their impact on chemical processes in the atmosphere. Whereas acid rain and photochemical oxidant pollution are somewhat localized environmental problems, modification of stratospheric ozone is a global phenomenon. Recognition of this fact has led to the implementation of the Montreal Protocol and its subsequent amendments.

Ozone, particularly its absorption of solar ultraviolet radiation, has a profound impact on the structure of the atmosphere. Figure 9.1 shows a typical mid-latitude temperature profile. Most ozone occurs in the stratosphere, and it is the

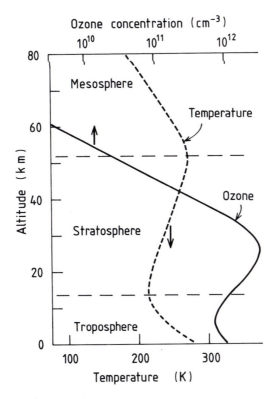

**Figure 9.1** *Ozone distribution and temperature profile in the atmosphere*

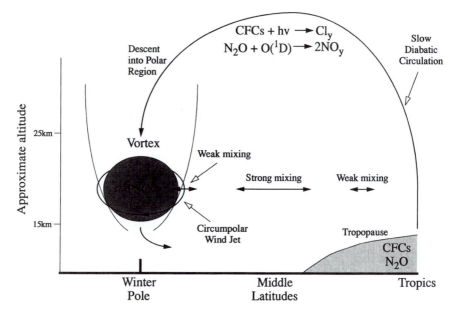

**Figure 9.2** *Schematic of the mean diabatic circulation in the wintertime hemisphere (in this instance, the south). Source gases enter the stratosphere in the tropics and are transported slowly to high altitudes where they are broken down by short wavelength light or the electronically excited $O(^1D)$ atom. Descent into the polar regions brings down high mixing ratios of $NO_y$, $Cl_y$ and $H_2O$. Clouds in polar regions activate chlorine and so bring about ozone depletion. A strong jet of westerly winds is present at the edge of the polar vortex, preventing rapid horizontal mixing. A similar barrier to mixing exists in the subtropics. Figure courtesy of Adrian Lee (Centre for Atmospheric Science, Cambridge)*

presence of ozone which produces the characteristic temperature inversion and static stability (*i.e.* hot, light air above colder, denser air) of the stratosphere. The temperature structure, and implied stability, of the stratosphere distinguishes it from the troposphere below, where heating at the surface causes more rapid vertical mixing. The stratosphere extends to a lower limit of 8 km to 16 km, depending on latitude (Figure 9.1), and extends to an upper limit of 45 km to 60 km, again depending on latitude. It should be noted that, due to the stability of the stratosphere against vertical mixing, pollutants in the stratosphere have a lifetime for removal by transport of several years, and hence can build up to globally damaging levels.

The circulation in the winter stratosphere is shown schematically in Figure 9.2. There is a slow overturning[3] that brings air upwards, from the troposphere, in the tropics and downwards, from the mesosphere and upper stratosphere, at

[3]M. E. McIntyre, 'Atmospheric Dynamics: Some Fundamentals, with Observational Implications', in Proc. Int. School Phys., 'Enrico Fermi' VXV Course, The Use of EOS for Studies of Atmospheric Physics', eds. J. C. Gille and G. Visconti, International School of Physics, North Holland, 1992, pp. 313–386.

the pole. At a given altitude, therefore, air is 'older', in the sense that it has been in the upper atmosphere for longer, near the winter pole and the air is 'younger' near the equator. Since CFCs are destroyed in the upper stratosphere and mesosphere, air near the winter pole is richer* in the breakdown products of CFC photodissociation than is air near the equator. This gradient in the 'age' of stratospheric air is particularly steep in the subtropics and at the edge of the polar vortex, where barriers to horizontal mixing exist. The barrier at the polar vortex edge is an important reason for the exacerbated ozone loss in the polar lower stratosphere. Between the mixing barriers, in the mid-latitude *surf zone*, quasi-horizontal stirring and mixing, on potential temperature† surfaces, is rapid.

## 9.2 STRATOSPHERIC OZONE CHEMISTRY

Over 100 years ago, Hartley[4] pointed out that ozone is a normal constituent of the higher atmosphere and is in a larger proportion there than near the earth's surface. We now know that ozone occurs in trace amounts throughout the atmosphere with a peak concentration in the lower stratosphere between about 20 and 25 km altitude (Figure 9.1). About 90% of all the ozone in the atmosphere resides in the stratosphere. Observations of the total amount of ozone in a vertical column have established the average column abundance and the general pattern of latitudinal, longitudinal, seasonal and meteorological variations. Total ozone, per unit area, is usually stated in *Dobson Units* or matm cm, related to the thickness of an equivalent layer of pure ozone at standard pressure and temperature. If the ozone in a column of the atmosphere were concentrated into a thin shell surrounding the earth at atmospheric pressure, it would be about 3 mm thick *i.e.* the average total amount of ozone is 300 DU. Seasonal and latitudinal variations at subpolar latitudes are about $\pm 20\%$ of this value. The annual average total ozone is a minimum of approximately 260 DU at equatorial latitudes and increases poleward in both hemispheres to a maximum at sub-polar latitudes of about 400 DU. The high latitude maximum results from transport of ozone from the equatorial middle and upper stratosphere, the region of primary production, to the polar lower stratosphere (Figure 9.2).

---

[4]W. N. Hartley, 'On the Absorption Spectrum of Ozone', *J. Chem. Soc.*, 1881, **39**, 57–60.

*Two measures of local abundance are used in this chapter. Absolute abundances are reported as *partial pressures*, $P_x$, in nbar. These are related to the other common unit of concentration, molec cm$^{-3}$, by the Ideal Gas Law. Relative abundances are reported as *volume mixing ratios*, $P_x/P_{air}$, expressed as parts per million (ppmv, vmr $\times$ 10$^{-6}$), parts per billion (ppbv, vmr $\times$ 10$^{-9}$) or parts per trillion (pptv, vmr $\times$ 10$^{-12}$). Because pressure decreases exponentially with height, the absolute concentration of chemicals in air parcels moving with the mean circulation will change, but the relative abundance will be conserved in the absence of chemical reactions.
†Potential temperature, $\theta$, is defined as $\ln \theta = 0.286 \times \ln(T(1000/p))$, where T is the temperature in Kelvin, and p the atmospheric pressure (in mbar), at the altitude of interest. $\theta$ is a measure of the entropy of air and so is conserved in adiabatic motion. The stable stratification of the stratosphere is reflected in the fact that air tends to move on *isentropic* surfaces of constant $\theta$.

## 9.2.1   Gas Phase Chemistry

How is this ozone produced? Chapman[5] proposed a dynamically static, pure oxygen, photochemical steady state model that agreed well with the observations available at the time. The reactions were:

$$O_2 + hv(\lambda < 243 \text{ nm}) \rightarrow O + O \tag{1}$$
$$O + O_2 + M \rightarrow O_3 + M \tag{2}$$
$$O_3 + hv(\lambda < 1180 \text{ nm}) \rightarrow O + O_2 \tag{3}$$
$$O + O_3 \rightarrow O_2 + O_2 \tag{4}$$

(a fifth reaction, the self-reaction of oxygen atoms, is too slow to be important in the stratosphere). Reaction 2 becomes slower with increasing altitude, since M, *i.e.* pressure, is decreasing, while Reactions 1 and 3 become faster, since the intensity of radiation, particularly short wavelength radiation, is increasing. Hence, O predominates at high altitudes and $O_3$ is favoured at lower altitudes. The reaction rate for loss of $O_3$ (Reaction 4) was unknown at the time of Chapman's proposal. Laboratory measurements have since revealed that Reaction 4 is too slow to destroy ozone at the rate it is produced globally. Therefore, for ozone to attain steady state, there must exist chemical processes which provide more efficient routes for the loss of odd oxygen. The reaction sequence:

$$X + O_3 \rightarrow XO + O_2 \tag{5}$$
$$XO + O \rightarrow X + O_2 \tag{6}$$
$$net \ O + O_3 \rightarrow 2O_2$$

achieves the same result as Reaction 4. The species X is a catalyst – is not consumed in the process – and so can effectively destroy many ozone molecules before being removed by some other chemical process. Several species have been suggested for X, including H, OH, NO, Cl and Br. The catalytic cycles are kinetically labile and will be able to compete with the direct Reaction 4, even when concentrations of X and XO are two or three orders of magnitude less than concentrations of the oxygen species. Because the interconversion of the various X/XO couples can be very rapid, it is often convenient to group the species together in *chemical families*. This is a particularly important concept in the numerical modelling of stratospheric chemistry and transport, since it allows equilibrium assumptions to be made between members of each chemical family, greatly reducing the number of reactive species to be integrated forward in time, and greatly increasing the timestep* which can be used in the integration. The

[5]S. C. Chapman, 'A Theory of Upper Atmospheric Ozone', *Mem. Roy. Met. Soc.*, 1930, **3**, 103–125.

*Numerical models of atmospheric chemistry are comprised of a set of simultaneous ordinary differential equations, one for each chemical species or family. The models solve for the chemical concentrations at a time, t, by integrating forward from initial conditions at time, $t_0$. To do this, the differential equations are approximated by difference formulae which apply over a timestep dt. The larger dt is, the faster the solution is reached, but if dt is too large relative to the rate of change of the chemical species, then the difference formulae will not be good approximations to the differential equations. Using families removes the fastest rates of change from the set of equations and so allows the use of larger dt.

various short-lived nitrogen compounds are grouped together as $NO_x (= N, NO,$ $NO_2, NO_3)$, the various short-lived chlorine species are grouped together as $ClO_x$ $(= Cl, ClO, Cl_2O_2)$, short-lived hydrogen species are grouped as $HO_x (= H,$ $HO, HO_2)$, and so on. By extension, the total amount of potentially reactive nitrogen, is known collectively as $NO_y (= NO_x, HONO, N_2O_5, HNO_3, HNO_4,$ $ClONO_2)$, and the total amount of potentially reactive chlorine, the *inorganic chlorine*, is known collectively as $Cl_y (= ClO_x, OClO, HOCl, HCl, ClONO_2)$.

In addition to Reactions 5 and 6, ozone may also be destroyed in cycles not involving atomic oxygen. Such, ozone-specific, cycles are important in the lower stratosphere because concentrations of atomic oxygen are low there. Variations on the general form are: (i) formation of an XO species which itself reacts with ozone, for example:

$$OH + O_3 \rightarrow HO_2 + O_2 \tag{7}$$
$$HO_2 + O_3 \rightarrow OH + 2O_2 \tag{8}$$
$$net\ 2O_3 \rightarrow 3O_2$$

and

$$NO + O_3 \rightarrow NO_2 + O_2 \tag{9}$$
$$NO_2 + O_3 \rightarrow NO_3 + O_2 \tag{10}$$
$$NO_3 + hv \rightarrow NO + O_2 \tag{11}$$
$$net\ 2O_3 + hv \rightarrow 3O_2$$

or (ii) formation of a compound from two XO species, leading to the recombination of oxygen, for example, elimination of $O_2$ in a thermal reaction:

$$Br + O_3 \rightarrow BrO + O_2 \tag{12}$$
$$OH + O_3 \rightarrow HO_2 + O_2 \tag{7}$$
$$HO_2 + BrO \rightarrow HOBr + O_2 \tag{13}$$
$$HOBr + hv \rightarrow OH + Br \tag{14}$$
$$net\ 2O_3 + hv \rightarrow 3O_2$$

or elimination of $O_2$ by photolysis of a different bond to the bond formed in the XO/YO combination:

$$Cl + O_3 \rightarrow ClO + O_2 \tag{15}$$
$$NO + O_3 \rightarrow NO_2 + O_2 \tag{9}$$
$$ClO + NO_2 + M \rightarrow ClONO_2 + M \tag{16}$$
$$ClONO_2 + hv \rightarrow Cl + NO_3 \tag{17}$$
$$NO_3 + hv \rightarrow NO + O_2 \tag{11}$$
$$net\ 2O_3 + 2hv \rightarrow 3O_2$$

A very important example of this last kind of ozone-specific cycle is the *ClO dimer cycle* which is responsible for the majority of the lower stratospheric ozone loss inside the polar vortices (see below):

$$2(Cl + O_3 \rightarrow ClO + O_2) \tag{15}$$
$$ClO + ClO + M \rightarrow Cl_2O_2 + M \tag{18}$$
$$Cl_2O_2 + hv \rightarrow Cl + ClOO \tag{19}$$
$$ClOO + M \rightarrow Cl + O_2 + M \tag{20}$$
$$net\ 2O_3 + hv \rightarrow 3O_2$$

Species X and XO may be involved in *null cycles* that do not remove odd oxygen. For X = NO we have:

$$NO + O_3 \rightarrow NO_2 + O_2 \tag{9}$$
$$NO_2 + hv \rightarrow NO + O \tag{21}$$
$$net\ O_3 + hv \rightarrow O_2 + O$$

This cycle is in competition with the catalytic cycle (Reactions 9, 10 and 11) and the $NO_x$ tied up in this cycle is ineffective as a catalyst. All the ozone destroying cycles above are in competition with null cycles and termination reactions that occur simultaneously. The termination of the catalytic and null cycles occurs by chemical conversion of radicals to more stable oxidation products. For example:

$$OH + NO_2 + M \rightarrow HNO_3 + M \tag{22}$$

The $HNO_3$ may be transported down into the troposphere and removed in rain or may be photolysed to regenerate $OH + NO_2$. This latter process is relatively slow, and $HNO_3$ is said to act as a *reservoir* of $NO_x$. Typically, in the wintertime lower stratosphere, more than 90% of the stratospheric load of $NO_y$ is stored in this $HNO_3$ reservoir.[6] Under the same conditions, about 40–80% of stratospheric $Cl_y$ is sequestered into $HCl$[7] via:

$$Cl + CH_4 \rightarrow CH_3 + HCl \tag{23}$$

The chlorine tied up as stable $HCl$ may be released by:

$$OH + HCl \rightarrow H_2O + Cl \tag{24}$$

or by heterogeneous reactions (see below). This $ClO_x$ can then again participate in catalytic cycles leading to the removal of ozone. Extensive research has emphasized the importance of more temporary reservoir species such as $ClONO_2$, $HOCl$, $N_2O_5$ and $HO_2NO_2$, which act to lessen the efficiency of $ClO_x$ and $NO_x$ species in destroying ozone in the lower stratosphere. These temporary reservoir species may be formed by reactons 10, 16 and:

[6] A. J. Weinheimer, J. G. Walega, B. A. Ridley, B. L. Gary, D. R. Blake, N. J. Blake, F. S. Rowland, G. W. Sachse, B. E. Anderson and J. E. Collins, 'Meridional Distributions of $NO_x$, $NO_y$, and Other Species in the Lower Stratosphere and Upper Troposphere during AASE II', *Geophys. Res. Lett.*, 1994, **23**, 2583–2586.

[7] C. R. Webster, R. D. May, L. Jaegle, H. Hu, S. P. Sander, M. R. Gunson, G. C. Toon, J. M. Russell, R. M. Stimpfle, J. P. Koplow, R. J. Salawitch and H. A. Michelsen, 'Hydrochloric Acid and the Chlorine Budget of the Lower Stratosphere', *Geophys. Res. Lett.*, 1994, **21**, 2575–2578.

$$NO_3 + NO_2 + M \rightarrow N_2O_5 + M \tag{25}$$
$$ClO + HO_2 \rightarrow HOCl + O_2 \tag{26}$$
$$HO_2 + NO_2 + M \rightarrow HO_2NO_2 + M \tag{27}$$

These reactions emphasize the coupling between the $HO_x$, $NO_x$ and $ClO_x$ families, and mean that the effects of the families are not additive. Members of one family can react with members of another to produce null cycles, chiefly by the production of $NO_2$:

$$NO + ClO \rightarrow NO_2 + Cl \tag{28}$$
$$HO_2 + NO \rightarrow NO_2 + OH \tag{29}$$

which can be followed by Reaction 21. The recycling of $HO_x$, $NO_x$ and $ClO_x$ from the reservoirs $HNO_3$, $N_2O_5$, $HOCl$ and $ClONO_2$ is by photolysis or heterogeneous reaction. Removal of $HO_x$ in the lower stratosphere occurs mainly by:

$$OH + HNO_3 \rightarrow H_2O + NO_3 \tag{30}$$
$$OH + HO_2NO_2 \rightarrow H_2O + NO_2 + O_2 \tag{31}$$

These reactions also release $NO_x$ from the reservoirs. The water formed may be physically removed from the stratosphere, or converted back into $HO_x$ by reaction.

Termination reactions become less efficient with increasing molcular weight in the halogen series, *i.e.* from Cl to I. The reaction of Cl with $CH_4$ (Reaction 23), to produce HCl, limits the abundance of active chlorine in the stratosphere, but the analogous reaction of Br with $CH_4$ is endothermic and can be neglected. The temporary reservoirs for bromine – HOBr and $BrONO_2$ – are very photolabile. As a result, BrO is the major form of $Br_y$ in the stratosphere. Therefore, on a molecule for molecule basis, bromine is a much more efficient sink for $O_3$ than is chlorine. Measurements of BrO show mixing ratios around $5 \times 10^{-3}$ ppbv at 18 km. These are sufficient to account for about 25% of the ozone depletion in the Antarctic ozone hole,[8,9] and to make bromine cycles significant contributors to mid-latitude ozone loss. Reactions involving iodine radicals have also been suggested recently.[10] As yet there are no measurements of iodine in the contemporary stratosphere, but models suggest that as little as $10^{-3}$ ppbv could have a significant effect on ozone concentrations.

Figure 9.3 shows a model calculation of the ozone loss rate due to the various chemical families ($HO_x$, $NO_x$, $ClO_x$, $BrO_x$) as a function of altitude, for vernal equinox at $40°N$.[11] These results show a smaller direct influence of $NO_x$-related

[8]M. B. McElroy, R. J. Salawitch, S. C. Wofsy and J. A. Logan, 'Reductions of Antarctic Ozone due to Synergistic Interactions of Chlorine and Bromine', *Nature*, 1986, **321**, 759–762.
[9]WMO (World Meteorological Organization), 'Scientific Assessment of Ozone Depletion: 1991', Global Ozone Research and Monitoring Program Report 25, WMO, Geneva, Switzerland, 1991.
[10]S. Solomon, R. R. Garcia and A. R. Ravishankara, 'On the Role of Iodine in Ozone Depletion', *J. Geophys. Res.*, 1994, **99**, 20491–20500.
[11]R. R. Garcia and S. Solomon, 'A New Numerical Model of the Middle Atmosphere, 2, Ozone and Related Species', *J. Geophys. Res.*, 1994, **99**, 12937–12952.

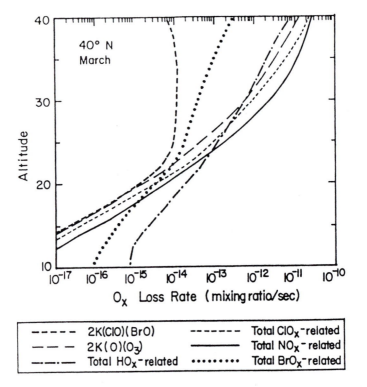

**Figure 9.3**   *Calculated 24-hour average $O_x$ loss rates from various chemical cycles for 40° N in March for non-volcanic aerosol loading. In these circumstances the dominant ozone loss process below 22 km is due to reactions involving $HO_x$, with $NO_x$ dominating between 23 km and 40 km. Under higher aerosol loading, coupled $HO_x$-halogen cycles become more important.* From the two dimensional model simulations of Garcia and Solomon (1994)[11]

cycles in the lower stratosphere than had been calculated previously (*cf.* for instance, Figure 4.6 in Wayne, 1991, p. 128[12]), due to the inclusion of heterogeneous processes in the model (see below).

Hydrocarbon oxidation is closely related to all other reactive trace gas species and hence ozone photochemistry. Methane is the dominant stratospheric hydrocarbon and is a chemical source of water in the stratosphere, *via*:

$$OH + CH_4 \rightarrow CH_3 + H_2O \qquad (32)$$

which is analogous to the Cl loss process (Reaction 23) above. The $CH_3$ radical is further oxidized to $CO_2$ through intermediate products $CH_3O_2$, $CH_3O$, HCHO and CO. Below 35 km there is sufficient NO for $CH_4$ oxidation to be a net source of $O_x$ (O + $O_3$) by the 'photochemical smog' reactions:

[12]R. P. Wayne, 'Chemistry of Atmospheres', 2nd ed., Oxford University Press, Oxford, 1991.

$$CH_3O_2 + NO \rightarrow CH_3O + NO_2 \tag{33}$$
$$NO_2 + hv \rightarrow O + NO \tag{21}$$
$$O + O_2 + M \rightarrow O_3 + M \tag{2}$$

Although Reaction 32 is a sink for OH, methane oxidation is also a source of $HO_x$ through the reactions:

$$CH_3O + O_2 \rightarrow HO_2 + HCHO \tag{34}$$
$$HCHO + hv \rightarrow H + HCO \tag{35}$$
$$H + O_2 + M \rightarrow HO_2 + M \tag{36}$$
$$HCHO + hv \rightarrow H_2 + CO \tag{37}$$

and further oxidation of CO.

## 9.2.2 Heterogeneous Chemistry

Aerosol particles provide sites for reactions which would otherwise not occur in the stratosphere. There are many types of aerosol in the stratosphere, ranging from meteoritic mineral debris to water droplets.[13] Of primary importance in our current understanding of the chemistry of the stratosphere are the sulfuric acid aerosol and polar stratospheric clouds (PSCs).

Sulfuric acid aerosol occurs in a wide altitude band, between about 15 and 30 km, throughout the globe. The aerosol is formed by inputs of $SO_2$ from volcanoes and COS from microbes. The volcanic influence makes the amount of aerosol highly variable in time: the eruption of Mt. Pinatubo in June 1991 increased the mass of aerosol in the stratosphere by more than an order of magnitude (from a total loading of around 2 Tg to around 30 Tg). The timescale for removal of the aerosol is of the order of 11 months,[2] so that these volcanic injections can have a profound global effect for some time (see below). Above temperatures of about 215 K, at 50 mbar, the aerosol exists as concentrated sulfuric acid droplets. Below this temperature solid sulfuric acid hydrates are the thermodynamically favoured phase, but there is a considerable kinetic barrier to freezing which enables liquid particles to remain until the temperature approaches the frost point[14] taking up water and nitric acid as they cool. The frost point for stratospheric air at 50 mbar is about 188 K.

Polar stratospheric clouds fall into two categories: type 1 PSCs are composed of nitric acid and water in volume ratios between 1:2 and 1:5, and form at temperatures about 2–5 K above the local frost point; type 2 PSCs are composed of water ice crystals and form, therefore, below the local frost point. PSCs are thought to form on the background sulfuric acid aerosol, which may, or may not, be frozen at the time. A complex hysteresis cycle of particle formation and growth results. It is important to understand the details of particle evolution in the lower

[13]O. B. Toon and N. H. Farlow, 'Particles Above the Tropopause', *Ann. Rev. Earth Planetary Sci.*, 1981, **9**, 19–58.

[14]K. S. Carslaw, B. P. Luo, S. L. Clegg, Th. Peter and P. Brimblecombe, 'Stratospheric Aerosol Growth and $HNO_3$ Gas Phase Depletion from Coupled $HNO_3$ and Water Uptake by Liquid Particles', *Geophys. Res. Lett.*, 1994, **21**, 871–874.

stratosphere because the different particles have different reactivities. The thermodynamically favoured solid phase for sulfuric acid aerosol is sulfuric acid tetrahydrate (SAT), and that for nitric acid aerosol is nitric acid trihydrate (NAT).

Reactions involving particles are more complicated to include in models than the simple molecular collisons occurring in the gas phase. Diffusion is important, and reactants partition between the gas and condensed phases according to their solubilities. In liquid particles, if reaction is slow, the reactants will have time to continuously adjust to their equilibrium partitioning (*e.g.* Cox *et al.*[15]). If the rate coefficient for reaction is faster than about $10^5$ $M^{-1}$ $s^{-1}$, the rate of reaction becomes limited by transport of reactants into the particle and the effective volume available for reaction is reduced[16] until, in the limit, the rate is dependent on the aerosol surface area rather than the volume. For solid particles, reaction rates depend linearly on the particle surface area.

The overall effect of heterogeneous reactions is to convert chlorine and hydrogen reservoirs into more active forms whilst converting $NO_x$ into nitric acid. Many of the heterogeneous reactions in the sulfuric acid aerosol are strongly temperature dependent, due to the variation of aerosol composition, and Henry's Law coefficients, with temperature. For aerosol loadings typical of a volcanically quiescent period, heterogeneous reactions begin to activate chlorine at temperatures below about 200 K. This activation is relatively slow, however. When aerosol loadings are much increased following a volcanic eruption, the temperature at which chlorine activation begins is increased by up to 5 K and aerosol reactions below 200 K can effectively compete with reactions on type 1 PSCs. One heterogeneous reaction which is not temperature dependent is the hydrolysis of $N_2O_5$, and it is this reaction which has led to a revision of the importance of $NO_x$-related catalytic cycles in the lower stratosphere (Figure 9.3).

In the polar regions, when temperatures fall below about 195 K, PSCs can form. These have a much larger total volume than the background aerosol, and so can convert chlorine reservoirs much more rapidly. At PSC temperatures, $Cl_2$ partitions into the gas phase, whereas $HNO_3$ remains in the particles. This re-partitioning of the chemical families is known as *chlorine activation*. If the temperatures remain cold, the condensed phase can sediment out, leading to irreversible *denitrification* and *dehydration* of the air. Denitrification prevents the reformation of $ClONO_2$, and so increases the effective chain length of chlorine catalytic cycles. In sunlight, $Cl_2$ is photodissociated to release atoms which may then be converted to ClO. The small amount of $NO_2$, produced by photolysis of $HNO_3$, can convert the ClO back to $ClONO_2$. Hence we have:

$$Cl_2 + h\nu \rightarrow Cl + Cl \qquad (38)$$
$$Cl + O_3 \rightarrow ClO + O_2 \qquad (15)$$
$$ClO + NO_2 + M \rightarrow ClONO_2 + M \qquad (16)$$

[15] R. A. Cox, A. R. MacKenzie, R. H. Müller, Th. Peter and P. J. Crutzen, 'Activation of Stratospheric Chlorine by Reactions in Liquid Sulfuric Acid', *Geophys. Res. Lett.*, 1994, **21**, 1439–1442.

[16] D. R. Hanson, A. R. Ravishankara and S. Solomon, 'Heterogeneous Reactions in Sulfuric Acid Aerosols: A Framework for Model Calculations', *J. Geophys. Res.*, 1994, **99**, 3615–3630.

This chlorine nitrate is then available to react with any remaining HCl, to release more $ClO_x$. Hence within the polar vortices we would expect to find high concentrations of ClO (see below). Over the wintertime and springtime pole, methane concentrations are low as a result of downward transport from the methane sink region (Figure 9.2) and $NO_x$ concentrations can be low as a result of denitrification. Hence the usual chain-breaking steps of Cl reaction with $CH_4$ or ClO reaction with $NO_2$, are particularly slow. In the Antarctic, the reduced rates of termination reactions, together with continued cold temperatures and PSCs, mean that all the ozone destroying catalytic cycles are able to continue throughout September. In the Arctic, however, sporadic cold temperatures and denitrification mean that there is more continuous competition between radical producton and radical termination.

## 9.3 NATURAL SOURCES OF TRACE GASES

Anthropogenic sources occur on top of natural background concentrations of many trace gases, which have a profound effect on the ozone layer. From a knowledge of the natural sources it is then possible to say how man's activities are perturbing the atmosphere.

Nitrous oxide is the dominant precursor of stratospheric $NO_x$. It is emitted predominantly by biological sources in soils and water. Tropical forest soils are probably the single most important source of $N_2O$ to the atmosphere. Tropical land-use changes and intensification of tropical agriculture indicate significant, growing sources of $N_2O$. The oceans, especially the upwelling regions of the Indian and Pacific Oceans, are also a significant source of $N_2O$. The recent IPCC (1995) report[17] estimated $N_2O$ source strengths for oceans, tropical soils and temperate soils as 3, 4 and 2 $Tg(N) \, yr^{-1}$, respectively.

Natural sources of $Cl_y$, providing enough to have a marked influence on atmospheric ozone, are few. Only three compounds, methyl chloride ($CH_3Cl$), dichloromethane ($CH_2Cl_2$) and chloroform ($CHCl_3$) are thought to have substantial natural sources. Methyl chloride, with an atmospheric abundance of approximately 600 pptv, is the dominant halogenated species in the atmosphere. A production rate of around 3.5 $Tg \, yr^{-1}$ is required to maintain this steady-state mixing ratio[18] most of which comes from the oceans, with a smaller fraction from biomass burning.

There are a large number of bromine compounds in the atmosphere. The most abundant organobromine species is methyl bromide ($CH_3Br$), which has both natural and anthropogenic sources. The main natural sources are oceanic biological processes where it is formed with other hydrogen containing molecules such as $CH_2Br_2$, $CHBr_3$ and $CH_2BrCl$, although these latter molecules are

[17] IPCC, 'Climate Change 1994 – Radiative forcing of Climate Change and An Evaluation of the IPCC IS92 Emission Scenarios', eds. J. T. Houghton, L. G. M. Filho, J. Bruce, H. Lee, B. A. Callander, E. Haites, N. Harris and K. Maskell, Cambridge University Press, Cambridge, 1994.
[18] R. Koppman, F. Johnen, C. Plass-Dulner and J. Rudolph, 'Distribution of methyl chloride, dichloromethane, trichloroethane and tetrachloroethane over the North and South Atlantic. *J. Geophys. Res.*, 1993, **98**, 20517–20526.

unlikely to reach the stratosphere. The ocean is believed to contribute between 30–70% of atmospheric methyl bromide.[19,20]

## 9.4  ANTHROPOGENIC SOURCES OF TRACE GASES

The production of nitrogen and chlorine compounds, in amounts equalling or exceeding natural sources, has led to numerous investigations aimed at assessing the impact of human activities on the ozone layer. Johnson[21] postulated that $NO_x$, emitted in the exhausts of supersonic aircraft (SST), would result in large ozone depletions. Early models predicted depletions of approximately 10% in total ozone. The altitude of greatest depletion occurred between 20 and 30 km and incremental decreases tended to be larger with each additional unit of $NO_x$ injection. As laboratory measurements improved chemical kinetic data, the predicted depletion gradually fell and by mid-1978 the total ozone was forecast to increase. In the event only very small fleets of SSTs were built, while plans for a US fleet were cancelled. There is once more increasing concern over the possible impact of aircraft flights, both supersonic and subsonic, on ozone levels in the lower stratosphere.[22–24]

Firstly, a new generation of SSTs is being proposed by the aircraft industry. Two-dimensional model simulations of the impact of a projected fleet (500 aircraft, each emitting 15 grams of $NO_x$ per kilogram of fuel burned at a cruise altitude of 20 km) in the stratosphere with a chlorine loading of 3.7 ppbv imply additional (*i.e.* beyond those from halocarbon losses) annual average ozone column decreases of 0.3–1.8% for the Northern Hemisphere.

Secondly, for many subsonic aircraft part of the flight operation actually occurs within the stratosphere. Subsonic aircraft flying in the North Atlantic flight corridor emit 44% of their exhaust emissions into the stratosphere. Models predict an ozone decrease in the lower stratosphere of less than 1%, but modelling the lower stratosphere is particularly difficult since a number of chemical processes are of comparable importance to each other, and to the transport processes.

The exhaust products of rockets contain many substances capable of destroying ozone. The possible impact of such emissions was recognized first in the early 1970s. Attention has focused on the chlorine compounds produced from solid fuel rockets. Rockets that release large amounts of chlorine per launch into the stratosphere include NASA's Space Shuttle (68 tons) and Titan IV (32 tons)

[19]H. B. Singh and M. Kanakidou, 'An Investigation of the Atmospheric Sources and Sinks of Methyl Bromide', *Geophys. Res. Lett.*, 1993, **20**, 133–136.

[20]C. E. Reeves and S. A. Penkett, 'An Estimate of the Anthropogenic Contribution to Atmospheric Methyl Bromide', *Geophys. Res. Lett.*, 1993, **20**, 1563–1566.

[21]H. S. Johnston, 'Reduction of Stratospheric Ozone by Nitrogen Oxide Catalysts from Supersonic Transport Exhaust', *Science*, 1971, **173**, 517–762.

[22]D. L. Albritton, W. H. Brune, A. R. Douglass, F. L. Dyer, M. K. W. Ko, R. C. Miake-Lye, M. J. Prather, A. R. Ravishankara, R. B. Rood, R. S. Stolarski, R. T. Watson and D. J. Wuebbles, 'The Atmospheric Effects of Stratospheric Aircraft: Interim Assessment of the NASA High-speed Research Program', NASA reference publication 1333, Washington DC, 1993.

[23]R. S. Stolarski and H. L. Wesoky, 'The Atmospheric Effects of Stratospheric Aircraft: A Second Program Report', NASA reference publication 1293, Washington DC, 1993.

[24]R. S. Stolarski and H. L. Wesoky, 'The Atmospheric Effects of Stratospheric Aircraft: A Third Program Report', NASA reference publication 1313, Washington DC, 1993.

rockets, and the European Space Agency's Ariane-5 (57 tons). The substances emitted from rocket exhausts are initially confined to a small volume of the atmosphere and results indicate that local ozone may be reduced by as much as 80% at some heights for up to three hours. Recovery of ozone is rapid with all but a fraction of a percent restored within 24 hours. Globally, little impact is predicted and the TOMS ozone record shows no detectable changes in column ozone immediately following each of several launches of the Space Shuttle.[9]

CFCs, unlike most other trace gases, are not chemically broken down or removed in the troposphere but rather, because of their exceptionally stable chemical structure, persist, mix throughout the troposphere, and are transported up to the stratosphere. Depending on their individual structure, different CFCs can remain intact for many decades, or even several centuries. Molina and Rowland[25] suggested that free chlorine could be introduced into the upper atmosphere as a result of the photodissociation of CFCs, *e.g.*:

$$CF_2Cl_2 + h\nu \rightarrow CF_2Cl + Cl \tag{39}$$
$$CFCl_3 + h\nu \rightarrow CFCl_2 + Cl \tag{40}$$

with the Cl released subsequently taking part in Reaction 5. It is now firmly established that anthropogenic chlorine plays a central role in stratospheric ozone depletion.[2,9] CFCs were used in a wide variety of industrial applications including aerosol propellants, refrigerants, solvents and foam blowing. The CFCs which have attracted most attention in ozone depletion are CFC-11 (CFCl$_3$) and CFC-12 (CF$_2$Cl$_2$). Despite controls on their production and consumption (see Section 9.8) the concentrations of CFCs are still rising. Recent data show that the growth rates of CFC-11 and 12 are slowing down. In particular tropospheric chlorine grew by about 60 pptv (1.6%) in 1992 compared to 110 pptv (2.9%) in 1989.[26] Peak total tropospheric chlorine loading was expected to occur in 1994, with the stratospheric peak lagging by three to five years.

Bromine is estimated to be about 50 times more efficient than chlorine in destroying stratospheric ozone on an atom-for-atom basis (see Section 9.2). Wofsy *et al.*[27] suggested that trace amounts of bromine might efficiently catalyse ozone destruction. Bromine is carried into the stratosphere in various forms such as halons and substituted hydrocarbons, of which methyl bromide is the predominant form. Three major anthropogenic sources of methyl bromide have been identified: soil fumigation; biomass burning; and the exhaust of automobiles using leaded petrol. Recent measurements have shown that there is three times as much methyl bromide in the Northern Hemisphere as in the Southern Hemisphere. Halon 1211 (CBrClF$_2$) and 1301 (CBrF$_3$) have been widely used in fire protection systems although their production has now ceased. Global back-

[25]M. J. Molina and F. S. Rowland, 'Stratospheric Sink for Chlorofluoromethanes: Chlorine Atom Catalysed Destruction of Ozone', *Nature*, 1974, **249**, 810–812.
[26]P. Fraser, S. Penkett, M. Gunson, R. Weiss and F. S. Rowland, in 'NASA Report on Concentrations, Lifetimes and Trends of CFCs, Halons and Related Species', eds. J. Kaye, S. Penkett and F. Ormond, NASA reference publication 1339, Washington DC, 1994, Ch. 1.
[27]S. C. Wofsy, M. B. McElroy and Y. L. Ying, 'The Chemistry of Atmospheric Bromine', *Geophys. Res. Lett.*, 1975, **2**, 215–218.

ground levels are about 2.5 pptv (H-1211) and 2.0 pptv (H-1301) and are growing at 3% and 8% per year, respectively. These rates have slowed significantly in recent years, consistent with reduced emissions.[2]

Carbon dioxide concentrations are increasing, largely as a result of burning fossil fuels. Increased $CO_2$ leads to lower stratospheric temperatures, and hence a slowing down of temperature dependent $(O + O_3, NO + O)$ ozone-destruction reactions. This mechanism is of potential importance in the middle to upper stratosphere. In the lower stratosphere a decrease in temperature could result in an increased frequency and duration of PSCs, leading to a greater activation of reactive chlorine and hence ozone depletion. Atmospheric methane concentrations are growing as a result of human activities. When methane is destroyed in the stratosphere, the chlorine and nitrogen cycles that destroy ozone are suppressed, thus increasing stratospheric ozone abundances. The destruction process also produces water vapour, which not only acts as a greenhouse gas but can also indirectly destroy ozone.

## 9.5  ANTARCTIC OZONE

The latest reports of ozone depletion in the Antarctic region (WMO, 1994, Chapter 3)[2] show that the ozone hole continues to appear every austral spring, and that it has been particularly severe in the 1990s. Figure 9.4 shows the monthly mean values of the ozone column over Halley Bay, Antarctica, for October. Prior to the mid-1970s the monthly mean column of ozone was approximately 300 DU. By 1985, the mean had fallen to below 200 DU and in the next 5–6 years appeared to be levelling off. However, the springs of 1992 and 1993 brought new record lows, with spot measurements below 100 DU for the first time, and the mean well below half that of the 1950s. The variation of ozone with altitude above Halley Bay is shown in Figure 9.5a. In mid-August the profile is near normal but in less than two months 97% of the ozone between 14 and 18 km in altitude has been destroyed.

The major chemical causes for the severe ozone depletion over Antarctica have now been established beyond reasonable doubt, and have been discussed in Section 9.2, above. It is now recognized that the severe nature of the Antarctic depletion is caused by man-made chlorine pollution of the atmosphere in combination with unique wintertime meteorological conditions. The circulation of the winter stratosphere over Antarctica and the surrounding oceans is dominated by the polar vortex (Figure 9.2). This is a region of very cold air surrounded by strong westerly winds. Air within the vortex is largely sealed off from that at lower latitudes and the vortex acts as a *containment vessel* in which chemistry occurs in near isolation.[28–30] In the lower stratosphere, temperatures within the vortex fall to well below 190 K; cold enough in the lower half of the

---

[28]M. E. McIntyre, 'On the Antarctic Ozone Hole', *J. Atmos. Terr. Phys.*, 1989, **51**, 29.

[29]M. E. McIntyre, 'The Stratospheric Polar Vortex and Sub-vortex: Fluid Dynamics and Mid-latitude Ozone Loss', *Phil. Trans. Roy. Soc.*, Series A, 1995, **352**, 227–240.

[30]R. L. Jones and A. R. MacKenzie, 'Observational Studies of the Role of Polar Regions in Mid-latitude Ozone Loss', *Geophys. Res. Lett.*, 1995, **22**, 3485–3488.

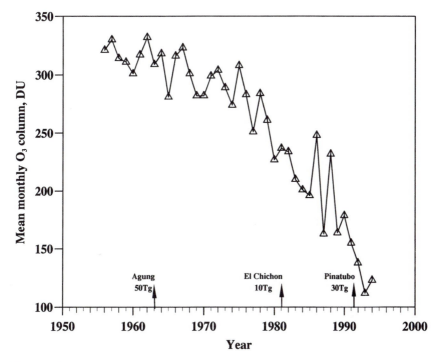

**Figure 9.4** *Mean October value of column ozone over Halley Bay (76°S) since 1957. The dates of major volcanic eruptions which injected sulfur into the Southern Hemisphere are also marked.* Ozone data courtesy of J. Shanklin and A. E. Jones (British Antarctic Survey, Cambridge)

stratosphere for type 1 and type 2 PSCs to form. Chlorine activation, denitrification and dehydration result.

In the decade following the discovery of the Antarctic ozone depletion, extensive measurements of $O_3$, ClO and other compounds have been made.* Figure 9.6 shows the evolution of ozone and related compounds on a single isentropic surface (the 465 K surface, about 19 km) as measured by the MLS instrument on board UARS.[31] In late autumn (28 April) descent of air at the pole has brought down higher mixing ratios of ozone, nitric acid and water vapour from the middle and upper stratosphere. Other measurements have shown that higher mixing ratios of $Cl_y$ are also present in this air. The formation of the polar vortex isolates the descended air from its surroundings. PSC formation has yet to begin and so ClO mixing ratios in the vortex are the same as outside the vortex,

[31]M. L. Santee, W. G. Read, J. W. Waters, L. Froideaux, G. L. Manney, D. A. Flower, R. F. Jarnot, R. S. Harwood and G. E. Peckham, 'Interhemispheric Differences in Polar Stratospheric $HNO_3$, $H_2O$, ClO and $O_3$', *Science*, 1995, **267**, 849–852.

*Most notably during the 1987 US Airborne Antarctic Ozone Experiment (AAOE, see *J. Geophys. Res.*, **94**, D9 & D14, 1989) and the 1994 US Airborne Southern Hemisphere Ozone Experiment (ASHOE). Satellite measurements, particularly those of the Upper Atmosphere Research Satellite (UARS, see *Geophys. Res. Lett.*, **20**, No. 12, 1993; *J. Atmos. Sci.*, **51**, No. 20, 1994), are also now available.

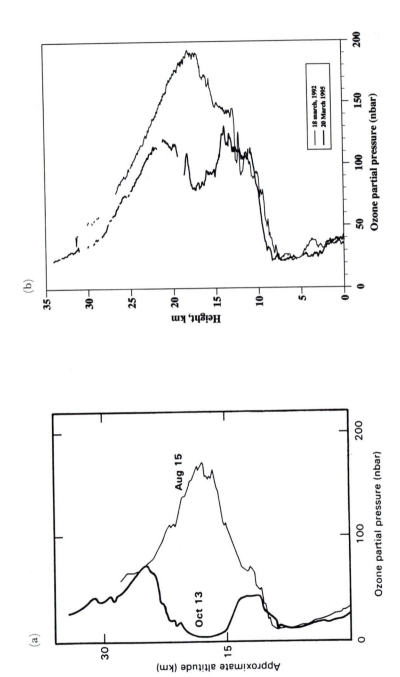

**Figure 9.5**  (a) *Vertical ozone profiles above Halley Bay for 15 August, 1987 and 13 October, 1987. (b) Vertical ozone profiles above Ny Alesund (78°N, 11°E) for 18 March 1992 and 20 March 1995. Both profiles are taken inside the polar vortex*

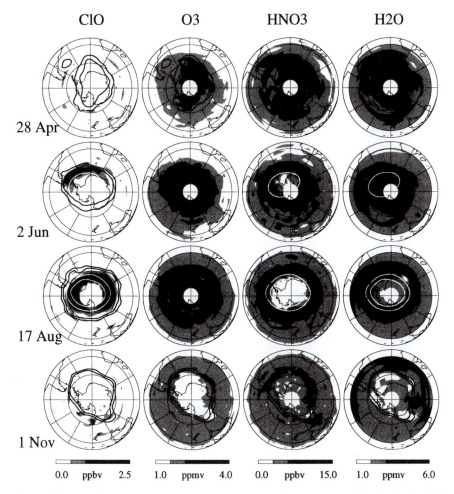

**Figure 9.6** *MLS satellite maps of ClO, O₃, HNO₃ and H₂O for selected days during the 1992 southern winter, interpolated onto the 465 K isentropic surface. The maps are polar orthographic projections extending to the equator, with the Greenwich meridian at the top, and the 30° S and 60° S latitude circles marked. No measurements were obtained poleward of 80° S, and, for H₂O, equatorward of 30° S. The edge of the polar vortex is shown by the two black contours. The black contour near S. America on 28 April surrounds air which has come from the polar vortex. Temperatures below which type 1 and type 2 PSCs can form are shown as white contours: the outer white contour is for type 1 PSCs.* Figure courtesy of Michelle Santee (Jet Propulsion Lab, California)

near zero on this scale. By 2 June, temperatures have fallen below the threshold for type 1 PSC formation, as can be seen from the reduced mixing ratios of gas phase nitric acid in the MLS measurements. Chlorine activation takes place on the PSCs and the ClO formed is blown downwind (Figure 9.6), but there has not yet been enough exposure to sunlight for large scale ozone depletion to be evident. By 17 August, temperatures are cold enough for type 1 and 2 PSCs to be present over large parts of the Antarctic continent, leading to reduced gas phase nitric acid and water vapours and to elevated ClO. Ozone depletion has begun on the sunlit outer rim of the vortex. By 1 November, temperatures are too high for PSC formation but nitric acid and water vapour are both reduced inside the vortex. This is due to denitrification and dehydration. ClO mixing ratios have returned to background levels almost everywhere since PSCs are no longer present. Ozone mixing ratios are now much reduced throughout the vortex, over an area larger than the Antarctic continent. At its greatest extent in 1992 and 1993, the area with ozone columns less than 220 DU approached 10% of the area of the Southern Hemisphere.[2] When the vortex breaks down, the ozone depleted air becomes distributed over the hemisphere, leading to ozone reductions at lower latitudes.

## 9.6   ARCTIC OZONE

Could an 'ozone hole' occur elsewhere? The possibility of an Arctic ozone hole is of such social and political importance that a number of scientific campaigns have been undertaken specifically to provide detailed analyses of the state of the Arctic for input into the political debate about control of CFCs.*

The evolution of ozone and related compounds in the Arctic polar vortex during the winter of 1992/93 is shown in Figure 9.7, which can be compared to the evolution in the Antarctic vortex of 1992, shown in Figure 9.6.[31] The initial descent into the vortex is less pronounced on 28 October in the Northern Hemisphere than on 28 April in the Southern Hemisphere, and the northern vortex is less fully developed than its southern counterpart. By 3 December the vortex has developed and descent, bringing down high mixing ratios of ozone, nitric acid and water vapour, has taken place. ClO mixing ratios are only elevated in a small region downwind of a patch of cold air over southern Finland. By 22 February there have been sufficient cold temperatures to activate chlorine throughout the vortex, but neither denitrification nor dehydration have taken place. Ozone mixing ratios remain high. By 14 March, ClO mixing ratios have again returned to background levels, nitric acid and water vapour mixing ratios remain high, and no ozone hole has formed. However, that is not to say that chlorine-catalysed ozone destruction has not taken place: ozone mixing ratios in March are lower than those in February. Given the near-continuous descent of ozone-rich air onto this isentropic surface in the northern vortex during winter, this decreased ozone implies chemical loss. Further evidence of chemical loss in the northern polar vortex has been adduced from the careful matching up of ozonesondes that have sampled the same air parcel.[32] A statistically significant

[32]P. van der Gathen, M. Rex, N. R. P. Harris, D. Lucic, B. M. Knudsen, G. O. Braathen, H. Debacker, P. Fabian, H. Fast, M. Gil, E. Kyro, I. S. Mikkelsen, M. Rummukainen, J. Stahelin and C. Varotsos, 'Observational Evidence for Chemical Ozone Depletion over the Arctic in Winter 1991/1992', *Nature*, 1995, **375**, 131–134.

*Those campaigns were: the 1989 US Airborne Arctic Stratospheric Experiment (AASE, *Geophys. Res. Lett.*, 1990, **17**); its 1991/92 follow-up (AASE 2, *Science*, 1993, **261**; and *Geophys. Res. Lett.*, 1994, **21**); the 1991/92 European Arctic Stratospheric Ozone Experiment (EASOE, *Geophys. Res. Lett.*, 1994, **21**); and the 1993/94–1994/95 Second European Stratospheric Arctic and Mid-latitude Experiment (SESAME). Satellite measurements, principally from UARS, are also now available.

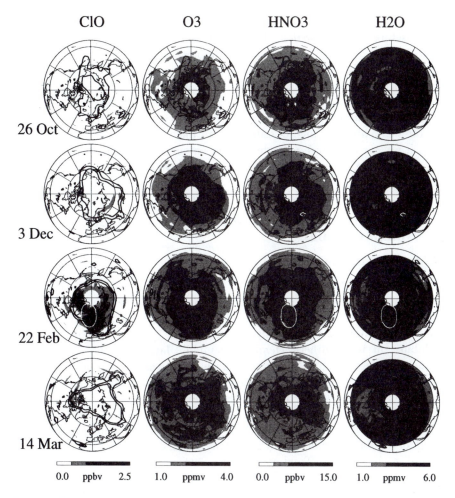

ClO          O3          HNO3          H2O

26 Oct

3 Dec

22 Feb

14 Mar

0.0   ppbv   2.5      1.0   ppmv   4.0      0.0   ppbv   15.0      1.0   ppmv   6.0

**Figure 9.7**   *As Figure 9.6, but for selected days during the 1992–1993 northern winter, with the Greenwich meridian at the bottom of each map.* Figure courtesy of Michelle Santee (Jet Propulsion Lab, California)

correlation is shown between ozone loss and hours of sunlight experienced between samplings.

Figure 9.7 demonstrates that the Arctic has a warmer and weaker vortex than the Antarctic. The greater wave activity in the Northern Hemisphere produces more rapid downward motion in the northern polar vortex relative to its southern counterpart. This downward motion heats the air by compression. Dissipating waves deposit momentum at the vortex edge, thus weakening it. The behaviour of the northern vortex can vary greatly from year to year. For winters which are cold early on, 1991/92 for example, large amounts of ClO can be formed, but it is formed when there is insufficient sunlight to cause severe ozone depletion.[33] Winters which are cold later, 1992/93 for example, produce more severe ozone depletion. The winter of 1994/95 has produced the most severe ozone depletion yet detected in the northern polar vortex. Figure 9.5b shows results from a pair of ozonesondes, launched from Spitsbergen, into the polar vortex, at about the same time of year in 1992 and 1995. About half the ozone has been destroyed in the 18 to 20 km region in 1995 relative to 1992. This is a situation which has been likened to that in early spring in the Antarctic. Whilst one might hesitate to call this amount of depletion an ozone *hole*, it is clear that, even with the present amounts of chlorine and bromine in the atmosphere, severe ozone depletion is possible in the Arctic in winters which are cold late on.

## 9.7 MID-LATITUDE OZONE, GLOBAL OBSERVATIONS AND GLOBAL MODELS

### 9.7.1 Mid-latitude Ozone

Analyses of global ozone records, from satellite and ground-based instruments, have shown that ozone loss is not confined to polar regions. Table 9.1 summarizes the trends in total ozone from 1979 to 1994. Trends in the Northern Hemisphere middle latitudes are significantly negative in all seasons. Similar trends are exhibited in the Southern Hemisphere middle latitudes. The trends in the Equatorial region (20°S–20°N) are slightly negative, but are not statistically significant. Greatest confidence is placed on the trends deduced from the ground-based network of Dobson instruments, although the geographic coverage is patchy, with most stations situated in the Northern Hemisphere. The Total Ozone Mapping Spectrometer (TOMS) data are the most reliable satellite-based monitor of total ozone because it gives daily global coverage and has a 14.5 year record of observations. Total ozone is also determined by the Solar Ultraviolet

**Table 9.1** *Trends (annual averages in % per decade) in total ozone.* Adapted from WMO $(1994)^2$

|  |  | Latitude | | |
|---|---|---|---|---|
|  |  | *Mid South* | *Equatorial* | *Mid North* |
| *Recent*: |  |  |  |  |
| Jan 1979 to May 1994 | SBUV + SBUV/2 | $-4.9 \pm 1.52$ | $-1.8 \pm 1.4$ | $-4.6 \pm 1.8$ |
| Jan 1979 to Feb 1994 | Dobson network | $-3.2 \pm 1.3$ | $-1.1 \pm 0.6$ | $-4.8 \pm 0.8$ |
| *Pre-Pinatubo*: |  |  |  |  |
| Jan 1979 to May 1991 | SBUV + SBUV/2 | $-4.9 \pm 2.3$ | $-0.8 \pm 2.1$ | $-3.3 \pm 2.4$ |
| Jan 1979 to May 1991 | TOMS | $-4.5 \pm 2.1$ | $+0.4 \pm 2.1$ | $-4.0 \pm 2.1$ |
| Jan 1979 to May 1991 | Dobson network | $-3.8 \pm 1.3$ | $+0.2 \pm 1.2$ | $-3.9 \pm 0.7$ |

Uncertainties ($\pm$) are expressed at the 95% confidence limits

---

[33]M. P. Chipperfield, 'A Three Dimensional Model Comparison of PSC Processing during the Arctic Winters of 1991/1992 and 1992/1993', *Ann. Geophysicae*, 1994, **12**, 342–354.

Backscatter (SBUV) instrument. Both TOMS and SBUV have shown a drift in calibration and full details of data quality evaluation techniques are given in WMO (1994).[2]

Various techniques have been used to measure vertical profiles of ozone. These include ground-based methods, such as Umkehr, ozonesondes and satellite instruments such as SBUV and the Stratospheric Aerosol and Gas Experiment (SAGE). The measurements indicate that at altitudes of 35–45 km, during 1979 to 1991, ozone declined 5–10% per decade at 30°N–50°N and slightly more at southern middle latitudes. Over the same period there was no significant depletion at any latitude between 25 and 30 km whilst sizeable reductions were observed in the 15 to 20 km region in middle latitudes. SAGE yields reductions of $20 \pm 8\%$ per decade at 16 to 17 km, while the average ozonesonde data shows smaller negative trends of $7 \pm 3\%$ per decade. Integration of the ozonesonde data gives total ozone trends consistent with total ozone measurements. SAGE measurements are subject to large uncertainties at low altitudes.

For the first two years after the Mt Pinatubo eruption, anomalously large downward ozone trends were observed in both hemispheres (see also Figure 9.4). Global total ozone values in 1992/93 were 1 to 2% lower than those expected from the long-term trend. It has been speculated that radiative, dynamical and chemical perturbations resulting from the Mt Pinatubo aerosol were responsible.

### 9.7.2 *In situ* Measurements

The detection of statistically significant downward trends in column ozone in the middle latitudes has provided the spur for extensive measurements of $O_3$, ClO and other compounds.* As well as accurate measurements, models, which couple the transport and the chemistry of long-lived trace species, are required if global ozone depletion is to be quantified. The importance of transport in establishing the conditions necessary for the ozone hole has been discussed above. In the middle latitudes, transport of air from both high and low latitudes determines the chemical composition. Transport from the polar vortex into the middle latitudes can produce ozone depletion by exporting ozone-poor vortex air, or by exporting chemically activated ClO-rich air. For example, ozone depletion may be observed as a result of air sinking through the vortex during the period of ozone destruction and then being transported to the middle latitudes. The polar vortex 'containment vessel' does not extend to the tropopause and so exchange can occur, in the very low stratosphere, between the middle latitudes and the polar regions, which are still cold enough at this height to support PSCs.[29]

---

*The 1992/93 US Stratospheric Photochemistry, Aerosols and Dynamics Experiment (SPADE), and the 1993/94–1994/95 Second European Stratospheric Arctic and Mid-latitude Experiment (SESAME). Measurements were also made as part of the polar campaigns mentioned in earlier sections, and as part of the UARS satellite measurement programme.

### 9.7.3 Global Modelling

The advent of supercomputing has made possible an integrated approach to the modelling of the stratosphere. Models are required ultimately for the prediction of the atmospheric effects of future emissions, so that legislation can be drafted in a timely fashion. Models also perform several, scientifically important, intermediate tasks, and the choice of model is determined by the task that it is required to carry out.

Two dimensional, in latitude and height, models can be integrated forward for many years, allowing an assessment of likely future ozone destruction to be made. They use the mean circulation, combined with eddy diffusion coefficients,* to transport chemicals. Unfortunately, the 'assessment' models do not resolve polar dynamics and chemistry sufficiently well to produce the observed chlorine activation and ozone destruction at both wintertime poles. Different assessment models also predict very different lower stratospheric abundances of trace gases.[2] Many of the problems with these models stem from the fact that the eddy diffusion concept does not accurately represent the essentially one-sided isentropic transport, associated with mechanisms such as wave breaking, occurring at the edge of the polar vortices and the sub-tropical mixing barrier.[3] Fully three-dimensional simulations of the chemistry and dynamics of the stratosphere are also possible (*e.g.* Lary *et al.* 1994[34]). However, such 3D *General Circulation Model* (GCM) integrations are usually limited to a few weeks, for reasons of economy, but also because the dynamics in the models are not sufficiently accurate to keep the model atmosphere close to observations.

For the interpretation of chemical observations, using prescribed meteorological fields to transport chemical tracers in 3D, is a very useful alternative to GCM integrations. These *Chemistry and Transport Models* (CTMs) also cannot be integrated forward for as long as the two dimensional assessment models, but low resolution seasonal runs are now possible.[35] Chemistry and transport models can be run on a single isentropic surface (making use of the fact that flow across isentropic surfaces is small, with a timescale of about a week). Even without capturing the details of the diabatic flow, single-layer models are a useful test-bed for chemical mechanisms. Figure 9.8 shows results from a single-layer CTM, run on the 475 K isentropic surface with a horizontal resolution of $5.6° \times 5.6°$, and which was initiated on 1 January 1992. The results are for 22 February 1992, 53 days into the simulation and a day for which there are measurements of the Northern Hemisphere (Figure 9.7). Although the winds and temperatures of the model are prescribed, the model chemistry is free-running after initialization. Clearly, model and measurements are in good agreement. Notice that on this day

---

[34]D. J. Lary, J. A. Pyle and G. Carver, 'A Three Dimensional Model Study of Nitrogen Oxides in the Stratosphere', *Q. J. Roy. Meteor. Soc.*, 1994, **120**, 453–482.

[35]M. P. Chipperfield, J. A. Pyle, C. E. Blom, N. Glatthor, M. Hopfner, T. Gulde, C. Piesch and P. Simon, 'The Variability of $ClONO_2$ and $HNO_3$ in the Arctic Polar Vortex', *J. Geophys. Res.*, 1995, **100**, 9115–9129.

*Eddy diffusion coefficients arise from a treatment of atmospheric mixing and stirring which is analogous to the treatment of molecular diffusion in Fick's Laws.

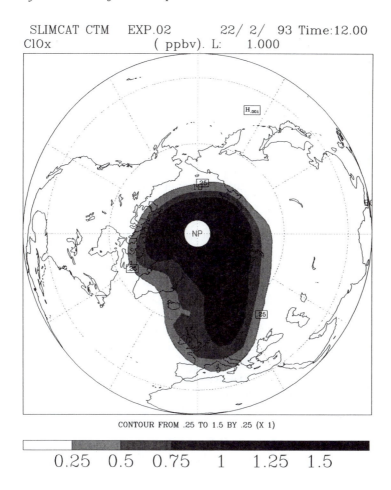

SLIMCAT CTM    EXP.02        22/ 2/  93  Time:12.00
ClOx             ( ppbv). L:    1.000

CONTOUR FROM .25 TO 1.5 BY .25 (X 1)

0.25   0.5   0.75   1   1.25   1.5

**Figure 9.8**  *ClO$_x$ mixing ratios (in ppbv) from a single-layer CTM, run on the 475 K isentropic surface at 5.6° × 5.6° horizontal resolution, initialized on 1 January 1992 and run out to 22 February 1992. The map is a polar orthographic projection. Cf. Figure 9.7.* Data courtesy of Martyn Chipperfield (Centre for Atmospheric Science, Cambridge)

the polar vortex was distorted and covered a large part of Europe, bringing ClO-rich air into sunlight, where ozone-destroying catalytic cycles can occur.

A variant of the single layer model follows the chaotic stirring of dynamical features by advecting material contours.[36,37] Results from the single-layer models, both CTMs and contour advection models, have proved to be particularly useful in the interpretation of high resolution *in situ* data. In particular, high

[36]W. A. Norton, 'Breaking Rossby Waves in a Model Stratosphere Diagnosed by a Vortex Following Coordinate System and A Technique for Advecting Material Contours', *J. Atmos. Sci.*, 1994, **51**, 654–673.
[37]D. W. Waugh and R. A. Plumb, 'Contour Advection with Surgery: A Technique for Investigating the Fine-scale Structure in Tracer Transport', *J. Atmos. Sci.*, 1994, **51**, 530–540.

resolution single-layer models have established the possible importance of the mixing out of chemically-perturbed air from the edge of the polar vortex into the middle latitudes.

## 9.8  LEGISLATION

On September 16, 1987, a treaty was signed that has been hailed as 'the most significant international environmental agreement in history', 'a monumental achievement' and 'unparalleled as a global effort'.[38] It had been long believed that this particular agreement would be impossible to achieve because the issues were so complex and arcane and the initial positions of the negotiating parties so widely divergent.

The Montreal Protocol on Substances that Deplete the Ozone Layer was adopted in 1987 by 25 countries, and entered into force on 1 January 1989. The Protocol required each party's production and consumption of CFCs -11 ($CFCl_3$), -12 ($CF_2Cl_2$), -113 ($C_2F_3Cl_3$), -114 ($C_2F_4Cl_2$) and -115 ($C_2F_5Cl$) first to be frozen at 1986 levels and ultimately reduced to 50% of 1986 levels by 1998. Production and consumption of halons 1211 ($CF_2BrCl$), 1301 ($CF_3Br$) and 2402 ($C_2F_4Br_2$) were to be restricted to 1986 levels.

Ozone depletion had, from the start, captured the US public imagination. The subject featured prominently in the media and in congressional hearings, and soon began to influence consumer behaviour. In 1978 the US, followed by Canada, Norway and Sweden, banned the use of CFCs as aerosol spray can propellants in nonessential applications.[39] Even before this ban, the US market for spray cans had fallen by nearly two-thirds because of public commitment to the protection of the environment. It is interesting to note that in the UK such a ban was not introduced. Concerns about ozone depletion were defused by ministerial assurances that any terrestrial effects would be inconsequential.[40]

In the light of increasing scientific and public concern over ozone depletion, empirical evidence of global depletion, confirmation that CFCs and other man-made ozone depletion substances were the major factor in creating the Antarctic ozone hole, and generally improved prospects for replacing such substances, significant changes to the Protocol were approved in London (1990) and Copenhagen (1992). In London, the control measures were adjusted to provide for the phase-out of CFC and halon production and consumption by the year 2000. An intermediate cut of 50% of the 1986 level by 1995 was also agreed for CFCs and halons, together with a 85% cut, for CFCs but not halons, by 1997. Additionally carbon tetrachloride and fully halogenated CFCs were to be phased-out by 2000 and methyl chloroform by 2005.

The 1990 Protocol Amendment introduced the concept of transitional substances, such as HCFCs. These are chemical substitutes for CFCs and other

---

[38]R. E. Benedick, 'Ozone Diplomacy', Harvard University Press, Cambridge, Massachusetts, 1991.

[39]S. C. Zehr, 'Accounting for the ozone hole: scientific representations of an anomaly and prior incorrect claims in public settings', *Sociol. Quart.* 1994, **35**, 603–619.

[40]M. Purvis, 'Yesterday in parliament: British politicians and debate over stratospheric ozone depletion, 1970–92', *Environ. Plann. C: Government and Policy*, 1994, **12**, 361–379.

controlled substances, but unlike CFCs, are reactive in the lower atmosphere and so, tonne for tonne, transport less chlorine to the stratosphere. They are necessary in some applications, in the short to medium term, to enable a rapid phase-out of the controlled substances to take place. A nonbinding resolution was approved with a view to replacing HCFCs by non ozone depleting alternatives by no later than 2040.

Further revisions to the Protocol were agreed in Copenhagen in 1992. The phase-out dates were brought forward and controls on several new substances were imposed. For CFCs, carbon tetrachloride and methyl chloroform, the new phase-out date was 1/1/96. Provision was made to allow production of these substances for 'essential use'. Within the European Union, Member states have adopted even tighter controls. The transitional substances were introduced into the Protocol and controls on HCFCs agreed. These controls set a cap on consumption which will be reduced stepwise, leading to a phase-out by 2030. The numerical value of the cap is the sum of the quantity of all HCFCs which were produced and used during 1989 and an amount equal to 3.1% of the calculated CFC consumption during 1989. Contributions to the 3.1% cap from individual HCFCs are adjusted by their ozone depletion potentials.

Based on assumed compliance, by all nations, of the revised Protocol stratospheric chlorine abundances will continue to grow from their current levels (3.6 ppbv) to a peak of approximately 3.8 ppbv around the turn of the century.

## 9.9 SUMMARY

Concern about ozone depletion from supersonic transport and CFCs from anthropogenic sources, were factors in the USA's decisions to cancel a proposed SST fleet and, in 1978, impose a ban on aerosol cans. Notwithstanding these control measures, dramatic depletion of ozone has been observed every spring since the early 1980s in the Antarctic polar vortex. Long-term depletion has been detected in the high and middle latitudes of both hemispheres, with most of the ozone loss occurring in the lower stratosphere. In the northern spring of 1995, ozone depletion similar to that observed each Antarctic spring was observed in the Arctic.

The destruction of the ozone layer is a prime example of a serious environmental problem of global extent. The Antarctic ozone hole has forced the international community to cooperate and has resulted in the world's first global treaty for the protection of the environment. It has also spurred investigators to study atmospheric chemistry and dynamics in new detail.

*Acknowledgements.* The task of writing a short description of stratospheric ozone is made immeasurably easier by the existence of the WMO Scientific Assessments. We thank the many scientists who have laboured long and hard on the latest Assessment. We thank Deb Fish, Jamie Kettleborough and Adrian Lee for many helpful comments and corrections. We thank Martyn Chipperfield, Anna Jones, Roland Neuber, Ross Salawitch, Michelle Santee and Jonathan Shanklin for supplying figures and data for figures. We thank Joe Farman for contributing to previous editions of this chapter. The Centre for Atmospheric Science is a joint initiative between the Department of Chemistry and the Department of Applied Mathematics and Theoretical Physics, at the University of Cambridge. RMK gratefully acknowledges funding from the Isaac Newton Trust.

CHAPTER 10

# Atmospheric Dispersal of Pollutants and the Modelling of Air Pollution

M. L. WILLIAMS

## 10.1 INTRODUCTION

Considerable resources are often devoted to the measurement of air pollutant concentrations in the ambient atmosphere but measurements on their own provide little information on the origin of the pollutants in question, on the dispersal process in the atmosphere and on the impact of new sources or the benefits of controls. There is frequently the need, therefore, for detailed knowledge of the characteristics and quantities of pollutants emitted to the atmosphere and on the atmospheric processes which govern their subsequent dispersal and fate. This knowledge must then be built into an appropriate dispersion model, whether the problem to be addressed is the emission from a single chimney or the emissions from a large multi-source urban/industrial area, or, on a larger scale, from a region or country.

The trend within the EEC towards formal air quality guidelines or standards is creating the need for formal air quality management systems and for a more strategic approach to air pollution control. Neither can be accomplished without the use of atmospheric dispersion modelling techniques. A wide variety of techniques are available, ranging from the most simple 'box' model through to numerical solutions of the basic equations of fluid flow, *etc.* For the most chemically reactive pollution, it is also necessary to incorporate the relevant atmospheric chemistry. The spatial and temporal resolution and accuracy of the model output ideally must match the questions being posed. Where emissions vary greatly, both in space and time, or it is necessary to predict the time series of concentrations at specified locations, then the modelling task is extremely difficult, even without the complications of a very chemically reactive pollutant species or substantial topographical effects on the dispersal pattern. On the other hand, if it is not necessary to know when a specified concentration will occur but rather it is the probability of occurrence during a given period (*e.g.* a year) which is of interest, then the modelling task is generally much less demanding. However in general, long-period (*e.g.*, annual) average concentrations can be modelled more accurately than can shorter averaging times where the turbulent fluctuations in the atmosphere can result in agreement within factors of 2 to 3 with observed values.

## 10.2  DISPERSION AND TRANSPORT IN THE ATMOSPHERE

A pollutant plume emitted from a single source is transported in the direction of the mean wind. As it travels it is acted upon by the prevailing level of atmospheric turbulence which causes the plume to grow in size as it entrains the (usually) cleaner surrounding air. There are two main mechanisms for generating atmospheric turbulence. These are mechanical and convective turbulence, and will be discussed in Sections 10.2.1 and 10.2.2 below.

### 10.2.1  Mechanical Turbulence

This is generated as the air flows over obstacles on the ground such as crops, hedges, trees, buildings and hills. The intensity of such turbulence increases with increasing wind speed and with increasing surface roughness and decreases with height above the ground. If there is only a small heat flux, either to or from the surface, in the atmosphere so that most of the turbulence is mechanically generated the atmosphere is said to be neutral or in a state of neutral stability. In this case the wind speed will vary logarithmically with height $z$:

$$u(z) = (u_*/k) \ln (z/z_0) \qquad (1)$$

where $k$ is von Karman's constant ($\sim 0.4$), $z_0$ is the so called surface roughness length ($\sim 1$ m for cities and $\sim 0.3$ m for 'typical' countryside in the UK), and $u_*$ is the friction velocity and is a measure of the flux of momentum to the surfaces.

### 10.2.2  Turbulence and Atmospheric Stability

As solar radiation heats the earth's surface, the lower layers of the atmosphere increase in temperature and convection begins, driven by buoyancy forces. The motion of air parcels from the surface is unstable as a parcel in rising finds itself warmer than its surroundings and will continue to rise. Convective circulations are set up in the boundary layer and this form of turbulence is usually associated with large eddies the effects of which are often visible in 'looping' plumes from stacks. At night when there is no incoming solar radiation and the surface of the earth cools, temperature increases with height, and turbulence tends to be suppressed. During calm clear nights, when surface cooling is rapid and little or no mechanical turbulence is being generated, turbulence may be almost entirely absent.

**Table 10.1** *Pasquill's stability categories*\*

| Surface wind speed m s$^{-1}$ ($\equiv U_{10}$) | Insolation | | | Night | |
|---|---|---|---|---|---|
| | Strong | Moderate | Slight | Thickly overcast or ⩾4/8 low cloud | ⩾3/8 Cloud |
| <2 | A | A–B | B | — | G |
| 2–3 | A–B | B | C | E | F |
| 3–5 | B | B–C | C | D | E |
| 5–6 | C | C–D | D | D | D |
| >6 | C | D | D | D | D |

Strong insolation corresponds to sunny midday in midsummer in England. Slight insolation in similar conditions in midwinter. Night refers to the period from 1 hour before sunset to 1 hour after dawn. A is the most unstable category and G the most stable. D is referred to as the neutral category and should be used, regardless of wind speed, for overcast conditions during day or night.
\*Based on Figure 6.10 in Reference 1.

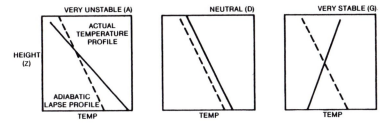

**Figure 10.1** *Typical atmospheric temperature profiles and corresponding stabilities*

When considering most dispersion problems it is convenient to classify the possible states of the atmosphere into what are usually referred to as stability categories. The typing scheme developed by Smith from the original Pasquill formulation[1,2] is widely used because of its relative simplicity yet dependence on sound physical principles. Stability is classified according to the amount of incoming solar radiation, wind speed and cloud cover. A semi-quantitative guide is given in Table 10.1 and Figure 10.1 shows typical temperature profiles corresponding to the unstable, neutral and stable cases. The adiabatic lapse profile in Figure 10.1 is the vertical temperature gradient for the atmosphere in a stage of adiabatic equilibrium when a parcel of air can rise and expand, or descend and contract, without gain or loss of heat; the temperature of the air parcel is always the same as that of the level surrounding air and the conditions correspond to neutral stability. The numerical value of the adiabatic lapse rate is ~1°C/100 m. For wind speeds in excess of about 6–8 m s$^{-1}$ mechanical

[1] F. Pasquill and F. B. Smith, 'Atmospheric Diffusion', Ellis Horwood, Chichester, 1983.
[2] 'A Model for Short and Medium Range Dispersion of Radionuclides Released to the Atmosphere', First Report of a UK Working Group on Atmospheric Dispersion, ed. R. H. Clark, NRPB Report R91, HMSO, London, 1979.

**Table 10.2**  *Typical annual frequency of occurrence
of stability categories in Great Britain*

| Stability category | Frequency of occurrence % |
| --- | --- |
| A | 0.6 |
| B | 6.0 |
| C | 17.0 |
| D | 60.0 |
| E | 7.0 |
| F | 8.0 |
| G | 1.4 |

turbulence dominates irrespective of the degree of insolation and neutral stability prevails. Table 10.2 gives typical annual frequencies of occurrence of the different stability categories in Great Britain. For other regions quite different frequencies might apply. In central Continental regions at lower latitudes, for example, the greater incidence of solar radiation would probably result in smaller incidence of neutral conditions and increased frequencies of unstable and stable categories.

### 10.2.3  Mixing Heights

The stable atmosphere depicted in Figure 10.1 is an example of a ground based temperature inversion, *i.e.* the temperature increases with height unlike the normal decrease. An elevated inversion is often observed where a region of stable air caps an unstable layer below. Pollutants emitted below the inversion can be mixed up to, but not through, the inversion, the height of which is referred to as the mixing height. This term can be used more generally to describe the height of a boundary between two stability regimes. In the case of an emission above an elevated inversion, the pollutant will be prevented from reaching the ground so that for both surface and elevated sources, inversions can have a significant effect on ground level concentrations. The variation of mixing heights throughout the day due to solar heating and atmospheric cooling can have profound effects on ground level concentrations of pollutants. At night the atmosphere is typically stable with a shallow ($\sim$1–300 m) layer formed by surface cooling. As the sun rises the surface heating generates convective eddies and the turbulent boundary layer increases in depth, reaching a maximum in the afternoon at a depth of $\sim$1000 m. As the solar input decreases and stops, the surface cools and a shallow stable layer begins to form again in the evening. In this idealized day, concentrations from surface sources will thus be at a maximum in the periods when the stable layers (with low wind speeds and mixing heights) are present and minimized during the afternoon, emission rates remaining the same. Sources emitting above the stable overnight layer will not contribute to ground-level concentrations until the height of the growing convective layer reaches the plume and brings the pollutants to ground level, a process known as fumigation. The patterns of ground-level concentrations from elevated sources

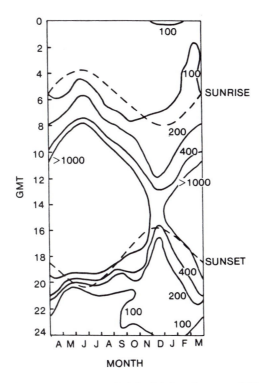

**Figure 10.2**  *Annual and diurnal variation of mixing height at Stevenage, 1981–3 (metres)*

can therefore be quite different from those of surface releases. In reality, the diurnal pattern of emissions can, of course, play a significant role.

In assessing air quality impacts, particularly of elevated sources such as power stations, estimates of mixing heights and their frequency of occurrence and variability throughout the day are therefore essential. WSL has used acoustic sounding (SODAR) to determine mixing heights[3] and an example of the diurnal and seasonal variation of mixing heights measured at Stevenage from 1981–1983 is given in Figure 10.2. The broad features described above are apparent in this diagram.

### 10.2.4  Building and Topographical Effects

Hills or buildings can have significant adverse effects on plume dispersion if their dimensions are large in comparison with the dimensions of the plume or if they significantly deflect or disturb the flow of the wind. Figure 10.3 shows a simplified and idealized representation of the flow over a building. There is a zone on the immediate downwind (or leeward) side of the building which is to some extent

---

[3]A. M. Spanton and M. L. Williams, 'A Comparison of the Structure of the Atmospheric Boundary Layers in Central London and a Rural/Suburban Site Using Acoustic Sounding', *Atmos. Environ.*, 1988, **22**, 211–223.

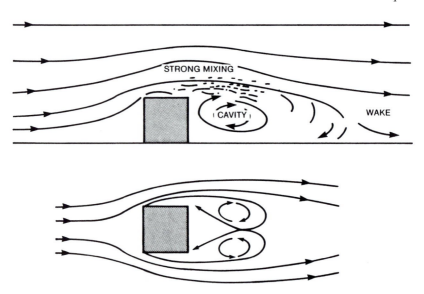

**Figure 10.3**  *Simplified schematic flow patterns around a cubical building*

isolated from the main flow and within which there is a reversal of the air flow. Further downstream the air flow is highly turbulent. Waste gases escaping through a relatively short chimney attached or adjacent to the building, will be entrained in this characteristic flow pattern and will not disperse according to the conventional Equations 2 and 3 (see Section 10.3.1). Recent wind tunnel studies demonstrate that up-wind buildings can have a significant effect on emissions from a chimney located within a few, say 5, building heights; for example, to maintain the same maximum ground level concentration the chimney height required in the presence of one building type studied would be between $1\frac{1}{2}$ and 2 times the height of the chimney required if the buildings was not present. Further downwind, beyond roughly 10 building heights, the near-field effects of the buildings can be incorporated into dispersion models in a parameterized way as discussed in Section 10.3.1.

Topographical features such as hills and sides of valleys can have similar effects on dispersion to those described above for buildings. Valleys are also somewhat more prone to problems arising from emission fairly close to the ground. The incidence of low level or ground based temperature inversions can be greater, either because solar heating of the ground is somewhat delayed in the morning or because during the night cold air drains down the valley sides (katabatic winds) thus creating a 'pool' of cold air on the valley floor. Any low level emissions will therefore disperse very slowly and may even accumulate to some extent. The relatively undiluted emissions can also drift along and across the valley thus affecting areas other than the immediate surroundings of the source. Emissions from high chimneys located on the valley floor may not be detected at all on the valley floor while the ground based inversion persists. However, considerable horizontal spreading of the plume aloft can occur and during the morning fumigation period large parts of the valley may experience relatively high concentrations at much the same time.

The effects of hills on the flow of pollutant plumes and the resulting concentrations are complex and will not be discussed here.

### 10.2.5 Removal Processes – Dry and Wet Deposition

When considering pollution impacts from nearby sources, *e.g.* within say 10 km or so, the various losses of pollutants are generally not important (unless one happens to be interested specifically in such issues as the short range washout of HCl or the deposition of a particularly toxic species in the near field). However, in considering impacts over long ranges and especially on the international scale then consideration of the removal processes is essential.

Dry deposition takes place continously in a turbulent boundary layer as a result of turbulent flux towards the surface. The efficiency of the process is determined by the deposition velocity which in general is a function of the prevailing level of turbulence (high levels of which result in increasing deposition other things being equal) and of the nature of the gas and the surface (for example a reactive gas such as $HNO_3$ will be deposited more readily than a less reactive species such as NO). The flux to the surface is given by the product of the surface concentration and the deposition velocity, $v_g$, so that the process is linear. The equivalent first-order rate constant for this process is given by $(v_g/H)$ where $H$ is the depth of the mixing layer through which deposition is taking place. Typical half-lives for this process are $\sim 1$–2 days for species such as $SO_2$ and $NO_2$, but $\sim 5$ days or more for sulfate aerosols, which is one reason why 'acid rain' is a continental rather than purely a local phenomenon.

Wet deposition is the term given to the removal of gases or particles from the atmosphere in clouds and/or in rain. For species such as $SO_2$ and aerosols this process is relatively efficient. Typical lengths of dry and wet periods in the UK are such that if transport times are of the order of the dry period duration ($\sim 70$ hours in the UK), wet removal processes should be included in the model.

### 10.3 MODELLING OF AIR POLLUTION DISPERSION

In Section 10.2 we discussed the underlying physical principles of air pollution dispersion, transport and transformation in the atmosphere. In this Section we summarize the methods used to apply these principles in a quantitative way to model the processes mathematically. The techniques used depend on the distance scales involved in the transport from the source to the receptor or 'target' area. If this distance is small, say of the order of tens or hundreds of metres, then very often buildings and local topography are important and the mathematical description of the ensuing complex flows and turbulence may not be tractable. In such cases (and others such as the dispersion of dense gases or longer distance problems in complex topography) a physical model in a wind tunnel may be the only practicable solution. These complications will be neglected in all that follows and we will deal with situations of ideal flat terrain.

It is fairly clear that as a plume is transported in the direction of the mean wind, it grows through the effect of atmospheric turbulence producing, very roughly, a cone shaped plume with the apex towards the stack. Now clearly the plume will continue to expand until, in the vertical, it fills the atmospheric boundary layer ($\sim 1$ km deep in neutral conditions). Beyond this point vertical dispersion has no further effect; concentrations are thence reduced only by horizontal dispersion, and by the deposition processes and, if appropriate, by chemical reactions. It can be shown that in neutral conditions this point is reached at downwind distances from a source of very roughly 50–100 km, so for source–receptor distances less than this value, vertical dispersion should be included in a model for an accurate representation of the dispersion. Beyond this region, plumes generally fill the mixing layer and uniformly mixed 'box-models' can be used with some confidence.

We will firstly discuss modelling on scales where vertical dispersion is important, before discussing problems involving longer range transport.

### 10.3.1 Modelling in the Near Field

In this section we will discuss modelling of pollutant dispersion from 0 to $\sim 100$ km, using the Gaussian plume approach. This is not to condemn more sophisticated methods, but for most practical applications, the quality of the available emission and meteorological data does not justify the increased resources required to set up and run more complex models. In many cases a sound knowledge of meteorology and aerodynamics can be used to parameterize the Gaussian model to simulate adequately, for example, the effects of buildings on dispersion.

In the Gaussian plume approach the expanding plume has a Gaussian, or Normal, distribution of concentration in the vertical ($z$) and lateral ($y$) directions as shown in Figure 10.4. The concentration $C$ (in units of $\mu$g m$^{-3}$ for example) at any point $(x, y, z)$ is then given by:

$$C(x, y, z) = \frac{Q}{2\pi\sigma_y\sigma_z U} \exp\left[-\frac{y^2}{2\sigma_y{}^2}\right]\left\{\exp\left[-\frac{(z - H_e)^2}{2\sigma_z{}^2}\right] + \exp\left[-\frac{(z + H_e)^2}{2\sigma_x{}^2}\right]\right\}$$

$$(2)$$

where $Q$ is the pollutant mass emission rate in $\mu$g s$^{-1}$, $U$ is the wind speed, $x, y$ and $z$ are the along wind, crosswind and vertical distances, $H_e$ is the effective stack height given by the height of the stack plus the plume rise defined below. The parameters $\sigma_y$ and $\sigma_z$ measure the extent of plume growth and in the Gaussian formalism are the standard deviations of the horizontal and vertical concentrations respectively in the plume. When $y = z = 0$, this equation reduces to the familiar ground level concentration below the plume centreline:

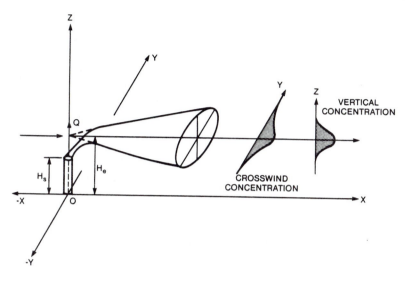

**Figure 10.4** *Gaussian plume distribution*

$$C(x) = \frac{Q}{\pi \sigma_y \sigma_z U} \exp\left(-\frac{H_e^2}{2\sigma_z^2}\right) \tag{3}$$

Equation 3 is rather less cumbersome to deal with and several points of interest emerge. Firstly, concentrations are directly proportional to the emission rate, $Q$, so it is essential that this is known accurately in any practical application. Secondly, unless $H_e = 0$ (*i.e.* unless the source is at ground level) the maximum concentration will occur at a point downwind and this downwind distance will increase with increasing $H_e$ and furthermore the value of $C_{max}$ will decrease with increasing $H_e$. In fact, $C_{max}$ is roughly proportional to $H_e^{-2}$. This is the mathematical statement of the so-called 'tall stacks' policy which underpinned air pollution control in the UK until relatively recently.

The specification of $H_e$ involves calculating the plume rise, which is the height above the point of emission reached by the plume due to its buoyancy (if it is warmer than the surrounding air, as most combustion emissions are) or momentum (plumes may be driven up stacks at relatively high velocities). For most plumes, buoyancy is the dominating force and there have been a large number of studies of methods of determining plume rise. A widely used method however is that due to Briggs,[4] where the plume rise $\Delta H$ is given by:

$$\Delta H = 3.3(Q_H)^{1/3}(10H_s)^{2/3}U^{-1} \quad \text{for} \quad Q_H \geqslant 20 \text{ MW}$$

or

$$\Delta H = 20.5(Q_H)^{0.6}(H_s)^{0.4}U^{-1} \quad \text{for} \quad Q_H \leqslant 20 \text{ MW} \tag{4}$$

[4] G. A. Briggs, 'Plume Rise', US Atomic Energy Commission, Washington, DC, 1969.

where $Q_H$ is the sensible heat emission from the stack and $U$ is the wind speed at the stack height $(H_s)$. The dependence on $Q_H$ is not strong and, if measured values of this quantity are not available, an approximation often used is to assume $Q_H$ is equal to one-sixth of the total heat generated in combustion of the fuel. An expression for $\Delta H$ due to Moore[5] has been developed for power stations in the UK and is:

$$\Delta H = aQ_H^{1/4}/U \tag{5}$$

where $a = 515$ in unstable and neutral conditions and $a = 230$ in stable atmospheres, with $Q_H$ in MW and $U$ in m s$^{-1}$. More recent (and more complex) formulations of Moore's and Briggs' formulae have been summarized by the UK Dispersion Modelling Working Group.[6]

The standard deviations of the plume in the vertical and lateral directions, $\sigma_z$ and $\sigma_y$, are extremely important quantities. They are determined by the prevailing atmospheric turbulence in the boundary layer.

Turbulent motions or eddies in the atmosphere vary in size and intensity; the greater their size and/or intensity the more rapid is the plume growth and hence the dilution of the pollutants. Small scale turbulent motions tend to dominate the plume growth close to the point of emission where the plume is still relatively small and the larger scale eddies dominate at greater distances. Furthermore, the small and larger eddies are associated respectively with short and longer time scales. Consequently, $\sigma_y$ and $\sigma_z$ increase in value with distance from the source (see Figure 10.4); also they increase with the time or sampling period over which they have been measured. This latter point means that it is essential to state the sampling period to which $\sigma_y$ and $\sigma_z$ apply, especially if comparisons are being made between calculated and measured concentration; ideally the two periods should be identical. Most values of $\sigma_y$ and $\sigma_z$ to be found in the literature are for sampling periods in the range 3–60 minutes. It is also evident that $\sigma_y$ and $\sigma_z$ are dependent on atmospheric stability, being smallest when the atmosphere is most stable (category $G$) *i.e.* when atmospheric turbulence is least, increasing to their greatest values in highly turbulent very unstable conditions (category $A$). The underlying surface roughness elements also play a part, $\sigma_y$ and $\sigma_z$ increasing with increasing surface roughness so that for a given distance downwind of a chimney $\sigma_y$ and $\sigma_z$ will be larger in, for example, an urban area than in an area of open, relatively flat agricultural land.

In general, lateral (horizontal) motion is less constrained than vertical motion with the result that there are larger scale eddies in the horizontal than in the vertical. Fluctuations in wind direction also become important for longer sampling periods. Consequently, $\sigma_y$ increases more rapidly with increasing

[5]D. J. Moore, 'A Comparison of the Trajectories of Rising Buoyant Plumes with Theoretical Empirical Models', *Atmos. Environ.*, 1974, **8**, 441–457.
[6]'Models to Allow for the Effects of Coastal Sites, Plume Rise and Buildings on Dispersion of Radionuclides and Guidance on the Value of Deposition Velocity and Washout Coefficients', Fifth Report of a UK Working Group on Atmospheric Dispersion, ed. J. A. Jones, HMSO, London, 1983, NRPB Report R157.

sampling or averaging period than does $\sigma_z$. This dependence of $\sigma_y$ on wind direction fluctuation also means that for longer sampling periods, say greater than one hour, $\sigma_y$ values can increase with increasing atmospheric stability because during low wind speed stable conditions plume meandering can be significant.

Ignoring for the moment the plume meandering component of $\sigma_y$ the parameters are often conveniently expressed in the form:

$$\sigma_y = \sigma_{yo} + ax^b$$
$$\sigma_z = \sigma_{zo} + cx^d \tag{6}$$

where $a$, $b$, $c$ and $d$ are constants dependent on atmospheric stability, $x$ is the downwind distance from the source and $\sigma_{yo}$, $\sigma_{zo}$ are the initial plume spreads generated by, for example, building entrainment. To incorporate plume meander into $\sigma_y$, an extra term is added so that:

$$\sigma_y^2 = \sigma_{yt}^2 + 0.0296 \ Tx^2/U \tag{7}$$

where $\sigma_{yt}$ is given by Equation 6 and $T$ is the averaging time in hours.

A simple expression for $\sigma_z$ based on Smith's[1] work is:

$$\sigma_z = \sigma_{zo} + 0.9(0.83 - \log_{10} P)x^{0.73} \tag{8}$$

which gives a good representation out to $\sim 30$ km. Here $P$ is Smith's stability parameter equal to 3.6 for neutral conditions and ranging from 0–1 (stability $A$) through to 6–7 in stability $G$.

Values of coefficients specifying $\sigma_{yt}$, the so-called microscale $\sigma_y$, i.e. not including any plume meander effects, are given in Table 10.3. This table also includes typical values of mixing heights in the stability categories A–G. The effect of the mixing height on vertical plume dispersion can be taken into account in the following modification of Equation 2 for ground level concentrations:

$$C(x) = \frac{Q}{\pi\sigma_y\sigma_z U} \exp\left[-\frac{y^2}{2\sigma_y^2}\right]\left\{\exp\left[-\frac{H_e^2}{2\sigma_z^2}\right] + \exp\left[-\frac{(2L - H_e)^2}{2\sigma_z^2}\right]\right\} \tag{9}$$

**Table 10.3**　*Typical mixing heights and coefficients in* $\sigma_y = cx^d$ *(x in km) for different stabilities*

|  | Stability | | | | | |
|---|---|---|---|---|---|---|
|  | *A* | *B* | *C* | *D* | *E* | *F/G* |
| Mixing height (m) | 1300 | 900 | 850 | 800 | 400 | 100 |
| c | 213 | 156 | 104 | 68 | 50.5 | 34 |
| d |  | 0.894 | 0.894 | 0.894 | 0.894 | 0.894 | 0.894 |

where $L$ is the mixing height. This equation is not valid for $H_e > L$ when the concentration is zero (*i.e.* the pollutant is emitted above the mixing height).

Where long period averages (*e.g.*, annual) are of concern the detailed dependence on $\sigma_y$ is of much less importance and the pollutant concentrations can be assumed to be uniformly distributed cross-wind within each wind sector. For a 30° sector the first two terms of Equation 9 become:

$$\frac{1.524Q}{U\sigma_z x} \tag{10}$$

The contribution of this wind sector to the overall annual average is then given by Equation 9 modified as in Equation 10 multiplied by the combined frequency of occurrence of that wind sector and stability category.

In Section 10.2.1 we introduced the concept of the logarithmic wind speed profile with height in conditions of neutral stability. In different atmospheric stability conditions the variation will be different, but, in general, the wind speed will increase with height because of the surface drag. As would be expected intuitively this variation is smallest in unstable conditions (since in such boundary layers there is a considerable degree of vertical mixing) and greatest in stable conditions (for the opposite reason). In general one can write:

$$U(z) = U_{10}(z/10)^{\alpha} \tag{11}$$

where $U_{10}$ is the 10 metre wind speed and $\alpha$ is $\sim 0.15$ in unstable conditions, $\sim 0.2$ in neutral and $\sim 0.25$ in stable conditions.

There are several interesting derivations from the standard Gaussian equation and a particularly useful one is the formula giving the concentrations from a line source (of infinite length) obtained by integrating, for simplicity, Equation 2 over $y$ to yield:

$$C(x) = \sqrt{\frac{2}{\pi}} \frac{Q}{\sigma_z U} \exp\left[-\frac{H_e^2}{2\sigma_z^2}\right] \tag{12}$$

which can be used to estimate the concentration downwind of roads, for example. Here, $Q$ is the mass emission rate per unit length of road ($\mu g \, m^{-1} \, s^{-1}$).

### 10.3.2  Emission Inventories

We have already seen how important it is to specify the emission rate of a single source in order to model concentrations with confidence. In single stack applications this is often relatively straightforward. However, in multiple source applications such as in the use of an urban air quality model, there can in principle be literally thousands of individual sources. It would be clearly impracticable to attempt to quantify the emission rate of every house, office, shop and car in, say, London so methods have to be devised of making the problem tractable yet retaining as accurate a description of reality as possible.

The usual way of achieving this is to apportion the area to be modelled into a grid and to combine all the numerous small emitters within each grid square (such as individual houses, cars, *etc.*) into so-called 'area sources'. Major sources are usually treated explicitly as individual point sources. The size of the grid square used will usually be determined by the size of the area, or domain, to be modelled and the computing resources available. Typical grid scales are 1 km (or smaller) for urban areas, 20 km for nationwide modelling in a country the size of the UK, and 50–100 km for European or other international scale long range transport.

The specification of emissions is therefore fundamental to modelling and, apart from single source problems, is a difficult task. The usual approach is to collect information on fuel consumption in particular sectors (such as power generation, domestic heating, *etc.*) and multiply this by appropriate emission factors which ideally will have been measured over a range of representative fuels, appliances and combustion conditions.

Very often such data on fuel consumption (or some other measure of industrial commercial activity) are available only on a large scale, *e.g.* at national level, when the area to be modelled is much smaller. The modeller then has to use some means of spatially disaggregating the total domain emissions over the individual grid square of the model. This usually involves the introduction of another level of uncertainty as surrogate statistics have to be employed – for example domestic heating emissions may be assumed to have the same spatial pattern over the model grid as population for which data are often fairly readily available. Other surrogates which can be used are office floor space for emissions from the commercial sector and population for motor vehicle emissions. If the domain is of an appropriate size then questionnaires and other, often labour intensive, techniques can be used to assess the magnitude and spatial pattern of emissions in a particular town or city.[7] Some examples of urban and national emission inventories are given in Figures 10.5 and 10.6. The emissions shown in Figure 10.5 formed part of a study of the London area by the London Research Centre[7] which collected data on fuel use and traffic activity and produced inventories of the emissions of six pollutants for 1990. Figure 10.5 illustrates the emissions of carbon monoxide in the London area (the data are displayed at 2 km resolution for clarity although the original study used a resolution of 1 km). Figure 10.6 is a map of total $NO_x$ emissions for the UK for 1993 at a resolution of $20 \times 20$ km grid squares and has been derived by NETCEN[8] on the basis of national and regional fuel use and other statistics, and forms part of the National Atmospheric Emissions Inventory.

We have already seen, in Section 10.3.1, how the averaging time inherent in the dispersion model structure and parameters should match that of the

[7]'Energy use and the Environment', eds. M. Chell and D. Hutchinson, London Research Centre, London Energy Study, 1993.
[8]H. S. Eggleston and G. McInnes, 'Methods for the Compilation of UK Air Pollutant Emission Inventories', Warren Spring Laboratory Report LR 634 (AP), Stevenage, UK, 1987.

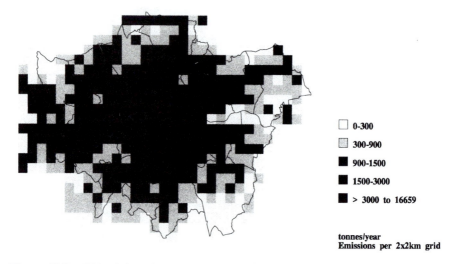

**Figure 10.5**   *1990 emissions of carbon monoxide in the London area (2 km squares)*

concentrations being modelled. Similarly it is vital that the emissions should be of the appropriate timescale. This is often straightforward in the case of annual averages when annual emissions must be used, but can be more difficult if seasonal or diurnal variations are being modelled. However, various approaches are possible.

Fuel requirements for space heating purposes (residential, commercial and a proportion of industry) depend on the ambient temperature and therefore vary with the season of the year. The Degree–Day principle[9] can be used to calculate these temperature dependent emission rates, $E(T)$, from the annual average emission rate $E$

$$E(T) = E_0[0.33 + 0.11(14.5 - T)] \quad \text{for} \quad T \leqslant 14.5°\text{C}$$
$$E(T) = 0.33 \, E_0 \qquad\qquad\qquad\quad \text{for} \quad T > 14.5°\text{C} \tag{13}$$

A further factor can be introduced if required to take account of the typical diurnal variation in emissions.

When modelling traffic pollution, diurnal variations in traffic flow are often available so that emissions can be scaled accordingly. One very important feature in dealing with traffic pollution is the variation of emissions with speed. This is particularly important for the pollutants carbon monoxide (CO) and hydro-carbons both of which, being products of incomplete combustion, are formed in the biggest quantities at low speeds. Recent work at WSL using on the road

[9]'Degree Days', Gas Council, London, Technical Handbook No. 101.

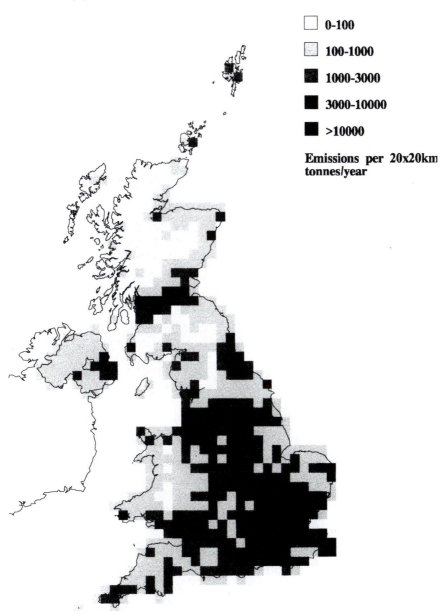

**Figure 10.6**   *Total $NO_x$ emissions in the UK*

measurements in actual driving conditions[10] has begun to quantify this effect. For example emissions of CO can vary from 25–30 g km$^{-1}$ at 20 kph to 5–10 g km$^{-1}$ at 100 kph.

[10]C. J. Potter and C. A. Savage, 'A Summary of Gaseous Pollutant Emissions from Tuned In-service Gasoline Engined Cars Over a Range of Road Operating Conditions'. Warren Spring Laboratory Report LR 447 (AP), Stevenage, UK, 1983.

### 10.3.3  Long Range Transport

With the increasing interest in the problems of acid rain and photochemical ozone in recent years, the modelling of these phenomena has become important. There are several important features which must be considered in modelling long (in this context, greater than 100 km) as opposed to shorter range transport. Firstly the time scales of transport such that the removal processes of wet and dry deposition must be incorporated. Secondly large scale meteorological features must be taken into account which involve specifying the movement of an air mass on a synoptic scale. Thirdly chemical reactions will generally be important and must be included. To incorporate all these effects in detail demands immense computing resources, as well as raising questions over how accurately the input data on emissions and meteorology can be known and how well one can describe the detailed physics and chemistry of the processes. In practice, therefore, simplified models have been used. These have generally been of two types, Lagrangian, where a series of air mass trajectories is followed, or Eulerian, where the governing equations are solved for every grid point in the domain at each time step. In many ways Lagrangian models are the simpler in concept and application, as Eulerian models generally require greater complexity of input data and computing requirements, as well as suffering from numerical 'pseudo-diffusion' if not appropriately constructed. The Lagrangian models in widest use in the UK take two forms, statistical models and trajectory/box models. The former are typically applied to long period averages of concentration and deposition fields of acidic pollutants such as $SO_2$ and sulfate, $NO_x$ and nitrate, and use of climatological data or annual frequencies of wind directions, speeds, stability categories, *etc.* Rainfall may be taken into account in several ways, the simplest but not necessarily the most satisfactory being to assume continuous constant rainfall (at an annual average rate). Alternatively the sporadic or stochastic nature of rainfall can be incorporated in a probabilistic way.[11] Trajectory/box Lagrangian models are usually applied to a succession of air mass trajectories arriving at a receptor at relatively short intervals (*e.g.* every six hours in the case of the most recent UN/ECE EMEP model[12]). The specification of the trajectories is a fundamental step in using these models and they are usually obtained from the detailed models used in national meteorological services. In practice back-track trajectories up to 96 hours are used; longer timescales would introduce unacceptable errors. As it is, errors in trajectories increase with time back along the path, and particularly in slack pressure areas these can be very large even at relatively short times. These models work by moving the box or air parcel along the trajectory and at each time step the appropriate emissions are introduced from the underlying grid, pollutant is lost by dry deposition at the rate given by $v_g c$ where the deposition velocity $v_g$ is appropriate to the underlying surface type. Wet deposition removes pollutant according to the rainfall field at

[11] H. Rodhe and J. Grandell, 'On the Removal Time of Aerosol Particles from the Atmosphere by Precipitation Scavenging', *Tellus XXIV*, 1972, **5**, 442–454.

[12] Ø. Hov, A. Eliassen and D. Simpson, 'Calculation of the Distribution of $NO_x$ Compounds in Europe', in 'Tropospheric Ozone, Regional and Global Scale Interactions', ed. I. S. A. Isaksen, Reidel, Dordrecht, 1988.

the particular location of the air parcel, and chemical transformations of reactive species are updated since the previous time step.[12] A good summary of the use of Lagrangian and Eulerian long range transport models of acid rain has been given by Pasquill and Smith.[1]

Much use has been made of Lagrangian box models in modelling photochemical ozone formation in the UK[13] and elsewhere.[14] Because the computational demands of the physical and meteorological aspects of the problems are relatively small, particularly if a one or two layer box model is used, quite complex chemical schemes can be used to describe the chemical processes involved. The model developed by Derwent,[13] for example, uses 339 chemical reactions describing the fate of 40 species. Explicit chemistry is used rather than so-called 'lumped' schemes where for example one hydrocarbon is used as a surrogate for its class (*e.g.* propene could be used to describe all alkenes, *etc*).

In attempting to simulate observed ozone concentrations with Lagrangian models, the correct specification of the air parcel's trajectory is of paramount importance. This may not be easy particularly, as we have noted, when anticyclonic conditions with slack pressure gradients exist. Then, although conditions may be optimal for ozone formation, the uncertainties in the calculated trajectories are often at their greatest. Large differences in the calculated ozone concentration can result, depending on the quantity of precursors (nitrogen oxides and hydrocarbons) picked up along each trajectory.

### 10.3.4 Operational Models

Although much basic research has been, and continues to be, carried out in dispersion modelling, an essential feature of models is their use in practical operational situations. Such uses are often made by non-specialists, and in the last decade or so there have been some significant developments in packaging models to facilitate their use. These developments have been made possible by the rapid expansion in personal and desk-top computing over this period, so that quite major calculations are now possible on PC-based systems without recourse to sophisticated computing facilities. It is worthwhile repeating, however, the crucial importance of high quality input data, particularly on emissions, as the numerical modelling calculations become easier through the use of standard packages.

Even without recourse to computers, some very useful estimates of air quality impacts of single sources can be made by the use of graphical workbooks, such as the well-known 'Workbook of Atmospheric Dispersion Estimates' by Turner,[15] and the NRPB report R91 referred to earlier.[2]

---

[13]R. G. Derwent and A. M. Hough, 'The Impact of Possible Future Emission Control Regulations on Photochemical Ozone Formation in Europe'. HMSO, London, Harwell Report AERE R 12919, 1988.

[14]A. Eliassen, Ø. Hov, I. S. A. Isaksen, J. Saltbones and F. Stordal, 'A Lagrangian Long-Range Transport Model with Atmospheric Boundary Layer Chemistry', *J. Appl. Met.*, 1982, **21**, 1645–1661.

[15]D. B. Turner, 'Workbook of Atmospheric Dispersion Estimates', Second Edition, Lewis Publishers, Chelsea, Michigan, 1994.

While such workbooks are probably most useful for screening calculations, more complex calculations require computer-based models, and the US EPA has produced a set of approved models, based on the Gaussian Plume approach, in its Users Network for Applied Modeling of Air Pollution (UNAMAP) system. A variety of models is available on diskette at low cost, covering single and multiple point source, area and line source applications. The models also have options to allow the influence of complicating factors such as building effects and the effects of complex terrain to be taken into account, in a relatively simple and approximate way. To assist the non-specialist in the use of these models, various companies operate training courses or consultancies. Other developments of the basic Gaussian plume model which have been produced by some companies in recent years have involved the addition of user-friendly software to assist in the development and handling of emission inventories and monitoring data for use with the models as part of an integrated package. While the input and output routines are often sophisticated, at heart these systems at present are generally built on the basic Gaussian plume model, although modular upgrading is often possible, to incorporate developments in dispersion theory.

Considerable developments in dispersion theory have been made over the past two decades, and the approximations inherent in the Gaussian plume approach have become more clearly understood. This has led to the development of a 'second generation' of operational models in the UK and in Europe and the USA. In the UK the most widely known of such models is the so-called UK-ADMS (Atmospheric Dispersion Modelling System) funded by a consortium of bodies including regulatory agencies, electricity generators, nuclear agencies and industrial companies. ADMS uses a numerical description of the boundary layer based on the Monin–Obhukov length scale, which essentially measures the relative contributions to boundary layer turbulence from convection and from mechanical wind shear. One practical consequence of the revised boundary layer description is that the standard deviations of the pollutant plume ($\sigma_y$ and $\sigma_z$ in the above discussion) can now vary with source height, a feature which accords more closely to observations. Equally, the boundary layer description also facilitates the incorporation of modules to treat building and topography effects in a more coherent manner.

Although these newer models offer the attraction that they incorporate more recent thinking on the properties of the atmospheric boundary layer and its effects on pollutant dispersion, there remains further work to be done in assessing the performance of such models in practical operational applications.

## 10.3.5  Accuracy of Models

In using models such as those described in Sections 10.3.2 and 10.3.3 for air pollution control purposes, for assessing air quality impacts or for elucidating chemical and transport mechanisms, it is important to assess the accuracy with which the model can reproduce observed concentrations. Detailed model valida-tion exercises can be expensive, involving considerable resources in measurements of pollutants at many locations and timescales, and the appropriate meteorological variables over the domain of the model. In general the more complex the model, the more complex is the validation required. However, the confidence one has in the output of a model depends very much on the questions one is attempting to answer. The prediction of a peak concentration at a specific location will place different demands on a model from answering a question such as whether or not controlling a particular category of source in a region would have a beneficial or an adverse effect.

In general, long-term (*e.g.* annual) average concentrations can be predicted with greater confidence than short-term (hourly or less) averages. Some assessments of the likely accuracy of dispersion models in predicting the concentrations of non-reactive pollutants have been given by Jones[16] for single source situations. In summarizing the work of several authors, he suggests that annual averages from a low level release can be predicted within a factor of about 2. At larger distances the factor increases with concentrations at about 100 km being predicted to within a factor of four with high probability. Factor of two accuracy for peak hourly concentrations is also suggested by Jones with the indication that if specification of the time and location of the peak are also required then the accuracy is likely to be worse.

In urban areas, where there are usually numerous sources on all wind directions around receptors, annual average concentrations of relatively inert pollutants, such as $SO_2$, can generally be predicted to better than a factor of two accuracy on most occasions. A summary of applications of a climatological Gaussian plume model to several urban areas of the UK[17] as shown in Figure 10.7. This diagram shows the frequency distribution of the percentage error of the calculation of annual average $SO_2$ concentrations. The percentage error here is defined as $100 \times$ (modelled value $-$ observed)/observed. For London, for example, the modelled results were within $\pm 30\%$ of observed for about 75% of the 20 receptors considered.

In the context of the evaluation of air pollution control strategies it should be noted that the Gaussian plume model and the behaviour of the primary pollutants to which it is usually applied are linear in emission rates so that such models will predict concentration reductions proportional to emission reductions which might arise from any postulated control technology. This is very straightforward in single source problems; however, in urban areas a larger number of sources will be present and only one sector (*e.g.* domestic sources under smoke control or motor vehicles under emission regulations) may be subject to controls. The accuracy of prediction of the effects of these controls will then depend on the accuracy with which the proportional contribution of the particular sources is predicted by the model, and this may be difficult to assess.

Turning to the larger scale effects such as acid deposition and photochemical oxidant formation, the question of the linearity, or proportionality, is more

[16]J. A. Jones, 'What is Required of Dispersion Models and Do They Meet the Requirements?' Paper to 17th NATO/CCMS International Technical Meeting on Air Pollution Modelling and its Applications, Cambridge, 1988.
[17]M. L. Williams, 'Models as Tools for Abatement Strategies', in 'Acidification and its Policy Implications', ed. T. Schneider, Elsevier, Amsterdam, 1986.

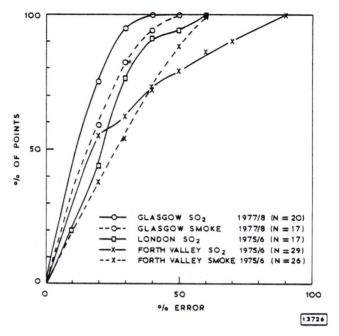

**Figure 10.7**  *Percentage of Receptor Points within percentages of observed concentrations for annual average $SO_2$ concentrations*

important, that is, whether or not for a given reduction in emissions of a particular species (such as $SO_2$ for example) there is likely to be a proportional reduction in deposition of sulfur. Total sulfur deposition over annual timescales has been shown to be approximately proportional[18] even though in some circumstances the wet deposition component may be non-linear.

The problem of the evaluation of photochemical ozone formation is more complex in that ozone is formed from the atmospheric reactions of nitrogen oxides and many individual hydrocarbon species. The governing reaction schemes are overall non-linear and can be further complicated by the fact that $NO_x$ (and some hydrocarbons) can act as both sources and sinks of ozone, on different scales.

With increasing interest in global problems such as the effects of so-called greenhouse gases on climatic change, and on the effects of man-made pollutants on stratospheric ozone, modelling techniques are now being used to advise pollution control policies over a very wide range of atmospheric problems.

[18]Acid Deposition in the UK 1981–1985', Second Report of the UK Review Group on Acid Rain, Warren Spring Laboratory, Stevenage, UK, 1987.

CHAPTER 11

# The Health Effects of Air Pollution

S. WALTERS and J. AYRES

## 11.1 INTRODUCTION

Worries that air pollution may have significant effects on health have recently been fuelled by publication of new evidence linking low levels of ambient air pollution with small public health effects. However this is a subject that also attracts a great deal of controversy, with strong views both in favour and against.

This chapter concentrates on health effects of air pollution in the general population at normal ambient levels. It will discuss important factors that need to be considered when reading about health effects of air pollution, enabling readers to appraise critically the extensive literature for themselves, and provide a brief summary of the effects on health of particulates, sulfur dioxide, nitrogen dioxide, ozone and carbon monoxide. It will also address the possible relationship between air pollution and cancer.

### 11.1.1 Exposure and Target Organ Dose

In order to suffer health effects, an individual must be exposed to a pollutant, and the pollutant must be able to reach those parts of the body which are vulnerable to its effect, which in most instances is somewhere in the respiratory tract. However the respiratory tract is very effective at dealing with noxious substances before they reach the lower airways and lung tissue. For example the nose and upper airways are excellent at filtering and removing coarse particulate material before it reaches the lower airways and lung tissue, and only a fraction containing the finest particles (under 10 $\mu$m) can reach these vulnerable sites. Therefore, although an individual may breathe a certain concentration of pollutant into the mouth or nose, this is not necessarily the concentration that reaches the target organ or tissue.

It may be easy to estimate the inhaled concentration of pollutant, but less easy to estimate the dose to which the target organ or tissue is exposed.

## 11.1.2  Factors Affecting Exposure

Target organ dose may be increased by exposure to greater pollutant concentrations, by exposure for a greater length of time or by behavioural factors, such as level of exercise and time spent in different microenvironments. For example, the concentration of particulates with an outdoor source is usually greater outdoors than indoors, although for ultrafine particles the difference in concentration may be very small, and the total particulate concentration including those of indoor origin may be higher indoors. Conversely, the concentration of nitrogen dioxide may be higher indoors, particularly if gas is used for heating or cooking.[1] Therefore people who work outdoors, or children who play outdoors, may be exposed to greater particulate doses, whilst those who remain indoors may be exposed instead to higher doses of nitrogen dioxide.

Dose may also be affected by exercise. This both increases the volume of air that is inhaled per minute, and decreases the effectiveness of the nasal filter due to increased mouth breathing. However faster breathing rates are associated with reduced ultrafine particle deposition due to a reduced time available for sedimentation.

Dose may also be enhanced by individual sources of exposure, notably cigarette smoking and occupational exposures, and possibly by co-exposure to other factors such as other pollutants or viral infections.

## 11.1.3  What is a Health Effect?

This is not a simple question. It is easy, for example, to measure small but subtle changes in lung function (for example changes in the amount of air that can be exhaled in one second, or the $FEV_1$) or bronchial reactivity (how sensitive the lung is to challenge with drugs that makes the airways constrict) when an individual is exposed to pollutants in the laboratory. However, these changes are usually transient and fully reversible in the experimental setting, and it may be argued that they are not lasting health effects. It may also be that a small change in bronchial reactivity is of no consequence to a normal individual, but makes a great deal of difference to a person suffering from asthma, who may suffer an attack as a result.

At the other end of the scale, we may observe an increase in deaths or hospital admissions on days following high levels of pollution. However this increase may represent in the majority of cases an effect on individuals who were already suffering from severe disease, brought forward maybe by just a few weeks or days. Figure 11.1 represents a pyramid of severity of effects that may be observed after exposure to pollutants. The strata are not meant to be quantitative, since quantification of the effects of air pollution is complex. The further question remains as to where effects cease to be reversible and become lasting health effects.

---

[1]F. E. Speizer, B. Ferris Jr, Y. M. Bishop and J. Spengler, 'Respiratory Disease Rates and Pulmonary Function in Children Associated with $NO_2$ exposure', *Am. Rev. Respir. Dis.*, 1980, **121**, 3–10.

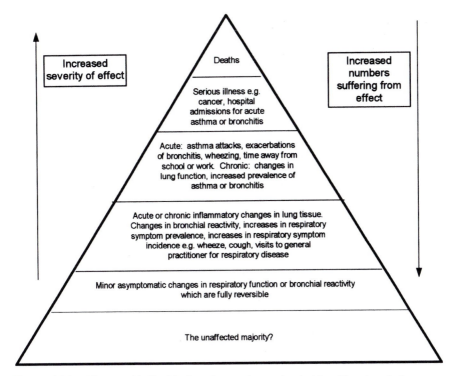

**Figure 11.1**  *The pyramid of health effects that may be associated with ambient air pollution*

### 11.1.4  Time Scales of Exposure–Effect Relationships

Some pollutants may have very rapid effects: for example sulfur dioxide causes constriction of the airways (bronchoconstriction) in sensitive individuals after a few minutes exposure.[2] However some pollutants may only affect individuals after prolonged or repeated exposure, which may take years. A good example is that of asbestos and pleural cancer, which occurs 20–30 years after exposure. It is much harder to establish a relationship for conditions with long latencies, such as cancer, since it is difficult to estimate exposure over a long period when both the concentration and the composition of ambient pollution has been changing, and when individuals move their residence, change their jobs and may change their smoking habits.

Pollutants may also interact with each other[3] or with other environmental factors such as temperature and airborne allergens,[4] to produce a health effect.

[2]J. R. Balmes, J. M. Fine and D. Sheppard, 'Symptomatic Bronchoconstriction after Short-term Inhalation of Sulfur Dioxide', *Am. Rev. Respir. Dis.*, 1987, **136**, 1117–1121.

[3]W. S. Linn, D. A. Shamoo, K. R. Anderson, R-C. Peng, E. L. Avol and J. D. Hackney, 'Effects of Prolonged, Repeated Exposure to Ozone, Sulfuric Acid and their Combination in Healthy and Asthmatic Volunteers', *Am. J. Respir. Crit. Care. Med.*, 1994, **150**, 431–440.

[4]N. A. Molfino, S. C. Wright, I. Katz, S. Tarlo, F. Silverman, P. A. McClean, J. P. Szalai, M. Raizenne, A. S. Slutsky and N. Zamel, 'Effect of Low Concentrations of Ozone on Inhaled Allergen Responses in Asthmatic Subjects', *Lancet*, 1991, **338**, 199–203.

**Table 11.1**   *Effects that have been postulated for exposure to pollutants in the short, medium and long-term*

---

*Short-term health effects (minutes to months)*
- Inflammatory cells in the lung
- Bronchoconstriction
- Changes in bronchial reactivity
- Minor respiratory symptoms – cough, wheeze, sore throat, sore eyes
- Hospital admissions for respiratory and circulatory diseases
- Deaths from respiratory and cardiac diseases

*Medium-term health effects (months to 10 years)*
- Increased prevalence of cough, wheeze, asthma, bronchitis
- Increased susceptibility to infection
- Reduction in lung function
- Reduction in lung growth in children
- Long-term inflammatory changes in bronchial walls
  (particularly smoke and $SO_2$ induced)

*Long-term health effects (10 or more years)\**
- Increased incidence of lung cancer
- Increased mortality from cardio-respiratory diseases

---

*These effects are more debatable but are likely to be real although the size of effect is very small

The exposure required to produce the effect may be either simultaneous or sequential, and the concentration of pollutant contributing to such an effect may vary from episode to episode. Table 11.1 shows health effects of exposure to pollutants in the short, medium and long-term.

### 11.1.5   Confounding Factors

In epidemiological studies of the relationship between exposure and disease, a possible cause may appear to be related to the potential effect, but this is only because they are both related to a third *confounding* factor. Confounding may occur either for the disease or the exposure, and may be in the short or long term.

For example, short-term exposure to high levels of particulates in winter is associated with still and cold anticyclonic weather conditions. However, cold snaps of this kind are also associated with increased mortality from cardiac and respiratory complaints.[5,6] Therefore an observed association between high levels of particulates and increased mortality may be due to confounding effects of temperature, unless this is taken into account in the analysis.

In the longer-term, we may observe a high prevalence of asthma and respiratory disease in residents near an industrial site. However, the population

[5]J. P. Mackenbach, C. W. N. Looman, and A. E. Kunst, 'Air Pollution, Lagged Effects of Temperature, and Mortality: The Netherlands 1979–87', *J. Epidemiol. Comm. Health*, 1993, **47**, 121–126.
[6]D. B. Frost, A. Auliciems, and C. de Freitas, 'Myocardial Infarct Death and Temperature in Auckland, New Zealand', *Int. J. Biometeorol.*, 1992, **36**, 14–17.

living in the zone of high exposure may live in poor social conditions, have high levels of occupational exposure, a high prevalence of cigarette smoking and so on. All these factors are also associated with respiratory disease, and unless they are taken into account the effects of the factory on health may appear exaggerated.

## 11.2 STUDYING THE HEALTH EFFECTS OF AIR POLLUTION

Bearing in mind the above difficulties in establishing a health effect, and determining exposure to pollutants for individuals and for target organs, the effects of air pollution on health can be studied in a variety of different ways ranging from simple experiments to complex epidemiological studies.

### 11.2.1 Experimental Studies

Experimental studies are designed to test a hypothesis, and therefore involve a measured exposure in a controlled system designed to test a specific effect. Experiments may be designed either to test the dose at which a toxic response occurs, or to observe more subtle mechanisms of effect. In general, experimental studies involve *in vivo* exposure in animals, or *in vitro* exposure of human or animal tissue or bacterial cultures. Experiments may involve exposure to higher concentrations than in ambient settings to produce a rapid result, or be specifically designed to look at the effect of low level concentrations. The toxicity of many atmospheric pollutants could only be determined in experimental settings. The experimental approach has also been used to determine the mutagenicity of potential carcinogens (cancer-causing agents), and the effects of combinations of pollutants, for example particulates coated with weak solutions of acid. They are very valuable in that confounding factors are eliminated by the controlled conditions, but can be misleading if different species have different susceptibility to a pollutant, or where specific combinations of exposure are needed to produce an effect.

### 11.2.2 Controlled-chamber Experiments

These consist of experimental exposures carried out under controlled conditions on human volunteers to examine physiological effects in terms of lung function or symptoms. By exposure to different concentration:time combinations it can be possible to establish a dose–response relationship. The volunteers can be normal individuals or those suffering from certain medical conditions such as asthma. Target organ dose may be increased by making subjects exercise. The outcomes measured usually include lung function, bronchial reactivity, markers of inflammation (in washings from the lung or nose) or the presence of symptoms. If an association between exposure and effect is found, causality may be established because of controlled exposure to a single or known combination of pollutants. They are also useful in establishing the range of inter-subject variability, and whether certain groups of people are more sensitive. However, these experiments can only consider short exposures (hours), and, because they are difficult and expensive to do, small numbers of subjects. People with severe respiratory disease may be unable to participate, and volunteers may not be typical of either the general normal population, or of all people with asthma or respiratory complaints.

### 11.2.3 Epidemiological Studies

Epidemiological studies estimate the effect on the health of whole populations, rather than individuals in experimental settings. They may deal with either short-term health effects resulting from short-term variations in exposure, or long-term health effects resulting from long-term average exposure.

*11.2.3.1 Short-term Studies.*   Short-term epidemiological studies are of two essential types:

1. *Ecological studies* – These examine the effects of day-to-day changes in air pollution levels on routinely measured health outcomes such as hospital admissions or mortality
2. *Panel studies* – In these studies, panels of individual volunteers measure their lung function and record the presence of symptoms every day, which are then related to ambient air pollution levels. These short-term studies generally refer to *incidence* or onset of disease *e.g.* asthma attacks, people developing a new cough on a particular day and so on.

Epidemiological studies are useful because they reflect real-life exposure conditions. Ecological studies cover the whole population, and therefore no groups are excluded, whilst panel studies can be specifically constructed to look at effects in vulnerable groups, or differences in individual and group sensitivity. Both types of study may be continued over relatively long time periods. However they do suffer from great problems with confounding due to concurrent fluctuations in other environmental factors (temperature, humidity, pollen count, virus infections), and cyclical fluctuations and trends in hospital admissions and mortality due to other causes. For a measured ambient level of pollution, the exposures of individuals in the population will differ greatly due to behavioural factors. It is also difficult to estimate the relative importance of different pollutants contained in mixtures which tend to rise and fall concurrently with the weather. For this reason statistical analysis needs to be complex, and it is not usually possible to infer causality from these studies. Finally, it is not possible to infer long-term health effects from the presence of short-term changes in response to changes in pollution. For example, we can conclude that asthma is made worse by air pollution exposure, but not that new cases of asthma are caused by air pollution exposure.

*11.2.3.2 Long-term Studies.*   In contrast to short-term studies, those in the long term examine disease *prevalence i.e.* the proportion of the population suffering from a particular disease or symptom. These studies are of three main types:

1. *Cross-sectional studies* – In these studies, the prevalence of disease in different communities is compared contemporaneously with the average ambient level of pollution in those communities,
2. *Cohort studies* – These studies follow up a group of individuals over a period of time, looking for changes in disease prevalence in relation to changes in average pollution exposure in this group,
3. *Migrant studies* – These study prevalence of disease in groups whose pollution experience has changed by virtue of migration to an area with markedly different ambient levels.

These studies also reflect real-life exposure conditions. However they suffer both from the effect of confounding factors (such as effects of cigarette smoking, exercise, occupation, race, socio-economic deprivation) on disease prevalence, and also the problem of estimating exposure over the whole latent period during which disease may develop. Very frequently it is not possible to take account of all such confounding factors in the design or analysis of cross-sectional studies. In addition, current disease patterns may reflect differences in exposure of individuals that were present 30 years ago but no longer exist today. Cohort studies allow more precise control of confounding factors, since these can be determined for all individuals enrolled, but such studies are expensive and time-consuming, and it may be difficult to both trace individuals and estimate their exposure over a long period. Migrant studies suffer from the problem that migants often differ fundamentally both from the population they left and the new population that they join, but can be very useful in establishing whether there is, for example, a critical age at which exposure needs to occur.

Finally, there is often a problem with statistical power. In order to be reasonably sure that a finding of no association is true, a large sample size is required, otherwise we may be missing a small but important effect. In this instance, this means comparing disease prevalence in many different sites, which is difficult. Consequently many studies compare only two or three places, making interpretation difficult.

## 11.2.4 Estimating Public Health Effect

Once a link between pollutant and health effect has been recognized, it is important to establish its potential effect on a population, in order to recommend air quality standards which adequately protect public health. Therefore we need to know the magnitude of the effect, the dose–response relationship, whether there is a threshold for effect and whether any groups are at particular risk.

These relationships may be different depending on the type of study (*controlled-chamber* or *epidemiological*), on the population to whom it refers (studies performed in the United States may not apply to the United Kingdom), on the average ambient levels of pollutants and on the proportion of the population falling into at-risk groups. For example, studies on effects of nitrogen dioxide appear to show effects in some ecological studies at concentrations 10 to 100 times lower than are required to produce effects in controlled-chamber experiments.[7] Some studies also appear to show no threshold for effect (this has been suggested to be the case for particulates, for example), but it is clearly not practicable to require a zero standard for airborne particulates because of natural sources of particulates such as wind-blown dust, pollen, spores and seawater aerosols, which cannot be controlled. Epidemiological studies are not often able to distinguish between the pollutant which produces an effect and those which simply co-vary with that pollutant. Finally, there are relatively few studies which provide *direct* comparison between normal individuals and vulnerable groups, or vulnerable groups which may contain a range of susceptibility.

People may be considered to be at risk either because they have greater *exposure* to pollutants (due to age, behaviour, occupation, activity levels), or because they are more sensitive to a given level of pollutant (*e.g.* some patients with asthma), or because the consequences of a given effect are greater for some individuals than others (*e.g.* increased bronchial reactivity is of greater consequence to somebody with asthma than to a normal individual). At-risk groups might therefore include children and infants, the elderly, people with existing respiratory disease (asthma, chronic obstructive pulmonary disease), people with ischaemic heart disease (angina pectoris), expectant mothers and non-smokers. The special problems of these groups must be taken into account when assessing appropriate air quality standards.

## 11.3  HEALTH EFFECTS OF PARTICULATES

Particulate air pollution was one of the first types of pollution demonstrated to have serious health effects, and for which there is greatest evidence of consistent short and long-term health effects at current low ambient levels in the United States and United Kingdom. There is now good evidence to suggest that short-term changes in mortality, hospital admissions, lung function and symptoms are associated with ambient levels of particulates, and there is some evidence to suggest that long-term prevalence of symptoms and mortality from respiratory diseases is associated with ambient particulate levels.

### 11.3.1  Experimental Studies

Because of the difficulties in generating particulates in experimental settings that adequately reflect the mix of particulates in the ambient air, few experimental studies have been published. Relatively little is therefore known about the mechanisms by which particulates produce either acute or chronic health effects.

There is some evidence that sulfuric acid[8] and ammonium bisulfate can induce airway narrowing in subjects with asthma. Whether acidic particles exert a permissive effect by enhancing the effect of other factors is uncertain.

---

[7] A. Ponka, 'Asthma and Low Level Air Pollution in Helsinki,' *Arch. Environ. Health*, 1991, **46**, 262–270.
[8] J. Q. Koenig, W. E. Pierson and M. Horike, 'The Effects of Inhaled Sulfuric Acid on Pulmonary Function in Adolescent Asthmatics', *Am. Rev. Respir. Dis.*, 1983, **128**, 221–225.

## 11.3.2 Ecological Epidemiological Studies

Following a severe episode of winter smog lasting four days in London during 1952 over 4000 excess deaths occurred, mainly from respiratory and cardiac conditions and largely in the elderly.[9] During this episode black smoke levels rose well over 4500 $\mu$g m$^{-3}$, along with high levels of sulfur dioxide and acid. This led to legislation to control urban pollution in the United Kingdom and elsewhere. In more recent years, concern has focused on the potential health effects of much lower levels of pollutants, particularly inhalable particulates (PM$_{10}$). Recent series of analyses from the United States have demonstrated consistent relationships between PM$_{10}$ (or other measures of particulates) and daily mortality, with approximately a 1% rise in all causes of mortality for every 10 $\mu$g m$^{-3}$ increase in PM$_{10}$.[10–16] The effect seems to be greatest for respiratory and circulatory causes of death and in the elderly, and for three–five day moving average particulate levels rather than hourly peak levels. These relationships appear to have no lower threshold, and to occur at levels well below current US air quality standards for particulates. These changes should be considered in the light of an average PM$_{10}$ level of 25–30 $\mu$g m$^{-3}$ in the United Kingdom.

There is also consistent evidence from the United States and Europe that hospital admissions and emergency room attendances for respiratory complaints and asthma are related to ambient levels of particulates, both smoke and PM$_{10}$ again without threshold and well below the ambient air quality standards.[17–19] In Birmingham, UK, hospital admissions for asthma and respiratory disease are significantly associated with levels of both black smoke[20] and PM$_{10}$.[21] A

[9]Ministry of Health, 'Mortality and Morbidity During the London Fog of December 1952,' HMSO, London, 1954.

[10]J. Schwartz and D. W. Dockery, 'Particulate Air Pollution and Daily Mortality in Steubenville, Ohio', *Am. J. Epidemiol.*, 1992, **135**, 12–19.

[11]J. Schwartz, 'Particulate Air Pollution and Daily Mortality in Detroit', *Environ. Res.*, 1991, **56**, 204–213.

[12]J. Schwartz and D. W. Dockery, 'Increased Mortality in Philadelphia Associated with Daily Air Pollution Concentrations', *Am. Rev. Respir. Dis.*, 1992, **145**, 600–604.

[13]C. A. Pope, J. Schwartz and M. R. Ransom, 'Daily Mortality and PM$_{10}$ Pollution in Utah Valley', *Arch. Environ. Health*, 1992, **47**, 211–217.

[14]J. Schwartz, 'Air Pollution and Daily Mortality in Birmingham, Alabama', *Am. J. Epidemiol.*, 1993, **137**, 1136–1147.

[15]J. Schwartz, 'Particulate Air Pollution and Daily Mortality in Cincinnati, Ohio', *Environ. Health Perspect.*, 1994, **102**, 186–189.

[16]P. L. Kinney, K. Ito and G. D. Thurston, 'A Sensitivity Analysis of Mortality/PM$_{10}$ Associations in Los Angeles', *Inhalation Toxicol.*, 1995, **7**, 59–69.

[17]C. A. Pope, 'Respiratory Hospital Admissions Associated with PM$_{10}$ Pollution in Utah, Salt Lake and Cache Valleys', *Arch. Environ. Health*, 1991, **46**, 90–97.

[18]J. Schwartz, D. Slater, T. V. Larson, W. E. Pierson and J. Q. Koenig, 'Particulate Air Pollution and Hospital Emergency Room Visits for Asthma in Seattle', *Am. Rev. Respir. Dis.*, 1993, **147**, 826–831.

[19]J. Sunyer, M. Saez, C. Murillo, J. Castellsague, F. Martinez and J. M. Anto, 'Air Pollution and Emergency Room Admissions for Chronic Obstructive Pulmonary Disease: A 5-year Study', *Am. J. Epidemiol.*, 1993, **137**, 701–705.

[20]S. Walters, R. K. Griffiths and J. G. Ayres, 'Temporal Association between Hospital Admissions for Asthma in Birmingham and Ambient Levels of Sulfur Dioxide and Smoke', *Thorax*, 1994, **49**, 133–140.

[21]J. Wordley, S. Walters and J. G. Ayres, 'Short-term Variations in Particulate Air Pollution and their Association with Hospital Admissions and Mortality in Birmingham,' *Thorax*, 1995, **50** (Supp. 2); A34.

10 $\mu$g m$^{-3}$ rise in PM$_{10}$ is associated with between 1.5 and 5% increase in hospital admissions or attendances. Where studies have considered other pollutants simultaneously, the association with PM$_{10}$ usually appears to be the strongest, and remains significant in multiple regression analysis.

### 11.3.3 Epidemiological Panel Studies

Panel studies in the United States and Europe have often, but not invariably, demonstrated a significant association between ambient levels of particulates and lung function, symptom incidence and use of treatment in children.[22,23] Children with symptoms appear in some, but not all, studies to be more sensitive to ambient particulates. The overall fall in lung function is small (less than 0.5% reduction in peak flow for every 10 $\mu$g m$^{-3}$ rise in particulates), and reversible. Recent studies have again suggested that there is no threshold, and that a relationship remains below existing air quality standards.

### 11.3.4 Long-term Epidemiological Studies

Many early studies did not adequately adjust for confounding factors such as smoking prevalence, and only compared two or three sites. In general, these suggested that the prevalence of respiratory symptoms was higher in polluted sites, but only a few studies have demonstrated a reduction in lung function. An association between particulates and prevalence of reduced lung function has been found in at least one large and well-conducted study[24]

The Harvard Six Cities Study carried out over a long period of time, has produced several important findings. They found that in children, prevalence of cough, bronchitis and chest illness was associated with particulates (PM$_{15}$ and PM$_{2.5}$, although less strongly with gaseous pollutants. There was no association between particulates and lung function. Prevalence of cough doubled over the measured range of pollutants (PM$_{2.5}$ 12–37 $\mu$g m$^{-3}$), and the effect on symptom prevalence in children with asthma was greater than for normal children. This, and other studies, revealed that the fine fraction of particulates and sulfate levels, showed strongest association with symptoms.[25,26] More recently, a fourteen year follow-up of a cohort of adults from each of the six cities showed a significant association between ambient particulate levels and mortality from cardiorespiratory disorders and lung cancer. This study carefully adjusted for confounding factors, although it could take no account of the possible effects of

[22] C. A. Pope and D. W. Dockery, 'Acute Health Effects of PM$_{10}$ Pollution on Symptomatic and Asymptomatic Children', *Am. Rev. Respir. Dis.*, 1992, **145**, 1123–1128.

[23] G. Hoek, B. Brunekreef and W. Roemer, 'Acute Effects of Moderately Elevated Wintertime Air Pollution on Respiratory Health of Children', *Am. Rev. Respir. Dis.*, 1992, **142**, A88.

[24] J. Schwartz, 'Lung Function and Chronic Exposure to Air Pollution: A Cross-sectional Analysis of NHanes II', *Environ. Res.*, 1989, **50**, 309–321.

[25] J. H. Ware, B. G. Ferris, D. W. Dockery, J. D. Spengler *et al.* 'Effects of Ambient Sulfur Oxides and Suspended Particles on Respiratory Health of Preadolescent Children', *Am. Rev. Respir. Dis.*, 1986, **133**, 834–842.

[26] D. W. Dockery, F. E. Speizer, D. O. Stram, J. H. Ware *et al.* 'Effects of Inhalable Particles on Respiratory Health of Children', *Am. Rev. Respir. Dis.*, 1989, **139**, 587–594.

pollution exposure in early life, before subjects were enrolled into the study. The closest association was observed between $PM_{2.5}$ and mortality, followed by sulfate and other particulate measures. Mortality was over 30% higher in the most polluted city compared to the least polluted city, although pollution levels were relatively low in all cities.[27]

## 11.4 HEALTH EFFECTS OF SULFUR DIOXIDE

Sulfur dioxide is a potent bronchoconstrictor at high levels, and patients with asthma are much more sensitive than normal individuals. Because levels of sulfur dioxide and particulates co-vary closely it has proved hard to demonstrate effects of sulfur dioxide that are independent from the effects of particulates in epidemiological studies. It is likely that sulfur dioxide contributes to respiratory symptoms, reduced lung function and rises in hospital admissions seen during pollution episodes.

### 11.4.1 Experimental Studies

Exposure to high levels of sulfur dioxide over a long period produces structural changes in the lung, with thickening of the lung lining, increase in glandular tissue, thickening of the protective mucus layer and reduction in mucus transport (which clear both mucus and other debris from the lung). Sulfur dioxide may also enhance sensitization to allergens and allergic response once challenged in animals.[28]

### 11.4.2 Physiological Studies

Sulfur dioxide produces bronchoconstriction in both normal and asthmatic individuals after exposure for only a few minutes. However the concentration required to produce an effect in asthmatic individuals is only one tenth of that required for normal individuals.[29] Exercise enhances the effect.[30] Both normal and asthmatic individuals vary in their sensitivity to the effect, so that although mean levels required to produce an effect in asthma patients may still be high in comparison with ambient levels (500–1000 $\mu g\ m^{-3}$, 200–400 ppb), the most sensitive individuals may respond at lower levels (100–300 ppb), or with greater falls in lung function.[31] Studies of the interaction between sulfur dioxide and

[27]D. W. Dockery, C. A. Pope, Xiping. Xu, J. D. Spengler *et al.* 'An Association between Air Pollution and Mortality in Six US Cities', *New England J. Med.*, 1993, **329**, 1753–1759.

[28]F. Riedel, S. Naujukat, J. Ruschoff, S. Petzoldt and C. H. Reiger, 'SO$_2$ Induced Enhancement of Inhalative Allergic Sensitization: Inhibition by Anti-Inflammatory Treatment', *Int. Arch. Allergy Immunol.*, 1992, **98**, 386–391.

[29]Department of Health, Advisory Group on the Medical Aspects of Air Pollution Episodes: Second Report, 'Sulfur Dioxide, Acid Aerosols and Particulates', HMSO, London, 1992, Chapter 6, pp. 71–100.

[30]D. Sheppard, A. Saisho, J. A. Nadel and H. A. Boushey, 'Exercise Increases Sulfur Dioxide Induced Bronchoconstriction in Asthmatic Subjects', *Am. Rev. Respir, Dis.*, 1981, **123**, 486–491.

[31]D. Horstman, L. J. Roger, H. Kehrl and M. Hazucha, 'Airway Sensitivity of Asthmatics to Sulfur Dioxide, *Toxicol. Ind. Health*, 1986, **2**, 289–298.

other pollutants, and the effect of sulfur dioxide on allergen responsiveness have been inconclusive.

### 11.4.3   Ecological Epidemiological Studies

Early ecological studies of the effect of particulates and sulfur dioxide on health did not attempt to separate the effects of particulates and sulfur dioxide on mortality, since these co-varied closely. In more recent studies, independent effects on mortality have been found for particulates but not sulfur dioxide. Ambient levels of sulfur dioxide were significantly associated with hospital admissions for respiratory conditions and asthma during summer in Canada.[32] A careful study of the effects of particulates and sulfur dioxide on emergency room visits for chronic obstructive pulmonary disease showed significant associations below current European guide levels in Barcelona, Spain.[19] An association was also found between sulfur dioxide levels and hospital admissions for asthma in Birmingham, UK, during the summer, with no association with particulate levels.[20] Although it is clear that ambient levels of sulfur dioxide may still be associated with health effects below current air quality standards, not all studies show this effect.

### 11.4.4   Epidemiological Panel Studies

As with ecological studies, the effects of sulfur dioxide are difficult to separate from those of particulates in most early studies. Panel studies of children following winter episodes of sulfur dioxide and particulate pollution in Europe have shown a reversible 5% fall in respiratory function following the episode, although which pollutant was responsible is not clear.[33] In some panel studies, both normal children and those with chronic respiratory symptoms have shown reductions in lung function with rises in ambient particulate and sulfur dioxide pollution, but the evidence for an effect on lung function at ambient levels is weaker for sulfur dioxide than for particulates. It is thought unlikely that any effect occurs at concentrations below 200 $\mu$g m$^{-3}$ (80 ppb), a level that is rarely exceeded in the United Kingdom nowadays.

### 11.4.5   Long-term Epidemiological Studies

Studies considering sulfur dioxide and the prevalence of respiratory disease have also apparently considered particulates and other pollutants. Some early studies did not use sufficiently sophisticated analysis to separate out the effects of individual pollutants. In several early studies the prevalence of respiratory symptoms was considered to be associated with ambient sulfur dioxide and particulate levels, at levels between 60 and 140 $\mu$g m$^{-3}$ (20–50 ppb) but in

[32]D. V. Bates, M. Baker-Anderson and R. Sizto, 'Asthma Attack Periodicity: A Study of Hospital Emergency Visits in Vancouver', *Environ. Res.*, 1990, **51**, 51–70.

[33]B. Brunekreef, H. Lumens, G. Hoek *et al.* 'Pulmonary Function Changes Associated with an Air Pollution Episode in January 1987', *JAPCA*, 1989, **39**, 1444–1447.

general, association has only been shown with respiratory symptom prevalence and not with lung function changes, at levels higher than current UK ambient levels.[34] These effects could not be distinguished from those of particulates, and at current much lower ambient levels the importance of $SO_2$ in this area must be regarded with caution.

## 11.5 HEALTH EFFECTS OF NITROGEN DIOXIDE

The rise in nitrogen dioxide emissions has led to concern about its health effects. However, evidence for significant short-term and long-term health effects of nitrogen dioxide is less consistent than for particulates and sulfur dioxide.

### 11.5.1 Experimental Studies

Nitrogen dioxide is an oxidizing agent and can damage lung tissue *via* its oxidizing properties. At very high doses it acts as a potent initiator of inflammation within the lung, preferentially affecting the small airways, close to the site of gas exchange in the lungs. Animal studies have demonstrated that nitrogen dioxide in high doses can impair ability to fight infection, although all studies involved nitrogen dioxide exposure to levels far exceeding ambient air, and exposure to high doses of infective agent. One study in humans showed no significant increase in infection rates, although it was higher in nitrogen dioxide exposed individuals.[35] Long-term exposure to high concentration may produce lung scarring (fibrosis) and emphysema in animals.[36]

### 11.5.2 Physiological Studies

Many controlled human exposure experiments have been carried out, often with conflicting results. This is because exposures took place under different conditions, in subjects with different characteristics, often with different measured endpoints. A recent meta-analysis[37] demonstrated that at exposures below 1880 $\mu g$ $m^{-3}$ (1000 ppb) only 47% of normal individuals showed increases in bronchial reactivity, whereas above this level 79% showed increases. In people with asthma, at exposures under 940 $\mu g\ m^{-3}$ (500 ppb) 69% showed changes in bronchial reactivity at rest, and 51% when exercising. The conclusions were that people with asthma were generally more sensitive than normal individuals to nitrogen dioxide, who were unlikely to show significant responses under 1880 $\mu g$ $m^{-3}$ (1000 ppb) and that exercise modifies response. The lowest level at which

[34]World Health Organization, 'Sulfur Oxides and Suspended Particulate Matter', Environmental Health Criteria No 8, World Health Organization, Geneva.

[35]S. A. Goings, T. J. Kille, L. R. Sauder *et al.*, 'Effects of Nitrogen Dioxide Exposure on Susceptibility to Influenza A Virus Infection in Healthy Adults', *Am. Rev. Respir. Dis.*, 1989, **139**, 1075–1081.

[36]K. Kubota, M. Murakami, S. Tanaka *et al.*, 'Effects of Long-term Nitrogen Dioxide Exposure on Rat Lung: Morphological Observations', *Environ. Health Perspect.*, 1987, **73**, 157–169.

[37]L. Folinsbee, 'Does Nitrogen Dioxide Exposure Increase Airways Responsiveness?', *Toxicol. Ind. Health*, 1992, **8**, 273–283.

nitrogen dioxide has been *consistently* shown to affect people with asthma is 564 $\mu$g m$^{-3}$ (300 ppb),[38,39] which is well above average UK ambient levels (56.4 $\mu$g m$^{-3}$, 30 ppb).

One study has shown an interaction between nitrogen dioxide and sulfur dioxide,[40] and another recent study from the UK has shown that low level (762 $\mu$g m$^{-3}$, 400 ppb) nitrogen dioxide exposure enhanced the response to allergen challenge in people with allergic asthma.[41]

### 11.5.3   Ecological Epidemiological Studies

In most studies of short-term variations in ambient nitrogen dioxide and health, no effect has been found on hospital admissions or mortality, even when these were found for other pollutants in the same study. Studies from Finland have shown an association between nitrogen dioxide and hospital admissions or clinic attendances for asthma, after taking other pollutants into account,[7,42] and a few other studies have shown an association at very low levels. Other studies which have claimed to show associations between nitrogen dioxide and mortality or hospital admissions have not considered whether these were independent of other pollutants. It is therefore likely that the association, if any, between nitrogen dioxide and hospital admissions or mortality is weaker than for other pollutants.

### 11.5.4   Epidemiological Panel Studies

Again, the majority of these studies have shown no association between ambient nitrogen dioxide levels and respiratory function. However, some well-designed studies have shown an association between nitrogen dioxide and respiratory symptoms in healthy individuals and people with asthma, the latter studies from Arizona also demonstrating a significant independent effect on lung function in asthmatic individuals.[43] One recent study from the United Kingdom found an association between nitrogen dioxide levels and lung function in patients with asthma, but particulates were not measured in this study.[44] In general, the

[38]L. J. Roger, D. H. Horstmann, W. F. McDonnell *et al.*, 'Pulmonary Function Airway Responsiveness and Respiratory Symptoms in Asthmatics Following Exercise in NO$_2$', *Toxicol. Ind. Health*, 1990, **6**, 155–171.

[39]M. A. Bauer, M. J. Utell, P. E. Morrow *et al.*, 'Inhalation of 0.30 ppm Nitrogen Dioxide Potentiates Exercise-induced Bronchospasm in Asthmatics', *Am. Rev. Respir. Dis.*, 1986, **134**, 1203–1208.

[40]R. Jorres and H. Magnussen, 'Airways Response of Asthmatics after a 30 minute Exposure at Resting Ventilation to 0.25 ppm NO$_2$ or 0.5 ppm SO$_2$', *Eur. Respir. J.*, 1990, **3**, 132–137.

[41]W. S. Tunicliffe, P. S. Burge and J. G. Ayres, 'Effect of Domestic Concentration of Nitrogen Dioxide on Airways Responses to Inhaled Allergen in Asthmatic Patients', *Lancet*, 1994, **344**, 1733–1736.

[42]O. V. I. Rossi, V. L. Kinnula, J. Tienari and E. Huhti, 'Association of Severe Asthma Attacks with Weather, Pollen and Air Pollutants', *Thorax*, 1993, **48**, 244–248.

[43]M. D. Lebowitz, L. Collins and C. I. Hodberg, 'Time Series Analysis of Respiratory Responses to Indoor and Outdoor Environmental Phenomena', *Environ. Res.*, 1987, **43**, 332–341.

[44]B. G. Higgins, H. C. Francis, C. J. Yates, C. J. Warburton, A. M. Fletcher, J. A. Reid, C. A. C. Pickering and A. A. Woodcock, 'Effects of Air Pollution on Symptoms and Peak Expiratory Flow Measurements in Subjects with Obstructive Airways Disease', *Thorax*, 1995, **50**, 149–155.

association between ambient nitrogen dioxide on panels of individuals appears to be weaker and less consistent than that for other pollutants.

### 11.5.5 Long-term Epidemiological Studies

Many of the published studies simply compare a polluted with a non-polluted area, and are unable to distinguish between effects of different pollutants. Many published studies have concentrated on indoor exposure to nitrogen dioxide. In general, several cross-sectional studies have demonstrated an association between ambient indoor or outdoor levels of nitrogen dioxide and prevalence of respiratory symptoms, but not lung function, in children.[45,46] The reverse seems true in adults, namely there is an association between nitrogen dixoide exposure and lung function, but not symptoms.[47] One large and well-controlled study from the United States found an independent association between nitrogen dioxide and prevalence of low lung function in young people.[24] It is therefore possible that a long period of nitrogen dioxide is required to affect lung function, these changes becoming manifest only in adults, but the evidence is conflicting.

## 11.6  HEALTH EFFECTS OF OZONE

There is very good, consistent experimental evidence that ozone has an effect on health, with a consistent dose–response effect on a number of lung function parameters at concentrations close to those seen in ambient air. The evidence from panel studies supports this, although there has been little published evidence of effects on mortality or hospital admission, particularly in the UK.

### 11.6.1  Experimental Studies

Ozone is a very powerful oxidizing agent, causing direct cellular damage by damaging the anti-oxidant mechanisms in cells lining the airway walls. It acts preferentially in the small airways and gas-exchange regions of the lung. Prolonged exposure of animals to high doses results in persistent inflammation of the small airways, similar to that induced by cigarette smoking. It may also induce scarring (fibrosis) in the lung. Acute exposure to ozone produces acute inflammation at quite modest levels of exposure (under 2000 $\mu$g m$^{-3}$, 1000 ppb).

There is also some evidence that prolonged exposure to ozone may impair cellular defences.

---

[45] F. E. Speizer, B. Ferris Jr, Y. M. Bishop and J. Spengler, 'Respiratory Disease Rates and Pulmonary Function in Children Associated with NO$_2$ Exposure', *Am. Rev. Respir. Dis.*, 1980, **121**, 3–10.

[46] L. M. Neas, D. W. Dockery, J. W. Ware *et al.*, 'Association of Indoor Nitrogen Dioxide with Respiratory Symptoms and Pulmonary Function in Children', *Am. J. Epidemiol.*, 1991, **134**, 204–219.

[47] P. Fischer, B. Remijn, B. Brunekreef and K. Biersteker, 'Associations between Indoor Exposure to NO$_2$ and Tobacco Smoke and Pulmonary Function in Adult Smoking and Non-smoking Women', *Environ. International*, 1986, **12**, 11–15.

## 11.6.2  Physiological Studies

There have been a large number of these studies carried out, which demonstrate a consistent curvilinear relationship between inhaled ozone concentration and respiratory function at all levels of exercise.[48] Increasing inhaled ozone concentration had a greater effect than either increasing duration of exposure or increasing ventilation (by exercise). Ozone affects a number of lung function parameters, including bronchial responsiveness. Recent studies have demonstrated effects in exercising individuals at concentrations which frequently occur in ambient air (160 $\mu$g m$^{-3}$, 80 ppb).[49] Individuals can develop tolerance to ozone exposure (diminishing effects with repeated challenge) but it is not known whether this is an adaptive response.[50] Individuals vary widely in susceptibility to inhaled ozone, with some showing great sensitivity. This sensitivity is not confined to those with respiratory disease, but occurs just as frequently in normal individuals.[51] For example people exercising in 240 $\mu$g m$^{-3}$ (120 ppb) ozone for 6 hours showed between a 4% and 38% reduction in FEV$_1$ (forced expiratory volume in one second). Ozone exposure also produces symptoms of cough, breathlessness and chest discomfort.

A recent study has shown that exposure to a low level of ozone can reduce the threshold at which allergic subjects respond to allergen challenge in humans, confirming previous findings in other species, and the effect of NO$_2$ described earlier.[4]

A review of these physiological studies at the population level suggested that at ambient levels of 200 $\mu$g m$^{-3}$ (100 ppb) some sensitive subjects would suffer a 10% decrement in lung function.[52]

## 11.6.3  Ecological Epidemiological Studies

Several studies from North America have shown an association between ambient levels of ozone and hospital admissions for asthma, although not all studies controlled for the effects of temperature and other pollutants.[53,32] Indeed, some have found no association between ozone and hospital admissions for respiratory complaints. More recent studies, with better control of confounding factors, have suggested that between 6 and 24% of the total daily variability in summertime

[48]M. J. Hazucha, 'Relationship between Ozone Exposure and Pulmonary Function Changes', *J. Appl. Physiol.*, 1987, **62**, 1671–1680.

[49]W. F. McDonnell, H. R. Kehri, S. Abdul-Saleem *et al.*, 'Respiratory Response of Humans Exposed to Low Levels of Ozone for 6.6 hours', *Arch. Environ. Health*, 1991, **46**, 145–150.

[50]S. M. Horvath, J. A. Gliner and L. J. Folinsbee, 'Adaptation to Ozone: Duration of Effect', *Am. Rev. Respir. Dis.*, 1981, **123**, 496–499.

[51]D. H. Horstman, L. J. Folinsbee, P. J. Ives, S. Abdul-Saleem and W. F. McDonnell, 'Ozone Concentration and Pulmonary Response Relationships for 6.6 hour Exposures with Five Hours of Moderate Exercise to 0.08, 0.10 and 0.12 ppm', *Am. Rev. Respir. Dis.*, 1990, **142**, 1158–1163.

[52]Department of Health, Advisory Group on the Medical Aspect of Air Pollution Episodes: First Report, 'Ozone', HMSO, London, 1991.

[53]R. P. Cody, C. L. Weisel, G. Birnboaum and P. J. Lioy, 'The Effect of Ozone Associated with Summertime Photochemical Smog on the Frequency of Asthma Visits to Hospital Emergency Rooms', *Environ. Res.*, 1992, **58**, 184–194.

asthma admissions may be due to ozone, alone or in combination with other components of acid summertime haze, but these studies nearly all come from North America.[54] There is no consistent evidence for an acute effect of ambient ozone on mortality. Asthma mortality did not rise following an acute ozone episode in England in 1976.

## 11.6.4  Epidemiological Panel Studies

The majority of these studies have taken place in children in summer camps in North America,[55–57] although some very good European studies have also been published.[58] These show consistent, reproducible and significant reductions in lung function at maximum ambient ozone levels below 500 $\mu$g m$^{-3}$ (250 ppb) often below 200 $\mu$g m$^{-3}$ (100 ppb) equivalent to a 3% fall in lung function for every 200 $\mu$g m$^{-3}$ (100 ppb) rise in ozone, a very small change. A careful analysis in Dutch children showed that there was wide variability in response between children, suggesting that there may be a sensitive sub-group, which is not confined to children with pre-existing respiratory complaints.[59] The characteristics which make these people more sensitive are not known.

## 11.6.5  Long-term Epidemiological Studies

There have been few published studies comparing more than two areas, or with good control for confounding factors. Such studies as have attempted to look at the question of whether long-term exposure to ozone increases the prevalence (as opposed to incidence) of asthma have generally been poorly designed. One study in Seventh Day Adventists (who are non-smokers) showed that frequent exceedance of the 200 $\mu$g m$^{-3}$ (100 ppb) threshold for ozone was associated with an increased prevalence of asthma in adult males only.[60] The balance of evidence is against the possibility that chronic ozone exposure causes a non-asthmatic individual to develop asthma.

[54] G. J. Thurston, K. Ito, P. L. Kinney and M. Lippmann, 'A Multi-year Study of Air Pollution and Respiratory Hospital Admissions in Three New York State Metropolitan Areas: Results for 1988 and 1989 Summers', *J. Expo. Anal. Environ. Epidemiol.*, 1992, **2**, 429–450.

[55] M. Lippmann, P. Lioy, G. Leikauf *et al.*, 'Effects of Ozone on the Pulmonary Function of Children', *Adv. Mod. Environ. Toxicol.*, 1983, **5**, 423–446.

[56] P. L. Lioy, T. A. Vollmuth and M. Lippmann, 'Persistence of Peak Flow Decrement in Children Following Ozone Exposures Exceeding the National Ambient Air Quality Standard', *J. Air Pollut. Control Assoc.*, 1985, **35**, 1068–1071.

[57] D. M. Spektor, M. Lippmann, P. J. Lioy *et al.*, 'Effects of Ambient Ozone on Respiratory Function in Active Normal Children', *Am. Rev. Respir. Dis.*, 1988, **137**, 313–320.

[58] G. Hoek, P. Fischer, B. Brunekreef *et al.*, 'Acute Effects of Ambient Ozone on Pulmonary Function of Children in the Netherlands', *Am. Rev. Respir. Dis.*, 1993, **147**, 111–117.

[59] B. Brunekreef, P. L. Kinney, J. H. Ware, D. Dockery *et al.*, 'Sensitive Subgroups and Normal Variation in Pulmonary Function Response to Air Pollution Episodes', *Environ. Health Perspect.*, 1991, **90**, 189–193.

[60] D. E. Abbey, P. K. Mills, F. F. Ptersen and W. I. Beeson, 'Long-term Ambient Concentrations of Total Suspended Particulates and Oxidants Related to Incidence of Chronic Disease in California USA Seventh Day Adventists', *Environ. Health Perspect.*, 1991, **94**, 43–50.

## 11.7  HEALTH EFFECTS OF CARBON MONOXIDE

Carbon monoxide exerts its toxic effect by binding very avidly to haemoglobin, thereby reducing the oxygen-carrying capacity of the blood. In very high doses it is fatal due to cerebral and cardiac hypoxia. In lower concentrations it may affect higher cerebral function, heart function and exercise capacity, all of which are sensitive to lowered blood oxygen content.

It is possible to obtain a direct measure of carbon monoxide exposure in humans by measuring carboxyhaemoglobin levels, normally around 1% of total haemoglobin. Carbon monoxide is present in very high concentrations in cigarette smoke, and cigarettes constitute by far the greatest source of exposure in smokers. This section therefore concentrates on the potential health effects of carbon monoxide in non-smokers exposed to ambient carbon monoxide. There has been relatively little recent research into the potential health effects of low level carbon monoxide exposure.

### 11.7.1  Experimental Studies

Neurobehavioural effects have been extensively studied in animals. In general these represent reduced ability to carry out complex tasks, although these effects require concentrations greatly in excess of normal ambient exposure. Chronic exposure to low-level carbon monoxide can affect brain structure. Cardiac effects of carbon monoxide include effects on the electro-physiological properties of the heart at quite low levels of carboxyhaemoglobin (5.5%), and may reduce the threshold at which cardiac arrhythmias or arrest can occur.[61]

High levels of carbon monoxide during pregnancy can reduce foetal growth and survival in animals,[62] and carbon monoxide has been shown to preferentially bind to foetal haemoglobin.

### 11.7.2  Physiological Studies

Neurobehavioural effects on humans suggest that effects on complex task performance (such as driving skills) are unlikely to occur below 5% carboxyhaemoglobin. Many of these studies are not very recent, and do not employ current research design such as double-blind procedures, making results difficult to interpret.

Impairment of exercise performance as measured by maximum oxygen uptake may occur at carboxyhaemoglobin levels between 3–4%, although these levels are still rare in non-smokers.[63] Minor changes in the electrocardiograph may be found in normal individuals but the significance is unclear. Of greater importance are the findings that people with existing ischaemic heart disease who have

[61]D. A. De Bias, C. M. Banerjee, N. C. Birkhead *et al.*, 'Effects of Carbon Monoxide on Ventricular Fibrillation', *Arch. Environ. Health*, 1976, **31**, 42–46.

[62]L. D. Longo, 'The Biological Effects of Carbon Monoxide on the Pregnant Woman, Fetus and Newborn Infant', *Am. J. Obstet, Gynecol.*, 1977, **129**, 69–103.

[63]S. M. Horvath, P. B. Raven, T. E. Dahms and D. J. Gray, 'Maximum Aerobic Capacity at Different Levels of Carboxyhaemoglobin', *J. Appl. Physiol.*, 1975, **38**, 300–303.

symptoms of angina pectoris may suffer these symptoms after a lesser degree of exertion following exposure to carbon monoxide. In some good experiments, effects were seen in angina patients at carboxyhaemoglobin levels of 2 to 4%, which may be seen in certain groups of non-smokers with high carbon monoxide exposures.[64,65] In people who have a tendency to irregular heart rhythm, this was enhanced after carbon monoxide exposure (6% COHb), although the clinical significance is unknown.[66]

### 11.7.3 Ecological Epidemiological Studies

Despite the evidence that patients with angina may be affected by high carbon monoxide levels, there have been relatively few studies of the effects of ambient carbon monoxide. Some studies have found no association between hospital attendances or mortality from ischaemic heart disease and ambient carbon monoxide concentrations. Some studies have found a significant association between carbon monoxide and hospital admissions for myocardial infarction, or case-fatality (proportion of cases admitted to hospital who subsequently died of myocardial infarction once they had reached hospital).[67,68] However these did not take into account the potential confounding effect of temperature, and low temperature is a potent cause of coronary artery constriction.

A recent study found an association between ambient carbon monoxide and hospital admissions for ischaemic heart disease and heart failure. After adjusting for particulate exposure, the relationship with heart failure remained significant.[69]

### 11.7.4 Long-term Epidemiological Studies

The majority of evidence comes from occupational studies looking at the risk of mortality from ischaemic heart disease in groups with high occupational exposure to carbon monoxide, such as bridge and tunnel workers, and drivers.[70,71] Small

[64]E. Anderson, R. Andelman, J. Strauch *et al.*, 'Effect of Low Level Carbon Monoxide Exposure on Onset and Duration of Angina Pectoris: A Study in Ten Patients with Ischaemic Heart Disease', *Ann. Int. Med.*, 1973, **79**, 46–50.

[65]E. N. Allred, E. R. Bleecker, B. R. Chaitman *et al.*, 'Short-term Effects of Carbon Monoxide Exposure on the Exercise Performance of Subjects with Coronary Artery Disease', *New Eng. J. Med.*, 1989, **321**, 1426–1432.

[66]A. Hinderliter, K. Adams, C. Price *et al.*, 'Effects of Low-level Carbon Monoxide Exposure on Resting and Exercise-induced Arrhythmias in Patients with Coronary Artery Disease and No Baseline Ectopy', *Arch. Environ. Health*, 1989, **44**, 89–93.

[67]S. Cohen, I. M. Deane, and J. R. Goldsmith, 'Carbon Monoxide and Survival from Myocardial Infarction', *Arch. Environ. Health*, 1969, **19**, 510–517.

[68]A. C. Hexter, and J. R. Goldsmith, 'Carbon Monoxide Association of Community Air Pollution and Mortality', *Science*, 1971, **172**, 265–267.

[69]J. Schwartz, and R. Morris, 'Air Pollution and Hospital Admissions for Cardiovascular Disease in Detroit, Michigan', *Am. J. Epidemiol.*, 1995, **142**, 23–35.

[70]G. Paradis, G. Theriault, and C. Tremblay, 'Mortality in a Historical Cohort of Bus Drivers', *Int. J. Epidemiol.*, 1989, **18**, 397–402.

[71]F. Stein, W. Halperin and R. Horning, 'Heart Disease Mortality among Bridge and Tunnel Workers Exposed to Carbon Monoxide', *Am. J. Epidemiol.*, 1988, **128**, 1276–1288.

excess mortality has been reported in these groups after adjusting for smoking prevalence, but it is not clear whether carbon monoxide is causally associated.

Poor foetal outcome is associated with acute episodes of high maternal carbon monoxide exposure,[72] although one case-control study of low birthweight babies failed to find an association between low birth weight and ambient carbon monoxide concentration in the area of residence of the mother.[73]

## 11.8  AIR POLLUTION AND CANCER

### 11.8.1  Problems in Studying Air Pollution and Cancer

Although it is relatively easy to establish whether an airborne chemical has the potential to cause cancer (a *carcinogen*), it is much harder to determine whether it actually does so at ambient concentrations in human populations.

Establishing the potential of a chemical to cause cancer can be done by looking at:

1. Its ability to damage the genetic material in cell cultures or bacterial cultures
2. Its ability to cause cancer in animals
3. Whether there is a higher incidence of cancer in people who have worked with very high concentrations of the particular chemical.

However, the term 'cancer' represents a variety of different diseases with a complex natural history. Development of cancer may depend on a initial triggering event (*initiation*), followed by exposure to other chemicals which promote the development of tumours or prevent the body from rejecting tumour cells (*promotion*). It may be necessary for these events to occur repeatedly, or in a particular sequence, at particular levels of exposure, and the process from initiation to development of cancer (*latency*) in a human may take decades. Therefore, people developing cancer today may be doing so as a result of exposure to ambient air pollution that no longer exists, and of which there are unlikely to be adequate records.

There is also the problem that inhaled carcinogens tend to be associated either with very common cancers (*e.g.* lung cancer) or very rare cancers (*e.g.* some types of leukaemia). With the former, it is difficult to associate a small effect of low level ambient air pollution from the very great effects of other factors, particularly cigarette smoking. With the latter, it may take a lifetime to accumulate sufficient cases at ambient levels of exposure to have adequate statistical power to detect an effect.

Finally, where a link with human cancer has been established in the occupational setting, this usually involves exposures several orders of magnitude higher

[72]C. H. Norman, and D. M. Halton, 'Is Carbon Monoxide a Workplace Teratogen? A Review and Evaluation of the Literature', *Ann. Occup. Hyg.*, 1990, **34**, 335–347.

[73]B. W. Alderman, A. E. Baron, and D. A. Savitz, 'Maternal Exposure to Neighbourhood Carbon Monoxide and Risk of Low Infant Birth Weight', *Public Health Rep.*, 1987, **102**, 410–414.

than those in ambient air. It is difficult to know if the dose–response curve is valid at much lower exposure levels, or whether there is a threshold for effect. Furthermore the latency may be much longer at lower exposure concentrations, so if epidemiological studies do not take this into account they may have false negative results.

For these reasons, we are still unclear about the potential role of ambient air pollution in the aetiology of cancer in populations. Two examples are explored in this section: the role of benzene in the aetiology of leukaemia, and the role of polycyclic aromatic hydrocarbons in the aetiology of lung cancer.

## 11.8.2 Airborne Carcinogens

There are many substances which have been shown actually or potentially to cause cancer. These have been classified by the International Agency for Research on Cancer into different categories according to their ability to cause cancer (Table 11.2).

**Table 11.2** *The International Agency for Research on Cancer classification of human carcinogens*

*Group 1 – proven human carcinogens*
Chemicals for which there is sufficient evidence from epidemiological studies to support a causal association between exposure and cancer.

*Group 2 – probable human carcinogens*
Chemicals for which evidence ranges from inadequate to almost sufficient.

> *Group 2A*: Limited evidence of carcinogenicity in humans and sufficient evidence for carcinogenicity in animals.
> *Group 2B*: Inadequate evidence for carcinogenicity in humans and sufficient evidence for carcinogenicity in animals.

*Group 3 – unclassified chemicals*
Chemicals which cannot be classified in humans, usually because of inadequate evidence.

Group 1 carcinogens present in ambient urban air include benzene. Group 2A carcinogens include benzo(a)pyrene, benzo(a)anthracene and other polycyclic aromatic hydrocarbons (PAH). Group 2B carcinogens include 2-nitrofluorene, 1,6-dinitropyrene and 1-nitropyrene. In general, these carcinogens are present in minute quantities, but there is concern over the potential for low level exposure to cause cancer in human populations.

## 11.8.3 Benzene and Leukaemia

Benzene is a group 1 carcinogen, with proven causal association with acute non-lymphocytic leukaemia in humans. The main toxic effects occur on the bone marrow, with toxic exposures producing bone-marrow suppression, and reductions in red cell, white cell and blood platelet production (pancytopenia) which may lead to bone marrow failure (aplastic anaemia).

It is important to adjust for smoking in epidemiological studies because benzene is present in high concentrations in cigarette smoke. Smokers may have up to ten times the exposure of non-smokers, particularly in rural areas. In the longer term, studies in workers exposed to benzene have clearly demonstrated an excess risk of acute non-lymphocytic leukaemia, but in general this was not detectable in workers exposed to less than 1.5 mg m$^{-3}$ (500 ppb) over a working lifetime.[74] There is evidence of chromosomal abnormalities in workers exposed to slightly lower levels (0.6 to 40 mg m$^{-3}$, 200 to 1300 ppb) over a long time period (over 11 years). This contrasts to ambient levels which are usually under 13 $\mu$g m$^{-3}$ (4 ppb) in the United Kingdom at the urban roadside and under 3 $\mu$g m$^{-3}$ (1 ppb) in rural areas.

Population epidemiological studies are extremely difficult to carry out because of the rarity of this type of leukaemia. Estimates of toxicity at low levels of exposure are therefore made from occupational studies. A combination of estimates of risk from a variety of studies suggest that for a lifetime exposure (70 years) to 1 $\mu$g m$^{-3}$ (0.3 ppb) benzene, the excess risk is between three to 30 cases per million population, with the World Health Organization consensus estimate being an excess risk of around four.[75]

However, acute non-lumphocytic leukaemia is extremely rare. There are only about six to seven cases per million per year in the United Kingdom, or 420–490 per million over a lifetime of 70 years. An additional four cases resulting from ambient levels of 1 $\mu$g m$^{-3}$ (0.3 ppb) would be almost impossible to detect in epidemiological studies, and the potential risk at ambient levels of benzene remains difficult to prove at the population level, or indeed in workers with a modest working lifetime exposures. In practice, the risk of leukaemia from ambient benzene exposure is so small as to be unmeasureable.

### 11.8.4   Polycyclic Aromatic Hydrocarbons and Lung Cancer

Polycyclic aromatic hydrocarbons (PAH) collectively describes a large number of chemicals, many of which, with their metabolites and nitro-derivatives, are known to be animal or human carcinogens. The majority derive from the combustion of organic fuels, including wood, coal, oil, petrol and diesel as products of incomplete combustion. The best studied PAH is benzo(a)pyrene (BaP), which along with others is present in cigarette smoke.

It is known that clearance of BaP is reduced if it is absorbed onto particles, and that the dose required to produce tumours in animals is reduced if absorbed onto particles.[76] This may therefore be relevant to the situation in

[74]Department of the Environment, Expert Panel on Air Quality Standards: First Report, 'Benzene', HMSO, London, 1994.
[75]World Health Organization, 'Air Quality Guidelines for Europe', WHO Regional Publications, European Series No 23, WHO, Copenhagen, 1987.
[76]A. R. Sellakumar, R. Montesano, U. Saffiotti *et al.*, 'Hamster Respiratory Carcinogenesis Induced by Benzo(a)pyrene and Different Levels of Ferric Oxide', *J. Natl Cancer Inst.*, 1973, **50**, 507–510.

ambient urban air. The fraction of urban particulates containing PAH is known to have a carcinogenic effect in animals, and one study demonstrated a dose–response relationship to concentration of BaP in extracts from urban particulates.[77] Human cell mutagenicity has also been demonstrated for urban air particulate extracts. Vehicle exhaust condensates are known to be carcinogenic in animals, and about 40% of this may be attributable to PAH.[78] Carcinogenicity of vehicle exhausts is demonstrable in inhalation experiments and is likely to be mainly due to the particulate fraction and specifically to the four–seven ring PAH-containing fraction.[79] However these experiments establishing carcinogenicity in animals used overwhelming doses of exhaust particulates.

The majority of evidence for carcinogenicity of benzo(a)pyrene comes from occupational studies in coal gasification workers, and coke production. These suggested a dose-dependent risk of lung cancer after some adjustment was made for smoking, but not all studies measured PAH or benzo(a)pyrene exposure directly, but used proxy measures such as where subjects worked in the plant, and the duration of employment. These workers were exposed to levels of benzo(a)-pyrene over ten times those which occurred in normal urban air during the 1950s and 1960s.[80,81] More recent studies have detected an excess of lung cancer (for example a relative risk around 1.5 in truck drivers) in workers exposed to high concentrations of vehicle exhausts which could not be attributed to other occupational exposure or smoking.[82] Small excesses have been found in other occupations (relative risks between 1.5 and 7), but some studies have been negative, and none have directly measured levels of PAH to which workers were exposed.

There have been a number of reported epidemiological prevalence studies in the general population demonstrating an excess of lung cancer in urban dwellers over rural dwellers, but many early studies failed to take into account potential confounding factors, particularly cigarette smoking, occupation and socio-economic status. Most, but not all, studies which adjust for smoking prevalence show a raised relative risk of between 1.5 and

[77]F. Pott, R. Tomingas, A. Brockhaus and F. Huth, 'Studies on the Tumorigenic Effect of Extracts and their Fractions of Atmospheric Suspended Particulates in the Subcutaneous Test of the Mouse', Zbl. Bakt. I. Abt. Orig. B, 1980, **170**, 17–34.

[78]J. Misfield, The Tumour-producing Effects of Automobile Exhaust Condensate and Diesel Exhaust Condensate', in 'Health Effects of Diesel Engine Emissions', eds. W. E. Pepelko, R. M. Danner and N. A. Clarke, Proceedings of an International Symposium, Vol 2, USEPA No EPA-600/9-80-057b.

[79]G. Grimmer, H. Brune, R. Deutsch-Wenzel *et al.*, 'Contribution of Polycyclic Aromatic Hydrocarbons to the Carcinogenic Impact of Gasoline Engine Exhaust Condensates Evaluated by Implantation into the Lungs of Rats', *JNCI*, 1984, **72**, 733–739.

[80]J. W. Lloyd, 'Long-term Mortality Study of Steelworkers: V. Respiratory Cancer in Coke Plant Workers', *J. Occup. Med.*, 1971, **13**, 53–68.

[81]J. F. Hurley, R. Archibald, McL. Collings *et al.*, 'The Mortality of Coke Workers in Britain', *Am. J. Ind. Med.*, 1983, **4**, 691–704.

[82]R. B. Hayes, T. Thomas, D. T. Dilverman *et al.*, 'Lung Cancer in Motor Exhaust Related Occupations', *Am. J. Ind. Med.*, 1989, **16**, 685–695.

2.0.[83–85] However small differences in the age at which smokers commence regular smoking might produce a relative risk of 1.5 for lung cancer.[86] Urban air pollution has fallen markedly and changed in the sources of particulates and probably the relative content of benzo(a)pyrene during the 30 year period during which ambient air pollution might be expected to contribute to current lung cancer rates. Recent studies suggest that only a very small proportion (under 5%) of the excess of lung cancer in urban areas may be attributable to ambient air pollution.[87,88]

It has been estimated that the lifetime risk from exposure to BaP varies from 0.3 to 1.4 deaths per year per 10 000 population per ng m$^{-3}$ BaP. This has been calculated to represent only 3% of lung cancer deaths in Sydney[88] and between 2 and 20% of lung cancer cases in the United States,[89] concurring well with other epidemiological studies suggesting that less than 5% of urban cases of lung cancer may be attributable at least in part to ambient air pollution, of which not all can be attributed to BaP.

## 11.9  CONCLUSIONS

Although there is clear evidence, at least for some pollutants, of health effects at both individual and population level, the size of these effects is very small.

There is a need for more research, particularly into the quantification of both individual and public health effects of air pollution to guide those who need to control pollution. Research is also needed into the more subtle ways, hitherto unsuspected, in which air pollution may exert effects on health.

## 11.10  SUGGESTIONS FOR FURTHER READING

1. Department of Health, 'Reports of the Advisory Group on the Medical Effects of Air Pollution Episodes:
   I    Ozone; HMSO, London, 1991.
   II   Sulfur Dioxide, Acid Aerosols and Particulates; HMSO, London, 1992.
   III  Nitrogen Dioxide; HMSO, London, 1993.
   IV   Health Effects of Exposure to Mixtures of Air Pollutants', HMSO, London, 1995.

[83] E. C. Hammond and L. Garfinkel, 'General Air Pollution and Cancer in the United States', *Prev. Med.*, 1980, **9**, 206–211.

[84] W. Haenszel, D. B. Loveland and M. G. Sirken, 'Lung Cancer Mortality as Related to Residence and Smoking History in White Males', *J. Natl. Cancer Inst.*, 1962, 947–1001.

[85] W. Haenszel and K. E. Tacuber, 'Lung Cancer Mortality as Related to Residence and Smoking History II White Females', *J. Natl. Cancer Inst.*, 1964, **32**, 803–838.

[86] I. C. T. Nisbet *et al.*, 'Review and Evaluation of the Evidence for Cancer Associated with Air Pollution', Clement Associates, Washington DC, 1983, EPA-450/5-83-006.

[87] P. A. Buffler, S. P. Cooper, S. Stinnett *et al.*, 'Air Pollution and Lung Cancer Mortality in Harris County, Texas, 1979–81', *Am. J. Epidemiol.*, 1988, **128**, 683–699.

[88] D. J. Freeman and F. C. R. Cattell, 'The Risk of Lung Cancer from Polycyclic Aromatic Hydrocarbons in Sydney Air', *Med. J. Aust.*, 1988, **149**, 612–615.

[89] E. Haemisegger, A. Jones, B. Steigerwald and V. Thomson, 'The Air Toxics Problem in the United States: An Analysis of the Cancer Risks for Selected Pollutants', Washington DC: Environmental Protection Agency, 1984, No EPA-450/1-85-001.

2. Department of Health, 'Reports of the Committee on the Medical Effects of Air Pollution:
   I   Asthma and Outdoor Air Pollution; HMSO, London, 1995.
   II  Non-biological Particles and Health', HMSO, London, 1995.
3. World Health Organization, 'Air Quality Guidelines for Europe', WHO Regional Publications, European Series No. 23, WHO, Copenhagen 1987.
4. D. W. Dockery, C. A. Pope, Xiping Xu, J. D. Spengler *et al.* 'An Association between Air Pollution and Mortality in Six US Cities', NEJM: 329; 1753–9, 1993.
5. Department of the Environment, 'Reports of the Expert Panel on Air Quality Standards:
   I    Benzene; HMSO, London, 1994.
   II   Ozone; HMSO, London, 1994.
   III  1-3, Butadiene; HMSO, London, 1994.
   IV   Carbon Monoxide; HMSO, London, 1994.
   V    Sulfur Dioxide; HMSO, London, 1995.
   VI   Particles; HMSO, London, 1995.

CHAPTER 12

# Effects of Gaseous Pollutants on Crops and Trees

T. A. MANSFIELD and P. W. LUCAS

## 12.1 INTRODUCTION

Research at the beginning of the 20th century showed quite clearly that the huge amounts of air pollution in industrial cities were deleterious to plants. The best of the early experiments were those conducted in Leeds by Cohen and Ruston.[1] They found that plants grew three–four times larger on the outskirts of the conurbation than in the foul air of the city centre, and they found a remarkably good correlation between the estimated annual deposition of $SO_3$ and the stunting of growth (Figure 12.1). Effects of this kind were not confined to the UK, but were later identified throughout Europe and N. America in regions where industrial activity or large urban centres produced localized sources of pollutants from fossil fuel combustion. In 1928 W. W. Pettigrew[2] delivered a lecture which included a graphic description of the problem of growing ornamental plants in Manchester: 'While a smoke-laden atmosphere is inimical to both animal and vegetable life, its effects are undoubtedly more apparent if not more deadly in the case of vegetation ...'. He noted that in order to maintain Philips Park 'in a presentable condition', it had to be planted up each year with over 7000 trees and shrubs.

In western Europe and N. America we now consider that the atmospheres of our cities are much cleaner than in the first half of this century. Air pollution is certainly less visible than it was then, and the exclusion of light by smoke – which was probably an important contributory factor in reducing plant growth – is rarely now a major problem. There are, however, still a lot of concerns about the influence of urban pollution on plants, and about the wider distribution of effects through suburbs and surrounding countryside. In the USA in the 1950s, it became apparent that topographic and climatic factors could combine to

[1] J. B. Cohen and A. G. Ruston, 'Smoke: A Study of Town Air', Edward Arnold, London, 1912.
[2] W. W. Pettigrew, 'The Influence of Air Pollution on Vegetation', Lecture to Smoke Abatement League of Great Britain, Harrogate, 1928.

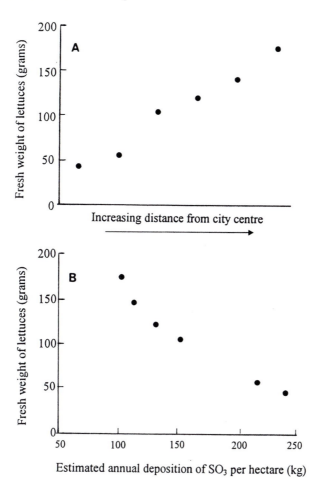

**Figure 12.1** *A. Fresh weights of lettuces grown at six different sites in Leeds and its suburbs between July and September 1911. B. Estimated annual deposition of SO₃ at the six sites showing a nearly linear relationship with the growth of the crop.* Plotted from data in the book by Cohen and Ruston[1]

produce photochemical smogs, and ozone damage to trees was well documented for areas such as the San Bernardino Forest to the east of the city of Los Angeles.

In the mid-1980s, researchers at Imperial College carried out experiments in London that were similar in concept to those performed by Cohen and Ruston in Leeds in the first decade of the century. They grew peas, clover and barley along a transect extending from central London in a south-westerly direction towards Ascot. There were improvements in plant performance further away from the city centre, and multiple regression analyses showed statistically significant relationships with concentrations of measured air pollutants ($NO_2$, $SO_2$ and $O_3$). Concentrations of $NO_2$ showed a steeper gradient along the transect than those of $SO_2$, and in Figure 12.2 we show the data for $NO_2$ alongside the dry

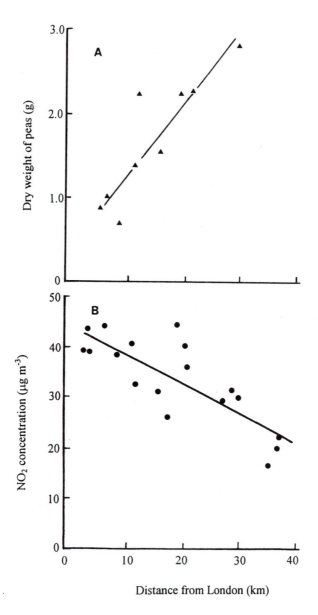

**Figure 12.2**   *A. Dry weights of garden peas (variety 'Progeta') grown at various distances from London in 1984. B. Mean $NO_2$ concentrations over the same transect during the summer months.* Compiled from data published by M. R. Ashmore and colleagues[3,3a]

[3]M. R. Ashmore, J. N. B. Bell and A. Mimmack, *Environ. Pollut.*, 1988, **53**, 99.
[3a]M. R. Ashmore and C. Dalpra, *London Environ. Bull.*, 1985, **3**, 4.

weights of peas (*Pisum sativum* L. cv. Progeta) produced at the different locations. Ashmore *et al.*[3] were correctly cautious in interpreting their findings, pointing out that experiments of this type with multiple environmental variables across sites can never provide conclusive proof of a causal link between an air pollutant and plant performance. Nevertheless, the close resemblance between Figures 12.1 and 12.2 is very striking, and it is clear that the air of a modern city remains toxic to plants despite efforts to regulate emissions of smoke. It seems likely, though unproven, that $NO_2$ may have replaced $SO_2$ as the main phytotoxic agent. Alternatively there may be synergistic effects of different pollutants, and we shall mention some possibilities later.

Apart from changes in the nature of pollutants in urban air, in the last few decades contaminants have become more widely distributed and the possibility that plants in rural areas may be affected has been much studied. In the 1960s and 1970s the research into effects of air pollutants on plants was mainly concerned with crops and grasslands, but during the 1980s the emphasis shifted from herbaceous plants to trees, particularly the economically important conifers grown in Europe and N. America. This shift occurred because of worrying reports of damage to forests in central Europe, Scandinavia and N. America, apparently worsening year by year, from about 1980 onwards.[4] Research on trees poses many problems and it was a long time before useful experimental results emerged, but we are now able to identify some effects of individual pollutants, especially ozone, that may contribute to aspects of 'forest decline'. There are however, many facets to this problem and much still needs to be done.

## 12.2  CURRENT KNOWLEDGE OF THE MECHANISM OF ACTION OF SOME MAJOR POLLUTANTS

### 12.2.1  Sulfur Dioxide

Sulfur is one of the important mineral nutrients of plants, and it is normally taken into the roots from the soil in the form of sulfate, then transported to the leaves. Enzymes for the reduction of sulfate are located in the leaves, and reduced forms of sulfur are used in the synthesis of amino acids. During the reduction of sulfate ions there is transient formation of sulfite, and there is unlikely to be any significant accumulation of this ion. Gaseous forms of sulfur can also be taken up directly by the leaves. Exchanges of carbon dioxide and water vapour between leaves and the atmosphere are fundamental to many physiological processes such as photosynthesis, respiration and transpiration. These gaseous exchanges are regulated by the opening and closing of stomata, and the uptake of pollutants such as $SO_2$ is also thought to occur primarily *via* these pores at the surface of the leaf.[5] There is, however, also evidence that some gaseous pollutants may be sorbed by the leaf cuticle which could have important implications in reducing cuticular integrity, leading to enhanced water loss.[6] There is, however, little

[4]L. W. Blank, *Nature*, 1985, **314**, 311.
[5]T. A. Mansfield and P. H. Freer-Smith, in 'Gaseous Air Pollutants and Plant Metabolism', eds. M. J. Koziol and F. R. Whatley, Butterworth, London, 1984, p. 131.
[6]K. J. Lendzian, *Aspects of Appl. Biol.*, 1988, **17**, 97.

likelihood that this sorption of $SO_2$ is accompanied by its significant permeation into the leaf through the cuticle.[7]

When $SO_2$ enters the leaves *via* the stomata, it dissolves in a film of water at the surface of the mesophyll cells inside the sub-stomatal cavity, forming sulfite and bisulfite ions. There is little evidence that $SO_2$ as a dissolved gas persists within the leaf.[8] The aqueous layer of the mesophyll cells is part of the pathway by which water and minerals from the soil are distributed to individual cells of the leaf. Thus sulfate ions will normally be present, but not sulfite or bisulfite in significant concentrations. The proton-pumping activities of the outer cell membrane (the plasmalemma), which are part of normal cellular functioning, alter the pH of the extracellular environment appreciably at different times of the day. Thus the equilibrium between the solution products of $SO_2$ may change, the ratio between sulfite and bisulfite increasing as the pH increases, but the balance is normally towards bisulfite.

We do not know to what extent the plasmalemma of the cells in a leaf can tolerate the presence of particular concentrations of sulfite and bisulfite, but damage to membranes is known to be one feature of cellular injury by $SO_2$. The permeability of the plasmalemma is known to be affected by exposure to $SO_2$ and essential ions such as potassium can then leak out of cells.[9] Nevertheless potassium-leakage rate for leaves are not well correlated with differences in $SO_2$ sensitivity between cultivars or species, and consequently it is not believed that the plasmalemma is the most susceptible cellular component to $SO_2$.[10] One reason for this may be the presence of the enzyme sulfite oxidase in the apoplastic (extracellular) water layer, which can act to detoxify sulfite by converting it to sulfate.[11]

The thylakoid membranes of chloroplasts seem to be more sensitive to the presence of $SO_2$. Swelling of these structures, which are the locations of the light-harvesting complexes and electron transport components, is seen very soon after the commencement of fumigation with $SO_2$.[12] This may explain why the process of photosynthesis is inhibited by $SO_2$. Despite this obvious sensitivity of the chloroplasts there is strong evidence that the primary sites of injury within leaves are located elsewhere. Noyes[13] fed bean (*Phaseolus vulgaris*) leaves with $^{14}CO_2$ and found that the translocation of $^{14}$C-labelled products out of the leaves was inhibited by $SO_2$ concentrations too small to reduce the rate of photosynthesis. Subsequent studies with several different species have confirmed the high sensitivity of assimilate translocation to $SO_2$.[14]

[7] K. Lendzian and G. Kerstiens, *Revs. Environ. Contam. Toxicol.*, 1991, **121**, 65.

[8] V. Black and M. H. Unsworth, *J. Exp. Bot.*, 1980, **31**, 667.

[9] E. Nieboer, D. H. S. Richardson, K. J. Puckett and F. D. Tomassini, in 'Effects of Air Pollutants on Plants', ed. T. A. Mansfield, Cambridge University Press, Cambridge, 1976, p. 61.

[10] D. T. Tingey and D. M. Olszyk, in 'Sulfur Dioxide and Vegetation', eds. W. E. Winner, H. A. Mooney and R. A. Goldstein, Stanford University Press, Stanford, California, 1985, p. 178.

[11] H. Rennenberg and A. Polle, in 'Plant Responses to the Gaseous Environment' eds. R. G. Alscher and A. R. Wellburn, Chapman & Hall, London, 1994, p. 165.

[12] A. R. Wellburn, O. Majernik and F. A. M. Wellburn, *Environ. Pollut.*, 1972, **3**, 37.

[13] R. D. Noyes, *Physiol. Plant Pathol.*, 1980, **16**, 73.

[14] S. B. McLauchlin and G. E. Taylor, in 'Sulfur Dioxide and Vegetation', eds. W. E. Winner, H. A. Mooney and R. A. Goldstein, Stanford University Press, Stanford, California, 1985, p. 227.

Long-distance transport of assimilates in plants involves their movement from sites of photosynthesis (principally the mesophyll cells) into the sieve tubes of the phloem. The final step is known as phloem-loading, and this can take place against a very large concentration gradient, *e.g.* 1:40. The energy required to transport sucrose against such a gradient is almost certainly generated by the consumption of ATP and extrusion of protons, which then return across the membrane 'carrying' sucrose molecules. Presumably the membrane site of this sucrose/proton co-transport may be the point of particular sensitivity of the solution products of $SO_2$. Minchin and Gould[15] used the very short-lived isotope [11]C to show that phloem-loading was almost immediately reduced after exposure of wheat (a $C_3$ plant) to $SO_2$. This was not the case, however, with maize, a $C_4$ plant in which there is a tight ring of cells protecting the phloem from the intercellular air spaces of the leaf. There is no such protective layer in wheat, and consequently $SO_2$ in the intercellular spaces may have almost direct access to the phloem tissue.

Studies of the effects of $SO_2$ on plants have often been performed in fumigation chambers subjected to natural illumination and ambient temperatures, and in some of these experiments the growth responses to the pollutant seemed to be affected by climatic changes. A study under controlled conditions with the grass *Phleum pratense* (Timothy) showed that when growth was rapid, in high irradiance and long days, 120 ppb $SO_2$ (higher than usually found in the most polluted situations nowadays) had no detectable effect on the plants. On the other hand the same concentration applied to plants in low irradiance and short days reduced growth by about 50 percent.[16] Similar changes in sensitivity were found when rate of growth was reduced by dropping the temperature.

The results of these and other similar experiments have led to the conclusion that there is no critical concentration which can usefully be regarded as the threshold for injury. Different amounts of $SO_2$ can clearly be tolerated under different climatic conditions, a factor which has not been considered in many of the models so far produced which attempt to evaluate the effects of pollutants on cultivated or natural vegetation. Perhaps our attention should be focused on those effects that occur under the least favourable conditions, particularly if they involve reductions in survival ability under extreme climatic conditions. There is now quite a considerable amount of evidence that plants exposed to $SO_2$ become more sensitive to frost injury. There is, for example, a substantial reduction in survival of $SO_2$-polluted ryegrass upon subsequent exposure to sub-zero temperatures.[17] It is not uncommon to see frost injury on pasture grasses during winter in the more polluted parts of the UK. The impact on the productivity of an established sward is, however, probably quite small because of recovery that can occur in spring and summer. The enhancement by $SO_2$ of frost sensitivity in woody plants is more likely to be of concern. There is evidence that even dormant deciduous plants may be affected by $SO_2$ uptake into the shoots in winter, with

[15]P. E. H. Minchin and R. Gould, *Plant Science*, 1986, **43**, 179.
[16]T. Jones and T. A. Mansfield, *Environ. Pollut.* Ser. A, 1982, **27**, 57.
[17]A. W. Davison and I. F. Bailey, *Nature*, 1982, **297**, 400.

subsquent death of the terminal buds.[18] This topic is covered in more detail in Section 12.4.2.1.

In some agricultural areas the soil is deficient in sulfur, and it is possible that $SO_2$ (and also $H_2S$) in the atmosphere might remedy this deficiency. The total sulfur deposition per unit area may exceed 6 g m$^{-2}$ per year in parts of the UK, and more than 12 g m$^{-2}$ per year in central and eastern Europe.[19] Such large inputs exceed the sulfur requirements of most plants. There is evidence that when supplies of sulfate in the soil are inadequate to support growth, then $SO_2$ in the atmosphere can make up the deficit.[20] When the plants are growing rapidly under favourable conditions, the $SO_2$ can be regarded as beneficial, but in most industrialized countries in Europe and N. America, $SO_2$ pollution is greater in winter when plant growth is slowest. To benefit from $SO_2$ the leaves must be able to use the supply of sulfur as it enters the cells. If metabolism is proceeding too slowly for this to happen then toxic ions such as $SO_3^{2-}$ and $HSO_3^-$ may accumulate with resulting damage to cells and cellular processes. It is thus possible to put mistaken emphasis on the beneficial effects of atmospheric $SO_2$. Sulfur deficiencies in the soil can be remedied by increases in the sulfate content of fertilizers at small cost to farmers, and the economic advantage of sulfur supplied by $SO_2$ pollution is therefore very small.

## 12.2.2 Nitrogenous Pollutants

It has long been recognized that nitrogen, which is a major constituent of all proteins and nucleic acids, is required by all plants for normal growth. As a consequence, plants have developed a range of diverse mechanisms to enhance their acquisition of this important nutrient, for example fixation of atmospheric nitrogen by symbiotic bacteria in legumes, associations with mycorrhizal fungi and insectivory. The majority of plants obtain nitrogen through root absorption of the inorganic ions ammonium $(NH_4^+)$ and nitrate $(NO_3^-)$ from the soil solution and have developed pathways of nitrogen assimilation which are now well characterized (Figure 12.3). (Soil processes such as nitrification and mineralization which influence the availability of nitrogen are complex and space does not permit a description of these processes here). Considerably less is known, however, about how plants that are adapted to low nitrogen inputs, particularly those growing in sensitive ecosystems such as forests or heathlands, react to additional nitrogen deposition from anthropogenic sources, either as gaseous nitrogen compounds $(NO, NO_2$ and $NH_3)$ or from wet deposition in rain or mist $(NO_3^-$ and $NH_4^+)$. The amount and forms of deposited nitrogen vary depending on the specific region under consideration.[21,22] Inputs of between 10 and 85 kg ha$^{-1}$ yr$^{-1}$ largely as $NH_3$ and $NO_2$, have been suggested for low

[18]T. Keller, *Environ. Pollut.*, 1978, **16**, 243.
[19]'Acid Rain', Report no 14, Watt Committee on Energy, London, 1984.
[20]D. W. Cowling and M. J. Koziol, in 'Effects of Gaseous Air Pollution in Agriculture and Horticulture', eds. M. H. Unsworth and D. P. Ormrod, Butterworth, London, 1982, p. 349.
[21]F. T. Last, J. N. Cape and D. Fowler, *Span*, 1986, **29**, 2.
[22]D. Fowler, in 'Air Pollution Effects on Biodiversity', eds. J. R. Barker and D. T. Tingey, Van Nostrand Reinhold, New York, 1992, p. 31.

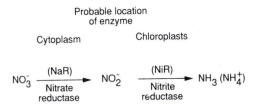

**Figure 12.3** *The reduction pathway from nitrate to ammonia which occurs in plants prior to amino acid synthesis*

altitude areas of western Europe, and more than 20 kg ha$^{-1}$ yr$^{-1}$ (predominantly as $NO_3^-$ and $NH_4^+$ in cloud and rainfall) for some high altitude regions of Britain and W. Germany. Vitousek[23] has estimated that human activities have altered the global biogeochemical cycle of nitrogen to such an extent that 'natural' N fixation has now been overtaken by anthropogenic N fixation. The long-term effects of high inputs of nitrogen to sensitive ecosystems are, however, only poorly understood, especially on soil processes,[24] the availability of other nutrients[25] and the possible interaction with environmental stresses such as frost.[26,27] In the southern Pennines in northern England the atmospheric concentrations of $SO_2$ have declined considerably since the 1950s but there has been no recovery of the *Sphagnum* moss, the decline of which was attributed to pollution from the nearby urban areas. This is thought to be because the *Sphagnum* is now primarily affected by nitrogenous pollutants which have increased in concentrations over the period during which $SO_2$ has declined.[28] Over the past ten years there has been a great deal of emphasis on studies of the impacts of $NH_3$ pollution, amounts of which have risen alarmingly in countries such as the Netherlands where farm animal production is very intensive. Duyzer *et al.*[29] estimated that at Speulderbos in the central Netherlands, the average annual flux of $NH_3$ amounted to about 50 kg N ha$^{-1}$ yr$^{-1}$ by dry deposition alone, which is appreciably higher than the critical load for coniferous forests. The precise reasons for the toxicity of $NH_3$ are not fully understood, but it is likely that the protons released when $NH_3$ assimilation becomes intensive can cause cellular acidosis.[30] There is also evidence that uptake of excess N into leaves can cause mineral nutrient imbalances in plants.[31,32]

*12.2.2.1 Responses to gaseous $NO_x$.* Most research has concentrated on the effects of $NO_2$ rather than NO, although the latter is often a major component of $NO_x$,

[23]P. M. Vitousek, *Ecology*, 1994, **75**, 1861.
[24]I. Kottke and F. Oberwinkler, *Trees*, 1986, **1**, 1.
[25]G. G. Gebauer and E. D. Schulze, *Aspects of Appl. Biol.*, 1988, **17**, 123.
[26]B. Nihlgard, *Ambio*, 1985, **14**, 2.
[27]A. J. Friedland, G. J. Hawley and R. A. Gregory, *Plant Soil*, 1988, **105**, 189.
[28]M. C. Press, S. J. Woodin and J. A. Lee, *New Phytol.*, 1986, **103**, 45.
[29]J. H. Duyzer, F. H. L. M. Verhagen, J. H. Westrate and F. C. Bosveld, *Environ. Pollut.*, 1992, **75**, 3.
[30]J. Pearson and G. R. Stewart, *New Phytol.*, 1993, **125**, 283.
[31]J. B. Aber, K. J. Nadelhoffer, P. Steudler and J. M. Melillo, *BioScience*, 1989, **39**, 378.
[32]E.-D. Schulze, *Science*, 1989, **244**, 776.

especially near sources such as busy roads. Anderson and Mansfield[33] found that NO was five times more soluble in xylem sap than in distilled water. The xylem sap is continuous with the extracellular water in the leaf mesophyll, and hence it is likely that the uptake of NO into leaves may be greater than would be predicted from its water solubility. We know little about the fate of NO after it has entered the leaf, but some research has suggested that it can be more toxic than $NO_2$.[34] This toxicity is very difficult to explain on the basis of our present knowledge of the solution products of the two gases, and it may depend on the direct action of NO *per se*. Petrouleas and Diner[35] found that NO may bind to a site on the Photosystem II reaction centre, making it capable of competing directly with $HCO_3^-$ or $CO_2$. During the last eight years NO has been a major focus of attention for animal physiologists, with roles so diverse and important for it to be designated 'molecule of the year' by *Science* in 1993. It certainly needs to be taken more seriously as an agent that may perturb the cellular functions of plants.

The nitrate and nitrite ions formed when $NO_2$ enters into solution might be expected to enter normal metabolism *via* the pathway that reduces them to ammonia (Figure 12.3). Nitrate and nitrite are the substrates for two reductase enzymes normally present in leaves, *viz.* nitrate reductase and nitrite reductase. The usual source of initial substrate is the nitrate (but not nitrite) transported into the leaves from the roots *via* the xylem.

Nitrite ions are known to be toxic to plant tissues and there is metabolic regulation within the reduction pathway to ammonia to prevent them accumulating. The necessary rate control is thought to be provided at the stage of nitrate reduction, the enzyme for which is located inside the cells but outside the chloroplasts[36,37] while nitrite reductase is within the chloroplasts. There must be mechanisms for keeping the toxic $NO_2^-$ ions away from sensitive sites during their movement within the cells, but we know little about these. It is, however, likely that the arrival of $NO_2$ at the cell surface, leading to $NO_2^-$ ions in the extracellular water, may pose problems because the movement of substantial numbers of these ions across the plasma membrane is probably not a normal requirement.

Figure 12.4 shows diagrammatically the locations of nitrate and nitrite reduction, and the incorporation of $NH_3$ into amino acids in chloroplasts. This is regarded as the normal pattern in the majority of plants but there are exceptions, and these may be important in determining differences in sensitivity to $NO_2$ between species.

It has been suggested[38,39] that the susceptibility of different plant species to $NO_x$ may depend on the precise location at which they normally carry out

[33]L. S. Anderson and T. A. Mansfield, *Environ. Pollut.*, 1979, **20**, 113.

[34]H. Saxe, *New Phytol.*, 1986, **103**, 185.

[35]V. Petrouleas and B. A. Diner, *Biochim. Biophys. Acta*, 1990, **1015**, 131.

[36]R. M. Wallsgrove, P. J. Lea and B. J. Miflin, *Plant Physiol.*, 1979, **63**, 323.

[37]P. J. Lea, J. Wolfenden and A. R. Wellburn, in 'Plant Responses to the Gaseous Environment', eds. R. G. Alscher and A. R. Wellburn, Chapman and Hall, London, 1994, p. 279.

[38]R. G. Amundson and D. C. Maclean, in 'Air Pollution by Nitrogen Oxides', eds. T. Schneider and L. Grant, Elsevier, Amsterdam, 1982, p. 501.

[39]A. R. Wellburn, *New Phytol.*, 1990, **115**, 395.

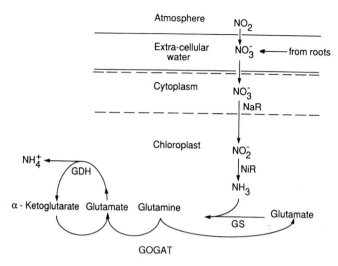

**Figure 12.4** *The relationship between the synthesis of amino acids in chloroplasts and the nitrate formed from $NO_2$ entering the leaf. NaR and NiR are nitrate and nitrite reductase, respectively*

nitrate/nitrite reductions. Some woody plants, for example, might be particularly sensitive to $NO_x$ because the reduction pathway is absent or poorly developed in the leaves. These species carry out the reductions of soil-derived nitrates in the roots, and hence nitrogen in an already reduced form is transported to the shoots. Research has so far failed to indicate precisely how such plants are affected by $NO_x$ entering the leaves from the atmosphere.

The exposure of plants to $NO_x$ pollution has been shown to lead to changes in the activities of nitrate and nitrite reductases in leaves.[40] There are usually marked increases in nitrite reductase, a clear indication that the plant has some ability for metabolic adjustment to detoxify the nitrite ions as they enter the cells. In contrast the activity of nitrate reductase may decline in leaves exposed to $NO_x$. This is probably because the supply of nitrite ions from the atmosphere reduces the need for the production of nitrite by nitrate reductase. A consequence may be loss of the normal rate of the pathway, which could produce serious imbalances in subsequent metabolism. Figure 12.4 shows the routes for the incorporation of ammonium into amino acids and it is these areas of metabolism that may be of much concern in future research.

### 12.2.3 Ozone

Ozone is highly toxic to both plants and animals, and it is the most thoroughly investigated of a group of phytotoxic compounds that are 'secondary' air pollutants, produced when 'primary' pollutants take part in reactions in the

[40]A. R. Wellburn, J. Wilson and P. H. Aldridge, *Environ. Pollut.*, 1980, **22**, 219.

atmosphere. If, as some authorities predict,[41,42] tropospheric $O_3$ concentrations increase by up to 50 percent over the next three–four decades, there could be serious impacts on crop production and on natural plant communities.

Considerations of the toxicity of $O_3$ are not complicated by its providing a source of an essential nutrient, as is the case with $SO_2$ and $NO_x$. We can therefore direct attention simply to the properties of the pollutant itself.

At high concentrations (above 100 ppb for most plant species) ozone causes a characteristic brown or white flecking to appear on the foliage, but there is some variation in visible symptoms and sensitivity between species, and even between cultivars of the same species. This variation is particularly wide between different cultivars of the tobacco plant, one of which, the variety Bel-W3, can be injured by hourly mean ozone concentrations above 40 ppb. Because of its high sensitivity, Bel-W3 is frequently used as a biomonitor for $O_3$ to provide a cheap and convenient substitute for chemical monitoring methods and it has been used in several studies in the UK.[43]

At lower concentrations, when there is no visible damage to foliage, ozone may reduce the growth rate of physiological activity of plants. Again, however, there is considerable variation from species to species, or between cultivars of one species. There are also marked differences in toxicity in different environmental conditions. Initially this variation caused problems for researchers, but more recently it has also been looked upon as an opportunity to identify those characteristics of a plant, or the physiological states determined by the environment, that lead to enhanced susceptibility.[44] This approach can provide important clues to the mechanisms behind injury.

Cells inside the leaf, especially the photosynthetic mesophyll, have been identified as primary sites of injury. Differences in the susceptibility of these cells do not, however, appear to be the most important cause of variation. It is the ability of a leaf to exclude $O_3$ from its intercellular spaces that emerges as one critical point of control.

Like other gases in the atmosphere $O_3$ gains access into leaves by diffusing through stomatal pores. There has been a great deal of research into the relationship between stomatal conductance and sensitivity to $O_3$, and there is general agreement that conductance is often the major determining factor. If the stomata are partially closed then some $O_3$ is excluded from the sensitive sites. Differences in stomatal opening can be inherent (*i.e.* genetically determined) or can result from responses to environmental conditions. Stomata tend to close partially when the atmosphere is dry or when the water content of the soil is low. Differences in $O_3$ sensitivity are often well correlated with such changes.[45]

This correlation with stomatal conductance does not appear so clearly when we consider other pollutants such as $SO_2$ and $NO_x$. A possible reason for its

[41]A. M. Hough and R. G. Derwent, *Nature*, 1990, **344**, 645.
[42]A. M. Thompson, *Ozone Sci. Eng.*, 1990, **12**, 177.
[43]M. R. Ashmore, J. N. B. Bell and C. L. Reily, *Nature*, 1987, **327**, 417.
[44]D. T. Tingey and G. E. Taylor, in 'Effects of Gaseous Air Pollution in Agriculture and Horticulture', eds. M. H. Unsworth and D. P. Ormrod, Butterworth Scientific, London, 1982, p. 113.
[45]R. L. Heath, in 'Plant Responses to the Gaseous Environment', eds. R. G. Alscher and A. R. Wellburn, Chapman and Hall, London, 1994, p. 121.

**Table 12.1** *Ethylene production and leaf injury in pea seedlings*

| Treatment | (A) Three-week fumigation ($C_2H_4$ evolved nmol g$^{-1}$ dry weight hr$^{-1}$) | (B) One-day fumigation (7 hr) ($C_2H_4$ evolved nmol g$^{-1}$ dry weight hr$^{-1}$) | Visible leaf injury to plants in (B) (%) |
|---|---|---|---|
| Control | 2.5 (0.1)* | 2.6 (0.2) | none |
| 50 ppbv $O_3$ | 1.9 (0.2)* | 4.7 (0.2) | 0–10 |
| 100 ppbv $O_3$ | 0.3 (0.1)* | 7.0 (0.4) | 10–35 |
| 150 ppbv $O_3$ | 0.2 (0.1)* | 5.7 (0.5)** | >50 |

The data in column (A) are for plants grown for three weeks and exposed to $O_3$ for seven hours daily. Those in column (B) are for plants grown in clean air for three weeks, then exposed to $O_3$ for seven hours for one day only.
Results represent means (s.e.m. in parentheses) of at least seven replicates. Treatments were compared using Student's t-test and the 100 and 150 ppbv treatments were highly significant at $P < 0.01$.
*No visible injury; **also with raised rates of ethane evolution.
Reproduced from Mehlhorn and Wellburn[46] with permission from Macmillan Journals Ltd.

importance in relation to $O_3$ sensitivity became apparent as a result of the research of Mehlhorn and Wellburn[46] and Mehlhorn, O'Shea and Wellburn.[47] Pea seedlings were fumigated with 50–150 ppb $O_3$ for seven hours daily for their first three weeks of growth after germination. Even in 150 ppb $O_3$ (regarded as a high level of pollution) there was no evidence of any injury apart from a slight curling of the leaves. On the other hand, if three-week-old seedlings that had been grown in clean air were fumigated for just one day under the same conditions and in the same concentration of $O_3$, there was very severe damage to the leaves (Table 12.1).

Subsequent studies showed that the two sets of seedlings differed markedly in the amounts of ethylene they were producing (Table 12.1). Ethylene is normally generated by plants and it is known to be able to regulate various aspects of growth and development. For this reason it is regarded as a hormone and it is unique as the only known gaseous hormone in plants or animals.[48] The pathway for its biosynthesis from methionine is well established (Figure 12.5). The point of regulation may be at the level of synthesis of ACC (1-aminocyclopropane-1-carboxylic acid), and it is known that some of the other plant hormones such as the auxins can affect ACC production.

Increased production of ethylene is associated with many events in plant development such as release from dormancy, stem and root growth and differentiation, abscission of leaves and fruits, flowering and fruit ripening. In particular, ethylene production increases in response to environmental stresses of various kinds (*e.g.* physical wounding, flooding).

[46]H. Mehlhorn and A. R. Wellburn, *Nature*, 1987, **327**, 417.
[47]H. Mehlhorn, A. R. O'Shea and A. R. Wellburn, *J. Exp. Bot.*, 1991, **42**, 17.
[48]P. J. Davies, 'Plant Hormones: Physiology, Biochemistry and Molecular Biology', Kluwer, Dordrecht, 1995.

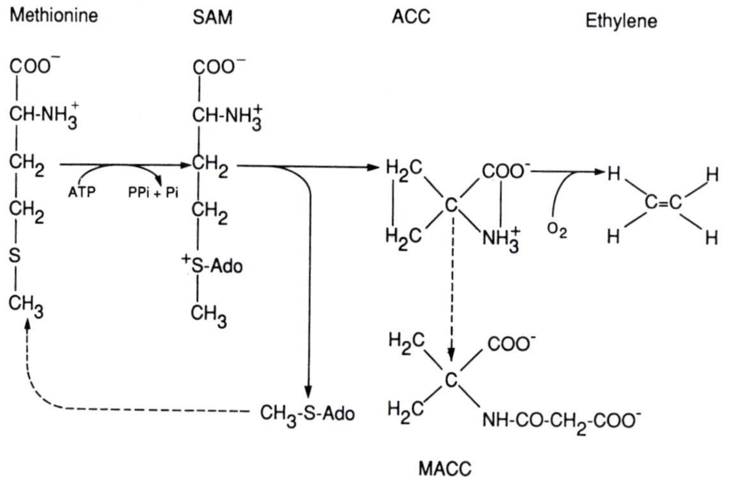

**Figure 12.5**  *The pathway of ethylene biosynthesis from methionine SAM = S-adenosylmethionine, ACC = 1-aminocyclopropane-1-carboxylic acid, MACC = 1-malonylaminocyclopropane-1-carboxylic acid*

In the experiments of Mehlhorn and Wellburn[46] the rates of ethylene evolution were more than doubled in the three-week-old plants fumigated for seven hours with $O_3$. The increased production of ethylene after $O_3$ fumigation was already well known from previous studies, and it was proposed that the production of ethylene and its reaction with $O_3$ entering through the stomata was a key factor in determining injury by $O_3$. They tested this hypothesis by treating plants with an inhibitor of ethylene biosynthesis, aminoethoxyvinylglycine (AVG), prior to exposing them by $O_3$. AVG inhibits the synthesis of ACC (Figure 12.5). The treatment with AVG reduced ethylene production by 85% and also almost completely prevented the damage normally caused to the three-week-old plants by the short $O_3$ treatment.

Ozone can react with unsaturated hydrocarbons to produce water-soluble free radicals. Mehlhorn and Wellburn[46] suggested that their formation in the intercellular spaces of the leaf around the mesophyll cells could trigger peroxidative processes inside the cells, ultimately leading to damage. More recent studies have shown that there are free radicals present in leaf tissues, closely correlated with ethylene formation and ozone treatments as postulated.[49] Independent evidence also suggested that isoprene, another unsaturated hydrocarbon produced and emitted by leaves, can react with ozone in the same way.[50]

These important new findings might explain why high normal stomatal conductance may be of special importance in connection with injury to leaves by $O_3$: only when the endogenous unsaturated hydrocarbon makes contact with the incoming $O_3$ does the critical formation of free radicals take place.

[49] H. Mehlhorn, B. J. Tabner and A. R. Wellburn, *Physiol. Plant.*, 1990, **79**, 377.
[50] C. N. Hewitt, G. L. Kok and R. Fall, *Nature*, 1990, **344**, 56.

The parts of cells most vulnerable to free radicals in the intercellular air are likely to be the plasmalemma membranes. $O_3$ fumigation is known to cause damage to membranes so that the retention of solutes is reduced. The paper by Heath and Castillo[51] provides a valuable review of this topic.

### 12.2.4 Acid Rain and Acid Mist

The presence of $SO_2$, $NO_x$ and $NH_3$ in polluted air has an enormous impact on the chemistry of rain and mist. In north-west Europe, the concentrations of ammonium and sulfate can be 1000-fold higher than in unpolluted areas, such as the South Island of New Zealand.[32] The wet deposition of acidity has been sufficient to increase the leaching of base cations from most soils in affected areas, but this will only increase the acidity within soil types in which the reserves of base cations are low.

There is now quite good scientific evidence that the pH values of some soils in Scandinavia and central Europe have decreased over the past few decades.[52-54] Increased base leaching has also been recognized in parts of North America, but here there is little to suggest that soils have yet reached dangerously low pH values.

The inputs of acid (principally sulfuric and nitric) from the atmosphere are not the only way by which soil pH can be decreased. Various biotic processes and humus accumulation can also contribute.[55] There is still some uncertainty about the relative contributions of these different processes, but nevertheless most soil scientists now accept that acidity in rainfall is playing an important part in the changes that are occurring. The direct and indirect effects on plants in the affected soils are much more open to dispute. In the Federal Republic of Germany in the 1980s the yellowing (chlorosis) of older needles of Norway spruce and silver fir increased markedly on nutrient-poor soils. This phenomenon appears to have been associated with deficiencies in cations, especially magnesium. The problem occurred over large geographical areas, many of which are long distances from the main sources of primary pollutants. It is, however, recognized that acidic deposition can occur far away from such sources.

The hypothesis that deterioration of the soil is a major factor in the decline in the health of trees has had much support, but other ideas have also been extensively discussed. In particular, the increased frequency of episodes of $O_3$ pollution at the damaged sites has attracted attention, and $O_3$ has been suggested to be either the predominant agent in causing damage, or at least in inciting

[51]R. L. Heath and F. J. Castillo, in 'Air Pollution and Plant Metabolism', eds. S. Schulte-Hostede, N. M. Darrall, L. W. Black and A. R. Wellburn, Elsevier Applied Science, London, 1988, p. 55.

[52]E. Matzner and B. Ulrich, in 'Effects of Atmospheric Pollutants on Forests, Wetlands and Agricultural Ecosystems', eds. T. C. Hutchinson and K. M. Meema, Springer-Verlag, Berlin, 1987, p. 25.

[53]G. Abrahamsen, in 'Effects of Atmospheric Pollutants on Forests, Wetlands and Agricultural Ecosystems', eds. T. C. Hutchinson and K. M. Meema, Springer-Verlag, Berlin, 1987, p. 321.

[54]J. J. Colls and M. H. Unsworth, in 'Air Pollution Effects on Biodiversity', eds. J. R. Barker and D. T. Tingey, Van Nostrand Reinhold, New York, 1992, p. 93.

[55]S. I. Nilsson, H. G. Miller and J. D. Miller, *Oikos*, 1982, **39**, 40.

stress. The injury of membranes caused by $O_3$ could, it is suggested, be the primary factor in the leakage of nutrients and their subsequent leaching from foliage by acid mist.[56,57] Another suggestion is that the deposition of nitrogen to forests from the increasing atmospheric concentrations of $NO_x$ has caused an imbalance in the major nutrients required for growth[32] (see Section 12.2). Some authors have suggested that 'nitrogen saturation' is beginning to occur in forest ecosystems, but there is not a clear physiological or ecological definition of what 'saturation' means in this context. It is clear, however, that floristic and biodiversity changes can be induced by high rates of acidic and nitrogen deposition.[58]

Most rainfall chemistry studies rely on the use of rain gauges for the collection of samples. Apart from other problems[19] simple rain gauges can also underestimate the amount of water from the atmosphere that is captured by vegetation in mist or clouds. This additional input of water from the atmosphere is known as 'occult' precipitation.[59] Not only is the volume of precipitation likely to be underestimated when plants are in cloud, but also the amounts of dissolved materials that are deposited.

Collection of occult deposition separately from rain is difficult and special techniques were developed in the 1980s. On Great Dun Fell in the northern Pennines in England, it has been estimated that the concentration of ions such as $H^+$, $SO_4^{2-}$ and $NO_3^-$ are two to four times higher in cloudwater than in rain.[59]

There are occasions when cloudwater may simultaneously condense on foliage and evaporate from it, and this can lead to a considerable increase in the concentration of ions in the liquid film or droplets on exposed surfaces.[60] This is thought usually only to occur for short periods, for example when the weather is improving after a dense cloud cover, but in thin cloud when some solar radiation can reach the surface it may be more prolonged.

Experimental studies of the effects of simulated rain on leaf surfaces have generally shown that there is little direct damage unless the pH is below 3.0, with the exception of a few cultivars of field crops that appear to be specially sensitive.[61] In some cases the growth of the plants was increased probably because of the fertilizing effects of additional nitrogen and sulfur. Very low pH values are of infrequent occurrence in rain but it seems likely that they could occur much more frequently during occult deposition. For this reason, the effects of cloudwater on forests at high altitude are now being considered as a further contributory factor towards the damage to trees, and there may be impacts on the frost sensitivity of some species (see below).

[56]R. A. Skeffington and T. M. Roberts, *Oecologia*, 1985, **65**, 201.
[57]T. M. Roberts, R. A. Skeffington and L. W. Blank, *Forestry*, 1989, **62**, 179.
[58]T. V. Armentano and J. P. Bennett, in 'Air Pollution Effects on Biodiversity', eds. J. R. Barker and D. T. Tingey, Van Nostrand Reinhold, New York, 1992, p. 159.
[59]M. H. Unsworth and D. Fowler, in 'Air Pollution and Ecosystems', eds. P. Mathy, D. Reidel Publishing Co., Dordrecht, 1988, p. 68.
[60]M. H. Unsworth, *Nature*, 1984, **312**, 262.
[61]L. S. Evans, K. F. Lewis, E. A. Cunningham and M. J. Patti, *New Phytol.*, 1982, **91**, 429.

## 12.3 INTERACTIONS BETWEEN POLLUTANTS

Air pollutants rarely occur singly, yet much of the literature covers their separate effects rather than their joint action in realistic mixtures. There are two main reasons for this: mechanisms behind injury are necessarily studied in the simplest context, at least in the first instance; and the facilities available to researchers have until recently rarely allowed for the design of experiments involving different pollutants in factorial combinations, especially for long-term studies.

The situation that has emerged so far is complex. The effects of combinations of pollutants are sometimes greater and sometimes less than we would predict from the separate effects, and only rarely are responses simply additive. In the space available here we shall concentrate on two specific pollutant combinations whose effects may be specially important.

### 12.3.1 Mixtures of $SO_2$ and $NO_2$

Combinations of these two primary pollutants have received attention because they do occur together in many situations. They are often produced simultaneously by the same sources, for example during the combustion of S-containing fossil fuels. Joint action of $SO_2$ and $NO_2$ may be important in urban areas where there are regular episodes of high concentrations, or in nearby rural areas where there may be longer exposures to lower concentrations of both gases.

Concentrations of $NO_2$ in urban situations commonly rise to peaks between 30 and 50 ppb, whereas peaks of $SO_2$ are nowadays generally lower, ranging from 10 to 30 ppb. Motor vehicle emissions have caused $NO_2$ concentrations to rise steadily in rural areas over the last three decades, but here there is less likely to be a coincidence of peaks of $SO_2$ and $NO_2$.

Short-duration fumigations with $SO_2$ and $NO_2$ have often led to severe injury, far greater than that found with the pollutants separately. Tingey *et al.*[62] were the first to perform a detailed study, and they found surprisingly high toxicity of some mixtures of $SO_2$ and $NO_2$ in four-hour exposures, at concentrations around 10% of those required for inducing visible injury by the individual gases. Since this early work there has been a mixture of confirmatory and contradictory reports, but there have been more than enough observations of synergism between $SO_2$ and $NO_2$ to suggest that simultaneous exposure to the two pollutants may sometimes be a cause of unexpected damage in the field.

Studies of long-term effects of $SO_2$ and $NO_2$ on grasses and cereals have shown that there are much greater growth reductions during the winter than in spring and summer.[16,63] This suggests that slower rates of metabolism determined by environmental conditions may predispose plants to this form of injury.

As discussed above (Section 12.2.2.1), the enzyme nitrite reductase is considered to play a critical part in the metabolism of the solution products of $NO_2$ after its entry into cells. Yoneyama and Sasakawa[64] made use of $^{15}NO_2$ to demonstrate

[62]D. T. Tingey, R. A. Reinert, J. A. Dunning and W. W. Heck, *Phytopathology*, 1971, **61**, 1506.
[63]P. C. Pande and T. A. Mansfield, *Environ. Pollut.* Ser. A, 1985, **39**, 281.
[64]T. Yoneyama and H. Sasakawa, *Plant Cell Physiol.*, 1979, **20**, 263.

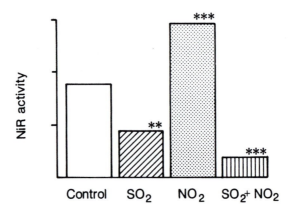

**Figure 12.6**    *Nitrite reductase activity in Phleum pratense ( Timothy grass) exposed for 230 weeks to clean air or atmospheres containing 69 ppb SO₂ and NO₂ alone or in combination. ** and *** indicate significant differences at* p < 0.01 *and* p < 0.001, *respectively. From Wellburn[65] and used with permission of Butterworth Scientific.*

the appearance in cells of nitrite and nitrate ions and to show that after their reduction the $^{15}$N was incorporated into amino acids. Analyses by Wellburn and his colleagues[39,65,66] of plastids extracted from leaves of grasses after exposure to $SO_2$, $NO_2$ or $SO_2 + NO_2$ revealed important differences in enzyme activities. Fumigation with $NO_2$ on its own led to a statistically significant increase in the activity of nitrite reductase, but no such increase was found when the $NO_2$ was accompanied by $SO_2$ (Figure 12.6). This obviously suggested that $SO_2$ prevented the induction of greater nitrite reductase activity, as normally occurs in the presence of $NO_2$. This means that the plants may have been unable to detoxify the solution products of $NO_2$. Thus there may be a fairly simple biochemical explanation of synergistic effects of $SO_2$ and $NO_2$ in damaging plants.

### 12.3.2   Mixtures of $SO_2$ and $O_3$

Experimental studies of this combination of pollutants have produced a much more confusing picture. The earliest studies involving short-term exposures provided evidence of synergism not unlike that often found with $SO_2$ and $NO_2$.[67,68] The types of foliar injury caused by $SO_2$ and by $O_3$ are quite different, and it was noted in these and in some later experiments that the enhanced damage in combinations of the two resembled that caused by $O_3$ rather than by $SO_2$. This might be explained by the enhanced opening of stomata that can occur in $SO_2$-polluted leaves (see next section). Thus $SO_2$ could assist the access of $O_3$ into the leaf's interior.

[65] A. R. Wellburn, in 'Effects of Gaseous Air Pollutants in Agriculture and Horticulture', eds. M. H. Unsworth and D. P. Ormrod, Butterworth, London, 1982, p. 169.

[66] A. R. Wellburn, in 'Gaseous Air Pollutants and Plant Metabolism', eds. M. J. Koziol and F. R. Whatley, Butterworth, London, 1984, p. 203.

[67] R. L. Engle and W. H. Gabelman, *Pro. Am. Soc. Hort. Sci.*, 1966, **89**, 423.

[68] H. A. Menser and H. E. Heggestad, *Science*, 1966, **153**, 424.

In more recent studies no consistent picture of the combined effects of $SO_2$ and $O_3$ has emerged. Reviews of the subject[69,70] have cited evidence for additive, less-than-additive and more-than-additive effects. The only consistency in the data from different sources was that most more-than-additive effects were found in short-term exposures. It seems unlikely that a simple description of the dose-response relationships for effects of these two pollutants on plants can be produced. This has serious implications for any modelling exercises aiming to predict the economic consequences of pollutants in the field.

The difficulties encountered by researchers in coming to a clear view of the combined effects of $SO_2$ and $O_3$ emphasize the complexity of the perturbations that can occur for plant life when we alter the atmospheric environment to even a small degree.

## 12.4 POLLUTANT/STRESS INTERACTIONS

The majority of experimental fumigations of plants with air pollutants have been conducted under conditions that are favourable for growth. There are good reasons for this: stress factors, biotic and abiotic, are often difficult to maintain and control at a predetermined level, hence their inclusion in experiments that are already technically very difficult has been seen as an unmanageable complication. Yet the evidence from a few well-designed experiments, supported by many observations in the field, suggests that such interactions can play an important role in determining plant growth.

### 12.4.1 Biotic Factors

One of the earliest known descriptions of the effects of air pollution on insects and plants was made by the diarist John Evelyn, who in 1661 said of the atmosphere of seventeenth-century London: 'It kills our bees and flowers abroad, suffering nothing in our gardens to bud, display themselves or ripen.' It was a further three hundred years before observational evidence of interactions between air pollutants and plant pathogens (insects or fungi) again began to appear in the literature (see Cramer[71] and Flückiger *et al.*[72]) and it is only in the last fifteen years that an experimental approach has been adopted to study these interactions.

Most of the evidence from these experiments suggests that exposure to pollutants such as $SO_2$ and $NO_2$ can predispose plants to attack by insects, with, in the majority of cases, an increase in insect performance being observed. For example, Warrington[73] produced a dose–response curve showing how precisely

[69]R. Kohut, in 'Sulfur Dioxide and Vegetation', eds. W. E. Winner, H. A. Mooney and R. Goldstein, Stanford University Press, Stanford, California, 1985.

[70]T. A. Mansfield and D. C. McCune, in 'Assessment of Crop Loss from Air Pollutants', eds. W. W. Heck, O. C. Taylor and D. T. Tingey, Elsevier Applied Science, London, 1988, p. 317.

[71]H. H. Cramer, *Forstw. Cbl.*, 1951, **70**, 42.

[72]W. Flückiger, S. Braun and M. Bolsinger, in 'Air Pollution and Plant Metabolism', eds. S. Schulte-Hostende, N. M. Darrall, L. W. Blank and A. R. Wellburn, Elsevier, London, 1987, p. 366.

[73]S. Warrington, *Environ. Pollut.*, 1987, **43**, 155.

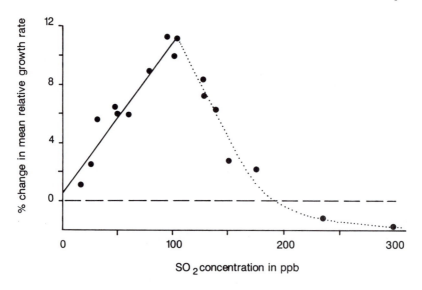

**Figure 12.7** *Effects of atmospheric SO₂ concentration on the mean relative growth rate of pea aphids* (Acyrthosiphon pisum). *The aphids fed for four days on garden peas growing in the various SO₂ concentrations, to which the plants had also been exposed for about 20 days prior to the placement of the aphids. Percentage changes compared with aphids on plants in clean air are shown.* From Warrington,[73] and used with permission of Applied Science Publishers Ltd.

the growth rate of aphids can be related to SO₂ concentration (Figure 12.7). There was a linear decrease from zero up to about 110 ppb SO₂ after which there was a steep decline. This was thought to be the concentration at which the SO₂ began to be directly toxic to the aphids. There is little evidence for direct effects of this nature occurring in the field under realistic concentrations and in general, observed responses appear to be due to an alteration in the quality of the host plant by the pollutants. Aphids, for example, feed specifically on the sap of the phloem in plants, which is the main transport pathway for sugars and amino acids. It is possible to raise some species of aphids on artificial diets providing the same nutritional factors as phloem. When such artificially maintained aphids are exposed to low concentrations of pollutants they remain unaffected. Thus there is little doubt that it is changes in the physiology of the plant which are responsible for the differences in aphid activity in polluted air.

The underlying mechanisms responsible for changes in the quality of the host plant have been extensively reviewed,[74,75] particularly for SO₂ and NO₂ where most studies have shown changes in the nitrogen balance of the host plant, especially amino acid composition. It does, therefore, seem likely that this is the main cause of the increase in the growth rate of aphids, an undesirable effect which may occur when the plants are themselves showing negative growth

[74]J. Riemer and J. B. Whittaker, in 'Insect–Plant Interactions, Vol. 1', ed. E. A. Bernays, CRC Press, Boca Raton, Florida, 1989, p. 73.
[75]J. B. Whittaker and S. Warrington, *Environ. Pollut.*, 1990, **65**, 363.

responses to the pollutants. There are thus likely to be greater final impacts on growth when exposure to pollutants is accompanied by aphid infestation. The precise economic consequences cannot, however, be determined from our present knowledge.

Whilst the impact of pollutants such as $SO_2$ and $NO_2$ on insects appears to be fairly clear, this is not the situation for $O_3$. Evidence from filtration experiments on crops in rural areas where $O_3$ is the main pollutant, suggests its presence can significantly improve insect performance. Conversely, experiments using indoor, closed-chamber systems reveal a more complex pattern. For example, chewing insects such as beetles and the Gypsy moth (*Lymantria dispar*) generally show an increase in performance, particularly in food consumption. However, the evidence is far less clear for aphids and both positive and negative effects have been reported (reviewed by Brown[76]). There is also more uncertainty about the mechanisms by which $O_3$ alters the host plant since no consistent changes have been shown in amino acid composition.

The possibility that fungal pathogens may sometimes show increased activity on polluted plants has also been studied. During the last 40 years the apparent absence in urban areas of *Rhytisma acerinum* Pers., the fungus responsible for tar spot on the leaves of sycamore (*Acer pseudoplatanus*) and its contrasting abundance in rural areas, has led to speculation that air pollution may be responsible, particularly $SO_2$.[77] However, in contrast to *Diplocarpon rosae* which is responsible for black spot on roses and which has been shown under laboratory conditions to be sensitive to $SO_2$[78] there is no direct evidence that $SO_2$ is the primary cause of the absence of tar spot infection of sycamore in towns and cities. A study by Leith and Fowler[79] in Edinburgh showed that present-day concentrations of $SO_2$ ($<20$ ppb) are unimportant in determining the distribution of the fungus. The major factor controlling its occurrence appears to be the abundance of overwintered sycamore leaves infested with tar spot, most of which in cities are removed either actively by man or passively by the wind. Clearing of dead leaves in this way would thus remove the source of infection and prevent the growth of the *R. acerinum* population in the city.

### 12.4.2 Abiotic or Environmental Factors

*12.4.2.1 Low Temperatures.* There is no simple relationship between temperature and individual physiological processes in plants. Different species are adapted to different environments and within a species individuals from different provenances can display distinctive behaviour. There are two main ways in

---

[76]V. C. Brown, in 'Insects in a Changing Environment', eds. R. Harrington and N. E. Stork, Academic Press, London, 1995, p. 219.

[77]G. N. Greenhalgh and R. J. Bevan, *Trans. Brit. Mycol. Soc.*, 1978, **71**, 491.

[78]M. Fehrmann, A. Von Tiedemann, L. W. Black, B. Glashagen and T. Eisenman, in 'How are the Effects of Air Pollution on Agricultural Crops Influenced by Interactions with Other Limiting Factors?' report of Working Party III, Concerted Action on 'Effects of Air Pollution on Terrestrial and Aquatic Ecosystems', p. 98, CEC, Brussels.

[79]I. D. Leith and D. Fowler, *New Phytol.*, 1987, **108**, 175.

which plants are normally damaged by low temperatures, *viz.* 'chilling' injury caused by low temperatures above 0°C and freezing injury. Some plants from warm climates cannot tolerate either chilling or freezing and so their geographical distribution is limited to areas where these stresses do not occur. Plants native to temperate regions can also be severely damaged or killed by chilling or freezing if they are not correctly acclimatized.[80] The process of acclimatization is usually called 'hardening' and it is achieved mainly by a period of prior exposure to temperatures slightly above the threshold for injury. In some plants, woody perennials for example, hardening may also require or be enhanced by decreasing photoperiods in the autumn, and it is therefore part of the dormancy cycle.

A full understanding of the mechanisms responsible for acclimatization has yet to be established. It is generally agreed, however, that during the autumn, perennial plants of temperate and Arctic latitudes acclimatize to low temperatures by adopting one or a combination of strategies, such as the accumulation of soluble solutes (proline, arginine and simple carbohydrates), changes in cell metabolism away from glycolysis towards the pentose phosphate pathway,[81] or modifications to the cell membranes which involve changes in phospholipids, sterols and glycolipids.[82] The ways in which pollutants may affect these processes are at present unknown but the hypothesis that alteration of membrane properties is involved appears to be plausible in view of the important physiological role of changes in membrane composition during cold hardening. The findings of Mehlhorn and Wellburn with regard to the likely vulnerability of plasmalemma membranes to free radicals are relevant here (Section 12.3).

Circumstantial evidence that exposure to air pollutants may reduce the frost resistance of woody plants first appeared in the Nordic countries.[83] Keller[18] was amongst the first to demonstrate experimentally that exposure of dormant beech seedlings to $SO_2$ during the winter led to increased concentrations of sulfur in the leaves that expanded in spring, and the death of some of the terminal buds which showed symptoms of the type associated with frost injury. Keller[84] also found that Norway spruce suffered frost injury in late winter if exposed to $SO_2$ from October to April.

Sulfur dioxide can also enhance frost injury in herbaceous plants. Davison and Bailey[17] found that the ability of perennial rye grass to undergo hardening against damage by sub-zero temperatures was reduced by prior exposure to 87 ppb $SO_2$ for five weeks, and Baker *et al.*[85] exposed winter wheat to $SO_2$ in a field-exposure experiment and noticed that a natural frost ($-9°C$) in mid-January caused increased injury to the fumigated plants. Freer-Smith and Mansfield[86] found that there were small but consistent increases in frost injury on the needles

[80]J. Levitt, 'Responses of Plants to Environmental Stresses, Vol. 1, Chilling, Freezing and High Temperature Stresses', Academic Press, New York, 1980.

[81]C. J. Andrews and M. K. Pomeroy, *Plant Physiol.*, 1979, **64**, 120.

[82]B. Yoshida and M. Uemura, *Plant Physiol.*, 1984, **75**, 31.

[83]S. Huttunen, in 'Air Pollution and Plant Life', ed. M. Treshow, John Wiley, Chichester, 1984, p. 321.

[84]T. Keller, *Gartenbauwissenschaft*, 1981, **46**, 170.

[85]C. K. Baker, M. H. Unsworth and P. Greenwood, *Nature*, 1982, **299**, 149.

[86]P. H. Freer-Smith and T. A. Mansfield, *New Phytol.*, 1987, **106**, 237.

of Sitka spruce cooled to $-5$ and $-10°C$ after exposure to 30 ppb $SO_2$ and 30 ppb $NO_2$ followed by a period of hardening in clean air.

Although atmospheric pollution has been implicated as a likely contributing factor to forest declines in Europe and N. America, as yet no clear consensus regarding the primary causes responsible has emerged. The concentrations of $SO_2$ and $NO_x$ in many locations where there is marked forest decline are very low throughout the year and it is thought unlikely that these pollutants acting alone are responsible. Of the hypotheses put forward as an explanation for tree injury, many involve ozone as a major factor, either acting alone or in combination with other pollutants or environmental stresses.[87] The involvement of $O_3$ is based on evidence[88] that in some of the affected areas, tropospheric concentrations in the summer have been found to be in excess of 50 ppb. This is especially true for high elevation sites ($>800$ m) where, due to a combination of meterological and topographic factors, these sites experience far fewer seasonal and diurnal fluctuations in tropospheric $O_3$ concentrations than occur at lower elevations.[3] In addition, winter stresses, such as frosts, will occur with a greater frequency at higher altitudes. Controlled fumigations with $O_3$ alone or in combination with acid mists, either at or in excess of 50 ppb, have failed to reproduce the symptoms of chlorosis or needle loss characteristic of the foliage of damaged trees which suggests that a complex combination of physical and chemical stresses may be involved.

Until the late-1980s there were few studies on the effects of ozone in relation to frost injury, presumably because the conditions necessary for the formation of this pollutant predominantly occur during the summer months (see Chapter 8). However, there is now sufficient evidence from controlled experiments to suggest that exposure to $O_3$ during the summer months can enhance the susceptibility of conifers to frost injury in the following autumn.[89–91] Although a clear effect of the photo-oxidant was observed in these experiments, the basic mechanism of winter acclimatization was not disrupted, because irrespective of treatment or species (Norway spruce, Sitka spruce) the majority of tree seedlings later developed full frost hardiness to temperatures below $-20°C$.

Apart from gaseous pollutants, it is now recognized that the frost hardiness of some coniferous tree species growing at high altitude sites can be reduced by exposure to acid mists containing $SO_4^{2-}$, $NH_4^+$, $NO_3^-$ and $H^+$ ions. Such sites frequently receive exposure to mists on 70% of days for up to ten hours per day on average and on 50% of these occasions mist pH is less than 3.5. Detailed studies by groups in the UK and the USA, summarized by Sheppard,[92] suggest that the uptake and accumulation of high concentrations of $SO_4^{2-}$ ions by younger foliage, in which the wax layer of the cuticle is not fully formed, lead to protein

[87] S. B. McLaughlin, *J. Air Pollut. Control Assoc.*, 1985, **35**, 512.

[88] B. Prinz, G. H. Krause and H. Stratman, 'Forest Damage in the Federal Republic of Germany', Land Institute of Pollution Control of the Land North-Rhine Westphalia, Wessen, West Germany, 1982.

[89] P. W. Lucas, D. A. Cottam, L. J. Sheppard and B. J. Francis, *New Phytol.*, 1988, **108**, 495.

[90] K. A. Brown, T. M. Roberts and L. W. Blank, *New Phytol.*, 1987, **105**, 149.

[91] J. D. Barnes and A. W. Davison, *New Phytol.*, 1988, **108**, 159.

[92] L. J. Sheppard, *New Phytol.*, 1994, **127**, 69.

denaturation and loss of membrane integrity and as a result a reduction in frost hardiness. It is believed that the presence of $NO_3$-N can mitigate the toxic effects of $SO_4^{2-}$ but that high concentrations of $SO_4^{2-}$ and $H^+$ ions can act synergistically and lead to a disruption in the control of cellular pH. In the field these effects may be exacerbated by low concentrations of soil-available $Ca^{2+}$ relative to the availability of $Al^{3+}$ which could further weaken cell membrane integrity.

*12.4.2.2  Water Deficits.*  After temperature, water availability is the main environmental factor governing the distribution of plants. Evolution has provided them with several features and mechanisms for the acquisition of water and to improve the economy of its usage. Some effects of air pollutants on plants have been identified which appear likely to disturb a plant's water economy.

Two pollutants, $SO_2$ and $O_3$, and mixtures of $SO_2$ and $NO_2$, have often been found to cause a change in the allocation of material between roots and shoots. There is reduced translocation of newly manufactured photosynthates from the leaves to the roots and the eventual effect on the plant can be a considerable increase in the shoot:root dry mass ratio.[93] This change in the balance of growth could have important implications for plants growing in drying soil, where the increased penetration of roots is important for the maintenance of water supplies for the shoot. However, in contrast to many previous studies, Taylor and Davies[94] exposed beech seedlings to the ambient atmosphere in southern England during the summer (polluted predominantly by $O_3$) and found no evidence that the allocation of dry matter between roots and shoots was significantly altered. However, both root extension and specific root length (root length per unit root dry mass) were significantly greater for beech grown in chambers receiving unfiltered air, which suggests that the root system was not only longer but also made up of thinner roots. These changes in root anatomy did not, however, appear to make the seedling more susceptible to an artificially imposed drought.

Effects of pollutants on stomatal behaviour have also attracted a lot of attention, since in most plants the strategy for avoiding drought stress involves limiting the rate of water loss through a reduction in transpiration *via* decreases in stomatal conductance. The opening of stomata ideally achieves an acceptable compromise between the plant's need to acquire $CO_2$ from the atmosphere for photosynthesis and the loss of water by transpiration. Any alteration in stomatal functioning can be serious because of interference with carbon gain or water usage.

There is a considerable amount of evidence in the literature that $SO_2$ and mixtures of $SO_2$ and $NO_2$ can sometimes cause abnormal stomatal opening.[95,96] This is probably the result of damage to the cells surrounding the stomatal guard cells. The latter are known to be more resilient than the other epidermal cells to

[93]T. A. Mansfield, P. W. Lucas and E. A. Wright, in 'Air Pollution and Ecosystems', ed. P. Mathy, Reidel, Dordrecht, 1988, p. 123.
[94]G. Taylor and W. J. Davies, *New Phytol.*, 1990, **116**, 457.
[95]T. A. Mansfield and P. H. Freer-Smith, 'Gaseous Air Pollutants and Plant Metabolism', eds. M. J. Koziol and F. R. Whatley, Butterworth, London, 1984, p. 131.
[96]V. J. Black, in 'Sulfur Dioxide and Vegetation', eds. W. E. Winner, H. A. Mooney and R. A. Goldstein, Stanford University Press, Stanford, California, 1985, p. 96.

adverse treatments.[97] The stomatal pore is opened as result of turgor pressure generated in the guard cells, which push against the resistance offered by the turgor of the surrounding cells. If the surrounding cells lose turgor then greater stomatal aperture occurs and there may not be enough pressure exerted on the guard cells to bring about closure of the pore. Under conditions of severe water stress the stomata are required to close quickly to protect the leaf.

Changes in stomatal function can also be brought about by exposure to ozone. Studies by Heggestad *et al.*[98] found that soil moisture stress enhanced the deleterious effects of $O_3$ on yields of soybeans so long as the concentrations of $O_3$ were fairly low (below 80 ppb for seven hours a day). At higher concentrations of $O_3$ opposite effects were found, *i.e.* there was a reduced impact of water shortage on yield. At higher concentrations, $O_3$ is known to induce closure of stomata and this may have had the dominant effect by reducing water loss from the leaves.[99] Detailed measurements of the stomatal resistance of beech leaves by Pearson and Mansfield[100] in which young trees were either subjected to a period of water stress or were watered throughout the experiment showed that in the presence of ozone, the stomatal resistance of well-watered trees was increased, whereas in the droughted trees ozone reduced stomatal resistance as the water stress developed (Figure 12.8). These subtle changes in stomatal function are important in two respects: during periods when water supply is plentiful, an increase in stomatal resistance could reduce the uptake of carbon dioxide for photosynthesis; during a drought, which often coincides with conditions favourable for the production of elevated concentrations of $O_3$ at ground level, it may be difficult for trees to control their water use effectively. Such changes, if they were to occur in the field may, in the long term, ultimately lead to growth reductions and in some cases the death of trees.

The method used to determine stomatal behaviour and water loss used by Pearson and Mansfield involved detailed measurements on individual leaves of the trees. This makes it difficult to extrapolate from such measurements to determine water loss by the whole tree, due to factors such as leaf age, leaf position in the canopy and interactions between the stomata and the environment. Recently, however, methods for measuring whole-tree transpiration and water use have become available, for example the heat-balance technique for measuring sap flow.[101] This procedure was adopted by Wiltshire *et al.*[102] to study the effects of $O_3$ exposure on water use by ash trees (*Fraxinus excelsior*). The use of the sap-flow method enabled both daily and seasonal changes in transpiration to be monitored and this showed a clear trend for the ozone-exposed trees to use more water per day than the controls during the early part of the growing season (June–July). Later in the season, the reverse was the case. It seems likely that

[97]G. R. Squire and T. A. Mansfield, *New Phytol.*, 1972, **71**, 1033.
[98]H. E. Heggestad, T. J. Gish, E. H. Lee, J. H. Bennett and L. W. Douglas, *Phytopathology*, 1985, **75**, 472.
[99]N. M. Darrall, *Plant Cell Environ.*, 1989, **12**, 1.
[100]M. Pearson and T. A. Mansfield, *New Phytol.*, 1993, **123**, 351.
[101]J. M. Baker and C. H. M. Van Bavel, *Plant Cell Environ.*, 1987, **10**, 777.
[102]J. J. J. Wiltshire, M. H. Unsworth and C. J. Wright, *New Phytol.*, 1994, **127**, 349.

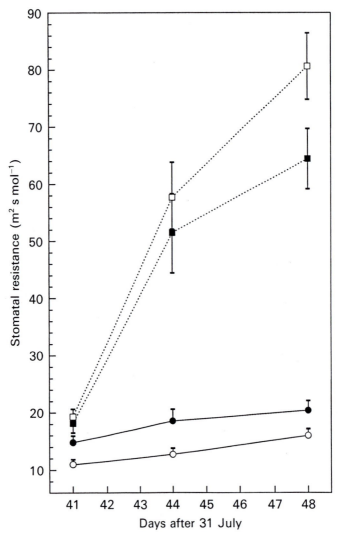

**Figure 12.8**   *The interaction between ozone and water stress on stomatal resistance in beech.*
*Measurements were made after the commencement of the water stress treatment (day*
*39). Each point is the mean of data collected at 08.15 and 11.00 hr. The vertical lines*
*represent standard errors of the means (N = 108). Ozone: well watered (●);*
*droughted (■). Control: well watered (○); droughted (□).*

these changes in water use were caused by increases in stomatal resistance during
the later stages of the experiment and it appears that $O_3$ could be a causal factor
in the decline of ash which has been observed in recent years in parts of
England.[103]

[103]S. K. Hull and J. N. Gibbs, 'Forestry Commission Bulletin 74, HMSO, London, 1991.

## 12.5 CRITICAL LOADS AND CRITICAL LEVELS

It is clear, from the evidence presented in the preceding sections of this chapter, that air pollutants can have extensive and varied effects on a wide range of vegetation. Over the last few years, attempts have been made to use this information to establish air quality control standards for the protection of both natural and managed ecosystems throughout Europe. The main forum for the development of these standards has been the United Nations Economic Development Commission for Europe (UNECE) through an international agreement termed the Convention on Long Range Transboundary Air Pollution. Originally, air quality standards agreed under this convention were based on fixed percentage reductions in emissions, for example the so-called '30% Club' for sulfur. More recently, a less arbitrary approach has been adopted based on the concepts of *critical loads* and *critical levels* which will be used to provide the scientific basis for the development of policies to control transboundary air pollutants. Both critical loads and critical levels are based on the precautionary principle with the aim of protecting the most sensitive elements of any ecosystem from long-term damage.

Critical loads refer to the total deposition of sulfur, nitrogen or hydrogen ions and are aimed primarily at preventing acidification or eutrophication of sensitive terrestrial or aquatic ecosystems. They are defined as 'the quantitative estimate of an exposure to one or more pollutants below which significant harmful effects on specified sensitive elements of the environment do not occur according to present knowledge'.[104] The method for estimating critical loads for vegetation for pollutants such as S and N rely on an empirical approach, since at present there does not seem to be a way of employing mass-balance models analogous to those used to set critical loads for soils and freshwaters. In the UK, a combination of experimental data, the National Vegetation Classification[105] and land-use maps have been used to identify sensitive vegetation types, and the results combined with pollutant deposition data to produce either critical load maps or exceedance maps for S and N (indicating where and by how much the estimated critical load is exceeded). Further information and critical load values for different types of vegetation, soils and freshwaters are provided by Hornung and Skeffington.[106]

In contrast to critical loads, critical levels refer to the *direct* effects of atmospheric pollutants on above-ground vegetation and are defined by the UNECE as 'the concentrations in the atmosphere above which direct adverse effects on receptors such as plants, ecosystems or materials, may occur according to present knowledge'. In certain areas of Europe, there is already circumstantial evidence that direct gaseous exposure to air pollution is at least as important as the indirect (critical load) pathway. At present the determination of critical levels for different types of vegetation, *e.g.* lichens and bryophytes, crop species, forest

[104]'Critical Loads for Sulfur and Nitrogen', eds. J. Nilsson and P. Grennfelt, Report of a Workshop held in Skokloster, Sweden, 19–24 March, Nordic Council of Ministers, Copenhagen, 1988.

[105]'British Plant Communities', ed. J. S. Rodwell, Cambridge University Press, Cambridge, 1991.

[106]'Critical Loads: Concepts and Applications', eds. M. Hornung and R. A. Skeffington, ITE Symposium No. 28, Proceedings of a Conference held on 12–14 February in Grange-over-Sands, Cumbria, HMSO, London, 1993.

**Table 12.2** *United Kingdom critical level values for the major air pollutants and vegetation categories*

| Pollutant | Receptor | Critical level |
|---|---|---|
| Sulfur dioxide | Agricultural crops | 30 $\mu$g m$^{-3}$ (11 ppb) (annual mean) |
| | Forestry | 20 $\mu$g m$^{-3}$ (8 ppb) (annual and winter mean) |
| | Natural vegetation | 20 $\mu$g m$^{-3}$ (8 ppb) (annual and winter mean) |
| | Lichens | 10 $\mu$g m$^{-3}$ (4 ppb) (annual mean) |
| Sulfate (particulate) | Forestry | 1 $\mu$g m$^{-3}$ (annual mean) |
| Oxides of nitrogen (NO$_x$) | Forestry and natural vegetation | 30 $\mu$g m$^{-3}$ (16 ppb) (annual mean, urban corrected) |
| Ammonia | Forestry and natural vegetation | 8 $\mu$g m$^{-3}$ (11 ppb) (annual mean) |
| Ozone* | Arable crops | 5300 ppb-hr (3 month growing season) |
| | Forestry | 10 000 ppb-hr (6 month growing season) |

*Critical levels for ozone are defined on the basis of cumulative exposure over a threshold concentration of 40 ppb (AOT40) rather than as a seasonal mean

trees and natural vegetation can only be estimated from experimental exposure experiments, most of which involve some form of chamber. The resulting critical levels determined from experimental data (Table 12.2) are therefore subject to various constraints and care needs to be taken in assessing how representative these results are of those found in the field. In addition, different types of vegetation, as well as different species and cultivars, may differ in their sensitivity to pollutants and in the type of effect which might be considered adverse. For example, a reduction in yield is of primary concern for arable crops, whereas both yield and conservation value are important for trees and the maintenance of biodiversity is of major importance for semi-natural ecosystems. Unfortunately, the information required to assign critical levels to a wide range of different species, representative of these different categories of vegetation, is not available and at present critical levels are set with the aim of protecting the species that are most sensitive in relation to the defined adverse effects. However, even with the limited information available, we do know that, apart from ammonia, the critical levels for all the major pollutants were exceeded in some part of the UK during the last five years and that the exceedance was greatest for ozone.

*Acknowledgement.* P. W. Lucas is grateful for the financial support of the UK Department of the Environment under their programme on Air Pollution: Environmental Impact Assessments, Critical Loads and Terrestrial Ecosystems.

CHAPTER 13

# Control of Pollutant Emissions from Road Traffic

C. HOLMAN

## 13.1 INTRODUCTION

### 13.1.1 Background

Air pollution is a problem in major cities throughout the world. International comparisons of urban air quality suggest that the concentrations found in conurbations in the United Kingdom are broadly similar to those found in similar sized conurbations elsewhere in northern Europe,[1] except where concentrations are influenced by local point sources. This is not surprising since the factors that influence ambient concentrations, such as the vehicle stock and meteorological conditions, are similar.

In most urban areas of the United Kingdom traffic generated pollutants – nitrogen oxides ($NO_x$), carbon monoxide (CO), volatile organic compounds (VOCs) and particulate matter (PM) – have become the dominant pollutants. Although little long-term monitoring data on these pollutants exists, it is likely that urban concentrations have increased over the past three or four decades. In contrast, early concerns about lead in the atmosphere from car emissions have now been effectively tackled by reducing the maximum permitted lead level and encouraging the use of unleaded petrol. As a result concentrations are now down to about 20% of those in the 1970s (see also Chapter 7).

Ozone is a regional scale pollutant formed from $NO_x$ and VOCs. The highest concentrations are typically found downwind of urban areas, although elevated levels are also observed in the summer in urban areas (see also Chapter 8).

In recent years carbon dioxide has become an important pollutant as concern about its role in changing climate has increased. Although road transport is not the major source of this pollutant, emissions from this sector are growing at a time when there is a Government commitment to stabilize emissions.

[1] C. D. Holman, 'Report on Current Air Pollution Due to Transport Activities (Regulated and Unregulated Exhaust Emissions) in Selected Cities in the European Community', Final Report, Prepared for DGXI/B3, Commission of the European Communities, Study Contract B4-3040/93/000359, Brussels, 1993.

This chapter aims to give a brief review of the contribution road traffic makes to total emissions and what factors influence emissions and then looks at strategies to reduce future emissions including technical and non-technical solutions.

### 13.1.2 Contribution of Road Transport (see also Chapter 7)

Table 13.1 shows the relative importance of different sources of sulfur dioxide ($SO_2$); black smoke (BS); $NO_x$; CO and VOCs in the United Kingdom in 1993. Black smoke is one constituent of airborne particulate matter.

At a national level, road transport is the single most important source of most of these pollutants. The exceptions are $SO_2$ and VOCs. In urban areas the contribution from road transport is likely to be greater than indicated by national emission data. This is because there is typically more traffic and less industry in urban areas. In addition, emissions from traffic have a greater impact on local air quality as they are at a lower height than those from industrial sources.

Cars are the dominant source of road transport emissions, contributing between 50 and 90 percent of the sector's share of CO, $NO_x$, VOC, $SO_2$ and $CO_2$ emissions. Cars are a less important source of PM and BS. The largest source of these pollutants is heavy duty vehicles (buses, coaches and lorries).

UK emissions from road transport increased rapidly during the 1980s.[3] This is shown in Table 13.2 and was largely the result of the increase in traffic. Since the introduction of cars with three way catalytic converters, emissions have begun to decline. In general, road traffic emissions peaked around 1990 and since then have shown a small decrease. Despite forecasts of traffic increasing by 75% to 160% (depending on economic growth) over the next 30 years[4] new technology is

**Table 13.1** *Sources of the principal pollutants 1993*

| Source | % of total emissions | | | | |
|---|---|---|---|---|---|
| | *Sulfur dioxide* | *Black smoke* | *Nitrogen oxides* | *Carbon monoxide* | *Volatile organic compounds* |
| Road transport | 2 | 51 | 49 | 91 | 38 |
| Power Stations | 66 | 5 | 24 | 1 | – |
| Other Industry | 24 | 4 | 14 | 2 | 55 |
| Domestic | 4 | 29 | 3 | 5 | 1 |
| Other | 5 | 11 | 9 | 1 | 5 |
| Total in k tonnes | 3188 | 444 | 2347 | 5641 | 2418 |

The term volatile organic compounds does not include methane
Assumes all solvent use is in the industrial sector. 28% of total is due to solvent use
Source: Department of the Environment, 1995[2]

[2] Department of the Environment, 'Digest of Environmental Statistics', No. 17, HMSO, London, 1995.
[3] H. S. Eggleston, 'Pollution in the Atmosphere: Future Emissions from the UK', Report No LR 888 (AP), Warren Spring Laboratory, Stevenage, 1992.
[4] Department of Transport, 'Transport Statistics Great Britain 1993', HMSO, London, 1993.

**Table 13.2** *Increase in UK estimated emissions from road transport 1982–1993*

| Pollutant | Percentage increase |
|---|---|
| Carbon monoxide (CO) | 31 |
| Nitrogen oxides (NO$_x$) | 61 |
| Volatile organic compounds (VOCs) | 8 |
| Black smoke (BS) | 85 |
| Sulfur dioxide (SO$_2$) | 27 |
| Carbon dioxide (CO$_2$) | 43 |

Source: Department of the Environment 1995[2]

expected to result in a reduction in emissions of most pollutants, until some time in the second decade of the next century. Thereafter, unless there are further technological improvements, the growth in traffic will result in an upturn in emissions. For PM the upturn is likely to occur sooner as the reduction in new vehicle emissions has been smaller than for the gaseous pollutants.

### 13.1.3 Factors Influencing Emissions

If complete combustion of the fuel was possible, vehicle exhaust would contain only carbon dioxide and water vapour. However, as a result of a number of factors including the short time available for combustion in the engine, poor mixing of the fuel and air (*e.g.* due to unburnt fuel getting trapped in crevices in the engine) and the high temperature of combustion, vehicle exhaust also contains CO, VOCs, PM and NO$_x$.

The emissions depend on a wide range of factors. The most important of these are discussed briefly in the following sections.

*13.1.3.1 Fuel Used.* Emissions depend on the fuel used to power the vehicle. For example, a car powered by petrol will emit more CO and VOCs and be less fuel efficient than a similar one powered by diesel. However, the diesel car will emit more NO$_x$ and PM.[5] Carbon dioxide emissions depend on fuel consumption and the carbon content of the fuel. Even though diesel is a more dense fuel, the carbon dioxide emissions from a diesel car are less than from a similar petrol car, but the benefit is smaller than the volumetric fuel consumption benefit. The relative differences in emissions between petrol and diesel cars are shown in Table 13.3.

A range of alternative fuels such as compressed natural gas and electricity have been used in trials in the UK. These have their own emissions characteristics. Electric vehicles, for example, while being clean (and quiet) in use, contribute to emissions at the power station. These in turn depend on the fuel used to power the generators.

[5]Quality of Urban Air Review Group, 'Diesel Vehicle Emissions and Urban Air Quality', 2nd Report, prepared for the Department of the Environment, Birmingham, 1993.

**Table 13.3**   *Comparison of emissions from petrol cars with three way catalysts and diesel cars*

| Pollutant | Petrol without catalyst | Petrol with three way catalyst | Diesel without catalyst | Diesel with oxidation catalyst |
|---|---|---|---|---|
| Nitrogen oxides ($NO_x$) | **** | * | ** | ** |
| Hydrocarbons (HC) | **** | ** | *** | * |
| Carbon monoxide (CO) | **** | *** | ** | * |
| Particulate matter (PM) | ** | * | **** | *** |
| Aldehydes | **** | ** | *** | * |
| Benzene | **** | *** | ** | * |
| 1,3-Butadiene | **** | ** | *** | * |
| Polycyclic aromatic hydrocarbons (PAHs) | *** | * | **** | ** |
| Sulfur dioxide ($SO_2$) | * | * | **** | **** |
| Carbon dioxide ($CO_2$) | *** | **** | * | ** |

Key: Asterisks indicate which type of car has typically the highest emissions
     *Lowest emissions; **/*** Intermediate; **** Highest emissions
     This table only indicates the relative order of emissions between the different types of vehicle. No attempt has been made to quantify the emissions. The difference in emissions between, say, **** and *** may be an order of magnitude, or much smaller
Source: Quality of Urban Air Review Group, 1993[5]

Currently there is much debate about the role of 'improved' conventional fuels. A number of studies have shown that certain fuel parameters can play an important role in controlling emissions.[6] This is discussed in further detail in Section 13.3.4 below.

*13.1.3.2   Engine Design/Pollution Control.*   Improvements to engine design and pollution control systems over the past twenty-five years has resulted in emissions from new vehicles now being more than an order of magnitude lower than those from unregulated vehicles.

For petrol engines one of the most important factors influencing emissions is the air to fuel ratio. Figure 13.1 shows that there is no ideal ratio at which all the main emissions are low and the engine power is at an acceptable level. Indeed, where CO and HC are at their lowest, the $NO_x$ emission is at its maximum. A good compromise is found in the lean burn region, and small lean burn engines have been produced. However, their emissions are not as low as conventional petrol engines with a three way catalyst.

The greatest step in controlling emissions from road vehicles was the introduction of closed loop (or controlled) three way catalysts for petrol vehicles. These remove 80 to 90 percent of the emissions of CO, VOCs and $NO_x$. The reactions taking place on the catalyst are shown below:

[6]European Commission, ACEA, UKPIA, Working Group European Commission – Industry Technical Group 1, 'Effect of Fuel Qualities and Related Vehicle Technologies on European Vehicle Emissions: An Evaluation of Existing Literature and Proprietary Data', Brussels, 1994.

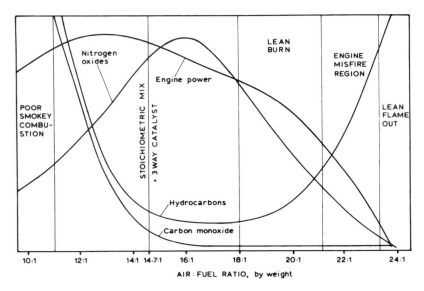

**Figure 13.1** *The effect of air/fuel ratio on engine operations*

*Oxidation reactions*

$$2CO + O_2 \rightarrow 2CO_2$$
$$HC + O_2 \rightarrow CO_2 + H_2O$$

*Reduction reactions*

$$2CO + 2NO \rightarrow 2CO_2 + N_2$$
$$HC + NO \rightarrow CO_2 + H_2O + N_2$$

Automotive catalysts are typically made from platinum and rhodium. For efficient removal of all three pollutants the air to fuel ratio needs to be close to the stoichiometric ratio (*i.e.* 14.7 to 1). Cars fitted with these catalysts require an oxygen sensor to monitor the exhaust gas composition and electronically controlled fuel management system to control the air to fuel ratio. The effect on catalyst efficiency of moving away from the stoichiometric air to fuel ratio (*i.e.* the equivalent ratio $\lambda = 1$) is shown in Figure 13.2.

This technology cannot be used in the oxygen rich exhaust of a lean burn petrol or diesel engine. For these engines CO and HC emissions can be reduced using a simple oxidation catalyst. Currently no catalyst exists that is capable of removing $NO_x$ from lean burn engines.

*13.1.3.3 Maintenance.* Poorly maintained vehicles consume more fuel and emit higher levels of CO and VOCs than regularly serviced ones. It has been shown in a number of studies that a relatively small number of vehicles contribute a disproportionate amount of pollution. For example, in 1991, the RAC found that

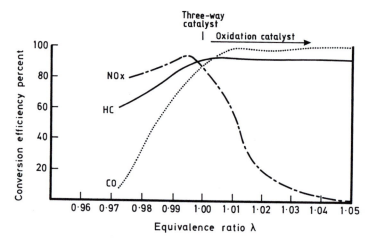

**Figure 13.2**   *The effect on catalyst efficiency of moving away from the stoichiometric air to fuel ratio*

**Table 13.4**   *Percentage of cars failing to comply with standards (1987–1992)*†

| Type of car | 'As received' |
|---|---|
| Non catalysts | 68% |
| Open loop catalysts | 74% |
| Closed loop catalysts[1] | 9% |
| Diesel | 0% |

†The standards relate to the Dutch incentive programme introduced in 1986
[1]Newest closed loop catalyst cars have better in-use performance
Source: Dutch Ministry for Housing Physical Planning and Environment, 1994[8]

in London and Leicester 12 percent of vehicles were responsible for half the CO emissions.[7]

In the Netherlands the government has systematically checked emissions of cars in service since 1987. Table 13.4 shows the percentage of different types of cars with excess emissions during the first five years of the programme. Most of the cars failing to meet the standards had excess carbon monoxide emissions. It should be noted that the data from the 6th year of the programme[8] showed that only 3% of cars with newer closed loop catalysts required tuning, suggesting that

[7]J. Vanke and J. F. S. Bidgood, 'Remote Sensing of Vehicle Emissions – Principles and Potential', Proceedings of the 24th FISITA Congress, The Vehicle and the Environment, Volume 2, held in London 7–11th June 1992, published by Institute of Mechanical Engineers, London, 1992.
[8]Ministry of Housing, Physical Planning and Environment, 'Project In Use Compliance: Air Pollution by Cars in Use: Annual Report 1992–1993', The Hague, 1994.

modern European catalyst cars are performing better in service than the early ones.

To ensure that in-use vehicles are well maintained a number of countries have introduced inspection and maintenance (I/M) programmes, in which emissions are periodically tested. In the UK this is now incorporated into the annual MOT test.

*13.1.3.4 Driver Behaviour/Traffic Congestion.* Emissions are not constant but vary depending on how the vehicle is being driven. $NO_x$ emissions increase when the engine is under load such as during rapid acceleration and when travelling at high speeds, while CO and VOCs emissions will increase when it is necessary to run rich, for example, when the engine is cold, and during accelerations. Thus, in general, emissions will be lowest when a car is driven at a steady speed. In the stop-start driving conditions that characterize congested urban areas emissions will be higher than at the same average speed but under free flow conditions.

A recent Dutch study has showed that average emissions can increase by a factor of 3.5 for CO and two for $NO_x$ by driving aggressively due to excursions from the ideal air to fuel ratio during the frequent changes from acceleration to deceleration and *vice versa*, compared with normal driving.[9]

Emissions also vary with vehicle speed. For petrol cars without catalysts emissions of CO and VOCs decrease with increasing speed. Emissions are highest at the slow driving speeds characteristic of urban driving. For $NO_x$ the opposite occurs; emissions increase with speed. Emissions from other types of vehicle (petrol with catalyst and diesel) typically have lowest values at medium speeds, with higher emissions at both low and high vehicle speeds. Particulate emissions from diesel vehicles show a less clear relationship with vehicle speed.

*13.1.3.5 Cold Starts.* Most car journeys are very short and are in urban areas. Emissions from cars are particularly high when first driven from a cold start. The emissions penalty is greater (relative to those when hot) for petrol cars with catalysts than for non catalyst petrol and diesel cars, as it takes a few minutes for the catalyst to become fully operational.

The cold start penalty is dependent on the ambient temperature. The colder it is the greater the penalty. For example, it has been estimated that during the first kilometre of a journey in a three way catalyst petrol car the emissions of CO and VOCs are 70% and 140% higher respectively when the temperature is $0\,°C$ compared to $10\,°C$.[10] Cold start penalties for $NO_x$ for petrol and diesel engines are small.

[9]D. M. Heaton, R. C. Rijkeboer and P. van Sloten, 'Analysis of Emission Results from 1000 In-use Passenger Cars Tested Over Regulation Cycles and Non-regulation Cycles', paper presented at the Symposium 'Traffic Induced Air Pollution', Graz University of Technology, 10th and 11th September 1992.
[10]C. Holman, J. Wade and M. Fergusson, 'Future Emissions from Cars 1990 to 2015: The Importance of the Cold Start Emissions Penalty', World Wide Fund for Nature UK, Godalming, 1993.

## 13.2  CONTROL MECHANISMS

### 13.2.1  Introduction

There are a number of pollution abatement measures available to governments. These include voluntary agreements between industry and the state, fiscal incentives to encourage consumers to buy cleaner products and to use them less and regulation. In recent years the last has been the most widely used control instrument to reduce air pollution from the road transport sector, although there is growing interest in the use of fiscal incentives in Europe and elsewhere.

Historically, air pollution from road vehicles in the European Union (EU) has been controlled by adopting the 'best available technology' approach to the control of emissions from new vehicles. Progressively more stringent limits have been introduced over the past twenty-five years. Recently, both the UK Government and the EU have moved towards an effects-based approach to air pollution control in which pollution abatement policy is determined within the context of air quality standards and other environmental criteria, and a wider range of measures are being considered.[11]

### 13.2.2  Regulation

*13.2.2.1  Vehicles.*  Originally emission limits for new vehicles were set by the United Nations Economic Commission for Europe (UNECE) and then introduced into EU legislation. However, in the last few years the EU has taken the initiative.

For cars and vans, emissions are tested on a chassis dynamometer over a test cycle that is designed to simulate typical European driving conditions, and measured in grams per kilometre. For heavy duty vehicles, emissions are measured from the engine (not the vehicle) using a bench dynamometer. The results are measured in grams per kilowatt hour, and the more powerful the engine the greater the permitted emissions.

Controls on emissions from vehicles were first introduced in the early 1970s. As environmental concerns grew vehicle emissions limits became progressively tighter. By 1983 there had been four amendments to the original UNECE regulation (Regulation 15) controlling emissions from cars. These had been mirrored by equivalent EC Directives.

In 1985 the Council of Ministers adopted standards equivalent to those in the USA for large cars. In 1991 it adopted the Consolidated Directive (Directive 91/441/EEC) which effectively mandated the fitting of closed loop three way catalysts to all petrol cars irrespective of size from 1992/1993 (the first date applies to new types of car; the second to all new cars). Yet more stringent emission limits were adopted in 1994 for introduction from 1996.

---

[11] Her Majesty's Government, 'This Common Inheritance: Second Year Report, Britain's Environmental Strategy', HMSO, London, 1992 and Commission of the European Communities, 'Towards Sustainability: A Community Programme of Policy and Action in Relation to the Environment and Sustainable Development', Luxemborg, 1993.

Emission limits for heavy duty vehicles have also been progressively tightened. The original UNECE regulation (Regulation 49) controlled emissions of CO, VOC and $NO_x$ from vehicles over 3.5 tonnes. In 1991, the EU introduced new limits values, incorporating PM for the first time, to be applied in a two stage process for introduction in 1992/93 and 1995/96.

In addition, emission limits are in the process of being agreed for the various categories of vehicle which are currently inadequately covered, such as vans and motorcycles.

Heavy duty vehicles have been subject to a regular emissions (smoke) test for a number of years in the UK and some other European countries. However, Directive 92/55/EC requires that all member states adopt an emissions test as part of the roadworthiness test for all types of vehicles by 1998. In the UK this has been incorporated in the MOT test for cars. These are known as inspection and maintenance (I/M) programmes.

However, even the stringent emission limits being introduced in the mid 1990s for new vehicles and the introduction of regular in-service emission tests, are not considered sufficient to solve Europe's air quality problems. The European Auto/ Oil programme has been set up with the aim of finding the least cost package of measures, for introduction in the year 2000, to meet air quality targets for the year 2010. The measures being considered fall into four broad categories: vehicle technology; fuel quality; inspection and maintenance and non-technical measures.

*13.2.2.2 Fuels.* In the past, fuel quality has been controlled by voluntary agreement under the auspices of the British Standards Institution. The European Union (EU) has a number of Directives controlling the permitted levels of lead, benzene and oxygenates in petrol and sulfur in diesel and these have been incorporated into the relevant British Standards.

In 1993 the European Committee for Standardization (CEN) agreed common specifications for diesel (EN 590) and unleaded petrol (EN 228) for adoption in EU and EFTA countries. These have been incorporated into new British Standards. The Government has the power to make the British standards mandatory but has only used them for leaded petrol.

The CEN standards largely relate to fuel parameters influencing vehicle performance and allow a large variation in the composition of the fuels. A survey of retail pumps has shown, for example, that the aromatic content of European petrol can vary from less than 20 percent to over 55 percent.[12]

### 13.2.3 Voluntary Agreements

Whilst the control of regulated emissions (*i.e.* CO, VOCs, $NO_x$ and PM) from motor vehicles has been dominated by legislation, controls over fuel economy and/or carbon dioxide emissions have, in Europe, been dominated by voluntary agreements. This is largely a result of the diversity of product on the market making the formulation of legislation difficult. Although a number of schemes

[12]M. J. Hawkins, Ford Motor Company, personal communication.

have been proposed within the EU to regulate carbon dioxide emissions from passenger cars based on a series of thresholds for different categories of cars, these have failed to win widespread acceptance.

There have been several voluntary agreements between industry and Governments to improve fuel consumption of new cars. In Europe the car manufacturers voluntarily agreed to improve fuel economy of new cars by 10 percent between 1978 and 1985 (on a sales weighted basis). In the event, the improvements were over 20 percent. It has been argued that the industry set a target which it knew would be easy to meet and that the agreement had no technology-forcing effect on the manufacturers.

In October 1991, the European Motor Manufacturers Association (ACEA) made a voluntary commitment to 'reduce the EC sales weighted $CO_2$ emissions of the new car fleet of each manufacturer by 10% on a voluntary basis within the period 1993–2005'. The industry has argued that this target is demanding, will be difficult to meet, and will require a shift towards more diesel cars. However, in 1995 a much more demanding target of 25% was agreed by the German Vehicle Manufacturing Association (VDA) and the German Government, with a commitment to investigate increasing this target.

### 13.2.4   Fiscal Measures

Fiscal incentives have been widely used in Europe to encourage the use of unleaded petrol. Several countries have also introduced tax incentives for the introduction of low sulfur diesel. In addition, a number of EU countries have successfully used fiscal incentives to encourage the purchase of cleaner vehicles. For example, in Germany and the Netherlands where fiscal incentives have been used, over 90% of the new petrol cars sold in 1990 had catalysts. This compares with less than 5% in Great Britain where no incentives were used.

The most comprehensive system of fiscal incentives for reducing the impact of traffic on air pollution exists in Sweden. In a major reform of the tax laws, the Swedish Government introduced a system of environmental taxation for all fuels and motor vehicles in the early 1990s. There is a carbon dioxide tax, a sulfur tax, a charge on emissions of nitrogen oxides (from large plant), an energy tax that favours 'clean' diesel and a purchase tax system that favours cleaner vehicles. From January 1991 the energy tax on diesel fuel was differentiated to encourage the use of improved diesel. Three classes were defined, and the improved grades (Class 1 and Class 2) were taxed at a lower rate than the standard grade (Class 3). Initially this was based on the sulfur and aromatics content and distillation range. A year later the classification system was extended to include other parameters. A similar scheme for petrol has been proposed but has yet to be fully implemented.

The effect of the tax differential on the diesel market in Sweden has been dramatic. Less than 15% of the diesel sold in 1990 would have met the current requirements for Class 1 or Class 2. By 1993 about 15% of all diesel fuel sold was Class 1 and about 60% was Class 2. Since then the proportion of Class 1 has increased as new production facilities have opened. The difference in tax between Class 1 and Class 3 is about 4 pence per litre.

The differentiated purchase tax for light duty vehicles (cars and vans) was introduced in 1993 and for heavy duty vehicles in 1994. The difference in tax between Class 1 and Class 3 is about £520 for cars; and £5600 for large lorries. Since Sweden joined the EU in January 1995 the scheme has been modified and there is some doubt as to whether it will be allowed, under EU law, to continue.

Economic instruments have also been considered as a good method to reduce the demand for road transport by, for example, increasing fuel prices, charging for use of roads (road pricing) and increasing parking charges. The Royal Commission on Environmental Pollution,[13] for example, recently suggested increasing the fuel duty to a level that will result in a doubling of fuel prices by the year 2005 as part of a recommended strategy for meeting a number of environmental targets.

## 13.3  TECHNICAL MEASURES

### 13.3.1  Introduction

In general, there is a trade off between emissions, performance and fuel economy. Automotive engineers strive to obtain the best balance recognizing the often conflicting demands of consumers and legislators with respect to environmental protection and vehicle performance. Experience has shown that there is initially resistance to new standards but that once legal requirements are agreed the technology develops quickly and costs are reduced compared to the first estimates.

In the past the emphasis of legislators was on improving emissions from new vehicles. However, it is now moving towards ensuring that vehicles remain clean throughout their life and that the best fuels are used.

### 13.3.2  New Vehicles

*13.3.2.1  Petrol.*  California has led the world in introducing clean petrol cars. In 1990 the Californian Air Resources Board (CARB) established four classes of emission standards for light duty vehicles: transitional low emission vehicles (TLEVs); low emissions vehicles (LEVs); ultra low emission vehicles (ULEVs) and zero emissions vehicles (ZEVs). The first three categories are scheduled to be phased in over a ten year period from 1994. The introduction of ZEVs is mandated such that by 2003 10% of each manufacturers' sales have to be ZEVs. Only electric vehicles currently meet the ZEV criteria.

The European standards due to be introduced in 1996 (Euro II) are broadly equivalent to the Californian TLEV standards, and many vehicles already meet these standards. In general terms the difference in technology between petrol cars meeting the Euro I (Directive 91/441/EEC) and Euro II standards (Directive 94/12/EU) is largely subtle optimization of the control system.

---

[13]Royal Commission on Environmental Pollution, Eighteenth Report, 'Transport and the Environment', HMSO, London, 1994.

It is likely the Euro III (year 2000) standards will be similar to those of the Californian ULEVs. To meet these standards motor manufacturers are addressing two factors: reducing emissions from the engine by greater control of the combustion process, and improving catalyst performance.

Most of the emissions from petrol cars occur during the first couple of minutes or so before the catalyst has fully reached its operational temperature. A number of solutions are being developed to reduce these emissions including using a rich fuel to air mixture together with secondary air injection to increase the temperature of the exhaust; moving the catalyst closer to the engine (close coupled catalyst) and external heating using electricity or a fuel burner.

The catalysts themselves are also being improved with new washcoat and precious metal formulations. New catalysts using palladium on its own or in combination with platinum and rhodium are now being developed and increasingly used. Very low levels of lead and sulfur in petrol would allow the wider use of palladium which is particularly effective in early light off catalysts but is poisoned by these elements. Lower emissions can also be achieved by using more accurate oxygen sensors and electronic control systems.

Further improvements in technology to reduce engine-out emissions remain possible. These include reducing pumping losses (through developments such as variable valve timing) particularly under part load conditions; changing the geometry of the combustion chamber to eliminate those areas where unburnt VOCs can be trapped and improving exhaust gas recirculation (EGR, *i.e.* recycling the exhaust gas through the combustion system to reduce the $NO_x$ emissions).

Several manufacturers have developed a new generation of lean burn engine. This type of engine was originally developed a decade ago for the fuel consumption benefits. However, their emissions remain higher than those of conventional engines with advanced emission control systems, and it remains to be seen whether this technology can meet the ULEV emission levels.

A rather different approach for the longer term is the use of a two stroke engine. The modern engine bears little relation to the two stroke engines used in Eastern European cars such as the Trabant. A Ford Fiesta with the new engine and a simple oxidation catalyst has been shown to be able to meet ULEV standards at low mileage. These engines also offer very good fuel economy. However, their ability to maintain low emissions throughout the life of the vehicle has yet to be proved.

Evaporative emissions, the lighter VOCs emitted from a vehicle's fuel system, are collected in a canister containing activated carbon, and recycled into the vehicle's combustion system when the vehicle is driven. The test procedure is currently being revised to better reflect real world conditions; however, it is likely that this method will continue to be the means of controlling these emissions.

*13.3.2.2 Diesel.* There are two major types of diesel engine. In the direct injection (DI) engine, fuel is injected and burnt directly in the cylinder. The indirect injection (IDI) engine has a separate pre-chamber (swirl chamber) where the air and fuel mixes, the fuel is ignited and combustion begins. The burning mixture then flows into the main combustion chamber and completes its combustion there.

IDI engines are able to work over a wider speed range than DI engines and have been more suitable for cars and vans. They are also less noisy. There is, however, a loss of efficiency compared to DI engines. DI engines have, until recently, been used on larger vehicles. However, the development of improved fuel injection technology, such as high pressure rotary pumps, has allowed the use of DI engines in passenger cars (*e.g.* the Rover Montego) and vans (*e.g.* the Ford Transit). It is expected that development of the DI engine will lead to its wider use in smaller vehicles.

In all diesel engines the main pollutants of concern are $NO_x$ and PM. IDI and DI engines have different emission characteristics with IDI engines emitting less $NO_x$ and PM. However, they are less fuel efficient and therefore emit more $CO_2$. Cars with DI engines may be up to 20% more fuel efficient than those with IDI engines, and up to 40% more efficient than those with petrol engines.

There are few diesel cars on the Californian market, and no manufacturer has announced a diesel car capable of meeting the ULEV standards. Small diesel engines are essentially a European product, developed and sold for the home market. In the UK there has been a rapid increase in the proportion of diesel cars sold over the past five years. Currently they make up over 20% of the new car market. In some other European countries, for example France, the proportion is even higher.

In diesel engines there is a trade-off in the emissions where measures to control PM emissions typically are those that promote high levels of $NO_x$. Thus, low emission engines rely on the precise control of the combustion process and increasingly are becoming electronically controlled with fuel injection becoming more precise.

Turbocharging on diesel vehicles is primarily intended to enhance the power output of the engine. The exhaust energy is used to drive additional air into the combustion chamber. This can reduce the emissions of PM, but results in an increase in $NO_x$ emissions. Intercooling, which cools the air between the turbocharger and the air intake system, increases the density of the air and removes the impact of turbocharging on $NO_x$ emissions. These technologies are standard on new large diesel engines and are increasingly being used on smaller ones. ERG is used to reduce the $NO_x$ emissions.

Whilst there remains some potential to further reduce engine-out emissions from diesel vehicles it is likely that after-treatment will be required. The three main types of after-treatment for diesel engines are:

- $DeNO_x$ catalysts
- Oxidation catalysts
- Particulate traps

Much effort has gone in recent years to developing a $NO_x$ catalyst capable of operating under lean conditions: the so called $DeNO_x$ catalyst. This could, when developed, also be used with lean burn petrol engines.

DeNO$_x$ catalysts developed so far have generally used zeolite. For these catalysts to work a reducing agent is required. Exhaust VOCs could be used, but their concentration is generally too low. It can be increased by deterioration of combustion or the addition of a hydrocarbon (such as ethylene) in the exhaust manifold. These catalysts have low conversion efficiencies and durability problems. In addition nitrous oxide (N$_2$O), a potent greenhouse gas, can be formed.

For stationary diesel engines selective catalytic reduction using a zeolite catalyst with ammonia (produced from urea) as a reducing agent is proven technology. The transfer of this technology to mobile sources is being investigated but there remain some problems particularly during transient operations. DeNO$_x$ catalysts are a promising, but immature, technology. It is likely that they will not be commercially available for up to five years.

Particulate matter from diesel engines comprise two components: the elemental carbon or soot and the soluble organic fraction (SOF). Oxidation catalysts remove the SOF associated with the PM, as well as CO and VOCs. These pollutants are associated with the odour and the mutagenicity of diesel exhaust. However, they have little effect on the elemental carbon. In total the mass of PM may be reduced by 20 to 30%[14] on an IDI engine and more on a DI engine. The efficiency of SOF removal is 50 to 70%.[14] Oxidation catalysts also oxidize any sulfur in the fuel to sulfate particulates. Without low sulfur diesel the increase in sulfate particulates can be greater than the decrease in the soluble organic fraction of the PM. However, from 1996 low sulfur diesel (less than 0.05% v/v) is required across the EU.

Particulate traps collect PM on a filter. After a short period the PM has to be removed otherwise the trap would become blocked and eventually exert an unacceptably high back pressure on the engine, adversely affecting engine performance, and increasing fuel consumption and emissions. The PM removal is known as trap regeneration.

A number of different traps have been developed. The most common is the extruded ceramic monolith (known as a ceramic wall-flow filter) which can trap over 80% of the exhaust PM, and has been shown to perform well over a long period. There is also a range of different methods for regenerating the traps. The collected PM burns at 550 to 600 °C and can be removed periodically using electric heaters or fuel burners. One commercially available system comes supplied with a mains regeneration unit to which a vehicle is connected for 20 minutes or so whilst regeneration takes place.

Alternatively the trap can be removed and replaced with a clean one so that no vehicle operation time is lost. These systems can be retrofitted to older vehicles to reduce emissions.

Another solution is to use a catalyst, either on the trap itself or as a fuel additive, to reduce the combustion temperature to less than 300 °C, which then allows the exhaust gases to remove the PM continuously. The lower operating temperature and continuous removal are less demanding, thus enhancing the

---

[14]Federal Environment Agency, 'Passenger Cars 2000, Requirements, Technical Feasibility and Costs of Exhaust Emission Standards for the Year 2000 in the European Community', Berlin, 1995.

durability of the system. However, as $NO_x$ is required for the operation of the catalyst there has to be a reasonable balance between the levels of $NO_x$ and PM in the engine exhaust and therefore this system may not be suitable for older engines.[15] The duty cycle of the vehicle should also regularly achieve exhaust temperatures above 275 °C. Different catalysts have been used including platinum on the trap and cerium in the fuel.

All particulate traps require low sulfur fuel (0.05%) for the same reasons as previously described for oxidation catalysts. Some, however, require even lower levels. For example, the Continuously Regenerating Trap (CRT) produced by Johnson Matthey requires a maximum of 0.01% sulfur.

Particulate traps have been widely tested in service. For example, a trial in Germany in the early 1990s involving 1200 heavy duty vehicles has shown that particulate traps can successfully remove 70 to 90% of the PM.[14] PM traps for heavy duty vehicles are commercially available, although they are expensive. They are not yet available for small vehicles although electrically heated and fuel additive systems are being developed. It is anticipated that this technology could become available in the next year or two.[14] However, if the technology is based on the use of fuel additives it may take longer for a full assessment of the environmental impacts to be undertaken to ensure that harmful levels of a new pollutant are not being emitted into the atmosphere.

### 13.3.3 In-Service

In the coming years there is likely to be increasing emphasis on ensuring that vehicles maintain their designed emission characteristics throughout their operational life and a number of measures introduced in the United States in recent years are likely to be modified for use in Europe. These include the following:

- more stringent Inspection and Maintenance (I/M) programmes;
- extended durability requirements;
- conformity of vehicles in circulation (recall) requirements and
- on-board diagnostics (OBD)

Typically I/M tests, such as in the British MOT test, involve a visual and idle test for petrol vehicles and a free acceleration test for diesel vehicles. Little analysis has been undertaken of the effectiveness of European programmes although a number of studies have been undertaken in the US.

Contrary to expectation, I/M programmes do not appear to have been very effective.[16] For example, comparison of average emissions from vehicles in areas of California with and without I/M programmes found that slightly more vehicles had high emissions in the areas with I/M programmes. Another survey showed that vehicles that had been subject to an I/M test within the previous 90 days had

---

[15] P. N. Hawker, 'Diesel Emission Control Technology: System Containing Platinum Catalyst and Filter Unit Removes Particulate from Diesel Exhaust', *Platinum Metals Review*, 1995, **62;39**, 2–8.
[16] A. Glazer, D. B. Kline and C. Lave, 'Clean on Paper, Dirty on the Road – Troubles with California's Smog Check', *J. Transport Econ. Policy*, 1995, **29**, 1.

higher emissions than those due for the test in the following 90 days.[16] It should be noted that these surveys used remote sensing to determine the effectiveness of the I/M programmes and the conclusions are not universally accepted.[17]

In the US, the Environmental Protection Agency has recommended the use of a short transient test cycle which requires the use of a chassis dynamometer at I/M stations. These are expensive, and the proposed test has met with opposition in some states including in California. A range of alternative testing systems, including remote sensing, are being explored in both the US and Europe.

Remote sensing devices were developed in the late 1980s to detect carbon monoxide levels in the exhaust of passing vehicles.[18] Subsequently they have been extended to include hydrocarbons. Essentially a beam of infrared light is passed through the exhaust plume of the vehicle which enables the concentration of pollution to be calculated. At the same time a video recording of the vehicle, focusing on the number plate, is made to enable its identification. The major advantage of remote sensing is its ability to measure the exhaust emissions from a large number of vehicles in a short period.

For most European countries with a mixture of catalyst and non catalyst petrol and diesel vehicles, it is unlikely that remote sensing systems could be used effectively to identify high emitting vehicles.[19] However, after the year 2000, when the majority of light duty vehicles will have low carbon monoxide emissions, remote sensing may play a useful role in an enhanced I/M programme.

Periodic emission tests are designed to ensure that vehicles are well maintained. However, in-use vehicles can be well maintained but give rise to high emissions due to poor design or quality control during the production process. Two European countries – Sweden and the Netherlands – and the United States undertake systematic in-use compliance testing programmes to ensure that well maintained vehicles meet the emissions requirements. Where faulty models have been identified, negotiations with the manufacturers/importers have led to faults being rectified. In Sweden, as in the US, the Environmental Protection Agency has the power to require manufacturers to recall and rectify faulty cars, although this is rarely used. With Sweden now a member of the EU it is expected that they will press for the adoption of similar legislation throughout Europe.

Under the current regulations, as part of the type approval process for cars, the manufacturer is required to show that the emissions after 80 000 kilometres are within defined limits. As a European car is typically driven much further than this during its lifetime, the extension of this to 160 000 kilometres is under consideration.

Simple on-board diagnostic systems (OBD) have been used on American cars for a number of years. These systems are designed to detect and alert the driver to the failure of an emissions control component. From the 1996 model year, 'second

[17]T. C. Austin and G. S. Rubenstein, 'An Analysis of Evidence Relating to the Effectiveness of Inspection and Maintenance Programs', prepared for the US EPA under contract 68-C1-0079, September 1994.

[18]S. H. Cadle and R. D. Stephens, 'Remote Sensing of Vehicle Exhaust Emissions', *Environ. Sci. Technol.*, 1994, **28**, 6.

[19]A. J. Hickman, Transport Research Laboratory, personal communication.

generation' systems (OBD II) will be required. When a malfunction occurs a light will illuminate on the dashboard and information will be stored in the vehicle's computer. The OBD system must also detect malfunction of the vapour recovery system. The intention is that soon after the light comes on the driver will take the car to be repaired, thus minimizing the period of excess emissions. Proposals for an European OBD system are being developed as this book is being written.

In the future telematics could be used to link OBD systems to roadside detectors, which could be used to aid enforcement. It has been suggested that an effective OBD system could remove the need for every car over a certain age (three years in the UK) to be subject to a regular I/M test.

### 13.3.4  Improved Fuels

Improving the quality of fuel can reduce emissions significantly as the benefits are seen immediately over the whole vehicle fleet. Improved fuels can also be used selectively to address local air quality problems.

A major US study – the Air Quality Improvement Research Programm (AQIRP) – was set up jointly by the oil and motor industries in 1989 to identify which petrol parameters have the most effect on car emissions.[20] In Europe, the two industries are co-operating in a similar research programme – EPEFE – the European Programme on Emissions, Fuels and Engine Technologies. This study, part of the European Auto/Oil Programme (see Section 13.2.2.1), aims to complement the work already undertaken in the US.

For some pollutants the relationship between fuel properties and emissions is clear (*e.g.* lead and benzene in petrol) while for others the relationship is less certain. There may be an interaction between different parameters. The effects of fuel parameters on emissions from petrol vehicles can depend on whether the car has a catalyst or not, and include the following:[6,21]

- Even very low levels of lead and sulfur in unleaded petrol can reduce catalyst efficiency.
- Oxygenates (such as MTBE) reduce CO emissions and to a lesser extent VOCs, but may increase $NO_x$ emissions and aldehydes.
- Reducing the aromatic content can reduce CO and to a lesser extent VOCs, but increases $NO_x$ emissions. However, there is an interaction with distillation characteristics of the fuel ($E_{100}$ – the percentage of the fuel that evaporates at 100 °C).
- Around half the benzene emissions comes from the benzene in the fuel. Reducing benzene from 3 to 2% would reduce benzene emissions by about 17%. The other half comes from the other aromatics in the fuel.
- Reducing olefins in the fuel reduces $NO_x$ and 1,3- butadiene emissions and the ozone formation potential of the evaporative emissions.

[20]Air Quality Improvement Research Programme, Phase 1 Final Report, US Society of Automotive Engineers, 1993.
[21]ACEA and EUROPIA, 'European Programme on Emissions, Fuels and Engine Technologies – Executive Summary', Brussels, July 1995.

- Increasing $E_{100}$ reduces the VOCs emissions but increases $NO_x$ emissions. CO emissions are at their lowest with a moderate $E_{100}$ (50% v.v).
- For cars without small canisters there is an exponential increase in evaporative emissions with Reid Vapour Pressure (RVP) and ambient temperature. For controlled cars (*i.e.* those with canisters) evaporative emissions only occur when the canister becomes overloaded.

The effects of fuel parameters on emissions from diesel vehicles include:[6,21]

- Reducing sulfur content of diesel reduces particulate emissions, particularly when oxidation catalysts are used. The effect is greatest on heavy duty vehicles.
- Reducing the density of diesel reduces $NO_x$ emissions from heavy duty engines and PM, CO and VOCs emissions from light duty engines. However, it also increases $NO_x$ emissions from light duty engines and CO and VOCs emissions from heavy duty engines.
- Reducing the polycyclic aromatics content of diesel reduces $NO_x$ and PM emissions from both heavy and light duty engines and VOCs from heavy duty engines. However, it increases CO and VOCs emissions from light duty engines.
- Increasing the cetane number reduces CO and VOCs emissions from both light and heavy duty engines and $NO_x$ emissions from heavy duty engines. However, it also increases PM emissions from light duty engines.
- Reducing $T_{95}$ (the temperature at which 95% is evaporated) increases CO and VOCs but reduces $NO_x$ emissions from heavy duty engines. It also increases $NO_x$, but reduces PM emissions from light duty engines.

EPEFE results[22] suggest that overall density and cetane number are the two most influential parameters on light diesel vehicle emissions. For heavy diesel vehicles cetane number and polyaromatics are the two most influential parameters. The sulfur effect was not investigated as part of EPEFE as it was felt that sufficient was known about the effect of this parameter. Reducing levels of sulfur reduces PM emissions but also, and probably more importantly, its removal allows the use of PM control technologies.

Some fuels contain detergents to stop the build up of sooty particles in the engine. These were first used in the US in the 1970s and in California their use in petrol has recently been mandated. Detergent additives are not, however, legally required in Europe, but are, nevertheless, now widely used in the UK.

A wide range of alternative fuels have been considered for use in the road transport sector, most of which offer both advantages and disadvantages in comparison with conventional fuels. Their use is largely at the demonstration stage and there is some uncertainty over the environmental benefits of many of them. To some extent this is because the vehicle performance depends on the fuel used, thus making direct comparisons difficult.

[22]Auto/Oil Programme, Informal Briefing, Brussels, March 1995.

Few alternative fuels have been exploited commercially due to two major problems. First there are the problems associated with the infrastructure requirements for the production, distribution and refuelling of new fuels. Secondly, cost is a major barrier when compared with conventional fuels. Certain fuels (natural gas and electricity) have production facilities and nationwide networks already in place and are more likely to overcome these market barriers. Nevertheless, additional facilities would need to be added if these energy sources were used on a large scale.

It is widely believed that conventional petrol and diesel will continue to be the dominant fuels for the next 20–30 years. Alternative fuels may find niche applications and gradually start to penetrate the market during the next five to ten years, but not significantly at the expense of petrol and diesel. Examples include urban buses and other urban vehicles operating on natural gas. Biomass fuels such as biodiesel and bioethanol are likely to be used mostly in developing countries and remote regions where conventional fuels are expensive or difficult to obtain. Hydrogen is a long term fuel that is unlikely to be exploited commercially for 30–50 years.

## 13.4  NON TECHNICAL MEASURES

### 13.4.1  Introduction

In the future non technical measures are likely to play an increasing role in local and national strategies to improve urban air quality. To tackle the problem of carbon dioxide emissions, for which there are no 'quick fix' technical solutions, non technical measures are likely to dominate reduction strategies.

Traffic management policies have some potential, particularly in the short term, to reduce traffic congestion and emissions by improving flow. However, improvements in traffic flow may release suppressed demand in congested areas, and recreate the very congestion they sought to combat, and therefore it is uncertain whether this approach will be effective. Alternatively they may divert traffic. Whilst this may help resolve acute city centre pollution problems, it will not resolve the problems of carbon dioxide emissions, which are independent of where they occur. To have an impact, traffic restraint policy will need to result in people moving from car use to walking, cycling or public transport and reducing the length and frequency of journeys.

There is a wide range of techniques available for promoting public transport, managing and limiting the use of private vehicles and encouraging other modes of transport such as walking and cycling. These range from regulatory and physical traffic controls to pricing and economic incentives, and include the use of public education campaigns and planning instruments. In this section some of these measures are briefly discussed. A strategy to reduce emissions from traffic will require a package of policies. It is unlikely that any single measure will have a large effect on its own. Disincentives to car use are likely to be unpopular with motorists, and are likely to be more acceptable if accompanied by incentives to use public transport, to walk or to cycle.

### 13.4.2  Parking

Limiting the availability of city centre parking and increasing the price, removes some of the advantages of cars over public transport, as travel time, cost and inconvenience for the car user increases. However, this approach could divert traffic to out of town facilities. Reductions in car commuting can be promoted by the provision of 'park and ride' schemes linked to an efficient public transport service.

### 13.4.3  Car Sharing

In theory car pooling and car sharing schemes could also reduce the number of cars used for commuting. However, these schemes have not been successful in Europe as few people choose to join them and there is a tendency for them to encourage a modal switch from public transport to car travel.

Where a city has an extensive motorway system there is scope for setting aside a lane for those vehicles with more than one person. It has been effective in North American cities, although, it is less appropriate in European cities which commonly have good public transport systems and are less likely to have a very extensive urban motorway network.

### 13.4.4  Area Bans

The use of area bans to restrict vehicles tends to divert traffic rather than restrain it. However, where good public transport is allowed access to large areas denied to car traffic, particularly in city centres, the use of area bans can result in a transfer from car to public transport, bicycle and walking.

The recent Corporation of the City of London restriction on the number of entry points to the City has reduced the number of vehicles entering the area by 25%.[23] However, the effects of the restrictions on the diversion of traffic were not studied.

In a number of cities rationing and permit systems have been introduced. The best-known form of rationing is the 'odds and evens' scheme, used in Athens and elsewhere, in which vehicles with odd numbers on their license plates are allowed into the restricted area on odd dates and vehicles with even numbers are allowed in on even dates. However, these systems have not been particularly efficient and there has been widespread abuse. In many attempts to make rationing schemes more sensitive to the needs of the travelling public an element of pricing has been introduced.

### 13.4.5  Road Pricing

Road pricing is a way of charging motorists for their use of the road system. It can be applied to a particular road, or to a wider area, and has been used both to

---

[23]Corporation of London, 'Vehicle Emissions in the City of London: Effects of the City's Traffic Management Scheme', 1994.

reduce congestion and to raise private funds for road building. Road pricing is seen as a potentially effective means of reducing car usage, as the price can be raised until the desired reduction in travel is brought about. However, it has met with considerable public opposition.

Area licensing has been introduced in a number of cities around the world, particularly in Scandinavia. The first such scheme was introduced in Singapore in 1975, where any vehicle driving into the designated central area and not displaying a pre-paid 'area licence' is fined. Electronic road pricing is a way of imposing different charges on drivers depending on the time of day and the particular roads they use. A pilot study undertaken in Hong Kong showed that the scheme is technically feasible. However, after the pilot study, electronic road pricing was rejected by the local people. The first British road pricing scheme is a pilot scheme due to open in Cambridge in the mid 1990s.

### 13.4.6  Public Transport

Public opinion surveys suggest that one of the major factors against using public transport is poor service. This is seen as more important, particularly for local journeys, than the level of fares. Where public transport is seen to be more convenient than driving patronage increases, for example in pedestrianized city centres where access has been denied to cars.

Modern electronics have opened up new opportunities for buses by giving them priority at traffic lights and computerized management systems. In a number of European cities such as Paris and Zurich, these have been introduced to identify and rectify problems such as 'bunching' and delays. They can also be used to keep passengers informed of when the next bus will arrive using, for example, electronic notice boards.

At the present time rail transport is undergoing something of a revival. Light railways, in particular, have been widely regarded as a way to ease urban congestion and increase public transport patronage. They fill the gap between conventional buses and trains and characteristically provide a fast and frequent service, with closely-spaced stops. Their major advantage is that they are cheaper to install than conventional rail. In Britain some 50 towns have seriously considered installation of such systems in the last few years. Light railways are already operational in Glasgow, Tyne and Wear, Blackpool, London and Manchester.

There is evidence that cheap integrated tickets can result in a modal shift from car to public transport, at least in the short term. For example, the Greater London Council's 'Fares Fair' scheme, introduced in the early 1980s, divided its area into zones, simplified the fares structure and introduced travel cards that could be used on buses, the Underground and British Rail. Traffic in London decreased by 16 percent between 1982 and 1984. Over the same period Underground patronage increased by 39 percent and bus use by 11 percent. Similar successes have been seen in other countries.

### 13.4.7  Walking and Cycling

Clearly a very large increase in the volume of cycling and walking would be needed to have any material impact on motorized traffic levels and it is difficult to see how a large increase could be achieved, especially as a significant proportion of journeys are longer than it is convenient to make by these non motorized modes. Nevertheless, short car journeys driven in a cold vehicle are proportionately more polluting than longer journeys undertaken in a fully warmed up vehicle. Cycling and walking are ideally suited to replace these journeys. To encourage cycling and walking the roads need to be made safer, and direct and attractive cycle lanes/ walkways provided. For longer journeys cycling can be combined with public transport through the provision of better bicycle parking facilities and improved provision of transport of bicycles by trains and express coaches.

### 13.4.8  Land Use Planning

The way in which land is used has important implications for transport demand and the ability of public transport to serve it. In compact cities, public transport patronage can exceed 400 trips per capita per year and up to 50% of journeys can be on foot. At the other extreme, in low density cities, public transport trip rates can be below 50 trips per capita per year and only 20% of journeys can be on foot.[24]

Towns and cities take many years to take shape. In the past there has been little concern over the effect of development on transport demand as illustrated by the rapid development of out-of-town shopping, leisure and business facilities. However, this is beginning to change. In the long term, land use planning has a vital role to play in reshaping towns and cities to reduce the need for motorized transport, and to ensure that a substantial proportion of it is well served by public transport.

### 13.4.9  Fiscal Measures

There are a range of fiscal measures that can be used to influence both the choice of vehicle (and fuel) and the amount it is used. These include increasing fuel prices, annual ownership taxes (vehicle excise duty) and purchase taxes. The use of fuel prices and purchase tax has been discussed in Section 13.2.4.

With the exception of the United States, most industrialized nations have long imposed large fuel taxes to raise revenue. These taxes have served as an incentive to consumers to purchase more fuel efficient cars and this is reflected in the smaller cars on Europe's roads compared to the US. However, fuel prices are lower in real terms now than they were before the first oil crisis in 1973 and fuel is only a small fraction of the cost of buying, operating and maintaining a modern passenger car. The UK Government is committed to a 5% per annum increase in real terms to reduce $CO_2$ emissions from traffic.

[24] J. Bayliss, 'Urban Traffic Management' in Transport Policy and the Environment, European Conference of Ministers/OECD, Paris, 1990.

However, to be effective it has been argued that fuel prices need to be increased further.[13]

Increasing fuel prices should encourage the purchase of smaller, more fuel efficient cars, stimulate manufacturers to produce vehicles that incorporate advanced technology and also be an incentive to drive economically, reduce mileage and encourage regular servicing. However, studies suggest that if the price change is maintained for a sufficiently long period for the car stock to change its fuel consumption characteristics these changes could offset some of the increase in fuel price. Thus if new cars use half the fuel motorists will be less sensitive to price increases and the fuel price increases would have to be even greater to have the same effect.

In some countries a graduated car purchase tax gives preferential treatment to certain classes of passenger car. However, as a small car using outdated technology can have higher emissions than a much larger vehicle of more modern design, a vehicle purchase taxation structure that directly links purchase tax to emissions would be better than one that uses a proxy *e.g.* size, power output, maximum speed *etc.*

It is difficult to devise a scheme graduated enough to ensure that the most expensive cars do not end up paying less tax than at present and therefore giving a market incentive to purchasers of these vehicles. Thus a revenue neutral scheme, over and above the purchase tax, in which cars that emit the most pay a charge and those that emit the least get a rebate appears to be an appropriate approach.

High purchase tax has also been used in a number of countries as a way of reducing car ownership levels. However, the disadvantage of such taxation is that it acts as a disincentive to buy new cars, and hence slows down the introduction of advanced vehicles. To encourage old (polluting) cars to be scrapped, several European countries, such as France and Spain, have introduced tax rebates for the early disposal of vehicles (scrappage fee).

Most industrialized countries have a progressive annual ownership tax. While some countries set the tax according to vehicle weight, others use engine size or power output, or a combination of these. All these taxation methods encourage car buyers to purchase more fuel efficient models since the less weight, power output and engine size, generally the better the fuel economy. In the UK the car excise duty is set at a fixed rate.

## 13.5 SUMMARY/CONCLUSIONS

Road traffic is a major source of air pollution. Despite the introduction of a series of more stringent emission controls for new vehicles over the past twenty-five years, further measures are likely to be needed to ensure that air quality meets recognized air quality standards. Given the current status of knowledge on the medical effects of air pollutants, it is likely that meeting desirable levels of ozone and PM will be particularly difficult.

In the past the emphasis was on reducing emissions from new vehicles. There remains some potential for further technical improvements to new vehicles but it is becoming increasingly apparent that a wider approach to controlling traffic pollution is required. A series of measures are expected to be introduced into the EU around the turn of the century that will not only reduce emissions from new vehicles, but will also ensure that emissions remain low throughout the life of the vehicle, and that better fuels are used. However, these technological solutions may not be sufficient, or even appropriate, for some air pollution problems. For example, it is likely that the control of $CO_2$ emissions from traffic will require changes in purchase patterns and a halt to the increase in car use.

Behavioural change is more difficult to achieve than technical change, as typically it cannot be mandated. It will require a wide range of non technical measures designed to encourage people to use their cars less, some of which have been described in Section 13.4. This will probably only be achieved by a 'carrot and stick' approach, that is, disincentives to car use linked to improvements to public transport, cycling and walking facilities. These measures are likely to be initially unpopular and will probably require a considerable period of time for the benefits to be seen.

For petrol vehicles the greatest potential for reducing emissions lies in cutting the cold start emissions through the use of early light-off catalysts. Emissions of $NO_x$ and PM from advanced diesel vehicles are at a much lower level than their predecessors, but still require emerging technologies, particularly the $DeNO_x$ catalyst, to become commercially available. For cars, particulate traps are not yet available, whilst for buses and lorries it is a proven, but expensive technology. Their use may be most suited to buses and other urban vehicles. Alternative fuels, particularly natural gas, may be a better approach to addressing emissions from these vehicles. However, in general, alternative fuels are unlikely to significantly penetrate the automotive fuels market until the next century.

Improved fuel quality is likely to play an important role in reducing emissions from vehicles and new minimum standards for petrol and diesel, based on emission performance rather than vehicle performance, are likely to be introduced in the EU from the turn of the century. Improved fuels were first introduced in the UK in 1995. However, due to their price premium they are unlikely to have mass appeal. In the absence of new standards it is probable that only with a tax incentive will these fuels obtain a large enough market share to have an impact on air quality. The introduction of improved fuels will not only reduce emissions from current vehicles but will also open the door to new advanced pollution control technologies.

Increasingly the emphasis is being put on identifying the gross polluters. As failure of an emissions related component may have no impact on the vehicle performance the driver may be unaware of a problem. The introduction of OBD systems onto European cars will not only alert the driver but will also store information for the repair mechanic to rapidly identify the problem. It is not known how long it will take the average European motorist to take the vehicle for repair after the warning light comes on and the use of telematics, linking the OBD system with a roadside detector, offers enforcement potential. However, there is likely to remain a need for some form of I/M testing, either using remote sensing or regular tests. The regular I/M test is likely to be modified and become more stringent in the medium term. In the longer term, when the vast majority of cars are fitted with OBD systems, the mechanic may only need to interrogate the on-board computer.

The type approval procedures are likely to be modified to ensure that manufacturers build their vehicles for low lifetime emissions. In particular European manufacturers face the expensive prospect of legally enforceable recall for failure of emission related components as well as being required to meet defined emission levels for the first 160 000 kilometres rather than the 80 000 kilometres as in the current legislation.

As the potential for further technological improvements declines in the coming years, non technical measures will have to play an increasing role in ensuring good air quality and reductions in emissions of greenhouse gases from the road transport sector.

*Acknowledgements.* Several sections of this chapter are based on previous work by the author. In particular, her contributions to the first two reports of the Quality of Urban Air Review Group and 'Greening Urban Transport – Environmentally Improved Grades of Petrol and Diesel' written for the European Federation for Transport and the Environment.

CHAPTER 14

# Soil Pollution and Land Contamination

B. J. ALLOWAY

## 14.1 INTRODUCTION

Soil is an essential component of terrestrial ecosystems because the growth of plants and biogeochemical cycling of nutrients depend upon it. Of the total area of the world's land mass ($13.07 \times 10^9$ ha), only 11.3% is cultivated for crops; permanent grazing occupies 24.6%, forest and woodland 34.1% and 'other land' including urban/industry and roads, accounts for 31%.[1] From a resource perspective, soil is vitally important for the production of food and fibre crops and timber and it is therefore essential that the total productive capacity of the world's soils is not impaired. Pollution, along with other types of degradation, such as erosion and the continuing spread of urbanization, poses a threat to the sustainability of soil resources. Soil pollution can also be a hazard to human health when potentially toxic substances move through the food chain or reach groundwater used for drinking water supplies.

In comparison with air and water, the soil is more variable and complex in composition and it functions as a sink for pollutants, a filter which retards the passage of chemicals to the groundwater and a bioreactor in which many organic pollutants can be decomposed. As a consequence of its occurrence at the interface between the land and the atmosphere, soil is the recipient of a diverse range of polluting chemicals transported in air. Further inputs of pollutants to the soil occur as a result of agricultural and waste disposal practices but, in general, the most severe pollution usually results from industrial and urban uses of land.

It is generally accepted that most of the soil in technologically advanced regions of the world is polluted (or contaminated), at least to a slight extent.[2] However, in many cases the relatively small amounts of pollutants involved may not have a significant effect on either soil fertility or animal and human health. More severe 'chemical pollution' which poses a greater hazard has been

[1] World Resources Institute, 'World Resources 1994–95: A Guide to the Global Environment', Oxford University Press, New York, 1994.
[2] K. C. Jones, *Environ. Pollut.*, 1991, **69**, 311–325.

estimated by a recent Global Assessment of Soil Degradation ('GLASOD') to affect a total of $21.8 \times 10^6$ ha of land in Europe, Asia, Africa and Central America.[3] Realistic estimates of areas affected by soil pollution are difficult owing to unreliable official figures and inadequate data for many parts of the world. Industrially contaminated land tends to contain higher concentrations and a greater possible range of pollutants than other sources of pollution. There are between 50 000 and 100 000 contaminated sites in the United Kingdom which occupy up to 100 000 ha.[4] In the USA, 25 000 contaminated sites have been identified; 6000 sites are being cleaned-up in the Netherlands, there are known to be at least 3115 sites in Denmark and 40 000 suspect areas have been identified on 5000–6000 sites in the former western part of Germany.[5]

Old industrial sites are generally characterized by being very heterogeneous, both with regard to the distribution of pollutants and also to the properties of the soil materials that control the behaviour of these chemicals. In contrast, atmospherically deposited pollutants tend to have a more even distribution with gradual changes in concentrations, which tend to decrease with distance from the source. The upper horizons of the soil are contaminated to the greatest extent by atmospheric deposition. In general, pollution, in common with other environmental impacts resulting from human activities, is linked to size of the population, its relative affluence and the level of technological development:[6]

$$Environmental\ impact = Population \times Affluence \times Technology$$

All contamination/pollution situations comprise the following components: (i) a source of pollutant, (ii) the pollutant itself, (iii) a transport mechanism by which the pollutant is dispersed and (iv) the target where the transport phase terminates. Transport can be by moving air or water, by gravity movement downslope or by direct conveyance and placement such as the haulage and spreading of waste materials. Although this is a very simple conceptual model, which does not take account of variations in time and quantity, it does provide a useful basis from which to consider the pollution of soils and other environmental media.

## 14.2 SOIL POLLUTANTS AND THEIR SOURCES

### 14.2.1 Heavy Metals

Heavy metals have a density of $>6\,\mathrm{g\,cm^{-3}}$ and occur naturally in rocks but concentrations are frequently elevated as a result of contamination. The term 'heavy metal' is imprecise but is widely used although others such as 'toxic metals', 'potentially toxic elements' and 'trace metals' are possible alternatives.

[3]L. R. Oldeman, R. T. A. Hakkeling and W. G. Sombroek, 'World Map of the Status of Human-Induced Soil Degradation', CIP-Gegevens Koninklijke Bibiotheek, Den Haag, 1991.
[4]House of Commons Committee, 'Contaminated Land', First Report, HMSO, London, 1990.
[5]E. M. Bridges, *Soil Use Management*, 1991, **7**, 151–158.
[6]D. H. Meadows, D. L. Meadows and J. Randers, 'Beyond the Limits', Earthscan Publications, London, 1992.

Heavy metals belong to the group of elements geochemically described as 'trace elements' because they collectively comprise $<1\%$ of the rocks in the earth's crust. All trace elements are toxic to living organisms at excessive concentrations, but some are essential for normal healthy growth and reproduction by either plants and/or animals at low but critical concentrations. These elements are referred to as 'essential trace elements' or 'micronutrients' and deficiencies can lead to disease and even death of the plant or animal. The essential trace elements include: Co (for bacteria and animals), Cr (animals), Cu (plants and animals), Mn (plants and animals), Mo (plants), Ni (plants), Se (animals) and Zn (plants and animals). In addition, B (plants), Fe (plants and animals) and I (animals) are also essential trace elements but are not classed as heavy metals.

Other elements, including Ag, As, Ba, Cd, Hg, Pb, Sb and Tl, have no known essential function and, like the essential trace elements, cause toxicity above a certain tolerance level. The most important heavy metals with regard to potential hazards and occurrence in contaminated soils are: As, Cd, Cr, Hg, Pb and Zn.[7]

*14.2.1.1 Sources of Heavy Metals.* *(a) Metalliferous mining:* This is an important source of contamination by a wide range of metals, especially As, Cd, Cu, Ni, Pb and Zn, because ore bodies generally include a range of minerals containing both economically exploitable metals (in ore minerals) and uneconomic elements (in gangue minerals). Most mine sites are contaminated with several metals and accompanying elements (*e.g.* sulfur). Wind-blown tailings (finely ground particles of ore and country rock) and ions in solution from the weathering of ore minerals in heaps of tailings tend to be the major sources of pollution from abandoned metalliferous mine sites.

*(b) Metal smelting:* This is the process of producing metals from mined ores and so can be a source of many different metals. These pollutants are mainly transported in air and can be in the form of fine particles of ore, aerosol-sized particles of oxides (especially important in the case of the more volatile elements such as As, Cd, Pb and Tl) and gases ($SO_2$). In some cases, pollution is directly traceable in soil $<40$ km downwind of smelters.

*(c) Metallurgical industries:* Pollution can include aerosol particles from the thermal processing of metals and solid wastes, effluents from the treatment of metals with acids and solutions of metal salts used in electroplating.

*(d) Other metal-using industries:* These can be a source of metals in gaseous/ particulate emissions to the atmosphere, effluents to drains and solid wastes. These include: the electronics industry where metals are used in semiconductors, contacts, circuits, solders and batteries (Cd, Ni, Pb, Hg, Se and Sb); pigments and paints (Pb, Cr, As, Sb, Se, Mo, Cd, Co, Ba and Zn); the plastics industry (polymer stabilizers such as Cd, Zn, Sn and Pb) and the chemical industry which uses metals as catalysts and electrodes including: Hg, Pt, Ru, Mo, Ni, Sm, Sb, Pd and Os.

*(e) Waste disposal:* Municipal solid waste, special wastes and hazardous wastes from many sources can contain varying amounts of many different metals.

---

[7]'Heavy Metals in Soils', ed. B. J. Alloway, Blackie Academic and Professional, Glasgow, 2nd edn, 1995.

*(f) Corrosion of metals in use:* Corrosion and chemical transformation of metals used in structures *e.g.* Cu and Pb on roofs and in pipes, Cr, Ni and Co in 'stainless steel', Cd and Zn in rust preventative coatings on steel; Cu and Zn in brass fittings and the deterioration of painted surfaces (Pb, Cr).

*(g) Agriculture:* This mainly includes As, Cu and Zn which are (or have been) added to pig and poultry feeds, Cd and U impurities in some phosphatic fertilizers, metal-based pesticides (historic and current) such as As, Cu, Pb and Zn.

*(h) Forestry:* Wood preservatives containing As, Cr and Cu have been widely used for many years and have caused contamination of soils and waters in the vicinity of timber yards.

*(i) Fossil fuel combustion:* Trace elements present in coals and oils include Cd, Zn, As, Sb, Se, Ba, Cu, Mn and V and these can be present in the ash or gaseous/particulate emissions from combustion. In addition, various metals are added to fuels and lubricants to improve their properties: Se, Te, Pb, Mo and Li.

*(j) Sports and leisure activities:* Game and clay pigeon shooting involves the use of pellets containing Pb and Sb (recently substitutes using steel, Mo and Bi are being introduced).

## 14.2.2 Hydrocarbon Pollutants

Hydrocarbon pollutants from petroleum mainly comprise a range of saturated alkanes from methane ($CH_4$), ethane ($C_2H_6$) and propane ($C_3H_8$) through straight and branched chains to $C_{76}H_{154}$. Aromatic hydrocarbons and organic components containing nitrogen and sulfur can also be important constituents of some petroleum deposits. The hydrocarbons derived from coal and petroleum tend to form the main group of organic macropollutants in soils. Organic solvents can be important soil pollutants at industrial sites. Apart from any toxicity hazard associated with the ingestion or inhalation of volatile hydrocarbons, there is also a high risk of fires and explosions.

*14.2.2.1 Sources of Hydrocarbon Pollutants.* *(a) Fuel storage and distribution:* Leaking underground storage tanks and spillages at distribution depots and from road accidents can lead to pollution of soils and aquifers with petrol and diesel fuels. It is possible that around 30% of filling stations in the United Kingdom may be causing some subsurface pollution through leakages from underground storage tanks. In view of the very large volumes of petroleum fuels used, this source must account for a high proportion of soil pollution by hydrocarbons. However, despite their ubiquitous occurrence, hydrocarbons are more readily degraded in soils and pose less of a toxicological risk than organo-micropollutants such as PAHs, PCBs, dioxins and many pesticide derivatives. However, Pb-containing petrol will continue to pose a long-term Pb contamination hazard.

*(b) Disposal of used lubricating oils:* In addition to hydrocarbons and abraded particles of metal, old lubricating oils contain PAHs and other products of partial combustion. Some domestic gardens, land around motor repair garages, farm

yards and sites of crashed and vandalized cars can often be polluted with this material.

*(c) Leakage of solvents from industrial sites:* Hydrocarbon solvents are used widely in industry for cleaning and degreasing metals and electrical components, and leakages (from storage, distribution and use) frequently result in contamination of soils and aquifers.

*(d) Coal stores:* Coal is a solid form of hydrocarbon and the main hazard associated with it is the risk of fires. Sites of coal stores at former industrial sites and distribution depots are the most likely to contain significant amounts of coal which could constitute a combustion hazard.

### 14.2.3   Toxic Organic Micropollutants (TOMPS) (see also Chapter 16)

The most commonly encountered toxic organic micropollutants include: polycyclic aromatic hydrocarbons (PAHs), polyheterocyclic hydrocarbons (PHHs), polychlorinated biphenyls (PCBs), polychlorinated dibenzodioxins (PCDDs), polychlorinated dibenzofurans (PCDFs) and pesticide residues and metabolites. Many of these organic pollutants are discussed in more detail elsewhere in this volume.

*Pesticides* – comprise a very large range of different types of organic molecules which are used with the intention of destroying pests of various types, including: insects, mites, nematodes, weeds and fungal pathogens. The types of compounds used as pesticides include:

| | |
|---|---|
| *Insecticides:* | Organochlorines (DDT, BHC) |
| | Organophosphates (Malathion, Parathion) |
| | Carbamates (Aldicarb) |
| *Herbicides:* | Phenoxyacetic acids (2,4-D, 2,4,5-T) |
| | Toluidines (Trifluralin) |
| | Triazines (Atrazine, Simazine) |
| | Phenyl ureas (Fenuron, Isoproturon) |
| | Bipyridyls (Diquat, Paraquat) |
| | Glycines (Glyphosate) |
| | Phenoxypropionates (Mecoprop) |
| | Translocated carbamates (Barban, Asulam) |
| | Hydroxyl nitriles (Ioxynil, Bromoxydynil) |

*Fungicides:*
        Non-systemic
                Inorganic and heavy metal compounds
                Dithiocarbamates (Maneb, Zineb)
                Phthalimides (Captan, Captafol)
        Systemic
                Antibiotics (Cycloheximide, Blasticidin-S)
                Benzimidazoles (Carbendazim, Benomyl, Thiabendazole)
                Pyrimidines (Ethirimol, Triforine)

As a consequence of the variety of compounds involved, there are major differences in soil behaviour and toxicity to plants, soil organisms and humans. Many pesticides break down into toxic derivatives and may cause phytotoxicity problems in sensitive crops. The most serious problems associated with pesticides are the contamination of surface and groundwaters and entry into the food chain through crops.

Typical rates of pesticide application in agriculture are 0.2–5.0 kg ha$^{-1}$ but frequently higher rates of some pesticides may be used for non-agricultural purposes, such as weed clearance on railway tracks and urban paths. In the United Kingdom, the total tonnage of pesticide active ingredient used decreased by 20% between 1980 and 1990 but the area treated with these substances increased by 9%.[8] Generally, less than 10% of the pesticide reaches its intended target; the remainder may reside in the soil, some will be volatilized and some will be leached through soils to groundwater or *via* field-drainage to water courses. Most pesticides have water solubilities greater than 10 mg l$^{-1}$ and are therefore highly prone to leaching through soils. The half-lives of many pesticide compounds in fertile soils range from 10 days to 10 years and so, in many cases, there is sufficient time for some leaching to occur. Atrazine with a half-life of 50–100 days gives rise to widespread groundwater contamination. The concentrations of soil-acting pesticides in the soil solution are thousands of times greater than the EC guideline concentration for potable waters (0.1 $\mu$g l$^{-1}$ per compound, total concentration 0.5 $\mu$g l$^{-1}$) and so there is a strong probability of groundwater contamination above EC limits.[9]

### 14.2.4 Other Industrial Chemicals (see also Chapter 1)

It is estimated that between 60 000 and 90 000 chemicals are in current commercial use. Although not all of these constitute potential toxicity hazards, many will cause pollution of soils as a result of leakage during storage, from use in the environment or from their disposal either directly, or of wastes containing them. Apart from industrial uses, a large number of chemicals are used in domestic products and so their use and disposal is less controlled than that of industrial chemicals (which are subject to strict regulations).

The total world production of hazardous and special wastes was 338 × 10$^6$ t in 1990.[10] Although the Red, Black and Grey Lists of hazardous chemicals contain a large number of priority substances which can pollute soils, only a few examples can be given in Table 14.1.

### 14.2.5 Nutrient-rich Wastes

*(a) Sewage sludges* – are the residues from the treatment of wastewater and large quantities are produced worldwide (6.3 × 10$^6$ t in the original 12 countries of the

---

[8]A. Brown, 'The UK Environment', Government Statistical Service, HMSO, London, 1992.

[9]S. S. Foster, P. J. Chilton and M. E. Stuart, *J. Inst. Water Environ. Managt.*, 1991, 186–193.

[10]Organization for Economic Co-operation and Development (OECD), 'The State of the Environment', OECD, Paris, 1991.

**Table 14.1** *Priority hazardous chemicals ( based on the UK Red List and EC Lists I and II)*

| | |
|---|---|
| Mercury and its compounds | Dichlorvos |
| Cadmium and its compounds | 1,2-Dichloroethane |
| $\gamma$-hexachlorocyclohexane (Lindane) | Trichlorobenzene |
| DDT | Simazine |
| Pentachlorophenol and its compounds | Organotin compounds |
| Hexachlorobenzene | Cyanide, Fluorides |
| Hexachlorobutadiene | Trifluralin |
| Fenitrothion | Azinphos-methyl |
| Aldrin, Dieldrin | Organophosphorus compounds |
| Endrin | Endosulfan |
| Carbon tetrachloride | Atrazine |
| Polychlorinated biphenyls | |
| Persistent mineral oils and | |
| hydrocarbons of petroleum | |

European Union in 1990 and $5.4 \times 10^6$ t in the USA). This sludge has usually been disposed of onto agricultural land (43% of total in UK and 22% in the USA), into the sea (30% in the UK), landfilled or incinerated. In the European Union, disposal at sea will no longer be permitted after 1998 and many other countries are also phasing out oceanic disposal. This implies that the other disposal routes will be used to a much greater degree and applications to land will greatly increase. Sewage sludge is a valuable source of plant nutrients (especially N and P) and of organic matter which has beneficial effects on soil aggregate stability. However, its value is somewhat diminished by its content of potentially harmful substances which include heavy metals, especially Cd, Cu, Ni, Pb and Zn and organic pollutants. The most important TOMPS in sewage sludges include: a) halogenated aromatics, *e.g.* polychlorinated biphenyls (PCBs), polychlorinated terphenyls (PCTs), polychlorinated naphthalenes (PCNs), polychlorinated benzenes and polychlorinated dibenzodioxins (PCDDs), b) halogenated aliphatics, c) polycyclic aromatic hydrocarbons (PAHs), d) aromatic amines and nitrosamines, e) phthalate esters and f) pesticides.[11]

*(b) Livestock manures* – these contain large amounts of N, P and K and are valuable sources of these nutrients for crops. However, the manures may also contain residues of feed additives which can include As, Cu and Zn and antibiotics such as sulfonamides fed to pigs and poultry.

### 14.2.6 Radionuclides (see also Chapter 17)

Nuclear accidents like those at Windscale (UK) in 1957 and Chernobyl (Ukraine) in 1986 resulted in many different radioactive substances being dispersed into the environment. The greatest long-term pollution problem with

---

[11] D. Sauerbeck, in 'Scientific Basis for Soil Protection in Europe', eds. H. Barth and P. L'Hermite, Elsevier, Amsterdam, 1987, p. 181.

radionuclides in soils is considered to be caused by $^{137}$Cs which has a half-life of 30 years and behaves in a manner similar to K in soils and ecosystems. Atmospheric testing of nuclear weapons dispersed large amounts of $^{90}$Sr which has a half-life of 29 years and behaves similarly to Ca in biological systems and poses a particularly dangerous hazard to humans because it is stored in the skeleton.

### 14.2.7  Pathogenic Organisms

Soils can be contaminated with pathogenic organisms (bacteria, viruses, parasitic worm eggs) from various sources, including the burial of dead animals and humans, manures and sewage and germ warfare experiments. The soil's own microbial biomass will destroy many of these pathogens but some appear to survive for relatively long times.

## 14.3  TRANSPORT MECHANISMS CONVEYING POLLUTANTS TO SOILS

Pollutants reach soils by four main pathways:

- atmospheric deposition of particulates (washout or dry deposition),
- sorption of gases (*e.g.* volatile organic compounds) from the atmosphere,
- fluvial transport and deposition or sorption from flood waters,
- placement (dumping, injection, surface spreading *etc.*).

The details of aeolian dispersion are covered in Chapter 10 but the critical factors determining the distance transported and the area of land affected are: the height of the emission above the ground, the windspeed, stability of the atmosphere and the size, shape and density of the particles and amounts of the pollutants emitted. Particles <0.05 mm diameter can be carried over large distances and those with diameters of less than 10 $\mu$m (0.01 mm) are classed as aerosols. These particles are small enough to remain suspended in turbulent air but are lost by impaction onto soil and vegetation, by gravity deposition and by washout in rain. They can also be inhaled into the respiratory system where the pollutants are more readily absorbed into the blood than from the digestive tract. Particles 0.05–0.5 mm in diameter are moved by saltation and this can be important in metalliferous mining areas where tailings particles can be transported from tailings heaps around the mine onto the surrounding soils.[12] Asbestos fibres are relatively aerodynamic and are readily dispersed by wind.

Fluvial transport is also dealt with elsewhere in this book (Chapters 1 and 4) but, with regard to soils, this mechanism is only important in land subject to flooding. This was an important pollution pathway in areas of metalliferous mining in the nineteenth century. Before pollution controls were introduced in 1876, Pb-Zn mines in the UK discharged waters from ore dressing operations directly into streams and rivers. This led to the alluvial soils in most flood plains of

[12]G. Merrington and B. J. Alloway, *Appl. Geochem.*, 1994, **9**, 677–687.

rivers draining mining areas being severely contaminated with Pb, Zn and other metals.[12] Soils on the flood plains of many major rivers in the world which drain industrial and urbanized areas have been significantly contaminated with a diverse range of substances from flood waters.

Volatilization involves a substance changing from a liquid to a gas and this is a very important mechanism by which many organic compounds become dispersed in the atmosphere and conversely, sorbed from the atmosphere onto soils or plants.

Placement of pollutants can occur in many ways; the most obvious being the spreading of wastes, such as sewage sludges or metal-rich manures from pigs or poultry. Phosphatic fertilizers can contain significant concentrations of Cd and U and have been at least partially responsible for the significantly elevated concentrations of these elements in many parts of the world.

## 14.4 THE NATURE AND PROPERTIES OF SOILS RELATED TO THE BEHAVIOUR OF POLLUTANTS

### 14.4.1 The Nature of Soils

Soil is the geochemically and biochemically complex material which forms at the interface between the atmosphere and the earth's crust and is highly heterogeneous in both composition and spatial distribution. It comprises a mixture of mineral and organic solids, permeated by voids containing aqueous and gaseous components and a microbial biomass. Soils are usually differentiated vertically into a series of distinctive layers, called 'horizons', which differ both morphologically and chemically from the layers above and below them. These horizons collectively form the 'soil profile' (or pedon) which is the unit of classification of soils. Soil formation results from interactions between: the weathered geological material on which the soil has formed, climate, vegetation cover, landscape position and the time over which the soil has been forming. Soil formation is a dynamic process and major changes in any of the environmental factors (such as climate, drainage or vegetation) will result in changes in the nature of the soil horizons. As a result of the wide range of rock types and environmental conditions around the globe, soils can differ markedly in physical, chemical and biological characteristics. Nevertheless, there are several properties which most soils have in common which relate to the behaviour of pollutants.

All soils contain humus which is highly polymerized organic material synthesized from the decomposition products of dead plant material. The organic matter contents of most soils lie in the range 0.1–10% but peaty soils can contain more than 70% organic matter. Soils in hotter climates tend to contain much lower amounts of organic matter than soils in cooler humid regions.

Soils contain varying amounts of different primary and secondary minerals. Primary minerals occur in unweathered fragments of igneous rock either from the parent material or erratic stones deposited by ice or water and their weathering provides plant nutrients and gives soils distinctive colour and chemical properties. Secondary minerals have been synthesized from the products of weathered rocks and can include hydrated oxides of Fe, Al and Mn (Fe oxides give soils their characteristic brown colour) and clay minerals which are thin layered forms of aluminium silicates. These secondary minerals and humus form the colloidal fraction which provides soils with sorptive properties which are very important in determining the fate of pollutants (and plant nutrients) in soils. Other secondary minerals, such as calcium carbonate, can be present in semi-arid and arid climates, but also occur as part of the parent material of soils which have developed on limestones in humid climates.

## 14.4.2 Chemical and Physical Properties of Soils Affecting the Behaviour of Pollutants

Space does not allow a detailed consideration of the chemistry related to the behaviour of pollutants in soils. However, the main considerations are the factors which control the sorption and desorption of ionic and non-ionic (uncharged) compounds. Most heavy metals exist in the soil predominantly as cations but important elements such as: As, B, Mo, V, Sb and Se occur as anions. Although some pesticides are ionic, most organic contaminants tend to be uncharged and hydrophobic.

Sorption of pollutants can be by several mechanisms, including:

*(i) Non-specific adsorption of cations and anions* (also referred to as Cation Exchange and Anion Exchange). Cations are adsorbed onto negatively charged surfaces in the soil colloidal fraction. This comprises the alumino-silicate clay minerals, hydrous oxides of iron and manganese and humic organic material. Anions are adsorbed on positively charged sites on the colloidal fraction which are mainly contributed by hydrous oxides of iron. The charge on these iron oxides depends on the pH of the soil. Below a pH of around 7.0 Fe oxides are positively charged and adsorb anions, but above this pH they are negatively charged and adsorb cations. Soil organic matter also has a pH-dependent surface charge but this is predominantly negatively charged above pH 2.5. Protonation of carboxyl and phenolic groups on the surfaces of humic polymers gives rise to negative charges. Clay minerals have permanent negative charges created by charge imbalances where isomorphous substitution of a major constituent (*e.g.* $Al^{3+}$ replacing $Si^{4+}$, or $Mg^{2+}$ replacing $Al^{3+}$) has occurred in the crystal lattice of the mineral during its formation.

The soil pH is the most important single physico-chemical parameter controlling the sorption–desorption of ions in soils. The normal range of pH in soils throughout the world is 4–8.5 owing to buffering by Al and $CaCO_3$ at the lower and upper ends respectively. In general, soils in humid regions, which are subject to leaching of bases, tend to have a pH range of 5–7 (although organic upland soils may have values of less than 4.0). Soils in arid regions tend to have pH values between 7–9 owing to the accumulation of $CaCO_3$ and other salts in the predominantly evaporating moisture regime.

Cation exchange occurs where a higher concentration of cations is held in the zone of attraction of the negative charges on the soil colloid surfaces than in the bulk of the soil solution. In general, it is found that the Cation Exchange

Capacity (CEC) of a soil increases with a rise in pH, at least up to pH 7.0. The adsorbed cations are in a dynamic state of flux dependent on the nature of the charged surface, the nature of the ion (its valency and hydrated size), its concentration and the relative concentrations of other ions in the soil solution. There is a general order of replacement whereby it is found that those ions which are most strongly attracted replace other cations in the zone of attraction. This order varies for different adsorbent materials, but for the clay mineral illite the order of increasing selectivity was given by Bittel and Miller[13] as:

$$Mg > Cd > Ca > Zn > Cu > Pb$$

Anion exchange involves the retention of anions on positively charged surface sites at pH values below 7.0 and by 'ligand exchange' where a surface complex forms between an anion and a metal, usually Fe or Al in a hydrous oxide or a clay mineral. The sorptive capacity of a soil is expressed in units of centimoles of charge per kg of soil $(cmol_c\ kg^{-1})$.

*(b) Specific adsorption.* This occurs where metals such as Cd, Cu, Ni and Zn form complex ions $(MOH^+)$ on surfaces that contain hydroxyl groups, especially hydrous oxides of Fe, Mn and Al. These complex ions do not undergo cation exchange but can be displaced by strong acids or complexing agents. Specific adsorption is strongly pH dependent and is responsible for the retention of larger amounts of metals than cation exchange. The general order of increasing strength of specific adsorption of heavy metals was given by Brummer[14] as:

$$Cd > Ni > Co > Zn >> Cu > Pb > Hg$$

*(c) Organic complexation of metals.* This occurs when the solid state humic material binds metals into a ring type structure (ligand molecule) most commonly a chelate. Humic compounds with hydroxyl, phenoxyl and carboxyl reactive groups can form coordination complexes with metals. The stability constants of chelates vary for different ligands and elements. In general, the stability constants of humic complexes tend to decrease in the order:

$$Cu > Fe = Al > Mn = Co > Zn.$$

Organic ligands can render many metals, especially Cu and Pb, relatively immobile. However, low molecular weight organic complexes of metals, not necessarily of humic origin, tend to be soluble and can prevent metals from being sorbed onto soil surfaces and thus render them more mobile and possibly more available for uptake by plant roots.

*(d) Sorption of organic contaminants on humic material.* This is the main mechanism by which non-polar, hydrophobic organic molecules are bound in soils. This may be by physical means or by chemical bonding.

[13]J. E. Bittel and R. J. Miller, *J. Environ. Qual.*, 1974, **3**, 243–244.
[14]G. W. Brummer, 'The Importance of Chemical Speciation in Environmental Processes', Springer Verlag, Berlin, 1986.

*(e) Chemisorption of elements.* This occurs when the element is incorporated into the structure of the compound. The most common example of this is when metals, such as Cd, replace Ca in the mineral structure of calcite ($CaCO_3$).

*(f) Co-precipitation of elements.* This is the simultaneous precipitation of an ion in conjunction with other elements. The elements typically found co-precipitated with secondary minerals in the soil were shown by Sposito[15] to include:

Fe oxides: V, Mn, Cu, Zn, Mo;

Mn oxides: Fe, Co < Ni, Zn, Pb;

Calcite: V, Mn, Fe, Co, Cd;

Clay minerals: V, Ni, Co, Cr, Cu, Pb, Ti, Mn, Fe

*(g) Precipitation.* This occurs when the concentrations of metal and accompanying ions exceed the solubility product of insoluble forms, such as $CdCO_3$, CdS and $Pb_5(PO_4)Cl$.

Those contaminants which are sorbed to soil solid surfaces tend to be held against leaching and are less readily available for uptake by crops than those remaining in unbound forms. Volatilization of organic molecules and the methylated forms of certain inorganic elements (As, Hg and Se) is also important. Some of the volatile compounds lost to the atmosphere may be decomposed by UV light (photolysis) or can also be sorbed onto the waxy cuticle of plant leaves and possibly enter the food chain.

The relative balance of reduction and oxidation (redox status) of a polluted soil plays an important role in the behaviour of some pollutants. Firstly, it will determine whether there will be an appreciable concentration of sorptive Fe and Mn oxides present. These are especially important for the sorption of As, Mo and Cd. Secondly, some elements such as Cd, which readily form insoluble sulfide precipitates (CdS) under strongly reducing conditions, will be very immobile in waterlogged soils. However, if these soils become aerated due to drainage and drying out, the sulfide will oxidize to form sulfuric acid and so the liberated $Cd^{2+}$ ions will be highly mobile and available for uptake. This can occur in contaminated paddy soils used for growing rice.

Organic pollutants bearing electrostatic charges will also be adsorbed onto oppositely charged sites on the soil colloids (*e.g.* the herbicides paraquat and diquat are strongly cationic). However, many organic pollutants are non-polar and uncharged and are normally bound to the soil organic matter by physical mechanisms, such as hydrophobic bonding. In soils which have received sewage sludge, the sludge material acts both as a source of several organic and inorganic pollutants and also as the major sorbent for them.

The sorptive properties of a soil for both inorganic and organic pollutants can be described mathematically; in most cases sorption is found to follow either the Langmuir or Freundlich adsorption isotherm equations but space does not permit its coverage here.

---

[15]G. Sposito, in 'Applied Environmental Geochemistry', ed. I. Thornton, Academic Press, London, 1983.

### 14.4.3   Degradation of Organic Pollutants in Soils

Organic pollutants can be degraded in soils and possibly, aquifers by either (a) non-biological mechanisms, including: hydrolysis, oxidation/reduction, photo-decomposition and volatilization or (b) microbial decomposition ('biodegradation').

When a pollutant chemical comes into contact with microbial colonies on the surfaces of soil solids ('biofilms') which line the voids within the soil matrix, various extracellular enzymes will be secreted. These may partially degrade the chemical which is then absorbed into the microbial cell where intracellular enzymes may catalyse further decomposition reactions which bring about the release of energy and nutrients. Several species of bacteria and different enzymes may be involved at the same time and bring about a sequence of degradation steps producing increasingly simpler compounds which are either used in 'anabolic' (cell building) or 'catabolic' (energy releasing) processes.[16]

The greatest energy yield is obtained when the catabolic decomposition of an organic substrate by microorganisms occurs in the presence of free oxygen, the next in order of decreasing energy yield is the reduction of Fe and Mn oxides, then the reduction of $NO_3^-$, followed by that of $SO_4^{2-}$ and finally the reduction of $CO_2$ to $CH_4$.

Although most organic molecules are ultimately biodegradable, the rate at which this takes place can vary greatly. The susceptibility of an organic molecule to biodegradation depends largely on its structure. Compounds which are most resistant to degradation tend to have halogen atoms in their structure, especially a large number of halogens, a highly branched structure, a low solubility in water or an atomic charge difference.[16] Straight chain aliphatic compounds are easily degraded, but unsaturated aliphatics are less degradable than saturated forms. Simple aromatic compounds are usually degradable by several ring-cleavage mechanisms, but the presence of halogens stabilizes the ring and makes the compound less readily degradable. In general, unless there are specifically adapted microorganisms present, there will be a tendency for the more readily degradable molecules to be catabolized first. Adaptation can occur in a population of microorganisms as a result of a selection pressure. Individual organisms with the ability to degrade or tolerate a toxic compound may arise as a result of mutations and gene transfer and these individuals will have a competitive advantage if they can utilize an abundant supply of an organic pollutant as a source of energy and nutrients.[16] There is a time lag while adaptation occurs before there are sufficient microorganisms present which have the ability to degrade the pollutant; this tends to give a 2-phase curve for concentration in the soil against time. The first phase is normally a steep decrease in concentration with time due to physical processes such as volatilization. The second phase is slower but goes on until the concentration approaches zero when the microorganisms have become adapted. 'Co-metabolism' can also occur and this is the term applied to the degradation of the pollutant (secondary substrate) by

[16]M. D. LaGrega, P. L. Buckingham and J. C. Evans, 'Hazardous Waste Management', McGraw-Hill, New York, 1994, p. 1146.

enzymes secreted by microorganisms to degrade another substance (the primary substrate).

Biodegradation requires appropriate conditions for the growth of the micro-organisms which include adequate moisture, a temperature between 10 °C and 45 °C, a pH which is preferably in the range of 6–8 and a supply of macro-nutrients (N and P). Some pollutant chemicals are highly toxic to soil micro-organisms and so inability to degrade them may be linked to a lack of tolerance.[16]

The organochlorine pesticides have been used for more than 50 years and are regarded as the most persistent of all groups of pesticides. The order for decreasing persistence is: DDT > dieldrin > lindane (BHC) > heptachlor > aldrin with half-lives of 11 years for DDT down to 5 years for aldrin.[17] The most persistent organic pollutants of all in soils are probably the more highly chlorinated PCBs and dioxins (PCDDs). However, the persistence of a chemical in any soil is determined by the over-all balance between its adsorption onto soil colloids (usually organic matter), extent of volatilization, uptake into, or binding on, plant roots and transformation/biodegradation processes. This balance depends on the nature of the pollutant, its concentration, the soil organic matter content, the soil pH and redox status, the available moisture, general soil fertility (level of nutrient supply, activity of soil microorganisms) and the time taken by the microorganisms to adapt to the new pollutant substrate.

## 14.5 THE CONSEQUENCES OF SOIL POLLUTION

Soil pollution can restrict the uses to which land is put because of the liability of pollutants to be hazardous to human health, harmful to living resources or to damage buildings and services.

### 14.5.1 Types of Hazards Associated with Contaminated Soil

The types of hazards and examples of the pollutants which cause them are given in Table 14.2.

### 14.5.2 Soil–Plant Transfer of Pollutants

The transfer of pollutants from soil into plants through the roots varies considerably due partly to the properties of the plants (genotypic factors) and to the extent to which the pollutant is sorbed in the soil in an unavailable form.

Kloke *et al.*[18] gave the general orders of magnitude of the transfer coefficients (also called concentration factors) for several heavy metals. These included values of 0.01–0.1 for As, Co, Cr, Hg, Pb and Sn, coefficients of 0.1–1.0 for Cu and Ni and values of 1.0–10 for Cd, Tl and Zn (which were the most readily accumulated by plants). Organic pollutants tend to have lower coefficients; a

[17]F. A. M. de Haan, in 'Scientific Basis for Soil Protection in Europe', eds. H. Barth and P. L'Hermite, Elsevier, Amsterdam, 1987, p. 181.
[18]A. Kloke, D. Sauerbeck and H. Vetter, in 'Changing Metal Cycles and Human Health', ed. J. O. Nriagu, Springer Verlag, Berlin, 1984.

**Table 14.2**   *Hazards and examples of the pollutants which cause them*

| Hazard | Pollutants |
|---|---|
| (1) Direct ingestion of contaminated soil (by gardeners, children, animals and on unwashed vegetables). | As, Cd, Pb, $CN^-$, $Cr^{6+}$, Hg, coal tars (PAHs), PCBs, dioxins, phenols, pathogenic bacteria, viruses, eggs of parasites |
| (2) Inhalation of dusts and volatiles from contaminated soil. | organic solvents (toluene, benzene, xylene) Radon, Hg, metal-rich particles, asbestos. |
| (3) Uptake by crop plants of pollutants hazardous to animals and humans through the food chain. | As, Cd, $^{137}Cs$, Hg, Pb, $^{90}Sr$, Tl, PAHs, various pesticides |
| (4) Phytotoxicity | $SO_4^{2-}$, Cu, Ni, Zn, $CH_4$, Cr, B |
| (5) Toxicity to soil microbial biomass | Cd, Cu, Ni, Zn |
| (6) Deterioration of building materials and services | $SO_4^{2-}$, $SO_3^{2-}$, $Cl^-$, tar, phenols, mineral oils, organic solvents |
| (7) Fires and explosions | $CH_4$, S, coal and coke dust, petroleum oil, tar, rubber, plastics, high calorific value wastes (old landfills) |
| (8) Contact of people with contaminants during demolition and site development | tars (PAHs), phenols, asbestos, radio-nuclides, PCBs, TCDDs, pathogenic bacteria and viruses |
| (9) Contamination of water | $CN^-$, $SO_4^{2-}$, metal salts, hydrocarbons, solvents, surfactants, sewage, farm wastes, pesticides. |

Adapted from Beckett and Sims,[19] ICRCL,[20] and Alloway.[21]

value of 0.033 has been suggested for dioxins and PCBs by Jackson and Eduljee.[22] The lower chlorinated congeners of PCBs tend to be more bioavailable (and biodegradable – see Section 14.4.3) than those with higher numbers of Cl atoms in their structure.

### 14.5.3   Ecotoxicological Implications of Soil Pollution

It has been recognized in recent years that, although soils mildly polluted with metals from sewage sludge to levels within EC guidelines do not cause obvious symptoms of phytotoxicity in most crops, some effects can occur in the soil microbial biomass. It is now recognized that Zn concentrations at close to the maximum permissible level of 300 $\mu g\,g^{-1}$ for European soils treated with sewage sludge can cause a marked decrease in the activity of the nitrogen-fixing bacteria of the species *Rhizobium leguminosum* bv *trifolii* which occur in nodules in the roots of white clover. Although it appears that the order of toxicity of heavy metals to

[19]M. J. Beckett and D. L. Sims, in 'Contaminated Soil', eds. J. W. Assink and W. J. van den Brink, Martinus Nijhoff, Dordrecht, 1986, pp. 285–293.

[20]Interdepartmental Committee on the Redevelopment of Contaminated Land, 'Guidance on the Assessment and Redevelopment of Contaminated Land', Guidance Note 59/83, Department of the Environment, London, 1987.

[21]B. J. Alloway, in 'Understanding Our Environment' ed. R. M. Harrison, Royal Society of Chemistry, 2nd edn., 1992.

[22]A. P. Jackson and G. Eduljee, *Chemosphere*, 1994, **29**, 2523–2543.

these bacteria is Cu > Cd > Ni > Zn, the greatest risk is posed by Zn. This is because many soils which have received several applications of sewage sludge may be approaching the maximum permissible concentration of 300 $\mu g\,g^{-1}$ for Zn, but remain well below the maxima for these other metals.[23]

## 14.6  SOIL ANALYSES AND THEIR INTERPRETATION

Analysis of soils for contaminants involves the collection of representative samples from suspected polluted sites and local controls. These samples are subsequently prepared for analysis, usually by drying and grinding followed by sieving. Analytical procedures involve either determination of the total concentration of the pollutant or a partial extraction procedure which can be correlated against a critical concentration. Space does not permit a detailed description of analytical procedures, but a brief summary is given in Table 14.3.

**Table 14.3**  *Summary of analytical procedures for soil pollutants*

| Pollutant | Analytical procedure |
| --- | --- |
| Heavy metals (in conc. acid digests or partial extractants) | Flame atomic absorption spectrophotometry (FAAS) or inductively coupled plasma atomic emission spectrophotometry (ICP-AES) or ICP-Mass spectrometry (ICP-MS). |
| As, Bi, Hg, Sb, Se, Sn and Te (in acid digests or partial extractants) | Hydride generation atomic absorption spectrophotometry (HGAAS) using sodium borohydride in NaOH. |
| Borate (water soluble) | ICP-AES (using quartz or plastic apparatus instead of Pyrex). |
| Organic pollutants (dissolved in appropriate organic solvents) | Gas chromatography with flame ionization detector (GC-FID) or with electron capture detector (GC-ECD) or GC combined with Mass spectrometry (GC-MS). |
| Cyanides | Colorimetrically – using pyridine pyrazalone (blue) reaction. |
| Sulfates (water soluble) | Elution through exchange column followed by titration with NaOH (or Ion Chromatography). |
| Sulfates (total) | Dissolution in HCl, precipitation of Al and Fe followed by gravimetric determination using $BaCl_2$. |
| Chlorides (total – in $HNO_3$) | Volhard's method – back titration with ammonium thiocyanate after initial precipitation with $AgNO_3$, (or Ion Chromatography). |

Adapted from Alloway, 1992[21]

[23]A. M. Chaudri, S. P. McGrath and K. E. Giller, *Soil Biol. Biochem.*, 1992, **24**, 625–632.

Having obtained concentrations of pollutants in soils, the interpretation of the data will be dependent upon national or international guideline concentrations or legal limits. These critical concentrations vary between countries and states. Perhaps the most widely known are the values used in the Netherlands but these are very conservative being based on ecotoxicological data and on the principle of 'multifunctionality of use' whereby soil should be maintained, or cleaned up to a standard which will allow it to be used for any purpose (including food crop production). This principle is not universally accepted, and other countries, such as the United Kingdom, use the principle of 'fitness for purpose'. This means that land to be used for domestic gardens where food crops will be grown should have much lower concentrations of hazardous chemicals than land to be built on for non-residential purposes. Values given in Table 14.4 are taken from the Soil and Water Quality Standards for the Netherlands.[24] This is not the complete list and the values are given for a 'standard' soil containing 10% organic matter and 20% clay. The A, B and C values, which were originally introduced in 1986, have been superseded by Target Values and Intervention Values and the latter have been revised (usually downwards).[24]

In the United Kingdom, the guideline values used in recent years have been based on the ICRCL provisional document of 1987.[20] These will shortly be replaced in 1995/96 by a completely new set of values based on probability risk assessment using the concept of exposure pathways. Other sets of guideline values which are sometimes used for comparative purposes are the Canadian Interim Environmental Quality Criteria for Soil (1991) (Canadian Council of Ministers of the Environment, 1991). Both the United Kingdom ICRCL 1987 values[20] and the Canadian Interim Values are based on fitness for purpose and have maximum values for domestic gardens of 3 and 5 $\mu g\,g^{-1}$ Cd, 130 and 100 $\mu g\,g^{-1}$ Cu, respectively and both give 500 $\mu g\,g^{-1}$ Pb for these soils. Space does not permit the various guideline values to be given in detail and readers wishing to find out more about this are recommended to consult the most recent values for their own countries.

## 14.7  SELECTED CASE STUDIES

### 14.7.1  Gasworks Sites

By the end of the 19th century, most towns and cities in North-west Europe and many other parts of the world had their own gasworks to produce coal gas for heating and lighting in industrial and domestic properties. When the total number of sites, including large municipal works and small units in some industrial premises are considered, there could be between 2000 and 5000 gas manufacturing sites in the UK. There are 234 former gasworks in the Netherlands which have been identified as being in need of major remediation.[25] The production of the gas was based on the heating of coal in a non-oxidizing

[24]Netherlands Ministry of Housing, Physical Planning and Environment, '1991 Environmental Quality Standards for Soils and Waters', The Netherlands, 1991.
[25]A. O. Thomas and J. N. Lester, *Sci. Total Environ.*, 1994, **152**, 239–260.

**Table 14.4** *The Netherlands' guideline values and quality standards for selected pollutants in soils*

| Substance | Critical values for soils ($\mu$g g$^{-1}$) | | | Target value |
|---|---|---|---|---|
| | A | B | C | |
| *Metals* | | | | |
| Arsenic | 20 | 30 | 50 | 29 |
| Cadmium | 1 | 5 | 20 | 0.8 |
| Copper | 50 | 100 | 500 | 36 |
| Mercury | 0.5 | 2 | 10 | 0.3 |
| Nickel | 50 | 100 | 500 | 35 |
| Lead | 50 | 150 | 600 | 85 |
| Zinc | 200 | 500 | 3000 | 140 |
| *Inorganic pollutants* | | | | |
| Cyanides (free) | 1 | 10 | 100 | 1 |
| Cyanides (complex) | 5 | 50 | 500 | 5 |
| Sulfur | 2 | 20 | 200 | |
| *Polycyclic aromatics* | | | | |
| PAHs (total) | 1 | 20 | 200 | |
| Naphthalene | 0.1 | 5 | 50 | 15 |
| Anthracene | 0.1 | 10 | 100 | 50 |
| Benzo(a)pyrene | 0.05 | 1 | 10 | 25 |
| *Chlorinated hydrocarbons* | | | | |
| CH (total) | 0.05 | 1 | 10 | |
| PCBs | 0.05 | 1 | 10 | |
| Chlorophenols (total) | 0.01 | 1 | 10 | |
| *Aromatic compounds* | | | | |
| Aromatics (total) | 0.1 | 7 | 70 | |
| Benzene | 0.01 | 0.5 | 5 | |
| Toluene | 0.05 | 3 | 30 | |
| Phenols | 0.02 | 1 | 10 | |
| *Other organic compounds* | | | | |
| Pyridine | 0.1 | 5 | 60 | |
| Gas oil | 20 | 100 | 800 | |
| Mineral oil | 100 | 1000 | 5000 | |

A = reference values, B = test requirement, C = intervention value (clean-up!), Target value is the concentration which ought to be aimed for in the longer-term[24]

atmosphere. The crude gas was then purified by being passed through separators to remove tar, condensers and wet purification to remove $NH_3$, HCN, phenols and creosols and dry purification with ferric oxide to remove S and cyanides. These purification processes have resulted in gasworks sites accumulating large amounts of some very hazardous materials.

The contamination problems associated with gasworks include contamination of surface waters by tarry wastes and phenols and the leaching of tars, phenols and cyanides into groundwaters. Atmospheric pollution by volatilization of naphthalene, lindane, benzene, acenaphthalene and cycloalkenes, thiophene, pyridine and hydrogen sulfide from contaminated soils at gasworks sites is also reported. In addition to the chemical pollution associated with gas manufacture, additional environmental pollution can also arise from asbestos used in insulation and Pb on old paintwork.[25]

### 14.7.2    Sites Contaminated with Solvents

Chlorinated hydrocarbon solvents such as trichloroethene, 1,1,1-trichloroethane, tetrachloromethane and tetrachloroethene are widely used in industry and are frequently found as soil pollutants at industrial sites. In most cases, pollution occurred as a result of spillages and this has resulted in the chlorinated hydrocarbons being found in the following four forms: (i) as isolated droplets within the pollution plume in the groundwater (aquifer) or trapped in pools in low permeability material, (ii) dissolved in porewaters, (iii) as vapour in the unsaturated zone (above the aquifer) and (iv) sorbed onto solid phase soil organic matter.[26] Soil gas monitoring has revealed that some derelict industrial sites with relatively low concentrations in soil have given rise to marked pollution of groundwater in boreholes below the sites at depths down to 187 m. The concentrations of chlorinated hydrocarbon solvents in the soil gas at derelict industrial sites were found to vary by seven orders of magnitude up to a maximum value of 2000 $\mu$g l$^{-1}$ trichloromethane.[26]

A fire at a solvent recovery works in the village of Carrbrook, in Cheshire, England, in 1981 caused severe contamination of soils in the local area with benzene, PCBs and dioxins.[27] Concentrations of up to 304 mg kg$^{-1}$ total solvents, 1160 mg kg$^{-1}$ PCBs and 168 $\mu$g kg$^{-1}$ dioxin (expressed as TEQ = toxic equivalents of 2,3,7,8-tetrachlorodibenzodioxin – TCDD) were found in soils at the site. Creaser *et al.*[28,29] reported the median concentration of dioxins in urban and rural soils in the UK as 1436 ng kg$^{-1}$ and 335 ng kg$^{-1}$ (sum of all tetra- to octa-chlorinated congeners) respectively. Domestic pets kept in houses and gardens near to the site were found to show abnormal health effects. Worst affected were guinea pigs which developed a terminal wasting disorder.[27] Concentrations of 100–1000 $\mu$g kg$^{-1}$ TCDD are considered typical for industrially contaminated sites but the levels at the solvent recovery works site were much higher than the maximum in this range. Perhaps the worst case of dioxin pollution occurred at Times Beach in Missouri where waste oil contaminated with distillation residues from organochlorine production had been used to reduce dust problems on dry soils. This contaminated oil gave rise to concentrations of up to 33 mg kg$^{-1}$ in the soils and caused the death of horses, cats, dogs, chickens and birds that came in contact with the polluted soil. Children playing

[26]P. R. Eastwood, D. N. Lerner, P. K. Bishop and M. W. Burston, *J. Inst. Water Environ. Managt.*, 1991, 163–171.
[27]T. Craig and R. Grzonka, *Land Contam. Reclam.*, 1994, **2**, 19–25.
[28]C. S. Creaser, A. R. Fernandes, A. Al-Haddad, S. J. Harrad, R. B. Homer, P. W. Skett and E. A. Cox, *Chemosphere*, 1989, **18**, 767–776.
[29]C. S. Creaser, A. R. Fernandes, S. J. Harrad and E. A. Cox, *Chemosphere*, 1991, **21**, 931–938.

in the area developed chloracne, a characteristic symptom of exposure to dioxins.[30]

### 14.7.3 Lead and Arsenic Pollution in the Town of Mundelstrup in Denmark

Severe Pb and As contamination was found in housing in the town of Mundelstrup Stationsby near Aarhus in Denmark in 1987. The pollution had arisen from the disposal of heavily contaminated fill from the site of a former fertilizer factory. Concentrations of up to 67 562 $\mu g\,g^{-1}$ Pb and 5481 $\mu g\,g^{-1}$ As were found in the soil of gardens around houses built on the fill. A comprehensive survey was carried out involving the analysis of approximately 1000 samples. The worst affected area was at the site of the original factory. In most cases the contamination was in the layer of fill which was up to 4 m deep. However, in areas of highly acid fill the pollutant metals had penetrated up to 3.5 m into the underlying intact deposits of clay till. The contaminated area covered 6700 m$^2$ and varied in depth between 0.5 and 8 m.

It was decided to remove all soil from around the houses which exceeded the Danish soil quality criteria of 40 mg kg$^{-1}$ Pb and 20 mg kg$^{-1}$ As; this involved a total of 30 gardens. A borrow-pit was dug to supply clean soil to replace the 50 000 m$^3$ of soil excavated from the contaminated gardens. The same pit was used as a special landfill to receive the contaminated soil after appropriate engineering of the base with layers of lime to prevent leaching and the whole landfill was later covered by a new motorway roundabout. This is an example of the ideal clean-up of polluted urban land. It was carried out at great cost (32 × 10$^6$ DKK) and this may not be possible in other countries with a legacy of a much greater number of historically polluted urban sites.[31]

### 14.7.4 Pollution from a Lead-Zinc smelter at Zlatna in Romania

Acid precipitation from a Pb-Zn smelter at Zlatna in Transylvania, Romania has led to obvious toxic effects on vegetation and damage to buildings within an area of at least 50 000 ha. The smelter lies in a valley and SO$_2$ fumes and metal aerosol particles which are often confined to the valley have killed many plant series but the effects are most pronounced in the 300 ha immediately down-wind of the smelter. The death of most of the vegetation cover has led to massive soil erosion and associated deterioration of the soil structure due to the loss of the surface organic matter. A layer of colluvium up to 46 cm thick has developed on the top of the soil profiles at the base of the slope. The metals deposited from the smelter fumes are rendered relatively highly mobile in the acid soils but the most destructive contaminant is the SO$_2$. The most immediate pollution control measures required are scrubbers to reduce the SO$_2$ concentrations in the smelter emissions. It is estimated that this would cost at least 525 × 10$^6$ Lei but the total

[30]C. D. Carter, R. D. Kimbrough, J. A. Liddle, R. E. Chine, M. M. Zack, W. F. Barthel, R. E. Koehler and P. E. Phillips, *Science*, 1975, **188**, 738–740.
[31]P. Clement, N. J. Olsen and P. Madsen, *Land Contam. Reclam.*, 1995, **3**, 39–46.

cost for emission control and remediation of the pollution damage is around 27 ×
$10^9$ Lei (*c.* £14 million).[32]

### 14.7.5 Cadmium Pollution in the Village of Shipham in England and in the Jinzu Valley in Japan

Zinc and lead mining was carried out around the village of Shipham in the
nineteenth century and the land has been left highly contaminated with Pb, Zn
and Cd. Garden soils used for growing vegetables were found to contain up to 360
$\mu g\,g^{-1}$ Cd, 37 200 $\mu g\,g^{-1}$ Zn and 6540 $\mu g\,g^{-1}$ Pb and the mean Cd concentration in almost 1000 samples of garden vegetables was 0.25 $\mu g\,g^{-1}$ which was
nearly 17 times the national average of 0.015 $\mu g\,g^{-1}$ Cd (dry matter). The
highest Cd concentrations occurred in leafy vegetables such as spinach, lettuce
and brassicas which contained up to 60 times more Cd than the same species
grown in uncontaminated soils. Although the vegetables contained relatively
high concentrations of Cd, no adverse health effects were found in the population
and this was ascribed to the garden produce only forming a low percentage of the
diet, a generally varied diet and a public water supply which conformed to
national and international standards.[33]

In Japan, rice-growing paddy fields in the Jinzu Valley had also become
contaminated with Cd and other metals from mining operations upstream but, in
contrast to Shipham, many people suffered ill health effects. Two hundred
elderly women who had given birth to several children were disabled by a Cd-
induced skeletal disorder known as 'itai-itai' disease and 65 died of this condition.
The disease became apparent during and immediately after the Second World
War when diets were more deficient in calcium and protein than normal.
Average Cd concentrations in rice from the contaminated paddy soils (0.7
$\mu g\,g^{-1}$ DM) were ten times greater than in local controls and the maximum Cd
content in the Jinzu Valley rice was 3.4 $\mu g\,g^{-1}$ DM. In the paddy soils the Cd
was present as insoluble CdS when the soils were flooded, but drainage in
readiness for harvest caused the sulfide to oxidize and release $Cd^{2+}$ ions and
sulfuric acid.[34] In the Shipham soils, a high content of $CaCO_3$ from mining waste
had helped to reduce the bioavailability of the Cd which was present at total
concentrations more than one hundred times higher than the Jinzu paddy soils.

### 14.8  CONCLUSIONS

Soils can become polluted by a wide range of chemicals arising from many
human activities, and in every case the bioavailability of the pollutant will
depend on: (a) the combined effects of the type and concentration of the polluting
substance, (b) the composition of the soil (especially its organic matter, clay and

[32]C. Răuță, C. Ciobanu, M. Dimitru, E. Dulvara, B. Kovacsovici, L. Latis and S. Cârstea.
Unpublished information.
[33]H. Morgan and D. L. Sims, *Sci. Total Environ.*, 1988, **75**, 135–143.
[34]T. Asami, in 'Changing Metal Cycles and Human Health', ed. J. O. Nriagu, Springer Verlag,
Berlin, 1984, pp. 95–111.

free carbonate contents), (c) the soil physico-chemical conditions (pH and redox status), (d) the genotype of the crops grown (species and varieties vary greatly in their capacity to accumulate pollutants) and (e) the climate (soil moisture status, temperature). Although these interactions can be relatively complex, there is enough information becoming available to allow predictive models to be developed at least for heavy metals. Much less is known about the behaviour of organic micropollutants in soils because their analysis is more difficult and more expensive. However, as analytical techniques for these compounds improve and become more widely used the results will allow models to be developed or refined. From the food safety aspect, in the United Kingdom, the Ministry of Agriculture, Fisheries and Food has stated that the priority organic pollutants of soils and crops are dioxins and PCBs, and As and Cd are the priority inorganic pollutants.[35]

Apart from the potential hazard of contaminated soils to human health, either by direct contact, inhalation or through the food chain, the link with contamination of groundwater is also very important and this will be very dependent on the hydraulic conductivity of soils, the balance between precipitation and evaporation and the nature of the aquifer. Serious water pollution problems occur in many industrial areas due to the infiltration of organic pollutants through soils to aquifers.

Although the long-term ideal would be to clean-up contaminated sites, economic feasibility dictates that in many cases clean-up will not be possible at least in the short-to-medium term. It will therefore be necessary to manage contaminated land in an appropriate manner to minimize the effects of the pollution.

[35]Ministry of Agriculture, Fisheries and Food, 'Surveillance and Applied R&D on Food: Requirements Document, 1996–97', MAFF, London, 1995.

CHAPTER 15

# Solid Waste Management

G. H. EDULJEE and D. A. ARTHUR

## 15.1 INTRODUCTION

Waste is not a unique material in terms of its constituents: the main distinguishing feature relative to the products from which it derives is its perceived lack of value. Dictionary definitions of waste include the descriptions 'useless' or 'valueless'. In the UK, the legal definition of waste as given in Section 75 of the Environmental Protection Act 1990 includes phrases such as 'scrap material or other unwanted surplus substance' and 'any substance which requires to be disposed of as being broken, worn out, contaminated or otherwise spoiled'.

Up to the 1970s, the perception of waste as unwanted, 'useless' material with no intrinsic value shaped society's approach to waste management. The ultimate disposal of waste was the overriding priority. Waste generators (domestic, commercial and industrial) sought disposal at the lowest cost (overwhelmingly in landfills) and had little or no incentive to 'manage' waste as opposed to merely disposing of it.

However, the past two decades have witnessed a sea change in attitudes to waste. Public sensitivity towards waste disposal outlets has put enormous pressure on operators and regulators alike; the siting of new facilities has at best proceeded after lengthy delays and in the teeth of intense opposition and at worst become all but impossible. Waste generators and policy makers have perforce turned their attention to upstream waste-related activities in an effort to minimize waste production and hence make more efficient use of ultimate disposal capacity.

Another key driver has been the concept of sustainable development,[1] defined as 'development that meets the needs of the present without compromising the ability of future generations to meet their own needs'. The slogan *more with less* encapsulates the new thinking: extracting the maximum value and benefit from products and services, using the minimum of energy and rejecting the minimum of waste materials or emissions to the environment. Arising out of this is the

---

[1]World Commission on Environment and Development, 'Our Common Future', Oxford University Press, Oxford, 1987.

concept of *sustainable waste management*.[2] In essence, waste is *given value*. Viewed against the principle of the conservation and nurturing of natural resources, the production of waste is in itself seen as a manifestation of the inefficient management of the earth's raw materials. Commencing with waste reduction during the production of goods and the provision of services, sustainable waste management calls for the recovery and reuse of materials so as to conserve raw materials, the use of waste as a source of energy in order to conserve non-renewable natural resources and finally for the safe disposal of unavoidable waste. A new policy approach is required to instil a sense of common responsibility towards waste, and to create overall strategies for waste management that do more than pay mere lip service to the concept of sustainability.

The aim of this chapter is to illustrate how the new approach has been applied to the management of municipal solid waste (MSW). For the purpose of this chapter, we define MSW as solid waste collected from households, commercial and industrial premises. In many ways MSW presents the greatest challenge to waste management. Waste arising from domestic, commercial and industrial premises is extremely heterogeneous in nature, and as a result components of potential value (such as glass, metals or biodegradables) cannot be beneficially used without significant effort being put into segregation and sorting schemes. Even with such systems, it is very difficult to achieve products of value due to cross-contamination and fragmentation of the constituents. In many countries, the poor standards of MSW disposal also leads to a significant threat to public health.

In keeping with the importance governments now place on sustainable development, this chapter emphasizes the preventative aspects and the resource potential of waste management as opposed to the environmental impact of MSW disposal *per se*: other publications have examined this latter aspect of waste management in some detail.[3,4] The chapter commences with an introduction to the so-called waste management hierarchy and the need to integrate various strands of policy into a coherent waste management strategy. Next, technical options for waste prevention and recycling are presented, followed by a discussion of the policy options that can be applied to encourage or to mandate waste prevention and reuse. Bulk waste reduction technologies, in particular incineration and composting, are then presented. Finally, some of the issues concerning the development of an integrated waste management strategy are discussed.

## 15.2  AN INTEGRATED APPROACH TO WASTE MANAGEMENT

### 15.2.1  The Waste Management Hierarchy

Options for waste management are often arranged in a hierarchical manner to reflect their desirability.[2] The first priority is waste avoidance, that is not

[2]Department of the Environment and the Welsh Office, 'A Waste Strategy for England and Wales', London, 1995.

[3]'Issues in Environmental Science and Technology No. 2 – Waste Incineration and the Environment', eds. R. E. Hester and R. M. Harrison, The Royal Society of Chemistry, Cambridge, 1994.

[4]'Issues in Environmental Science and Technology No. 3 – Waste Treatment and Disposal', eds. R. E. Hester and R. M. Harrison, The Royal Society of Chemistry, Cambridge, 1995.

producing the waste in the first place. If the waste must be produced, then the quantities should be minimized. Once that has been achieved, the next priority is to maximize recovery, reuse and recycling of suitable waste materials. Taken together, these three options are often called waste prevention, although strictly speaking only the first two are prevention whereas the third is already an end of pipe solution.

Once the possibilities for waste prevention have been exhausted, the next priority is to reduce the volume of residual wastes being passed on for final disposal, extracting resources in the form of products and/or energy in the process.

Figure 15.1 represents this hierarchy as an upright cone, with the most desirable option, waste avoidance, at the apex. By coincidence, the volume of each of the layers in the cone is also roughly proportional to the relative quantities of waste currently being managed by each of the options in most countries around the world. In other words, while there is general agreement on the order of desirability of the various options, in practice the current situation in terms of relative quantities is generally inversely proportional to desirability.

An alternative representation of the hierarchy, in which the volume of the layers is proportional to their desirability, is as an inverted cone (Figure 15.2) rather than as an upright cone. This reflects the relative quantities of waste which the introduction of a sustainable waste management strategy seeks to achieve.

A comparison of Figures 15.1 and 15.2 allows one to draw another analogy with waste management in practice. An upright cone is a stable structure, while the inverted cone is inherently unstable. To move from the current situation in which the majority of wastes are dealt with by final disposal or other end of pipe solutions to one where waste prevention becomes dominant, it is necessary for governments to provide some support through policy measures. Figure 15.3 illustrates that policies which encourage waste avoidance and waste minimiza-tion are to be preferred over those which focus purely on further encouraging present efforts in recycling, recovery and reuse.

The waste management hierarchy is discussed further in Section 15.6.

### 15.2.2   An Integrated Approach

If waste management is to change significantly, the behaviour of individuals and groups in society will have to change. Three groups in society are key to this process: government, industry and commerce, and individuals. A fourth group, lobby groups and NGOs, is effective as a conduit of ideas and energy. Policies need to be designed to change behaviour of all these groups in order to reduce or even reverse the growth in waste generation which accompanies increasing wealth.

Effective policies generally operate on a two-pronged approach. Where possible, waste reduction policies should be implemented on a voluntary basis, by all the above groups. In order to kick-start some initiatives, government can provide an incentive or '*carrot*' in the form of financial or other support. Combined, these form the first prong. For the second prong, governments

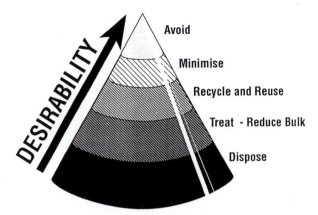

**Figure 15.1** *The waste management hierarchy in its cone shape*

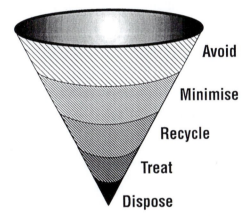

**Figure 15.2** *The waste management hierarchy in its inverted cone shape, indicating that the aim is for quantities of waste managed by each method to decrease as one moves down the hierarchy*

**Figure 15.3** *The waste management hierarchy illustrating the role of policy support in stabilizing the ideal hierarchy in its inverted cone shape*

implement policies to tackle the underlying causes relating to waste generation and positively discourage it. The UK Government's introduction of a landfill tax, which will focus attention on the costs of disposal, is such a policy. Some such measures need to be implemented alongside those from the first prong. Others may only be required if measures from the first prong, for whatever reason, prove to be ineffective. This second prong of policies can be viewed as the disincentives, sanctions or '*sticks*' which complement the '*carrots*' above.

In addition, it is essential that policies build on rather than undermine existing strengths. Where there are already efforts to reduce waste generation or recover materials for recycling and reuse on a voluntary basis or using current market mechanisms, policies must aim to reinforce these efforts and not be so invasive as to undermine such existing initiatives.

Finally, even with all these policies and measures in place and operational, some wastes will still be generated and require disposal. The provision of bulk waste reduction technologies will help to reduce the volume and weight of these remaining residues prior to final disposal.

An integrated strategy for waste reduction requires a combination of all the types of measure detailed above. The following sections address these measures, along with technical options for waste prevention and segregation.

## 15.3  TECHNICAL OPTIONS FOR WASTE PREVENTION AND RECYCLING

### 15.3.1  Opportunities for Waste Avoidance and Minimization

Waste avoidance and minimization are the most desirable options in the waste management hierarchy as discussed in Section 15.2.1 above. This section explores the ways in which waste avoidance and minimization have been or may be practised to reduce the amounts of domestic, commercial and industrial wastes that arise. Table 15.1 lists some options which may be appropriate and these are discussed in greater detail under each sector. Some of these waste avoidance and minimization options may be applicable to all waste generators while some would be applicable only to industry (or even to specific industry sectors).

*15.3.1.1  Domestic Sector.*  There are many ways in which individuals can avoid or minimize the amount of waste they put out for disposal. As consumers, they may select product types, packaging types and material types that would lead to the generation of less waste. For example, waste can be reduced by buying in bulk, utilizing reusable shopping bags, buying reusable and more durable products and by buying equipment that generates less waste (for example electrical equipment which runs off mains power or uses rechargeable batteries to reduce the quantity of primary cells used). A major factor is to change the public's perception of waste and how to deal with waste materials. Education and communication programmes therefore need to be built into any approach aimed at the domestic sector before significant take-up of the options offered can be achieved.

**Table 15.1** *Ways to achieve waste avoidance and minimization*

*Product Design Change*
– product design with less waste
– increase product life

*Package Change*
– product in bulk or concentrate form
– reusable or recyclable pack

*Material Change*
– substitution of less toxic materials
– use of reusable or recyclable materials

*Technological Change*
– improved/more efficient equipment
– cleaner technology

*Good Housekeeping/Management Practices*
– proper operating procedures and regular maintenance
– inventory control
– training and clear instructions
– waste segregation

Good housekeeping can also contribute to waste avoidance and minimization. For example, proper operation and regular maintenance of equipment is likely to significantly lengthen its useful life, reducing the need for replacement, and better household management can reduce the unnecessary purchasing of consumer durables and perishables, and hence waste. In a Dutch study on consumer habits relating to the purchase of milk and bread[5] it was found that about 15% of bread was wasted because it was stored too long in the home, resulting in 70 000 tonnes of bread being rejected as waste and hence landfilled. The accompanying wasted packaging amounted to some 6000 tonnes of plastic and paper.

Some waste avoidance and minimization measures taken within the domestic sector would not be possible without the supporting actions of both industry and commerce. The manufacturing, packaging and labelling of products and the provision of some services that would lead to the generation of less waste are necessary to enable householders to have some choice. This type of responsibility relating to producers is examined in later sections.

*15.3.1.2 Commercial and Industrial Sectors.* As manufacturers, companies may produce longer-life products, requiring less maintenance. Manufacturing processes could use less toxic substitutes and recycling materials and may be improved in terms of reduction in material wastage. Proper operating procedures and regular maintenance would also reduce wastage as well as reducing other emissions. Such measures are likely to be industry specific, *i.e.* their application and potential to reduce wastes will vary industry by industry. For example, in the plastics industry, internal recycling of segregated clean plastics may contribute

[5]M. Kooijman, *Environ. Managt.*, 1993, **17**, 575–586.

10% to 15% of total consumption of plastic materials. In the metal products fabrication industry, improved maintenance of cutting machines has been identified as a way of preventing contamination of scrap metal waste which may then be recycled. In the wearing apparel industry, use of a wider roll of cloth and the re-design of garments may significantly reduce cutting wastes. The experience of McDonald's in the US shows that, among other initiatives, a simple trimming back of one edge in their paper napkins reduced the amount of paper used by 21%. Over the past three decades producers have reduced raw materials usage in food packaging: metal drinks cans weigh half as much in the 1990s as they did in the 1960s. Improving management and control of processes can lead to significant waste reduction, which in turn can lead to considerable savings in costs due to less wastage and more efficient procedures. For example, between 1986 and 1990 the UK consumption of drinks increased by 20%, while industry used 7% less energy and generated 14% less solid waste per litre of drink produced.[6]

### 15.3.2 Collection and Sorting

A prerequisite for the cost-effective reuse and/or recycling of potentially valuable components in MSW is that these materials be separated out from the bulk waste. A number of options are available, as discussed below.

*15.3.2.1 Source Separation.* Collections of waste materials at source are termed 'kerbside collection systems'. This method involves the householder putting out recyclable materials for collection separate from the normal refuse. These recyclable materials may either be mixed, all materials being placed into one container for future sorting either by the collector at a reclamation facility, or separated into individual materials.

In the *mixed at source* scheme the householder places all recyclable materials into one container which is emptied into the collection vehicle and taken to a central facility for sorting. This central facility is usually referred to as a Materials Reclamation Facility (MRF). In MRFs, co-mingled materials are re-separated, stored and perhaps some initial processing is carried out, prior to selling on to a manufacturer or into the secondary materials markets. Two processing lines are usually set up; one for the separation of co-mingled containers (aluminium cans, steel cans, plastic bottles and glass bottles) and the other for co-mingled paper (newspaper, domestic paperboard, white paper). The process separates these and prepares them for onward sale to recycling companies. The initial processing of waste at a MRF (for example, washing, baling, *etc*) can add value to the separated materials.

In the *separation at source* scheme, the householder is either required to place recyclable materials into one container for sorting by the collector at the kerbside when the materials are collected, or the recyclables are placed in separate containers. An example of the former is the 'Blue Box' scheme in Sheffield, where in the first year of operation some 277 tonnes of paper, 78 tonnes of glass,

[6]I. Boustead, 'Resource Use and Liquid Food Packaging', Incpen, London, 1993.

39 tonnes of cans and 27 tonnes of plastic were collected from 3300 properties, resulting in a reduction of 17% in the quantity of materials entering the domestic waste scheme. The latter scheme has been in operation in Leeds, Bury and Milton Keynes. The UK's first purpose built MRF was opened in Milton Keynes in late 1993: it currently processes approximately 20 500 tonnes of material per annum, or about one fifth of its potential capacity.[7,8] Sorting from a co-mingled source can also be done at the kerbside using a specially adapted vehicle.

### 15.3.3 Bring Systems

*Bring systems* are very widely used in many parts of the world and are typically employed for the recovery of glass or paper. In the UK, the recycling centres at Civic Amenity sites and recycling facilities, such as bottle banks and paper banks outside supermarkets and in town centres, are examples of 'bring systems'. At most Civic Amenity sites, recycling centres have been set up to permit the deposit of recyclable materials in separate facilities. Typically, separate facilities are used for glass, metals, aluminium and steel cans, paper, cardboard, oils and textiles. Occasionally containers for hardcore, wood, car and domestic batteries, *etc.* are also provided. Refrigerators containing CFCs can also be recycled.

The overall diversion rate of materials from the waste stream destined for direct landfill varies widely, but is generally higher for kerbside collection, presumably because this involves less effort on the part of the householder.

## 15.4 POLICY OPTIONS TO MAKE WASTE PREVENTION AND RECYCLING WORK IN PRACTICE

### 15.4.1 Introduction

For recycling to become more effective, it is necessary to introduce measures to 'level the playing field' so that secondary materials can compete more fairly with virgin stock, and to provide direct support to increase the size of local markets. These measures can be grouped under the following policy options:

- voluntary participation;
- positive encouragement, generally in the form of some sort of government support (*i.e.* the *carrots* referred to in Section 15.2.2);
- persuasion or mandatory measures, enacted by government to more forcefully encourage the adoption of waste reduction measures (*i.e.* the *sticks* discussed in Section 15.2.2);
- provision of bulk waste reduction technologies, to reduce the weight and volume of remaining wastes prior to final disposal.

The central objective of the application of these policy measures is to encourage waste avoidance, minimization, reuse and recycling. In general, they each seek to

[7]S. Leadbeater, *J. Waste Managt. Resource Recovery*, 1994, **1**, 13–18.
[8]Environmental Resources Management, Unpublished data base of international waste management practice.

change behaviour by making waste avoidance, minimization, reuse and recycling more attractive options than disposal. The first three policy measures are discussed in this section. The fourth is addressed in the next section.

### 15.4.2   Voluntary Participation

Waste avoidance, waste minimization and the separation of waste materials at source to facilitate recycling, all require the active participation of the waste generators, including householders and commercial/industrial companies and their employees.

*15.4.2.1   Sectoral and Cross-Sectoral Collaboration.*   Two vital aspects of any successful waste reduction strategy are an information and education programme directed at the various sectors of the community and a series of voluntary co-operative programmes to implement waste reduction in practise. Education and other communications measures to raise awareness of the issues relating to waste reduction and the actions necessary by all to achieve the desired results, form a major aspect of any waste reduction strategy. For voluntary participation to have a major impact, and to reduce the need for more formal intervention by Government, sectoral programmes of waste prevention and recovery are often initiated. This involves, for example, particular commercial or industrial sectors working together, in liaison with government and perhaps also the voluntary sector, to devise and implement coherent waste reduction plans within a sector.

Cross-sectoral collaboration offers potential for removing the constraints resulting from a lack of information and practical experience, particularly among small businesses, and also to engender competition in terms of achieving waste reduction. International precedents include a number of US co-operative programmes to facilitate the sharing of new ideas and innovations in waste prevention and collection of recyclables, which involve public commitments by participating companies to achieving certain targets. Examples include the *Wastewi$e* initiative co-ordinated by the US EPA and the 33/50 waste reduction programme in which companies contract to successively reduce waste production by 33% and 50% by specified target dates.

A useful complement to such sectoral or cross-sectoral approaches is one which focuses on particular components of the waste stream, such as packaging wastes, newspaper and magazines, wood in construction wastes, used batteries, used electronic goods, used automobiles, *etc.*

*15.4.2.2   Producer Responsibility.*   In order to implement the 'Polluter Pays Principle', the concept of 'producer responsibility' is one which is rapidly becoming the norm rather than the exception around the world for the management of wastes. The concept is that the manufacturer or importer of the products giving rise to the waste should take responsibility for those wastes. These groups are thereby encouraged to consider the implications of disposal of their product and are given an incentive to investigate methods of reducing, reusing or recycling their wastes. The producer typically levies a charge on the product to

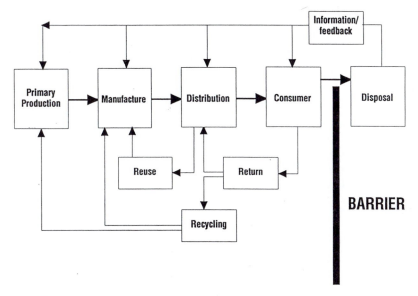

**Figure 15.4** *The concept of producer responsibility*

finance the cost of recovery and collection of materials. The concept is illustrated in Figure 15.4.

This concept has been implemented in a number of countries on a voluntary basis whereby industry negotiates agreed targets for waste prevention and recycling with government and is then left to implement these in the most cost-effective manner. All voluntary schemes worldwide have been negotiated between industry and government on the understanding that if a satisfactory agreement is not reached, or if agreed targets are not met, then a mandatory scheme will be introduced.

Voluntary agreements have been reached between government and industry groups or local authorities to achieve waste reduction, reuse and recycling targets. For example, in the Netherlands voluntary agreements have been negotiated for a total of 29 separate waste streams. The agreements in the form of covenants, which are formalized in the Dutch Packaging Ordinance, incorporate targets for waste reduction and recycling. The success of the voluntary system in the Netherlands is partly due to the fact that Dutch industry negotiates through industry organizations which have the mandate to establish regulations governing their members.

In other countries voluntary agreements have been made in a less formal way and often under the threat of stringent legislation. For example, in Victoria, Australia, industry agreed to fund the establishment of kerbside recycling schemes following the drafting of deposit refund legislation. In the UK, the Government gave the packaging industry's Producer Responsibility Group (PRG) a deadline of six months to develop a plan to recover 50–75% of all packaging materials. The resulting plan, *Real Value from Packaging Waste – A Way*

*Forward*, aimed for 58% recovery of all packaging waste by the end of the decade, overseen by a new organization, VALPAK.

*15.4.2.3  Preferential Purchase Schemes.*  A concern with many recycling initiatives relates to identification of reliable markets for the recovery materials. Many recycling initiatives in the past have failed either because markets for the recovered materials could not be found or because they collapsed shortly after the scheme was initiated. Importance must therefore be given to stimulating the demand side of the recycling process to ensure that sufficient markets exist to absorb the recovered materials. Preferential purchasing schemes is one of the options for achieving this: under a voluntary participation approach, such schemes will relate only to the private sector.

*15.4.2.4  Eco-labelling.*  An eco-labelling scheme for products may complement these types of voluntary initiatives and encourage other consumers to purchase environmentally friendly (including low waste) products. Labels may help to facilitate recycling for materials which would otherwise be difficult to segregate (*e.g.* different types of plastics).

Eco-labels seek to give a product a rating for environmental impact and are designed to help consumers make environmentally based purchasing decisions and to educate consumers about environmental issues. Some recent schemes in Europe have taken a 'cradle to grave' approach, evaluating a product from initial resource use through to final disposal.

Currently there are about thirty eco-labelling programmes operating in the world including those in Austria, Canada, France, Germany, Ireland, the Netherlands, the Nordic countries, Spain, Japan and Singapore.[8] For example, the Austrian eco-label identifies plastic bags as being HDPE and fully recyclable.

Eco-labelling helps to support and reinforce other policy measures. For example, if a product charge were introduced, an eco-label could briefly explain the purpose of the product charge. As in Austria, an eco-label could also facilitate the recycling of materials such as plastic bags.

### 15.4.3  Positive Encouragement

Three types of positive encouragement are examined in this section:

- Grants and subsidies;
- Recycling credits;
- Preferential purchase policies.

*15.4.3.1  Grants and Subsidies.*  Grants in the form of low interest loans or one-off cash payments and/or tax allowances for investments may be used to encourage innovative projects on waste avoidance, waste minimization, collection and processing of recyclables. This type of support is compatible with funding schemes for demonstration projects (such as the UK's DEMOS scheme operated by the Department of Trade and Industry) introduced at the inception stage before the innovation becomes a commercially viable option in the free market system. Grants and subsidies tend to be aimed at supporting the initial capital outlay for the introduction of new technologies or for 'kick starting' other waste reduction initiatives. Normally, grants would not exceed 50% of the total capital required.

Several countries use grants or subsidy systems to encourage the development of recycling schemes or initiatives. Subsidy systems can be used to encourage both supply (encouragement of collection schemes) and demand (support for processing facilities and innovative project development) for materials for recycling. Examples include the subsidized introduction of home composting units in Canada, grants towards the establishment of separate collection programmes for household recyclables in Luxembourg and grants to cover the capital cost of thermal recycling facilities for waste plastic in Japan. Ireland has a system of grants which cover up to 50% of the cost of approved recycling developments. In Taiwan the Government has allocated over $400 million to finance low interest loans for pollution abatement investments by private sector companies. The incentive is the interest rate of 2% below prime. Similar schemes have been set up in Germany, Japan, Norway, Spain and the USA.

*15.4.3.2 Recycling Credits.* A recycling credit can be defined as a payment to those who divert materials from final disposal for recycling. The recycling credit would reflect the saving in reduced collection, transfer and disposal costs. The level of recycling credits may also be set to reflect the weight and volume of the waste types, which may incur different collection, transport, and disposal costs (*e.g.* lower density materials such as plastics may incur higher collection, transport and disposal costs per unit weight).

The aim of recycling credits is to encourage the recovery of materials in situations where the economics of doing so would be marginal. Typically, these situations include operations where the amounts of materials recovered are small compared with the effort required and where the value of the recovered material is sufficiently low to make the recovery process marginal in financial terms. Recycling credits are not intended to support recovery activities that are already economically viable, those that are completely untenable (in that, for example, there is no market for the recovered material) or, indeed, recycling industries.

Recycling credit schemes have been introduced in the UK, Canada and Australia. In the UK, a scheme was introduced under Section 52 of the Environmental Protection Act, and implemented by Waste Disposal Authorities (WDAs) paying a credit to Waste Collection Authorities (WCAs) operating recycling schemes, and also to third parties collecting waste for recycling. WCAs in turn can make a discretionary payment to recyclers of a collection credit commensurate with the saving in household waste collection costs. A survey[9] indicated that in 1992/93 about £2.65 million was paid by WDAs on 445 000 tonnes of recycled waste. In 1992 the value of the credit was set at half the long-run marginal cost of disposal, but in April 1994 the disposal credit was raised to its full marginal value.

[9] R. L. Pocock and R. M. Thomas, *J. Waste Managt. Resource Recovery*, 1995, **2**, 117–124.

*15.4.3.3   Preferential Purchase Policies.*   Whilst preferential purchasing initiatives have already been discussed in Section 15.4.2 under voluntary participation, such schemes could also be implemented as part of *Positive Encouragement* policies by Government. For example, initiatives have been taken by the US Government in relation to preferential purchasing, initiated by President Clinton in 1993.[8] The need for a preferential purchasing policy implies that the products targeted cost more than other products. If the products receiving the preference could operate under normal market conditions, the preference would not be needed. For example, if recycled paper is more expensive than paper produced from virgin pulp because there is greater demand for the latter, then until the demand for recycled paper increases, the small market for recycled paper means that the prices are higher. The objective is to encourage greater purchases which will stimulate price reductions and, in the long run, encourage further purchases of recycled paper. Another example of demand-side initiatives is the subsidizing of energy from the incineration of waste under the Non Fossil Fuel Obligation (NFFO) scheme in the UK.

### 15.4.4   Persuasion Measures

Most successful schemes for waste prevention and recycling work by combining the 'carrot' of financial incentives and other measures to encourage positive behaviour with the 'stick' of either financial penalties or legal requirements to discourage negative behaviour. Policies and other measures may target either those who provide the goods (*e.g.* manufacturers and importers) or those who use and dispose of the goods (*e.g.* householders), or both.

*15.4.4.1   Producer Responsibility.*   Several alternatives for a mandatory producer responsibility scheme are available:

- Mandatory 'take back' requirements for producers
- Deposit refund systems on industry and consumers
- Product charges on producers and consumers
- Raw material charges on producers

In the 'take back' system, householders are required to return the waste to the retailers to enter a parallel, private waste collection/recycling/disposal system rather than the public system (*e.g.* the Austrian scheme); or it could be a compulsory deposit-refund system (*e.g.* as in Taiwan) or it could involve a charge levied on either raw materials used or on products sold.

The best known producer responsibility scheme is the German packaging waste system, introduced through the 1991 Packaging Ordnance.[10,11] The Ordnance obliged distributors and producers to take back packaging from consumers. This obligation was implemented in three stages, for transit

---

[10]D. Berndt and M. Thiele, 'Status des Dualen Systems und seine Kosten', Verpackungs-Rundschau, October 1993, pp. 84–88.
[11]B. K. Fishbein, 'Germany, Garbage and the Green Dot – Challenging the Throwaway Society', INFORM Inc, New York, 1994.

packaging (1991), secondary packaging (1992) and sales packaging (1993). Quotas were initially set for collection, sorting and recycling of different packaging materials and were scheduled to increase in 1995. However, a proposed amendment to the Ordnance will abolish the quota system, replacing it with recycling targets to take effect from 1996. A number of organizations were established to organize collection, sorting and recycling on behalf of the manufacturers and retailers:

- Duales System Deutschland (DSD) for sales packaging
- RESY for transport packaging
- Interseroh AG for transport and secondary packaging

This has resulted in the setting up of a co-operative, industry-wide 'dual' waste management system, paid for by industry through a levy on all packaged products which must bear a 'green dot'.

The German system has come under criticism because the collected materials flooded markets in other countries, distorting local markets. The scheme is being challenged in the European Court under competition rules. However, this is not related to the approach adopted but rather to the recycling targets set and the lack of measures to increase levels of demand for recycled materials. In response to these problems other countries have widened their definition of recycling (for example to include chemical recycling for plastics or incineration with energy recovery) and/or graduated the levies to encourage a reduction in the quantity of packaging material used, or to encourage a shift to more benign packaging types.

It has been estimated that in Germany in 1993 the DSD scheme diverted 4.3 million tonnes of materials from landfill, representing 60% of the total amount of 7.3 million tonnes of post consumer packaging. In the same year, about 54% of consumer packaging (3.9 million tonnes) were recycled. First estimates from Germany suggest that the per capita consumption of packaging material has decreased by about 10% between 1991 and 1993.

Other schemes, learning from the German experience, have been less draconian. For example, in France, industry is working actively with the existing municipality waste management system to establish new collection schemes which will meet their agreed recycling targets. France is currently conducting preliminary studies of Producer Responsibility for batteries. In Austria, a producer/retailer recovery obligation, which relied on the environmental awareness of the consumer to return batteries, has resulted in a 30% reduction in the battery-derived mercury content of domestic waste since the implementation of the Waste Management Act in 1990.

*15.4.4.2 Product Charges and Product Taxes.* Product charges are levied at the point of consumption on products that are harmful to the environment. The charges can be set at such a level as to achieve the desired reduction in usage, or to incorporate some or all of the costs of recycling or disposing of the product. The former is preferable.

Italy and Austria have introduced product charges on disposable carrier bags, in an effort to reduce waste and encourage consumers to use durable bags. In Italy, a mandatory charge of L150 on plastic bags was first introduced in 1988. The charge met with opposition from the plastic bag producers but reportedly little opposition from consumers who appreciated the environmental objectives of the charge. In January 1994, the mandatory charge was withdrawn but retailers have continued to charge for plastic bags. Charging is widespread across all retail types. The charge ranges between L150 and L200. Paper bags are charged a similar amount. Immediately after the introduction of the charge in 1988 the consumption of plastic bags declined by 20–30%. The long term reduction is estimated at between 20% and 30%. These reductions have been achieved through a combination of responses, namely: use of durable bags; reuse of plastic bags; substitution to paper bags and, to a lesser extent, use of the biodegradable corn-based bag. It is understood that where consumers purchase plastic bags, they reuse them several times on subsequent shopping trips.

The use of product taxes is growing in many European countries as a means of raising the price of disposal or non-recyclable goods relative to less environmentally demanding alternatives. Taxes are commonly used as an incentive to set up deposit/refund schemes, as is the case in Norway and Finland, where tax exemptions are allowed if a suitable return rate is achieved. Other countries considering product taxes as a way of influencing consumer behaviour and providing subsidies for recycled materials include Denmark (beverage containers and tableware), the Netherlands (PET soft drinks bottles), Belgium and Switzerland.

*15.4.4.3  Refunds.*   In the UK, the supermarket chain J Sainsbury plc launched an initiative in 1991 to encourage the reuse of plastic bags. The supermarket provides a refund of £0.01 when the consumer reuses a plastic or other type of bag. To claim the refund the consumer had to bring back and reuse a Sainsbury or, alternatively, another retailer's bag. Consumers are encouraged through the provision of in-store collection boxes to donate the refund to charity. Sainsbury refund, on average, 1.7 million pennies each week and they estimate that the scheme has saved over 60 million plastic bags per annum, which represents a reduction of about 13%. All of Sainsbury's bags contain 75% recycled material, which is obtained from post-use UK waste.

Deposit refund schemes are another policy instrument for waste reduction. For example, there have been several new initiatives in Scandinavia, while in Germany and the Netherlands deposit schemes for plastic beverage containers have been introduced. There are also examples of deposit refund schemes used for a number of other products including car bodies (Sweden, Norway), batteries (Denmark, Netherlands, the US) and disposable cameras (Japan). In Korea an industrial deposit refund scheme has been set up whereby manufacturers are required to pay a deposit to the government which is refunded if, after customer use, the company collects and treats the product itself.

*15.4.4.4  Compulsory Collection and Recycling.*   Legislation to force local authorities to collect and recycle materials is becoming a widely used means of achieving a reduction in the quantity of MSW which is sent for disposal. Measures of this type have proliferated in Europe during the early 1990s and usually focus on particular components of the MSW stream. For example, legislation has been enacted in both the Netherlands and Austria which obliges municipal authorities to set up source-separated organics collection programmes to collect and compost household organic waste. Similar legislation has been proposed in Denmark, Germany and Luxembourg.

Compulsory collection and recycling programmes usually have to achieve certain waste reduction or recycling targets (for example 50% diversion from landfill by the year 2000 in Ontario). Alternatively, disposal limits for recyclable waste can be set, for example in Austria, where the organic content of MSW sent to landfill must be below 5% since July 1994. The programmes are often funded by state or country-wide waste disposal levies.

## 15.5 BULK WASTE REDUCTION TECHNOLOGIES AND FINAL DISPOSAL

Whatever success is achieved in reducing waste and in separating materials for recycling, some waste will always remain. To achieve high waste reduction rates in terms of landfill demand, a technology component is required. Some options are listed in Table 15.2.

Other than the physical size and weight reduction technologies such as baling and separation, of the options listed in Table 15.2 waste fired power generation and composting are perhaps the most widely used. These two options are discussed in this Section, along with landfilling, which remains the most common waste disposal option.

### 15.5.1 Mass Burn Incineration

Mass burn waste incineration has been practised as a waste management and volume reduction technique since the 1890s. It is an extremely effective bulk waste reduction technology, typically reducing waste volume by 90% and mass by around 70%. In terms of waste processing, mass burn incineration is a relatively simple option, with unsorted waste being fed into a furnace and, by burning, reduced to one-tenth of its original volume. Typically the only materials removed from the waste stream prior to burning are large bulky objects such as refrigerators and mattresses, or potentially hazardous materials such as gas bottles.

The combustion gases then typically pass through a boiler system to recover energy. The most flexible means of recovering energy from the hot gases is to produce steam for direct use (at a lower temperature) or for electricity generation. To generate electricity, superheated steam is passed from the boiler system through a turbine generator. The gases are then cleaned prior to discharge to atmosphere through a tall stack, in order that the discharge conforms to the strict emission limits laid down by national governments. Gas cleaning strategies aim to remove the following components:

**Table 15.2**  *A list of bulk waste reduction technologies*

*Size Reduction Technologies*
- Baling
- Pulverization/shredding
- Homogenization/wet pulping

*Weight Reduction Technologies*
- Separation
- Materials separation facilities
- Waste derived fuels

*Waste Fired Power Generation (WFPG)*
- Mass burn incineration
- Fluidized bed incineration
- Combustion of prepared waste derived fuels

*Other Combustion Technologies*
- Aggregate/block production
- Cement firing
- Wood burning power/CHP stations
- Tyre burning power/CHP stations

*Biological Systems*
- Composting
- Vermiculture
- Hydrolysis
- Anaerobic digestion

*Others*
- Pyrolysis
- Gasification

- Acid gases such as nitrogen and sulfur oxides and hydrogen chloride. A variety of wet and dry scrubbing systems can be applied, but in essence the gas cleaning method involves neutralization of the acidity by dosing with an alkaline reagent such as lime. Chemical methods of controlling nitrogen oxide emissions (so called De-NOX systems) have also been developed.
- Particulate matter, often associated with trace metals and semivolatile organic micropollutants. Electrostatic precipitators (ESPs) and fabric filters are generally used for this purpose. The particulate matter is retained in the ESP or the filter and is periodically removed from the system as ash which requires land disposal.
- Organic micropollutants such as dioxins. Control of these emissions is achieved by control of combustion conditions, as well as by dosing in the gas cleaning train with activated carbon or other adsorbents. Simultaneous reduction of emissions of metals such as mercury and cadmium is also achieved.

The impact of prior materials separation and recycling on the calorific value of waste sent for incineration has been studied. UK research has indicated that the calorific value will fall if large quantities of only paper and plastics are recovered.

However, if putrescibles are also separated from the waste stream, there is little effect on the overall calorific value.

### 15.5.2  Composting

Composting is essentially the controlled aerobic decomposition of putrescible material. Of the various methods of aeration, windrowing (mechanical or manual turning of the material) and forced aeration of the static pile are the most common methods used. During composting, putrescible material is progressively broken down by microorganisms in a series of distinct stages. In the mesophilic stage, microorganisms begin to actively break down the organic material, the temperature of the composting material rising to around 50 °C in about two days. During the second, or thermophilic stage, temperatures begin to rise so that only the most temperature resistant microorganisms survive. As the microorganism population reduces, the composting material cools and anaerobic conditions may develop unless sufficient air is introduced. In the third stage, the material continues to cool and microorganisms begin to compete for the remaining organic material, in turn leading to breakdown of cellulose and lignin in the waste. During the final, maturation stage, levels of microbial activity continue to fall as the remaining organic material is broken down and the microorganisms die off as their food sources deplete.

Overall, maintaining the correct balance of oxygen, and therefore temperature, is crucial for the successful degradation of wastes. If the process becomes anaerobic, odour problems can be severe and microbial activity can cease. Temperature and moisture content can be monitored by electronic sensors. Depending on the type of organics and moisture levels in the material to be composted, volume and weight reduction levels for that waste fraction are typically in the range of 40–60% and 40–50% respectively.

Despite widespread use as a waste management strategy, in the past many composting schemes have been unsuccessful. Maintaining a consistently high quality and as such a marketable compost product, and producing it in an efficient and environmentally acceptable manner, were the key problems facing early schemes. However, composting is currently enjoying a resurgence of interest, and prospects for composting schemes look promising. The key to this change of fortune has been a switch to the composting of uncontaminated source separated organic wastes rather than of mixed MSW. Operational experience in the US and elsewhere with mixed MSW feedstock has indicated that the end product is often of poor quality and has limited end use. Recent composting schemes have therefore increasingly focused on processing source segregated kitchen and garden wastes, 'green' wastes from parks and gardens, and food wastes from the food processing industry and large commercial generators.

### 15.5.3  Landfilling

Landfills are the final destination for the residues from incineration and other treatment and processing options, as well as for the primary waste stream – in the

UK about 90% of domestic waste and 85% of commercial waste is dispatched directly to landfill.[2] It is therefore a critical element in a waste management strategy, since despite the best attempts at waste minimization, recycling and recovery, its use is unavoidable. The challenge is to design and manage the process of landfilling in a sustainable manner so as not to leave a long-term potential for environmental damage.

The process of landfilling consists of waste preparation (shredding, compaction, *etc.*), waste placement (involving deposition, compaction and covering of the waste within the landfill) and finally landfill completion, in which the completed landfill is capped with a low permeability layer to minimize the ingress of rainwater and covered with soil to return the area to the surrounding landscape. In practice, landfill restoration is a continuing process throughout the lifetime of the site: void space is typically utilized in a phased manner, with progressive contouring, capping and restoration.

Within the landfill, the constituents of the waste undergo biodegradation and stabilization. The infiltration of rainfall and surface waters into the waste mass, coupled with the biochemical and physical breakdown, produces a leachate which contains soluble components of the waste.

The breakdown products also include so-called 'landfill gas', which can be harnessed for its energy content. The maximum gas volume which is generated from the decomposition of organic matter is in the region of 350–400 $m^3$ per tonne of MSW, amounting to about 6 $m^3$ per tonne per year. The gas typically consists of 50–70% methane and 30–50% carbon dioxide with traces of nitrogen, hydrogen, oxygen, hydrogen sulfide and a range of trace organic compounds. Landfill gas utilization schemes can capture 50–60% of the gas released by the site: its combustion can potentially generate about 350–370 kWh of energy per tonne of MSW.

### 15.5.4 Environmental Considerations

*15.5.4.1 Waste Incineration.* All bulk waste reduction technologies have the potential to produce gaseous, liquid and solid contaminants. For waste incineration, the key environmental issues are:

- Dust and odour from waste handling and storage
- Ash management (grate ash and flyash from the gas cleaning system)
- Atmospheric emissions from the stack

The potential effects of dust and odour from waste handling can be controlled by the following measures:

- Location of the waste unloading area within an enclosed building and the use of dust suppression waste sprays at the waste tipping bays.
- Extraction of air from above the storage bunker for use as combustion air in the incinerator, where any odorous compounds and dust entrained in the air will be destroyed. This also results in a slight negative pressure within the

building which draws air inwards, thus minimizing the escape of dust or odours.

These measures are now standard in most modern incinerator plants, and are not identifiable as separate environmental mitigation within a scheme.

Grate ash and flyash are produced at a rate of 20–30% and 3–4% of the waste input respectively.[12] Grate ash is essentially inert; flyash has relatively high levels of heavy metals and is classed as a hazardous waste in some countries. Due to its inert characteristics, the disposal of grate ash is not generally considered a significant pollution risk and it is often used for beneficial purposes such as road building, in construction materials, for intermediate cover at landfill sites, *etc.* Prior to disposal, the ash is normally scavenged for ferrous metals by magnetic separation: up to 10% of the ash may be recoverable ferrous material.[13]

The cleaning of combustion gases prior to their release to atmosphere has been discussed in Section 15.5.1. Additional information of the environmental impact of atmospheric releases can be obtained from other references.[14,15]

*15.5.4.2 Composting.* The key environmental issues of the composting process are as follows:[16]

- Fugitive emissions of litter and dust. These emissions may arise from wind dispersal of the waste feedstock, especially paper, during preparation and from compost piles.
- Odour, arising from trace organics produced in anaerobic conditions are permitted to occur in the composting process.
- Leachate generation, since the requirement for moisture in the composting process may lead to excess water and run-off from the composting pile, which may be contaminated by materials present in the waste.

Dust, litter and odour from waste material preparation can be mitigated by enclosure of the operation, both with respect to individual items of equipment (shredders, *etc.*) and by locating the process within a building with appropriate ventilation systems. Dust, litter and odour from the composting process can also be mitigated by enclosure of the activity, or by in-vessel composting. However, since these measures are relatively expensive due to the large volume of material which needs to be processed, windrow composting is most commonly applied.

[12] A. Porteous, 'Prospects for MSW incineration', Waste Management Conference Proceedings, AEA Technology, Harwell, October 1992.

[13] C. R. Dobbie, 'Waste Management: A View from the Buyers' Side', ETSU Waste to Energy Conference, Winchester, May 1994, ETSU, Harwell.

[14] G. H. Eduljee, in 'Issues in Environmental Science and Technology No. 2 – Waste Incineration and the Environment', eds. R. E. Hester and R. M. Harrison, The Royal Society of Chemistry, Cambridge, 1994.

[15] P. T. Williams, in 'Issues in Environmental Science and Technology No. 2 – Waste Incineration and the Environment', eds. R. E. Hester and R. M. Harrison, The Royal Society of Chemistry, Cambridge, 1994.

[16] P. R. Bardos, 'Survey of Composting in the UK and Europe', Paper No. W93036, Warren Spring Laboratory, Stevenage, 1993.

This is normally conducted out of doors, although Dutch barn buildings are sometimes used (*i.e.* roofed buildings with open sides). Regular wetting of the windrows and proper management of the composting process assist in minimization of dust, odour and litter.

The requirement to keep the compost pile wet increases the potential for leachate generation from excess water and run-off. The control of this leachate is important for the protection of water resources. Typical design features include organization of the windrows or compost piles to retain water inflow and avoid run-off, or the use of a concrete plinth equipped with a controlled drainage system to collect run-off.[17]

*15.5.4.3 Landfilling.* Of the potential releases and environmental effects of landfilling[18,19] the following have raised most concern:

- *Leachate:* leachate arises from the moisture contained in the deposited waste, from the infiltration of water into the site and from the biodegradation process itself. Escape of leachate, for example due to engineering failures of landfill caps, covers and liners, has been linked to contamination of water resources.
- *Landfill gas:* landfill gas, a mixture of methane and carbon dioxide, can cause damage to vegetation, and is also an explosion hazard. Methane and carbon dioxide are also greenhouse gases.
- *Trace organics:* a variety of trace organic compounds can be entrained with landfill gas, for example vinyl chloride, benzene, toluene, alkanes, organosulfur compounds, *etc.* While many of these compounds are potentially toxic, their concentration in offsite air is generally too low to pose a threat to public health. Odour nuisance is potentially a more common problem.
- *Litter, vermin, noise*, etc: The nuisance aspects of landfills and their operation are potentially the most intrusive in terms of disturbance and disruption to the amenities enjoyed by the surrounding population. Careful consideration is given to operational work plans, traffic movements on and off site, the fitting of screens to reduce the visual intrusion and dispersal of litter, daily cover of the waste, *etc.* to mitigate the possibility of nuisance.

Landfilling provides a method of reclaiming existing excavations as well as the development of new landforms. Hence, it can be used to return unproductive land to beneficial use. However, after landfilling operations are complete and the site has been capped, the *in situ* processes of biodegradation continue for a significant length of time, measured in decades. As biodegradation is still active, the generation of leachate and landfill gas also continues, as does the potential for offsite migration of these releases. Post closure management of the landfill is therefore a key consideration: typically this takes the

[17]P. Stenbro-Olsen and P. Collier, *J. Waste Managt. Resource Recovery*, 1994, **1**, 113–118.
[18]K. Westlake, in 'Issues in Environmental Science and Technology No. 3 – Waste Treatment and Disposal', eds. R. E. Hester and R. M. Harrison, The Royal Society of Chemistry, Cambridge, 1995.
[19]D. J. Lisk, *Sci. Total Environ.*, 1991, **100**, 415–468.

form of the installation of a landfill gas utilization/control scheme, provision for leachate collection and treatment, and regular monitoring of releases from the site.

## 15.6 INTEGRATED WASTE MANAGEMENT STRATEGIES

### 15.6.1 Revisiting the Waste Management Hierarchy

In order to achieve an integrated approach to waste management, the following need to be included in any strategy:

- Policy and other measures to encourage the avoidance and minimization of waste.
- Policy and other measures to encourage the recovery, recycling and reuse of materials that would otherwise enter the waste stream.
- Adoption of bulk waste reduction technologies which will effectively reduce the volume of materials remaining in the waste stream after the above measures have been put into place, thus minimizing the amount of material requiring final landfill.

While there are options within each of these categories, a strategy which omits the inclusion of measures from any one complete category is likely to achieve much lower levels of overall waste reduction than one which uses a more integrated approach. In support of this conclusion, reference is made to a series of hypothetical scenarios in Figure 15.5, which illustrates how a combination of

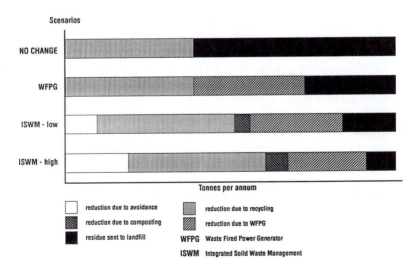

**Figure 15.5** *Combining policy and technical measures into a waste management strategy – potential waste reduction under four scenarios*

policy and technical measures could be combined to reduce the final tonnages of waste dispatched to landfill.

- The *first scenario* ('no change' or 'do nothing') shows an existing situation where approximately one third of the waste is recovered by recycling.
- The *second scenario* indicates a simple 'technical fix' in which no policy measures are introduced, reliance being placed instead on a comprehensive programme of building incineration facilities. This scenario cuts landfill demand by 50% but does nothing either to tackle waste growth (a continuing programme of building new incineration plants would be needed) or to reduce expenditure on waste collection, transfer, treatment and disposal.
- The *third and fourth scenarios* indicate 'low' and 'high' applications of waste avoidance, reduction, *etc.* policies and technologies. The benefit of the integrated approach is not only a substantial increase in waste reduction, but also a substantial potential to reduce the costs of collection and disposal.

Table 15.3 summarizes the combination of measures some countries are taking to strive towards sustainable waste management.

In conceptual terms, what an integrated strategy seeks to achieve is to convert a linear transference of materials through the producer-user-disposer chain, to a series of circular and as far as possible, closed systems where beneficial aspects of the waste (materials, energy, *etc.*) are drawn out at each stage. However, the phrase 'waste management hierarchy' may suggest that after waste prevention and minimization, the remaining options nearer the apex of Figure 15.2 invariably represent more environmentally acceptable solutions and conversely, options such as landfilling invariably represent less environmentally acceptable and less sustainable solutions. This view could lead to anomalous and unbalanced waste management strategies which take insufficient account of local situations and economic sustainability.[20-22] For example, in a predominantly rural area in which MSW is generated in sparsely populated and widely separated conurbations, long transport hauls to a central composting or waste to energy incineration facility may not represent an environmentally or economically preferable option relative to local landfilling if the cost, energy and emissions relating to transportation are taken into account. Reusable containers need to withstand repeated cleaning and handling and therefore require more material and energy in their manufacture than do single-trip containers.[20] The overall energy consumption of a bring system which involves the householder in a 2 km car ride to a Civic Amenity site to deliver recyclables is over double that of a basic system involving non-segregated kerbside collection followed by landfilling.[21]

[20] J. Bickerstaffe, *J. Waste Managt. Resource Recovery*, 1994, **1**, 91–96.
[21] P. R. White, M. Franke and P. Hindle, 'Integrated Solid Waste Management – A Lifecycle Inventory', Blackie Academic and Professional, London, 1995.
[22] R. Watkinson, 'Towards Sustainable Waste Management in Practice', Harwell Waste Management Symposium, AEA Technology, Harwell, 1995.

**Table 15.3**  *The use of waste reduction policy instruments in some countries*

| Country | Deposit/refund | Producer responsibility | Voluntary agreements (recycling targets) | User charges | Waste disposal levies/taxes | Product taxes | Recycling credits | Market support/preferential purchase | Compulsory collection/recycling | Grants/subsidies |
|---|---|---|---|---|---|---|---|---|---|---|
| Australia | √ | | √ | √ | √ | | √ | | | √ |
| Austria | √ | √ | √ | √ | √ | | | | √ | |
| Belgium | √ | A | √ | √* | √ | √ | | | | |
| Canada | | A | | √ | √ | A | √ | | √** | √ |
| Denmark | √ | √ | √ | √* | √ | √ | | √ | P | √ |
| Finland | √ | | | √ | P | √ | | | √ | √ |
| France | √ | √ | √ | √* | P | | | √ | | √ |
| Greece | | √ | | √* | | | | √ | A | |
| Ireland | | | | √* | | | | | | √ |
| Italy | | √ | | √* | | √ | | | | |
| Netherlands | √ | | √ | √* | √ | √ | | √ | √ | √ |
| Norway | √ | | | √* | | √ | | | | √ |
| Portugal | | | | √* | | | √ | √ | | √ |
| UK | | P | P | √* | P | | √ | √ | | √ |
| USA | √** | P | | √* | √** | | | √ | √** | √ |
| European Union | | | P | | | | | | A | |

*Status*: √ implemented; A Agreed; P Proposed
Notes: *Through general municipal revenue; **In some states only

In order to optimize the overall waste management strategy, with respect both to the environment and to economics, it is necessary to address the entire strategy in a holistic sense. The final section of this chapter discusses some comparative tools that are being developed for this purpose.

### 15.6.2   Optimization of Waste Management Strategies

The development of a sustainable waste management strategy involves a structured approach. A powerful analytical framework for the comparison of waste management options or combinations of options is provided by Life Cycle Analysis (LCA). The methodology comprises the following stages:[23]

- *Goal definition and scoping*, which defines the system's boundary. For example, the boundary for the assessment can be drawn at the point where products (for example, plastics) enter the waste management cycle (the *Foreground System*[24]). Alternatively, a wider perspective can be taken whereby the resource, energy, environmental and economic implications of the primary production process is also included in the assessment (the *Background System*[24]).
- *Inventory analysis*, in which data on resources used, energy consumed, released and recovered and emissions to atmosphere of the system under study are compiled in the form of an inventory table.
- *Impact assessment*, in which the inventory data is converted into a quantified indication of environmental impact.
- *Improvement assessment*, in which options for reductions in resource and energy consumption, environmental impact and/or waste generation are identified and assessed against the baseline situation.

The main focus is presently on the collection of sufficient data on the resource, energy and emissions characteristics of waste management systems; most LCAs terminate at the inventory stage, a process called *Life Cycle Inventory* (LCI) analysis.

Considerable literature exists on the LCA and LCI analysis of materials such as paper, plastics and glass, but less so for other components of MSW, or for waste management processes such as collection, incineration, composting and land-filling. The UK Government is currently developing a framework for applying LCI analysis to the management of non-hazardous waste and a preliminary systems definition has been completed.[25]

---

[23]Society of Environmental Toxicology and Chemistry (SETAC), 'Guidelines for Life-Cycle Assessment: A Code of Practice', Brussels, 1993.

[24]A. Doig and R. Clift, 'Apply Life Cycle Thinking to Solid Waste Management', in 'Life Cycle Inventory Analysis for Waste Management', Culham Laboratory, May 1995.

[25]D. Daley and J. Barton, 'Life Cycle Inventory for Waste Management – Stage 1 of the DoE Research Programme', in 'Life Cycle Inventory Analysis for Waste Management', Culham Laboratory, May 1995.

A hypothetical case study LCI analysis has been undertaken,[21] which illustrates the methodology. The following waste management options were examined:

(1) Unsorted collection of household waste followed by landfilling.
(2) Unsorted collection of household waste followed by mass burn incineration.
(3) Unsorted collection of household waste followed by composting and marketing of the compost.
(4) Separate collection and composting of putrescibles and landfilling of the rest.
(5) Separate kerbside collection of dry recyclables and incineration of the rest.

There is a net benefit in energy consumption and a reduction in global warming potential when waste is incinerated rather than landfilled (Scenario 1 *versus* Scenario 2). In the case of Scenario 3, there is no net reduction in energy consumption relative to direct landfilling of the waste, but this changes to a net benefit if account is taken of the savings resulting from product substitution. In general, Scenarios 2 and 5 (involving incineration) were preferable in terms of net energy consumption than the remaining scenarios. Other scenarios and variations thereof can also be assessed. For example, the collection of waste can be supplemented by sorting and recycling of glass, or it can be assumed that the compost generated in Scenario 3 has no market.[21]

Emissions of other pollutants can also be assessed on a comparative basis. For example, a comparison of emissions from the combustion of MSW *versus* the combustion of conventional oil or coal for the production of energy indicated significant decreases in emissions of acid gases and particulate matter were a replacement MSW incinerator to be installed.[26]

## 15.7 CONCLUSIONS

Waste management strategies are increasingly being developed as a combination of policy and other measures, with certain core policies driving the strategy forward. As shown in Table 15.3, these typically include mandatory producer responsibility combined with contract-bound voluntary agreements, refund schemes or public commitments such as in the Wastewi$e scheme. A strategy which relies more on voluntary measure and market forces is likely to be less effective in achieving the goals of sustainable waste management than one which also has strong legislative and mandatory countermeasures. For example it has been pointed out that in the UK, the Producer Responsibility Group's target for 58% recovery of packaging material by the year 2000 on a voluntary basis (see Section 15.4.2.2) would perhaps be more effectively attained by the simple

[26]H. F. Taylor, in 'Proceedings of the USEPA/AWMA Second Annual International Conference on MSW Combustion', Tampa, Florida, 1991, US EPA, Washington, DC.

expedient of introducing a mandatory ban on the landfilling of used transit packaging.[27]

Nevertheless the trend towards developing strategies with a view to achieving sustainable waste management is now firmly established. Some national systems (as in Germany and the Netherlands) are well advanced. Other countries such as Hong Kong and Thailand which have experienced explosive MSW growth rates, have embarked on a phased programme of improvements in waste management along the lines of sustainable development, drawing on the lessons learned in establishing initiatives such as the German DSD system to shape the formulation of similar indigenous systems. The revolution that commenced in the 1970s is now truly a global phenomenon.

[27]M. Rose, *Warmer Bulletin*, 1995, **44** (February), 22–23.

# CHAPTER 16

# The Environmental Behaviour of Toxic Organic Chemicals

S. J. HARRAD

## 16.1 INTRODUCTION

The last thirty years has seen tremendous growth in public, scientific and governmental interest in the environmental effects of toxic organic chemicals. This is attributable to a number of factors, principally: the increasing weight of scientific evidence related to their adverse effects, the dramatic development of the analytical technology required to measure such compounds and enhanced public awareness – the latter promoted by media coverage of incidents such as the 1976 explosion at a chemical plant in Seveso, Italy, alongside books such as Rachel Carson's 'The Silent Spring',[1] that publicized the potentially detrimental effects of pesticide use. This chapter addresses the environmental impact of toxic organic chemicals and the factors influencing that impact.

### 16.1.1 Scope

Given the tremendous array of organic compounds present in the environment, many of which may elicit adverse effects, the range of chemicals examined in this chapter is limited to a selection of key pollutants on the basis of their toxicity. Of primary interest, are the so-called semi-volatile organic compounds (SVOCs) – loosely defined as those possessing a vapour pressure of $< 130$ Pa at $25\,^\circ\text{C}$ – although a few of the more widely studied volatile organic compounds (VOCs), namely benzene, toluene, ethyl benzene and the three isomeric xylenes (collectively referred to as BTEX) will be dealt with briefly. Those SVOCs considered here are: polychlorinated dibenzo-*p*-dioxins (PCDDs) and polychlorinated dibenzofurans (PCDFs) (collectively known as PCDD/Fs), polychlorinated biphenyls (PCBs), polycyclic aromatic hydrocarbons (PAHs) and organochlorine pesticides (OCPs).

---

[1] R. Carson, 'The Silent Spring', Hamish Hamilton, London, 1963.

For each class of compounds, the following key areas pertinent to their environmental impact are addressed: their toxicology – with particular emphasis on their human effects; the methods used to monitor their presence in the environment; their major sources; their physicochemical properties and the influence of these on their environmental fate and behaviour, and, finally, a brief examination of the use of theoretical models to predict such behaviour.

### 16.1.2 Source Material

A wealth of review literature exists on the toxic organic chemicals examined in this chapter. An excellent general article covering their global distribution and fate is that of Ballschmiter,[2] whilst introductions to PCDD/Fs, PCBs, PAHs and VOCs are also available.[3–6] Readers seeking more in-depth treatment of both the chemistry of organic pesticides and the fugacity concept in environmental modelling are also referred elsewhere.[7,8]

### 16.1.3 Chemical Structure and Nomenclature

The environmental fate and behaviour of groups of toxic organic compounds is strongly structurally-dependent, and a brief discussion of their structures and nomenclature is deemed worthwhile.

*16.1.3.1 PCDD/Fs.* The basic chemical structures of PCDD/Fs are illustrated in Figure 16.1. The numbers indicate sites of chlorination – for example, 2,3,7,8-tetrachlorodibenzo-*p*-dioxin (2,3,7,8-TCDD) contains four chlorine atoms, one each at the 2, 3, 7 and 8 positions; in total, there are 75 possible PCDDs and 135 possible PCDFs. Each individual PCDD/F is referred to as a congener, whilst those congeners possessing identical empirical formulae are isomers of each other. Each group of isomers (there exists one for each degree of chlorination) constitutes a homologue group.

*16.1.3.2 PCBs.* Figure 16.2 illustrates the basic chemical structure of PCBs. As with PCDD/Fs, the numbers denote sites of chlorination – *e.g.* 3,3',4,4',5-pentachlorobiphenyl possesses five chlorines, three attached to one biphenyl ring at the 3, 4 and 5 positions, the remainder on the other ring in the 3' and 4' sites. 209 PCB congeners are possible, and the definitions and distinctions between congeners, isomers and homologues are identical to those given above

[2] K. Ballschmiter, *Angew. Chem. Intl. Ed. Engl.*, 1992, **31**, 487.

[3] S. J. Harrad and K. C. Jones, *Chem. Br.*, 1992, **28**, 1110.

[4] K. C. Jones, V. Burnett, R. Duarte-Davidson and K. S. Waterhouse, *Chem. Br.*, 1991, **27**, 435.

[5] M. L. Lee, M. V. Novotny and K. D. Bartle, 'Analytical Chemistry of Polycyclic Aromatic Compounds', Academic Press Inc., London, 1981.

[6] 'Chemistry and Analysis of Volatile Organic Compounds in the Environment', eds. H. J. Th. Bloemen and J. Burn, Blackie Academic and Professional, London, 1993.

[7] K. A. Hassall, 'The Chemistry of Pesticides', Macmillan Press, London, 1982.

[8] D. Mackay, 'Multimedia Environmental Models: The Fugacity Approach', Lewis Publishers Inc., Michigan, USA, 1991.

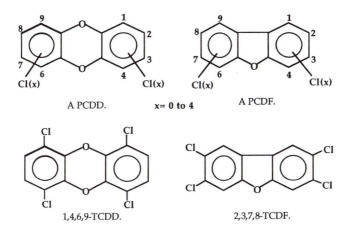

A PCDD.    x= 0 to 4    A PCDF.

1,4,6,9-TCDD.    2,3,7,8-TCDF.

**Figure 16.1** *PCDD/F nomenclature*

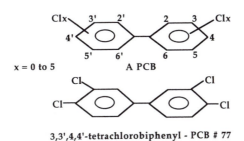

x = 0 to 5    A PCB

3,3',4,4'-tetrachlorobiphenyl - PCB # 77

**Figure 16.2** *PCB nomenclature*

for PCDD/Fs. Unlike PCDD/Fs, an IUPAC system[9] assigning a single unique number to each possible PCB congener is widely – indeed almost universally – utilized and this greatly simplifies reference to individual PCBs. For example, 3,3',4,4'-tetrachlorobiphenyl is referred to as PCB # 77.

*16.1.3.3 PAHs.* There are three possible PAH isomers consisting of three rings. When this number increases to six rings, 82 isomers are possible. When 12 rings are present, the possible total amounts to 683 101. It is thus unsurprising that only around 45% of these have been identified and that the complex chemistry of these compounds requires a complex nomenclature system. The standardization of chemical nomenclature is the responsibility of IUPAC but undergoes frequent changes and the interested reader is referred to Zander.[10] Figure 16.3 gives the chemical structures of some environmentally significant PAHs, which are an important sub-division of the wider group of compounds known as polyaromatic compounds (PACs). This encompasses a far wider range of compounds, including

[9] D. E. Schulz, G. Petrick and J. C. Duinker, *Environ. Sci. Technol.*, 1989, **23**, 852.
[10] M. Zander, 'The Handbook of Environmental Chemistry', 1980, **3(A)**, 109.

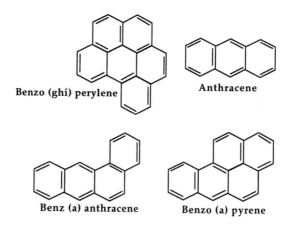

**Figure 16.3**   *Selected PAH structures*

heterocyclic derivatives such as dibenzothiophenes and carbazoles, along with substituted PAH such as nitro-PAH.

*16.1.3.4   OCPs.*   As with PAHs, a limit to the number considered in this chapter has been imposed and Figures 16.4, 16.5 and 16.6 show the chemical structures of aldrin, dieldrin, a toxicologically active toxaphene, DDT and metabolites (DDE and DDD) and the hexachlorocyclohexanes (HCHs).

*16.1.3.5   VOCs.*   The chemical structures of the limited number of VOCs chosen as illustrative of the environmental behaviour of this vast class of compounds; namely: benzene, toluene, ethyl benzene and the isomeric xylenes (BTEX), are considered sufficiently well known to preclude illustration here.

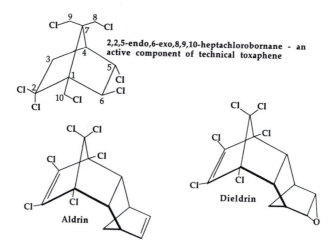

**Figure 16.4**   *Structures of chlorinated cyclodiene pesticides*

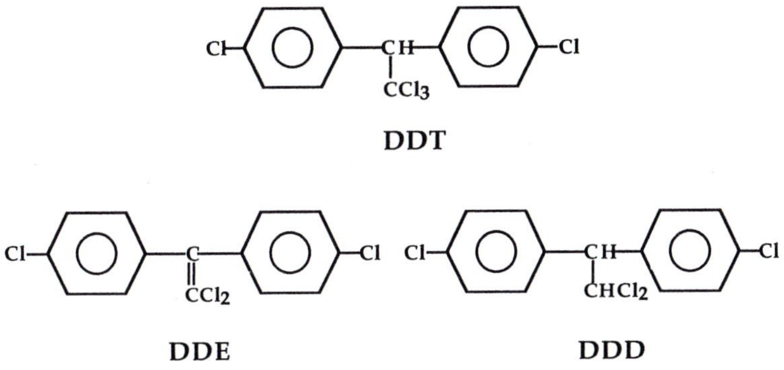

**Figure 16.5**   *Structures of the DDT 'family'*

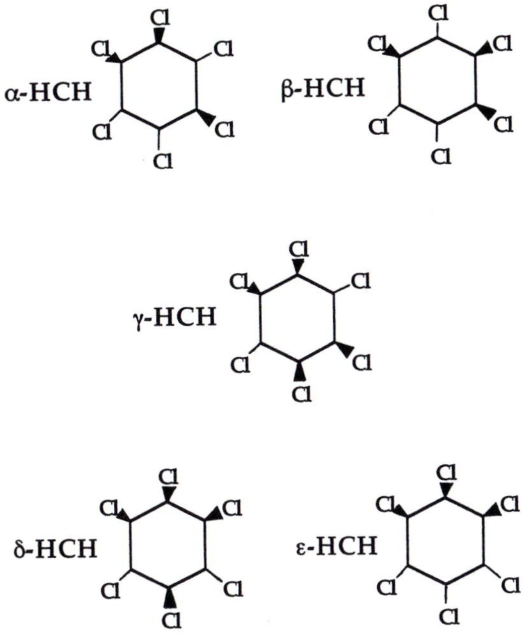

**Figure 16.6**   *Structures of selected HCHs*

## 16.2   ADVERSE EFFECTS (see also Chapters 11 and 18)

An in-depth treatment of the toxicology and ecotoxicology of toxic organic chemicals is beyond the scope of this chapter and only a brief summary of the adverse effects of each compound class is presented. A general consideration in relation to the toxic effects of compounds present at extremely low levels in the environment, is the difficulty in ascribing health or reproductive abnormalities to contamination by a specific chemical alone, in the presence of other similarly potentially toxic pollutants, often at much higher concentrations. In particular, one cannot discount the possibility of synergistic and/or antagonistic† toxicological effects arising from contamination by a 'cocktail' of pollutants like PAHs, PCBs and PCDD/Fs, all of which are known to evoke similar biochemical responses. Indeed, evidence exists for the existence of a synergistic effect accentuating the toxic effects of lindane and heptachlor.[11] This area, which has important health implications, is now receiving much closer scientific scrutiny. In addition to these difficulties, there are considerable problems with assessing the effects in humans, in the understandable absence of clinical data. Instead, evidence for adverse effects in humans is reliant on epidemiological surveys and extrapolations from animal studies, with all the problems inherent in such indirect measurements of human toxicity. A final general point, is the tendency of all of the compounds considered here to partition into fatty tissues. As a result, their adverse effects are compounded by their bioaccumulation in species at the head of food chains, such as marine mammals, birds of prey and humans.

### 16.2.1   Polychlorinated Dibenzo-*p*-dioxins and Polychlorinated Dibenzofurans (PCDD/Fs)

Probably no other group of chemicals has been subjected to greater scrutiny with respect to their environmental effects than PCDD/Fs. It is to the Seveso incident in 1976, that present interest may be traced. The episode occurred on July 10th, 1976 at the ICMESA chemical plant in Seveso, near Milan, Italy. The alkaline hydrolysis of 1,2,4,5-tetrachlorobenzene to produce 2,4,5-trichlorophenol (2,4,5-TCP) – an intermediate in the production of the bactericide, hexachlorophene – went out of control, and the resultant explosion distributed large quantities of 2,3,7,8-TCDD (formed *via* the self-condensation of 2,4,5-TCP) over the surrounding area. Although the exact environmental impact of the incident has yet to be fully evaluated – whilst widespread animal mortality occurred, the only undisputed adverse effect in humans remains the induction of chloracne (an extremely disfiguring skin complaint) – the psychological effect on the general public was immense. Widespread media coverage both in Italy and beyond – ranging from the responsibly factual to the hysterical (an assertion was made that dioxin could facilitate 'the end of Western civilization') – propelled the 'dioxin issue' onto the political agenda of the industrialized world. Since then, a number of incidents have maintained a high public profile for these compounds. In particular, there has been much controversy in the USA over the exposure of Vietnam War veterans to the defoliant Agent Orange (a mixture of esters of di- and trichlorophenoxy acetic acids and heavily contaminated with PCDD/Fs).

---

[11]The World Health Organization (WHO), 'Public Health Impact of Pesticides used in Agriculture', WHO, Geneva, 1990.

†Synergism is defined as a situation in which the combined effect of two or more substances exceeds the sum of their separate effects. Antagonism is the opposite phenomenon, whereby the combined presence of two or more substances lessens the effects of those substances acting independently.

**Table 16.1** *PCDD/F congeners for which TEFs are defined*

| Congener | TEF |
|---|---|
| 2,3,7,8-tetrachlorodibenzo-*p*-dioxin (2,3,7,8-TCDD) | 1 |
| 1,2,3,7,8-pentachlorodibenzo-*p*-dioxin (1,2,3,7,8-PeCDD) | 0.5 |
| 1,2,3,4,7,8-hexachlorodibenzo-*p*-dioxin (1,2,3,4,7,8-HxCDD) | 0.1 |
| 1,2,3,6,7,8-hexachlorodibenzo-*p*-dioxin (1,2,3,6,7,8-HxCDD) | 0.1 |
| 1,2,3,7,8,9-hexachlorodibenzo-*p*-dioxin (1,2,3,7,8,9-HxCDD) | 0.1 |
| 1,2,3,4,6,7,8-heptachlorodibenzo-*p*-dioxin (1,2,3,4,6,7,8-HpCDD) | 0.01 |
| octachlorodibenzo-*p*-dioxin (OCDD) | 0.001 |
| 2,3,7,8-tetrachlorodibenzofuran (2,3,7,8-TCDF) | 0.1 |
| 1,2,3,7,8-pentachlorodibenzofuran (1,2,3,7,8-PeCDF) | 0.05 |
| 2,3,4,7,8-pentachlorodibenzofuran (2,3,4,7,8-PeCDF) | 0.5 |
| 1,2,3,4,7,8-hexachlorodibenzofuran (1,2,3,4,7,8-HxCDF) | 0.1 |
| 1,2,3,6,7,8-hexachlorodibenzofuran (1,2,3,6,7,8-HxCDF) | 0.1 |
| 1,2,3,7,8,9-hexachlorodibenzofuran (1,2,3,7,8,9-HxCDF) | 0.1 |
| 2,3,4,6,7,8-hexachlorodibenzofuran (2,3,4,6,7,8-HxCDF) | 0.1 |
| 1,2,3,4,6,7,8-heptachlorodibenzofuran (1,2,3,4,6,7,8-HpCDF) | 0.01 |
| 1,2,3,4,7,8,9-heptachlorodibenzofuran (1,2,3,4,7,8,9-HpCDF) | 0.01 |
| octachlorodibenzofuran (OCDF) | 0.001 |

It is important to note that amongst the range of 210 compounds known as PCDD/Fs, there exist wide differences in toxicological potency. In summary, only those PCDD/Fs chlorinated at the 2, 3, 7 and 8 positions are considered of toxicological significance. The seventeen congeners fitting this criterion are listed in Table 16.1, along with their toxic equivalent factors (TEFs†). This structural-dependence of toxicity, is due to the fact that the toxic effects of PCDD/Fs are mediated by initial interaction with a cellular protein – the Ah-receptor. Such binding is subject to the 'lock and key' principle commonly associated with compound-receptor interactions and hence, only those compounds able to assume a coplanar molecular configuration similar to that of 2,3,7,8-TCDD are capable of binding to the Ah-receptor. This receptor-mediated theory of toxicity also lends a rationale to the considerable species-specific variations in the toxicity of 2,3,7,8-TCDD (guinea-pigs have an $LD_{50}$ of 0.6 $\mu$g kg$^{-1}$ body weight (bw), *c.f.* hamsters which have an $LD_{50}$ of 3500 $\mu$g kg$^{-1}$ bw),[13] as the precise nature of the Ah-receptor varies widely between species.

[12] 'International Toxicity Equivalency (I-TEF) Method of Risk Assessment for Complex Mixtures of Dioxins and Related Compounds', Pilot Study on International Exchange on Dioxins and Related Compounds, Report no 176, NATO/CCMS, 1988.

[13] F. W. Karasek and O. Hutzinger, *Anal. Chem.*, 1986, **54**, 309.

†TE (or TEQ) is an acronym for 2,3,7,8-TCDD equivalents. It is a means of expressing the toxicity of a complex mixture of different PCDD/Fs in terms of an equivalent quantity of 2,3,7,8-TCDD. Each of the seventeen 2,3,7,8-chlorinated congeners are assigned a Toxic Equivalency Factor (TEF) based on its toxicity relative to that of 2,3,7,8-TCDD, which is universally assigned a TEF of 1. Multiplication of the concentration of a PCDD/F by its assigned TEF gives its concentration in terms of TE and the toxicity of a mixture is the sum of the TEs calculated for all congeners. Whilst several different weighting schemes exist, the one adopted by the UK government and referred to in this chapter, is that devised by the NATO Committee on Challenges to Modern Society (NATO/CCMS).[12]

Scientific opinion concerning the exact toxicological potency of PCDD/Fs differs widely. Despite this, there is a significant body of opinion that PCDD/Fs may well present a human cancer hazard, and that they are potent toxins with the potential to elicit a range of non-cancer effects, starting with binding to the Ah-receptor. These effects may be occurring in humans at very low levels comparable with the upper limit of background (*i.e.* non-occupational) exposure and include reproductive, immunological and developmental effects. To put this into context, average UK background exposure to PCDD/Fs is estimated at around 1 pg $\Sigma$TE kg$^{-1}$ body weight per day, compared with the UK government's guideline exposure limit of 10 pg $\Sigma$TE kg$^{-1}$ body weight per day.

### 16.2.2  Polychlorinated Biphenyls (PCBs)

PCBs have been linked to a number of toxic responses, including the impairment of immune responses in biota. Other cited effects include hepatotoxicity, carcinogenicity, teratogenicity, neurotoxicity and reproductive toxicity. In evaluating the significance of these effects, it is important to recognize that – as for PCDD/Fs – considerable species- and congener-specific variations exist. In particular, there are indications that their potency in humans is markedly less than in other animals, and it is very important that caution is exercised when extrapolating toxicological data obtained from laboratory animals to humans, and that PCB behaviour is studied on an individual congener basis. Also of relevance, is the fact that some PCBs (especially those lacking chlorine substituents at the 2, 2′, 6 and 6′ positions), are capable of eliciting similar toxicological effects to the 2,3,7,8-chlorinated PCDD/Fs. This is due to their ability to adopt the coplanar molecular configuration necessary to interact with the Ah-receptor responsible for mediating 'dioxin-like' effects and has led to the tentative assignation of TEFs to such 'coplanar' PCBs. It should be noted, however, that such extension of the TEF scheme to non-PCDD/F compounds has yet to be accepted by any regulatory body.

### 16.2.3  Polycyclic Aromatic Hydrocarbons (PAHs)

The toxicity of PAHs, like PCBs and PCDD/Fs, varies widely on both a compound- and species-specific basis. Their adverse effects have been known for a considerable time, and hindsight allows us to associate them with one of the first recorded cases of occupational cancer, *viz* the high incidence of scrotal cancer in chimney sweeps reported by Percival Pott in 1775, which was subsequently attributed to the high concentrations of PAHs – in particular, benzo(a)pyrene – in chimney soot. Several PAHs have been shown to be acutely toxic. However, health concerns regarding these compounds centre on their metabolic transformation by aquatic and terrestrial organisms into carcinogenic, teratogenic and mutagenic metabolites such as dihydrodiol epoxides, which bind to and disrupt DNA and RNA, with possible resultant tumour formation. Such metabolic conversion and activation is essential if PAHs are to exhibit their latent carcinogenic properties. The most potent carcinogens of the PAH group – in addition to benzo(a)pyrene – include: the benzofluoranthenes, benz(a)anthracene, dibenzo(ah)anthracene and indeno(1,2,3-cd)pyrene. Such PAHs are recognized by regulatory agencies such as the EC and the United States Environmental Protection Agency (USEPA) as priority pollutants.

### 16.2.4  Organochlorine Pesticides (OCPs)

A common effect of all three of the compound groups considered – *i.e.* the DDT group, the HCH group and the chlorinated cyclodiene family (*e.g.* dieldrin and toxaphenes) is an ability to induce reproductive effects, including reduced fertility and eggshell thinning in birds (this is as a result of the insecticide's ability to affect calcium metabolism in birds and mammals), as well as direct toxic effects. The combined consequence of such effects, was a marked reduction in populations of birds of prey. Restrictions on the use of the insecticides responsible have since resulted in a welcome recovery in affected populations. However, their environmental persistence, coupled with continued 3rd World usage, means that the danger is considered far from over. Furthermore, as a result of the marked bioaccumulative properties of these compounds, their presence in human fat is of some concern. For example, at the peak of DDT usage, the typical American adult contained 12 mg kg$^{-1}$ DDT in their fat. The International Agency for Research into Cancer (IARC) regards both HCHs and technical toxaphene as 'possibly carcinogenic to human beings' – although there is some dissent with this view – whilst the evidence for human carcinogenicity for aldrin, DDT and dieldrin is considered 'inadequate'. The general lack of data from human studies is striking however, and the WHO deems further study imperative. In addition, other adverse human effects are known. For example, organochlorine insecticides in general, are associated with enhanced induction of drug-metabolizing enzymes in the liver, whilst both DDT and lindane ($\gamma$-HCH) are known to elicit skin sensitization. An important fact concerning the relative toxicity of the different HCH isomers (although five are shown in Figure 16.6, eight are possible), is that the $\gamma$-isomer is fifty to several thousand times more toxic to insects than the $\alpha$- and $\delta$-isomers, whilst the $\beta$- and $\varepsilon$-isomers are usually almost inert.

### 16.2.5  Volatile Organic Compounds (VOCs)

In general, the acute toxicity of the VOCs considered here is low to moderate, with LD$_{50}$ (lethal dose causing 50% mortality) values in excess of 200 mg kg$^{-1}$ body weight in laboratory animals. However, acute sub-lethal effects including anaesthesia and loss of co-ordination are of concern. With respect to long-term adverse effects, only benzene is a recognized human carcinogen and haemotoxin, and little evidence of mutagenicity or teratogenicity exists for any of the other BTEX compounds. Most concern, therefore, centres around the neurotoxicity of these VOCs and, in particular, their ability to exert harmful effects on the central nervous system, leading to symptoms such as dizziness and amnesia. Possible links between elevated VOC levels and building-related health problems such as the so-called 'sick building syndrome' (SBS), have also been reported,[14] although BTEX compounds are unlikely to be implicated in this.

## 16.3  MEASUREMENT TECHNIQUES

Table 16.2 indicates 'typical' levels of various toxic organic chemicals in soil, air, human milk and freshwater. From this, it is evident that sensitivity is an important prerequisite of any measurement technique for such compounds, particularly for PCDD/Fs, where the quantities involved may justifiably be described as 'ultra-trace'. Given the increasing level of legislation relating to the presence of such compounds in the environment, there is also a clear need for the techniques to be both as accurate and as precise as is possible. In achieving these aims, the crucial role of sampling methodology must be recognized, although it should be noted that the direct measurement of VOCs in air is possible, using open-path spectroscopic techniques like IR, which obviate the need for separate sampling and analysis. Such methodology is likely to find increasing favour, and the interested reader is referred elsewhere for a detailed description of such techniques.[15] Finally, the measurement techniques employed must be able to distinguish the target compounds from the complex 'soup' of other chemicals present in the sample – in short, they must be selective.

**Table 16.2**  *Typical environmental levels of toxic organic chemicals*

| | 'Typical' Concentration of | | | | |
|---|---|---|---|---|---|
| Compartment | PAHs[a] | PCBs[b] | PCDD/Fs[c] | Toxaphene[d] | Benzene |
| Soil | $2000 \ \mu g \ kg^{-1}$ | $30 \ \mu g \ kg^{-1}$ | $0.3 \ \mu g \ kg^{-1}$ | $200 \ \mu g \ kg^{-1}$ | $10 \ \mu g \ kg^{-1}$ |
| Air | $100 \ ng \ m^{-3}$ | $1 \ ng \ m^{-3}$ | $0.003 \ ng \ m^{-3}$ | $0.03 \ ng \ m^{-3}$ | $2 \ ppb^{e}$ |
| Human milk[f] | —[g] | $10 \ \mu g \ kg^{-1}$ | $0.015 \ \mu g \ kg^{-1}$ | $5 \ \mu g \ kg^{-1}$ | $0.1 \ \mu g \ dm^{-3h}$ |
| Freshwater | $10 \ ng \ dm^{-3}$ | $1 \ ng \ dm^{-3}$ | $0.003 \ ng \ dm^{-3}$ | $1 \ ng \ dm^{-3}$ | $30 \ ng \ dm^{-3}$ |

[a]Expressed as the sum of the 16 PAH required to be monitored by the United States Environmental Protection Agency (USEPA). These include naphthalene, anthracene, benzo (a) pyrene and benzo (ghi) perylene.
[b]Expressed as the sum of 44 of the most commonly occurring PCBs.
[c]Reported as the sum of all tetra- through octachlorinated congeners.
[d]Measured as the sum of all congeners present in technical toxaphene.
[e]Equivalent to $7000 \ ng \ m^{-3}$ at STP.
[f]Data on a whole milk basis.
[g]Data not available. Human metabolism of PAH considered too fast.
[h]Level in whole blood.

[14]L. A. Wallace in 'Chemistry and Analysis of Volatile Organic Compounds in the Environment', eds. H. J. Th. Bloemen and J. Burn, Blackie Academic and Professional, London, 1st edn, 1993, p. 1.
[15]W. A. McClenny in 'Chemistry and Analysis of Volatile Organic Compounds in the Environment', eds. H. J. Th. Bloemen and J. Burn, Blackie Academic and Professional, London, 1st edn, 1993, p. 237.

### 16.3.1 Sampling Methodology

Even using the most sophisticated analytical instrumentation, a measurement will be severely compromised if the sample taken for analysis is unrepresentative. Hence, a good sampling method provides a sample that accurately reflects the levels of the chosen analyte(s) in the matrix studied. It must also provide sufficient sample for the purposes of the study (in this respect, the sensitivity of the analytical instrumentation employed must also be taken into account). Another important consideration for both soil and water sampling, is that the glass containers for sample storage must be thoroughly cleaned prior to use, to minimize potential contamination.

In terms of sampling methodology, the major differences occur according to the nature of the matrix under scrutiny – *e.g.* soil, air or water – rather than on a compound-specific basis. As a result, the following sections will discuss the types of sampling strategies used for the determination of trace levels of organic chemicals in air, soil and water, with allusion to compound-specific refinements where necessary – in particular, the special requirements of VOC sampling.

*16.3.1.1 Air Sampling.* There are two principal categories of 'air' sampling. One involves sampling ambient air, *i.e.* that to which humans and other animals are exposed in their normal working and recreational lives, whilst the other involves the study of the gaseous and/or dust emissions from a pollutant source such as a waste incinerator. For the purposes of this chapter, we shall refer to the two as 'ambient' and 'source' respectively.

A general consideration for all air sampling categories is the monitoring of sampling efficiency – *i.e.* the proportion of analyte retained by the sampler. This may be achieved by the addition of 'sampling standards' to the sampler prior to commencement of the sampling campaign. The fraction remaining at the end of sampling affords a measure of the sampling efficiency.

*SVOC Ambient Air Sampling.* Figure 16.7 shows a schematic of the sampling apparatus used to sample SVOCs in ambient air. The equipment basically comprises a $PM_{10}$-selective impactor head – designed to remove particles above 10 μm diameter – followed by a smaller sampling head housing a filter paper – commonly glass fibre or PTFE – the latter is more expensive, but minimizes analyte reaction on the filter during sampling – followed by one or two cylindrical 'plugs' of polyurethane foam (PUF). SVOCs present in ambient air will exist in both gaseous and aerosol form (vapour and particulate phase respectively). The aerosol fraction is collected by the filter, with the vapour phase adsorbed onto the PUF 'plugs'. Owing to the low levels of SVOCs (particularly PCDD/Fs) in ambient air, very high volumes (100–3000 $m^3$) must be sampled in order to procure measurable quantities of the analyte(s) of interest.

*VOC Ambient Air Sampling.* Only a very small fraction of VOCs exist adsorbed to atmospheric particles. Furthermore, the atmosphere is an important repository for them, with the consequence that their concentrations in air are often elevated by comparison with SVOCs. As a result, VOC sampling in ambient air requires less powerful pumps, and there is little need to separate the aerosol and vapour phases. Small pumps pull air through a small tube (usually metal or glass) packed

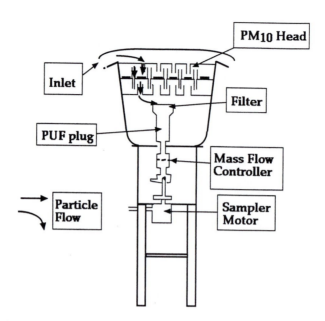

**Figure 16.7**   *Schematic of hi-vol apparatus for sampling SVOCs in ambient air*

with an adsorbent medium – usually fine beads of polymeric resin. The main potential problem with sampling air-borne VOCs, lies with their tendency to volatilize from the sampling medium and thus reduce sampling efficiency. Such losses may be minimized by avoiding excessive sampling rates, by employing the most efficient adsorbent medium for a specific analyte and by cooling the adsorbent trap during sampling. Alternative sampling approaches include so-called 'grab' sampling methods, whereby an air sample is trapped in an inert sampling container, usually either a plastic (Tedlar) bag or a chrome-nickel oxide coated stainless steel canister.

*SVOC Source Sampling.* Although the basic principles of ambient air and source sampling are similar, in that sampled gas is sucked through a sampling head or probe, fitted with a filter and adsorbent material (usually a polymeric resin rather than a PUF plug) to trap aerosol and vapour-phase pollutants respectively, there are some major differences. First, the higher concentration of analytes in the sampled gas stream means that smaller sample volumes are necessary. Second, the disturbance in the flow of the gas stream caused by the insertion of the sampler 'probe', will result in inaccurate sampling of the aerosol fraction, unless sampling is conducted under isokinetic conditions, *i.e.* the velocity at which air is pulled through the sampler must equal the velocity of the sampled gas stream. A detailed account of the theory and practice of isokinetic sampling is beyond the remit of this chapter, and the interested reader is referred elsewhere.[16] Finally, the gas must be sufficiently cool before reaching the filter and adsorbent if acceptable sampling efficiency is to be achieved.

[16]C. A. Pio in 'Handbook of Air Pollution Analysis', eds. R. M. Harrison and R. Perry, Chapman and Hall, London, 2nd edn, 1986, p. 1.

*VOC Source Sampling.* As with ambient air sampling, the principal difference between VOC and SVOC source sampling lies in the fact that VOCs are almost entirely associated with the vapour phase. As a result, the need for isokinetic sampling conditions is far less crucial, as is the requirement to separate the aerosol and vapour phase fractions. 'Grab' sampling, using either Tedlar bag or canister techniques is also widely practised.

*16.3.1.2 Soil Sampling.* Procuring soil samples is relatively straightforward. A cylindrical metal corer of *ca.* 10 cm diameter is sunk into the soil, with the material removed being placed into a clean glass jar, which is sealed to minimize adsorption and volatilization of analytes from and to the air.

An important aspect of any soil sampling campaign is the sampling depth. This will depend on the nature of the analyte (SVOCs like PCDD/Fs, PCBs and PAHs will be less prone to movement through the soil column than VOCs) and on whether the soil has been ploughed or otherwise recently disturbed, as disturbance will enhance analyte distribution throughout the soil column. Typical sampling depths for PCDD/Fs, PCBs and PAHs are 5–10 cm on undisturbed land, as concentrations fall dramatically below such depths. Obtaining a representative sample is also very important and is commonly achieved by combining sampled soil cores from various points in the vicinity of the sampling site. These points are situated along either an 'X' or a 'W' pattern – the latter is illustrated in Figure 16.8 – covering an area typically 10 m by 10 m. The combined samples are homogenized and appropriately sized sub-samples taken for analysis.

*16.3.1.3 Water Sampling.* Hydrophobicity is a common property of all of the chemicals dealt with in this chapter. As a result, the levels present in aqueous media are low (see Table 16.2), and much of the burden of such chemicals associated with 'water' samples, is in fact bound to particulate matter. Consequently, large sample volumes are necessary (typically 2 dm$^3$ for PCDD/Fs) and

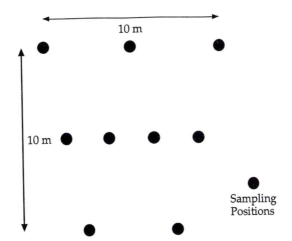

**Figure 16.8** *Schematic of a typical soil sampling strategy*

the particulate fraction must be separated from the aqueous fraction if the true 'dissolved phase' concentration is to be measured. Such separation is carried out by pulling the sample through a fine mesh filter (usually of 0.45 $\mu$m porosity), to trap particle-bound pollutants. The filtrate is subsequently either solvent extracted, or passed through an adsorptive resin-filled tube to concentrate the dissolved phase, which is subsequently eluted with a small volume of solvent.

### 16.3.2   Analytical Methodology

Once sampling is complete, a vast array of analytical techniques are available to the analyst. That chosen depends on both the nature of the matrix (the determination of organic chemicals in sewage sludge is extremely difficult owing to the high levels of potential interferences that are present), the expected concentrations of the target analyte, and the nature of the analyte itself. In this section therefore, the range of available techniques will be examined on a compound-specific basis.

*16.3.2.1   PCDD/Fs.* The universal method of choice for PCDD/F determination is GC/MS – a good introductory guide to which is available.[17] The selectivity afforded by mass spectrometry is essential, given that the levels of potential chemical interferences present in samples will usually be far in excess of those of PCDD/Fs, whilst the requisite sensitivity is partly achieved *via* use of selected ion monitoring (SIM). In general, and particularly for the determination of PCDD/Fs in ambient air – where individual congener concentrations may be as low as a few fg m$^{-3}$ (1 fg $= 10^{-15}$ g) – the use of high resolution mass spectrometers capable of providing 'on-column' detection limits of 25 fg is favoured.

Quantitation of PCDD/Fs is made *via* reference to $^{13}C_{12}$-labelled PCDD/F internal standards added to the sample prior to extraction. These standards – which are identical to the corresponding 'native' PCDD/F except for their mass – are used to correct analyte concentrations for losses throughout the extensive extraction and purification procedures employed prior to GC/MS analysis. The quantification method employed is referred to as 'isotope dilution'.

*16.3.2.2   PCBs.* There are two principal selectivity requirements for the analysis of PCBs. The first, the separation of individual PCB congeners from each other, is crucial, given the congener-specific variations in toxicity, and is achieved by the use of capillary GC. The second selectivity requirement – *viz* the ability to differentiate PCBs from co-eluting interferences – is less pivotal than for PCDD/Fs, given the fact that environmental concentrations of individual PCBs are much greater. As a result, the most common detection technique for PCBs is the electron capture detector (ECD), which offers excellent sensitivity ($\leqslant$1 pg 'on-column'), alongside a reasonable degree of selectivity. Mass spectrometric detection – especially when employed alongside the use of $^{13}C_{12}$-labelled PCB internal standards – offers advantages; *e.g.* in the resolution of congeners that

[17]R. K. Mitchum in 'Instrumental Analysis of Pollutants', ed. C. N. Hewitt, Elsevier Applied Science, Barking, UK, 1st edn, 1991, p. 147.

coelute on the GC column, but possess different numbers of chlorines and hence molecular ions. Mass spectrometric detection also offers enhanced selectivity in terms of differentiating PCBs from co-eluting interferences such as $\beta$-HCH and DDE.

*16.3.2.3 PAHs.* As environmental concentrations of PAHs are generally around two orders of magnitude greater than those of PCBs, the strict sensitivity and selectivity requirements associated with analytical techniques for the determination of the latter are less important. Perhaps the most commonly used technique for PAH analysis is reverse-phase HPLC, in conjunction with variable-wavelength UV and/or fluorescence detection. Other methods are capillary GC-based, with the flame ionization detector (FID) being the most common method of detection. However, the inherent universality of response of FID, means that more specific detectors like ECD and fluorescence are sometimes employed. GC/MS has also found widespread use, with isotopically-labelled internal standards (either deuterated or $^{13}$C-labelled) and SIM, enhancing accuracy and sensitivity.

*16.3.2.4 OCPs.* The almost universally favoured analytical technique for the determination of OCPs, is capillary GC/ECD, although detection of low levels in complex matrices such as sewage sludge, is greatly facilitated by the use of GC/MS. The analytical chemistry of toxaphenes remains a fertile area for research, with current methods centred on GC/ECD and GC/ECNIMS (electron capture negative ion mass spectrometry).

*16.3.2.5 VOCs.* The method of choice for BTEX is capillary GC, interfaced either with FID or MS. Given the volatility of these analytes, solvent extraction from the sample followed by concentration prior to GC analysis is not feasible. As a result, there are a number of techniques for transferring the analytes from the sample to the head of the GC column. The basic principle of each is the same; the sample (or adsorbent tube for air samples) is purged – usually at elevated temperatures – for one–two mins with the GC carrier gas. The analytes volatilized by this process, are introduced to the head of the GC column held at sub-ambient temperatures over the purging period, in order to 'trap' the purgate in a sharp band and optimize chromatographic resolution.

## 16.4 SOURCES

Source apportionment is an important research area, the purpose of which is to identify and rank the major sources of pollutants to the environment. Such ranking tables permit the identification of major release pathways and hence the prioritization of emission control strategies. The basic strategy of a source inventory is to derive an emission factor for a specific source activity (*e.g.* 10 $\mu$g per t of waste burnt) and subsequently to multiply this by an activity factor – *i.e.* the extent to which the activity is practised (*e.g.* 3 million t waste burnt per year). In this way, an estimate of annual pollutant emissions for a specific source can be derived – in the case illustrated, annual emissions would amount to 30 g. The degree of difficulty involved in the construction of a source inventory – and hence its accuracy – varies widely according to the nature of the emission source (*e.g.* quantifying pesticide releases using well-documented industry production figures is more reliable than estimating emission factors for unintentional releases of PAHs from fossil fuel combustion) and, as a result, the accuracy of such source inventories varies widely, with the primary benefit of many proving to be the identification of areas requiring further investigation. The section that follows summarizes present knowledge of the magnitude and sources of current and past releases of toxic organic chemicals.

### 16.4.1 PCDD/Fs

PCDD/Fs have never been intentionally produced – other than on a laboratory scale. Figure 16.9 illustrates the temporal variation of PCDD/F concentrations in archived soil taken from the Rothamsted experimental crop station in the south of England. This study – which covers the period 1844–1986 – shows there to have existed a low, essentially constant 'background' of PCDD/Fs throughout the latter part of the 19th century, after which, there has been a dramatic rise.[18] The inference from this, is that – whilst some natural sources exist, notably forest fires (which may arguably be of significance in some countries) – the origins of the environmental burden of these compounds are principally as by-products of anthropogenic activities, in particular the manufacture and use of organochlorine chemicals (such as chlorophenols) and combustion processes (such as waste incineration and coal combustion).

The mechanisms *via* which PCDD/Fs form during combustion activities are complex, and are still not wholly understood. It appears that the mechanism of formation is *via de novo* synthesis – *i.e.* formation from the basic chemical 'building blocks' of carbon, chlorine, oxygen and hydrogen. In summary, this means that, provided the conditions of temperature *etc.* outlined below are met, PCDD/F formation can occur from the combustion of any 'fuel', providing sources of these

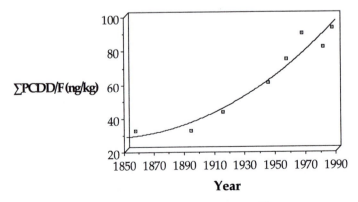

**Figure 16.9**   *Temporal trends in ΣPCDD/F soil concentrations*[18]

[18]L.-O. Kjeller, K. C. Jones, A. E. Johnston and C. Rappe, *Environ. Sci. Technol.*, 1991, **25**, 1619.

four elements are present, although formation is enhanced if levels of so-called 'precursor' compounds like chlorophenols and chlorobenzenes are present in the 'fuel'. The evidence of laboratory experiments and studies on working waste incinerators is that PCDD/F formation during combustion occurs in post-combustion zones – *i.e.* oxygen-rich regions where temperatures are in the region 250–350°C – such as electrostatic precipitators. In these regions, a series of reactions, catalysed by the presence of metal chlorides, occur on the surface of fly ash particles and as a result, PCDD/Fs are formed.

The presence of PCDDs in chlorophenols and products produced *via* chlorophenol intermediates (such as chlorophenoxy acetic acid derivatives; one of the principal constituents of 'Agent Orange' – a defoliant used in the Vietnam War, and heavily contaminated with 2,3,7,8-TCDD – was the *n*-butyl ester of 2,4,5-trichlorophenoxyacetic acid) is more easily explained. PCDD formation occurs *via* the inadvertent reaction of two chlorophenol molecules (or chlorophenate ions), which yields PCDD(s), the exact identity(ies) of which are dependent on the chlorination pattern of the reactants and the possibility of Smiles rearrangement products, as illustrated in Figure 16.10.

It should be noted that there exist many other known sources, though a dearth of data prevents estimates of emissions from these. Indeed, the continuing discovery of fresh sources in recent years (including metal smelting and chlorine bleaching of paper pulp), has made it clear that much work remains to be done before the origins of the present environmental burden of PCDD/Fs are fully clarified. A further illustration of this, is that source inventories of atmospheric emissions of PCDD/Fs conducted in the UK[19] and Sweden,[20] have both failed to

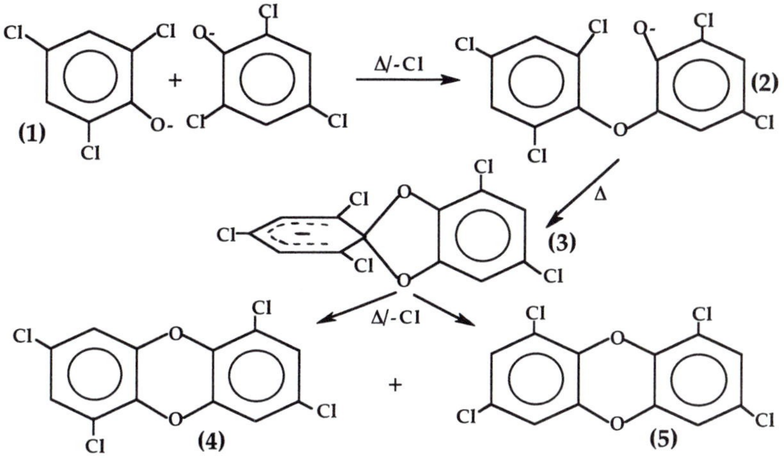

**Figure 16.10**   *Self-condensation of 2,4,6-trichlorophenate (1) via intermediate species (2) and (3) and Smiles rearrangement of (3) to give 1,3,6,8-TCDD (4) and 1,3,7,9-TCDD (5)*

[19]S. J. Harrad and K. C. Jones, *Sci. Total Environ.*, 1992, **26**, 89.
[20]C. Rappe, *Organohalogen Comp.*, 1990, **4**, 33.

account for more than 10–20% of deposition to the surface of these countries. One possible explanation for at least part of this discrepancy is the potentially significant contribution arising from recirculation of the existing environmental burden; specifically, the redeposition of PCDD/Fs released from topsoil (which bears >95% of the total UK burden) *via* both volatilization and soil erosion.

### 16.4.2 PCBs

In stark contrast to PCDD/Fs, the origin of the environmental burden of PCBs, is almost exclusively their deliberate manufacture – primarily for use as dielectric fluids in electrical transformers and capacitors, but also for use in *inter alia*, carbon-less copy papers and inks. Industrial manufacture commenced in the USA in 1929 – UK production in 1954 – and between then and the late 1970s, when production (although not use of existing stocks) ceased in most western nations, an estimated total of 1.2 million t were produced – 67 000 t in the UK. Figure 16.11 shows the temporal variations in PCB levels in soils over the period 1942 to 1992, from the so-called 'Woburn Market Garden' experiment. This study[21] revealed UK PCB levels to have essentially reflected temporal trends in use, with the restrictions on such use producing a precipitous decline in soil concentrations. The principal loss mechanism from soils – which has been estimated to hold >90% of the UK's burden of PCBs – is considered to be volatilization – degradation in soils is known to be slow – and hence it is unsurprising that re-circulation of the existing environmental burden (which includes resuspension of soil particles), has been identified as the major source of current PCB input to the UK. Other significant sources include leaks from PCB-filled transformers and capacitors still in use, the scrapping of redundant PCB-contaminated electrical goods such as refrigerators (fragmentizing) and volatilization of PCBs during the recycling of contaminated scrap metal.

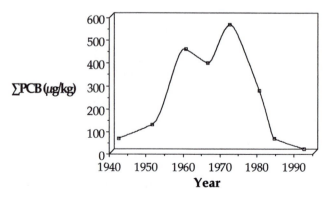

**Figure 16.11**   *Temporal trends in ΣPCB soil concentrations*[21]

[21]R. E. Alcock, A. E. Johnston, S. P. McGrath, M. L. Berrow and K. C. Jones, *Environ. Sci. Technol.*, 1993, **27**, 1918.

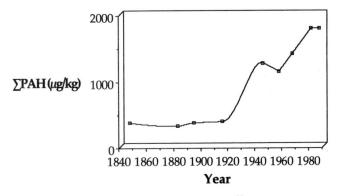

**Figure 16.12** *Temporal trends in ΣPAH soil concentrations*[22]

### 16.4.3 PAHs

Like PCDD/Fs, PAHs have never been intentionally produced. In contrast to PCDD/Fs however, it is widely accepted that natural sources can make a significant contribution to their environmental burden in countries where forest fires and volcanic activity are prevalent. Furthermore, the biogenic formation of perylene in anaerobic sediments is well-established, although the significance of similar mechanisms for other PAHs is unclear. However, the overwhelming bulk of releases to the atmosphere occur *via* incomplete combustion of organic materials, in particular fossil fuels. The major sources to the UK environment are domestic coal combustion and motor vehicle fuel combustion, with the volatilization of the existing burden in soils (especially those from contaminated sites – *e.g.* from disused coal gasworks) and from creosote-treated timber considered likely to play important rôles. The relative importance of anthropogenic and natural sources of environmental PAH are illustrated by Figure 16.12, which shows the variation in PAH levels over the period 1846–1986 in soils sampled from the Rothamsted experimental station.[22]

### 16.4.4 OCPs

Sources of organic pesticides in the environment are entirely anthropogenic. Production of Lindane, 'purified' and 'technical' HCH commenced during World War II and, despite the introduction of use restrictions in the last 25 years, they remain in use throughout the globe. It is difficult to accurately estimate production volumes of the toxicologically important γ-HCH, given the fact that, in addition to its widespread European and N. American usage in its pure γ form (lindane), it has found significant use in the Southern Hemisphere in the form of 'purified' HCH (60–98.9% γ-HCH) and 'technical' HCH, in which the proportion of γ-HCH ranges between 10 and 16%. However, cumulative

[22]K. C. Jones, J. A. Stratford, K. S. Waterhouse, E. T. Furlong, W. Giger, R. A. Hites, C. Schaffner and A. E. Johnston, *Environ. Sci. Technol.*, 1989, **23**, 95.

worldwide use of HCH products since their introduction (as insecticides and as a wood preservative), has been put in excess of $10^6$ t. The family of toxaphene compounds are also particularly worthy of mention, as, in 1971, technical toxaphene accounted for 80% of all chlorinated cyclodiene insecticides still used at that point by US farmers, chiefly to control cotton pests. Despite being banned in the USA in the early 1980s – up to which point, at least $3 \times 10^5$ t had been used in the USA alone – it is still manufactured and used in developing countries. Estimates of cumulative worldwide toxaphene use amount to a minimum of $5 \times 10^5$ t, although the true figure may be much higher, given the fact that between 1955 and 1960 Egypt alone used 54 000 t.

For DDT, figures relating to world production are only very approximate, but an estimate in excess of $3 \times 10^6$ t would not seem unreasonable. Global production of aldrin and dieldrin will be lower than for the other pesticides considered here, but is still significant, particularly in some countries, where their use may have exceeded that of DDT.

### 16.4.5   VOCs

Environmental inputs of benzene, toluene, ethyl benzene and the isomeric xylenes, are predominantly anthropogenic. The principal sources of BTEX include chemical industry uses such as solvents, oil refineries, fuel oil/petrol combustion for industrial, transport and domestic purposes, the production and use of paints and glues and emissions from coke production facilities. Estimated annual UK emissions of BTEX are currently as follows: benzene 13 000 t; toluene 69 000 t; xylenes 46 000 t and ethyl benzene 4000 t.[23] With regard to emissions from motor vehicles; BTEX compounds are released as unburnt components of the fuel and, additionally, as products from thermolytic dealkylation of higher molecular weight monoaromatic compounds. Other sources associated with the use of motor vehicles include refuelling of both individual vehicles and petrol stations.

### 16.5   IMPORTANT PHYSICOCHEMICAL PROPERTIES AND THEIR INFLUENCE ON ENVIRONMENTAL BEHAVIOUR

The following sections deal with the influence of a number of important physicochemical properties of toxic organic chemicals on their environmental behaviour; in particular, their availability for uptake by plants and animals, routes of human exposure, their relative partitioning between different environmental compartments and their transport throughout the environment. Note that the practical difficulties† of accurately measuring such properties, has

---

[23] N. R. Passant, 'Emissions of Volatile Organic Compounds from Stationary Sources in the UK', Warren Spring Laboratory Report LR990, Warren Spring Laboratory, Stevenage, UK, 1993.

†To illustrate, accurate measurement of $K_{ow}$ for highly hydrophobic compounds like PAHs, PCBs and PCDD/Fs is extremely difficult. Measurements of pollutant concentration in the octan-1-ol phase are relatively simple, but accurate determination of the extremely low levels in the water layer is severely hampered by the difficulty in achieving complete separation of the two layers. Given the much higher

resulted in considerable uncertainty surrounding values reported for the compounds studied here. As a result, a table of definitive values for physicochemical properties is not included, and selected values are instead quoted to illustrate general trends, and to indicate the order of magnitude of these properties for individual pollutants.

### 16.5.1 Log $K_{ow}$

Strictly, $K_{ow}$ is a measurement of the partition coefficient of a compound between water and octan-1-ol. The significance of this – at first sight, seemingly rather obscure – parameter is greater than might be expected, as it approximates to the lipid-water partition coefficient ($K_{lw}$). As a result, $K_{ow}$ (the log form is most commonly quoted for convenience) is considered an expression of a compound's hydrophobicity, with implications for its bioaccumulative tendencies and movement through the food chain. It is important to note that the 'solubility' of organic chemicals in octan-1-ol only varies within a very narrow range (*i.e.* 200 to 2000 mol m$^{-3}$). As a consequence, $K_{ow}$ for an organic chemical is predominantly dependent on water solubility, and the common assumption that $K_{ow}$ is an expression of lipophilicity is misleading, as most organic chemicals possess an equal affinity for lipids, whilst having very different affinities for water. As a general rule, $K_{ow}$ increases with increasing molecular weight for a given class of compounds and hence, OCDD (8.2) and benzo(a)pyrene (6.3) have higher log $K_{ow}$ values than 2,3,7,8-TCDD (6.8) and naphthalene (3.6) respectively. Log $K_{ow}$ values are significantly lower for BTEX, lying between 2.1 for benzene and 3.2 for ethyl benzene.

### 16.5.2 Log $K_{oc}$

The exact definition of $K_{oc}$ is the partition coefficient of a chemical between water and natural organic carbon. The wider implications of this parameter (which has units of dm$^3$ kg$^{-1}$), is as an indication of the tendency of a chemical to sorb to organic matter. This can have important consequences for – amongst other things – a chemical's ability to leach through soil columns, its availability for uptake by plants and animals from soil and sediments, and its tendency to volatilize from soil. As a general rule for a particular group of chemicals, $K_{oc}$ increases with increasing molecular weight and hence, OCDD and benzo(a)pyrene have higher $K_{oc}$ values than 2,3,7,8-TCDD and naphthalene respectively. As a very general approximation, $K_{oc} = 0.41 K_{ow}$, although the dependence of $K_{oc}$ on the chemical form of organic carbon must not be overlooked, and the margin of error associated with this approximation is about a factor of two.

---

levels present in the octan-1-ol phase, the presence of even a very small fraction of this phase in the aqueous layer, will lead to an exaggeration of the pollutant concentration in the latter. To illustrate, if 50 $\mu$g of a substance with a log $K_{ow}$ of 8, is partitioned between 1 dm$^3$ of both octan-1-ol and water, the equilibrium concentrations in the two layers will be 50 $\mu$g dm$^{-3}$ and 0.5 pg dm$^{-3}$ respectively. Inadvertent contamination of the aqueous layer with just 1 $\mu$l of the organic phase will raise the apparent aqueous phase concentration to 50.5 pg dm$^{-3}$, giving rise to an apparent log $K_{ow}$ value of 6.

### 16.5.3  Aqueous Solubility

As would be expected, an apolar organic compound's aqueous solubility is inversely related to its $K_{ow}$ and $K_{oc}$ and a good approximation is that solubility at neutral pH and a given temperature, will decrease with increasing molecular weight for a given class of compounds. To illustrate, the water solubilities at 25 °C of benzene, *o*-xylene, 2,3,7,8-TCDD and 1,2,3,4,7,8-HxCDD are, respectively: 0.02, $1.7 \times 10^{-3}$, $3.5 \times 10^{-8}$ and $3.2 \times 10^{-9}$ mol dm$^{-3}$. The environmental significance of this property is obvious; not least because more water-soluble chemicals can enter humans *via* water consumption, as they will be more susceptible to leaching through soil into groundwater supplies.

### 16.5.4  Environmental Persistence

An exact measure of environmental persistence is impossible to define, as persistence will vary tremendously between different environmental media and there are so many confounding variables. Even if environmental persistence is arbitrarily defined as the half-life in soil, measurements of this one parameter vary wildly, depending on the binding of the compound to the soil (which is, in turn, dependent on factors like soil organic content, available surface area, log $K_{oc}$ and soil moisture content) and the depth of incorporation below the surface (this influences the intensity of UV radiation received by the chemical and hence its photolysis rate). Even where a definitive half-life can be obtained for a compound, there are grave difficulties in reliably assessing the relative importance of each contributory loss mechanism – *e.g.* bacterial degradation, photolysis, leaching, chemical reaction, uptake by biota and volatilization. Despite these caveats, knowledge of the range of a compound's persistence is extremely important in assessing its environmental impact, not least because a longer soil half-life will allow more time for the chemical to transfer into the food chain, and elicit any adverse effects in biota. Illustrative values for soil half-lives range from a few days for BTEX, to months for lower molecular weight PAHs and OCPs and up to 10 years or more for PCDD/Fs, higher molecular weight PAHs and PCBs.

### 16.5.5  Henry's Constant

Henry's Constant ($H_c$ is essentially a measure of a compound's tendency to partition between a solution and the air above it and can be expressed in different ways. Its dimensionless form $H_c'$, relates the concentration of a chemical in the gas phase $C_{sg}$ to its concentration in the liquid phase $C_{sl}$,

$$H_c' = C_{sg}/C_{sl} \qquad (1)$$

Henry's Constant can also be written thus,

$$H_c = P_{vp}/S \qquad (2)$$

where $P_{vp}$ is the vapour pressure of the chemical (in Pa) and $S$ the aqueous solubility or saturation concentration in mol m$^{-3}$. H$_c$ therefore has units of Pa m$^3$ mol$^{-1}$ and can be related to H$_c'$ by,

$$H_c' = H_c/RT \qquad (3)$$

where $R$ is the gas constant (8.31 Pa m$^3$ mol$^{-1}$ K$^{-1}$) and $T$ is the temperature (K).

The utility of Henry's Constant is that it describes the tendency of a chemical to volatilize from water/plant/soil surfaces into the atmosphere, and compounds with a high H$_c$ value are more prone to volatilization from aqueous solutions. In practical environmental terms, knowledge of H$_c$ affords an insight into the movement of a chemical into and out of aquatic ecosystems such as ponds, lakes or oceans. Examples of H$_c$ values (expressed as Pa m$^3$ mol$^{-1}$) at 25 °C are: benzene (550), 2,2',3,3'-tetrachlorobiphenyl (22), 2,2',4,4',5,5'-hexachlorobiphenyl (43), 2,3,7,8-TCDD (3.3) and OCDD (0.68).

### 16.5.6 Vapour Pressure

As a general rule, the vapour pressure of a chemical decreases with increasing molecular weight and hence decachlorobiphenyl possesses a lower vapour pressure (5 × 10$^{-8}$ Pa) than 2,2',3,3'-tetrachlorobiphenyl (2.3 × 10$^{-3}$ Pa). Other illustrative measurements of vapour pressure are: toluene (3800 Pa), *p*-xylene (1200 Pa), 2,3,7,8-TCDD (2 × 10$^{-7}$ Pa) and OCDD (1.1 × 10$^{-10}$ Pa). As with all of the other properties dealt with in this section, the influence of vapour pressure on environmental behaviour cannot be considered in isolation, and it is strongly dependent on environmental factors such as temperature. The environmental significance of vapour pressure is tremendous; *inter alia* it influences a compound's partitioning between air and other environmental compartments, thus affecting its rate of atmospheric transport, availability for uptake by biota and susceptibility to photolytic degradation (compounds spending more time in the vapour phase are generally more susceptible to this). Vapour pressure also exerts a powerful influence on the relevance of inhalation as a human exposure route. This pathway is far more significant for BTEX compounds than for any of the SVOCs considered here, in particular the higher molecular weight species, for which inhalation makes only a negligible contribution to human exposure.

### 16.5.7 General Comments

The combined influence of the above physicochemical properties of individual chemicals, together with environmental factors such as meteorology (temperature, rainfall) and other factors like soil properties *etc.*, governs the environmental fate and behaviour of such chemicals in a predictable fashion. Thus, detailed knowledge of a compound's physicochemical properties and the range of prevailing environmental conditions which it may experience, affords important insights into how an individual compound may behave once released into the environment.

For example, most SVOCs display common characteristics in their environmental behaviour. They reside primarily in soil, accumulate through the food chain, and their principal route into humans is *via* food, specifically *via* the consumption of animal fats. In addition, they are capable of volatilizing from soils and water and subsequently undergoing long-range atmospheric transport either in the vapour phase or bound to aerosol. It is *via* such atmospheric transport, which occurs in a repeated cycle sometimes referred to as 'the grasshopper effect', that these chemicals distribute in an ubiquitous fashion. This universal distribution is illustrated by the presence of SVOCs in polar regions, which is a matter of concern, for although inputs to such locations are small, the low temperatures minimize volatilization losses, leading to a steady accumulation of the overall pollutant burden. This accumulation is compounded by the reduced biomass to surface area ratio of polar regions, with the result that the pollutant burden is distributed amongst a much smaller biomass than in industrialized areas. These factors account for the observations of PCB concentrations in the breast milk of Canadian Inuit mothers, that exceed those found in women from urban Canada.

## 16.6   MODELLING ENVIRONMENTAL BEHAVIOUR

The environmental behaviour of a chemical under uniform meteorological and other environmental conditions, is largely predictable in terms of a relatively small number of physicochemical properties. Without doubt, the most challenging aspect of predicting the environmental behaviour of a chemical lies in accounting for the uncertainties introduced by fluctuations in meteorological and other environmental parameters. For example, a hot, still and sunny day will result in higher vapour-phase concentrations, with concomitant susceptibility to photolytic degradation (enhanced by the greater sunlight intensity), greater water solubility resulting in greater leaching and enhanced uptake by aquatic biota, as well as more intensive biotic metabolism, leading to greater chemical uptake by animals and plants. Similarly, the influence of both the magnitude and chemical composition of the organic content of a soil will exert a profound influence on the environmental fate of a given pollutant. To illustrate, PCBs present in peat-like soils will bind much more strongly to the soil than if they were present in a sandy soil. As a result, movement of pollutants from peat soils – either by volatilization or leaching – is minimal; there is also evidence to suggest that uptake of pollutants from such soils by biota is similarly reduced – in other words, the 'bioavailability' of the chemical is diminished.

These simple examples, serve as illustrations of the complex requirements that a successful environmental modelling technique must meet. Whilst a completely satisfactory model has yet to be developed – there are too many real-world variables for that – the work of Prof. Donald Mackay and co-workers at the University of Toronto has provided an excellent framework within which the environmental fate of organic contaminants may be predicted with some

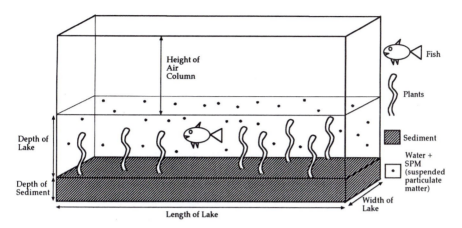

**Figure 16.13**  *A typical 'Model World' scenario*

accuracy. His approach is essentially to construct a 'model world'; for example a small lake of defined dimensions and with clearly defined values for properties like suspended particulate matter (SPM), the organic content of such matter, and the 'volume' of biota like fish and plants that are present. Assumptions must also be made concerning the depth and physical properties – such as organic content – of lake sediment, and the height of the air layer above the lake surface deemed relevant. Figure 16.13 illustrates such a 'model world'.

Combined with data on the 'model world' dimensions and composition, are mass loadings and calculated values of fugacity for the compound(s) of interest; fugacity essentially describing the potential for a substance to move from one environmental compartment to another. Incorporation of all these data into a single computational package, allows the construction of a reasonably realistic model, capable of predicting the distribution of a chemical throughout the various environmental compartments of the 'model world'. The model can be further enhanced by inclusion of transport factors relating to processes such as leachability and volatilization, along with reaction properties to provide information on loss processes such as biodegradation and hydrolysis, and this basic approach can be modified to consider smaller or larger 'model worlds' as appropriate, in whatever degree of detail necessary. To illustrate, the concept has been used to assess the global fate and distribution of toxaphenes – dividing the globe into nine separate 'model worlds' such as the Arctic, the Antarctic, Northern Hemisphere temperate, tropical *etc.* and modelling chemical distribution within and between each of the 'model worlds'.[24] One use for this global model has been to assess the relative propensity for chemicals to transfer from temperate zones and 'condense out' at the polar regions. On a smaller scale, the approach can be utilized to predict chemical movement within a soil column following amelioration with sewage sludge, and assess the potential for

[24] F. Wania and D. Mackay, *Chemosphere*, 1993, **27**, 2079.

leaching to groundwater or uptake by crops. In conclusion, it is important to note that although such models are extremely useful in comparing the behaviour of different chemicals under identical environmental conditions, they are not sufficiently reliable to provide accurate predictions of actual concentrations.

CHAPTER 17

# Radioactivity in the Environment

C. N. HEWITT

## 17.1 INTRODUCTION

Pollution of the natural environment by radioactive substances is of concern because of the considerable potential that ionizing radiation has for damaging biological material. Although both the benefits of controlled exposure for medical purposes and the catastrophic effects of large doses of radiation (for example, those received by the inhabitants of Hiroshima and Nagasaki in 1945 and in the vicinity of Chernobyl in 1986) are well understood, what is less clear are the effects of small doses on the general population. For this reason it is necessary to evaluate in detail the exposure received from each of the multiplicity of natural and man-made sources of radioactivity in the environment. In order to do this, an understanding is required of the actual and potential source strengths, the pathways and cycling of radioactivity through the environment and their flux rates and the possible routes of exposure for man.

## 17.2 RADIATION AND RADIOACTIVITY

### 17.2.1 Types of Radiation

Radiation arises from a spontaneous rearrangement of the nucleus of an atom. Whilst some nuclei are stable many are not and these can undergo a change, losing mass or energy in the form of radiation. Some unstable nuclei are naturally occurring while others are produced synthetically. The most common forms of these radiations are alpha particles, beta particles and gamma rays, the physical properties of which are shown in Table 17.1 and are described below.

(a)   Alpha ($\alpha$) particles consist of two protons and two neutrons bound together and so are identical to helium nuclei. They have a mass number of 4 and a change of $+2$. Because they are so large and heavy, alpha particles travel slowly compared with other types of radiation (at maximum, about 10% the speed of light) and can be stopped relatively easily. Their ability to penetrate into living tissue is therefore limited and

**Table 17.1**   *Types of radioactive particles and rays*

| Radiation | Symbol | Composition | Charge | Mass number | Approximate tissue penetration (cm) |
|---|---|---|---|---|---|
| Alpha | α | particle containing two protons and two neutrons | 2+ | 4 | 0.01 |
| Beta | β | particle of one electron | 1− | 0 | 1 |
| Gamma | γ | very short wavelength electromagnetic radiation | 0 | 0 | 100 |

damage occurs to animals only when alpha-emitting isotopes are ingested or inhaled. However, their considerable kinetic energy and the double positive charge which attracts and pulls away electrons from atoms belonging to tissue means that alpha particles can cause the formation of ions and free radicals and hence severe chemical change along their path. The emission of alpha particles is common only for nuclides of mass number greater than 209 and atomic number 82, nuclides of this size having too many protons for stability. An example of α-decay is:

$$^{210}_{84}\text{Po} \rightarrow {}^{206}_{82}\text{Pb} + {}^{4}_{2}\text{He}$$

Note that the sum of the superscripts (mass numbers or sum of neutrons and protons) and the sum of the subscripts (atomic numbers or sum of protons) remain unchanged during the decay.

(b) Beta ($\beta$) particles are simply electrons emitted by the nucleus during the change of a neutron into a proton. They have minimal mass ($1.36 \times 10^{-4}$ times that of an alpha particle) but high velocity (typically 40% the speed of light) and a charge of $-1$. They may penetrate through skin or surface cells into tissue and may then pass close to the orbital electrons of tissue atoms where the repulsion of the two negative particles may force the orbital electron out of the atom, ionizing the tissue and forming radicals. Because beta decay involves the change of a neutron into a proton the atomic number of the nuclide increases by one, but the mass number does not change. Beta decay is a common mode of radioactive disintegration and is observed for both natural and synthetic nuclides. An example is:

$$^{90}_{38}\text{Sr} \rightarrow {}^{90}_{39}\text{Y} + {}^{0}_{-1}\text{e}$$

(c) Gamma ($\gamma$) radiation is very short wavelength electromagnetic radiation. It travels at the speed of light, is uncharged, but is highly energetic and so has considerable penetration power. As it passes through biological tissue the electric field surrounding a gamma ray may eject orbital electrons from atoms and so can cause ionization of the tissue and formation of radicals along its path. The emission of gamma radiation does not lead to changes in mass or atomic numbers and may occur either on its own from an electronically excited nucleus or may accompany other types of radioactive decay. Examples of these two processes are:

$$^{125}_{52}Te^* \rightarrow \,^{125}_{52}Te + \gamma$$

the asterisk signifying the excited state of the nucleus and:

$$^{137}_{55}Cs \rightarrow \,^{137}_{56}Ba + \,^{0}_{-1}e \rightarrow \,^{137}_{56}Ba + \gamma$$

## 17.2.2 The Energy Changes of Nuclear Reactions

The energy changes associated with nuclear reactions are considerably greater than those associated with ordinary chemical reactions. The sum of the mass of the products of the nuclear reaction is invariably less than the sum of the mass of the reactants and the amount of energy released ($\Delta E$) is equivalent to this difference in mass ($\Delta m$). Most of this energy is released as kinetic energy, although some may be used to promote the nucleus to an excited state from where it will lose energy in the form of radiation and return to the ground state.

The energy equivalent of a given mass can be calculated by means of Einstein's equation:

$$\Delta E = (\Delta m)c^2$$

The energy equivalent of one atomic mass unit (1 u $= 1.660566 \times 10^{-27}$ kg) is:

$$\Delta E = (1.660566 \times 10^{-27} kg)(2.99792 \times 10^8 \text{ m s}^{-1})^2$$
$$= 1.49244 \times 10^{-10} \text{ J}$$

This is usually expressed in units of electron volts (eV) where:

$$1 \text{ eV} = 1.60219 \times 10^{-19} \text{ J}$$
$$\text{and} \quad 1 \text{ MeV} = 1.60219 \times 10^{-13} \text{ J}$$

The energy equivalent of 1 u is therefore 931.5 MeV.

The amount of energy released by a decay process can now be calculated. For example, in the alpha decay of polonium-210:

$$^{210}_{84}Po \rightarrow \,^{206}_{82}Pb + \,^{4}_{2}He$$

$$\Delta m = (\text{mass } ^{210}_{84}Po) - (\text{mass } ^{206}_{82}Pb + \text{mass } ^{4}_{2}He)$$
$$= 209.9829 \text{ u} - (205.9745 \text{ u} + 4.0026 \text{ u})$$
$$= 0.0058 \text{ u}$$

The energy released by the decay is therefore:

$$\Delta E = 0.0058\text{u} \times 931.5 \text{ MeV/u} = 5.4 \text{ MeV}$$

### 17.2.3 Rates of Radioactive Decay

The rates of decay of radioactive nuclides are first-order and independent of temperature. This implies that the activation energy of radioactive decay is zero and that the rate of decay depends only on the amount of radioactive substance present. If $N$ is the number of atoms present at time $t$ the rate of change of $N$ is given by:

$$\mathrm{d}N/\mathrm{d}t = -\lambda N$$

where $\lambda$ is the characteristic (or disintegration or decay) constant for that radionuclide. Integrating between times $t_1$ and $t_2$ gives:

$$N_2 = N_1 \exp\left[-\lambda(t_2 - t_1)\right]$$

where $N_1$ and $N_2$ are the number of atoms of the radionuclide present at times $t_1$ and $t_2$ respectively. If $t_1$ is set to zero then:

$$N = N_0 \exp\left(-\lambda t\right) \tag{1}$$

where $t$ is the elapsed time and $N_0$ is the number of atoms of the radionuclide present when $t_1 = 0$.

When $N/N_0$ is equal to 0.5 (*i.e.* half the atoms have decayed away) $t$ is defined as being the half-life, $t_{1/2}$. Then:

$$N/N_0 = 0.5 = \exp\left(-\lambda t_{1/2}\right)$$
$$t_{1/2} = 0.693/\lambda$$

Equation (1) can now be written in terms of the more readily available $t_1$, rather than $\lambda$, to give:

$$N = N_0 \exp\left(-0.693 t_1/t_{1/2}\right).$$

The half-lives of some selected radionuclides are given in Table 17.2.

### 17.2.4 Activity

The amount of radiation emitted by a source per unit time is known as the activity of the source, expressed in terms of the number of disintegrations per second. The unit of activity, the becquerel (Bq), is defined as being one disintegration per second. The activity of a source is proportional to the number of radioactive atoms present and so diminishes with time according to first order kinetics.

**Table 17.2** *Half-lives of some environmentally important radionuclides*

| Radionuclide | half-life |
|---|---|
| $^{131}I$ | 8.1 d |
| $^{85}Kr$ | 10.8 yr |
| $^{3}H$ | 12.3 yr |
| $^{90}Sr$ | 28 yr |
| $^{137}Cs$ | 30 yr |
| $^{239}Pu$ | $2.4 \times 10^4$ yr |
| $^{238}U$ | $4.5 \times 10^9$ yr |

## 17.2.5 Radioactive Decay Series

Some radioactive decay processes lead in one step to a stable product but frequently a disintegration leads to the formation of another unstable nucleus. This can be repeated several times, producing a radioactive decay series which only terminates on the formation of a stable nuclide. There are three naturally occurring decay series, headed by $^{232}Th$, $^{238}U$ and $^{235}U$. Each of these nuclides has a half-life which is long in relation to the age of the earth and each finally produces stable isotopes of lead, $^{208}Pb$, $^{206}Pb$ and $^{207}Pb$ respectively. The 14 steps of the $^{238}U$ series are shown in Figure 17.1 and it will be seen that at several points branching occurs as the series proceeds by two different routes which rejoin at a later point.

## 17.2.6 Production of Artificial Radionuclides

The first artificial transmutation of one element into another was achieved in 1919 when Ernest Rutherford passed $\alpha$-particles (produced by the radioactive decay of $^{214}Po$) through nitrogen:

$$^{14}_{7}N + ^{4}_{2}He \rightarrow ^{17}_{8}O + ^{1}_{1}H$$

Subsequently the first artificial radioactive nuclide, $^{30}P$, was produced:

$$^{27}_{13}Al + ^{4}_{2}He \rightarrow ^{30}_{15}P + ^{1}_{0}n \text{ (neutron)}$$
$$^{30}_{15}P \rightarrow ^{30}_{14}Si + ^{0}_{1}e \text{ (positron)}$$

Bombardment reactions of this type have been used to produce isotopes of elements which do not exist in nature and which are particularly important as pollutants of the environment, for example:

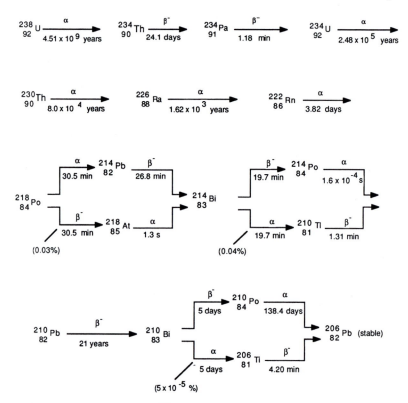

**Figure 17.1**   *Disintegration series of $^{238}_{92}U$ (half-lives of isotopes are indicated)*

$$^{238}_{92}U + ^{1}_{0}n \rightarrow ^{239}_{92}U + \gamma \qquad \text{uranium}$$

$$^{239}_{92}U \rightarrow ^{239}_{93}Np + ^{0}_{-1}e \qquad \text{neptunium}$$

$$^{239}_{93}Np \rightarrow ^{239}_{94}Pu + ^{0}_{-1}e \qquad \text{plutonium}$$

$$^{239}_{94}Pu + ^{2}_{1}H \rightarrow ^{240}_{95}Am + ^{1}_{0}n \qquad \text{americium}$$

### 17.2.7   Nuclear Fission

Many heavy nuclei with mass numbers greater than 230 are susceptible to spontaneous fission, or splitting into lighter fragments, as a result of the forces of repulsion between their large number of protons. Fission can also be induced by bombarding heavy nuclei with projectiles such as neutrons, alpha particles or protons. When the fission of a particular nuclide takes place the nuclei may split in a variety of ways, producing a number of products. For example, some of the possible fission reactions of $^{235}U$ are:

**Table 17.3** *Yields of some long-lived radionuclides following uranium and plutonium fission*

| Radioisotope | Half-life | Yield (%) $^{235}U$ | $^{239}Pu$ |
|---|---|---|---|
| Strontium-90 | 28 y | 5.8 | 2.2 |
| Iodine-131 | 8 d | 3.1 | 3.8 |
| Caesium-137 | 30 y | 6.1 | 5.2 |
| Krypton-85 | 10.3 y | 0 | — |
| Cerium-141 | 33 d | 6.0 | 5.2 |

$$^{235}_{92}U + ^1_0n \rightarrow ^{95}_{39}Y + ^{138}_{53}I + 3^1_0n$$

$$^{235}_{92}U + ^1_0n \rightarrow ^{97}_{39}Y + ^{137}_{53}I + 2^1_0n$$

$$^{235}_{92}U + ^1_0n \rightarrow ^{90}_{36}Kr + ^{144}_{56}Ba + 2^1_0n$$

$$^{235}_{92}U + ^1_0n \rightarrow ^{90}_{35}Br + ^{143}_{57}La + 3^1_0n$$

In each of these reactions neutrons are formed as primary products and these neutrons can, in turn, cause the fission of other $^{235}U$ nuclei, so causing a chain reaction of the fissile uranium.

The number of nuclei of a particular daughter product formed by the fission of 100 parent nuclei is defined as the yield of the process. Fission yields vary, fission induced by slow neutrons (as in a nuclear power plant) having a different set of yields to that induced by fast neutrons (as in many nuclear weapons). Table 17.3 shows examples of the percentage yields of the fast-neutron induced fission of $^{239}Pu$. It also shows some of the yields of $^{235}U$ fission, which are similar for both slow and fast neutron induced processes.

## 17.2.8 Beta Decay of Fission Products

The primary daughter products of nuclear fission are almost always $\beta$-radioactive and often these also quickly produce $\beta$-radioactive products. Only after several such decays are products with long half-lives formed. For example the fission of $^{235}U$ produces $^{90}Br$:

$$^{235}_{92}U + ^1_0n \rightarrow ^{90}_{35}Br + ^{143}_{57}La + 3^1_0n$$

which quickly decays to $^{90}Kr$ with a half-life of 1.4 seconds. This in turn has a half-life of 33 seconds and decays to $^{90}Rb$ ($t_{1/2} = 2.7$ minutes), and this to $^{90}Sr$:

$$^{90}_{35}Br \rightarrow ^{90}_{36}Kr + ^0_{-1}e \quad (t_{1/2} = 1.4 \text{ s})$$

$$^{90}_{36}Kr \rightarrow ^{90}_{37}Rb + ^0_{-1}e \quad (t_{1/2} = 33 \text{ s})$$

$$^{90}_{37}Rb \rightarrow ^{90}_{38}Sr + ^0_{-1}e \quad (t_{1/2} = 2.7 \text{ min})$$

Strontium-90 decays to yttrium-90:

$$_{38}^{90}\text{Sr} \rightarrow\ _{39}^{90}\text{Y} +\ _{-1}^{0}\text{e} \quad (t_{1/2} = 28 \text{ years})$$

and has a half-life of 28 years, which is sufficiently long for it to be widely circulated through the environment, in contrast to its short-lived precursors. After one more beta decay a stable product, $^{90}\text{Zr}$, is formed. It is therefore radioactive strontium rather than its precursors that is the important environmental pollutant in this series.

Examples of other environmentally-important nuclides formed by beta decay chains are iodine-131 and caesium-137. $^{131}\text{I}$ is formed from $^{131}\text{Sn}$, itself produced by the fission of $^{235}\text{U}$ and forms the stable nuclide $^{131}\text{Xe}$:

$$_{50}^{131}\text{Sr} \rightarrow\ _{51}^{131}\text{Sb} +\ _{-1}^{0}\text{e} \quad (t_{1/2} = 3.4 \text{ min})$$

$$_{51}^{131}\text{Sb} \rightarrow\ _{52}^{131}\text{Te} +\ _{-1}^{0}\text{e} \quad (t_{1/2} = 2.3 \text{ min})$$

$$_{52}^{131}\text{Te} \rightarrow\ _{53}^{131}\text{I} +\ _{-1}^{0}\text{e} \quad (t_{1/2} = 24 \text{ min})$$

$$_{53}^{131}\text{I} \rightarrow\ _{54}^{131}\text{Xe} +\ _{-1}^{0}\text{e} \quad (t_{1/2} = 8 \text{ days})$$

$^{137}\text{Cs}$ begins as the uranium fission daughter $^{137}\text{I}$ and ends as the stable barium isotope, $^{137}\text{Ba}$:

$$_{53}^{137}\text{I} \rightarrow\ _{54}^{137}\text{Xe} +\ _{-1}^{0}\text{e} \quad (t_{1/2} = 24 \text{ s})$$

$$_{54}^{137}\text{Xe} \rightarrow\ _{55}^{137}\text{Cs} +\ _{-1}^{0}\text{e} \quad (t_{1/2} = 3.9 \text{ min})$$

$$_{55}^{137}\text{Cs} \rightarrow\ _{56}^{137}\text{Ba} +\ _{-1}^{0}\text{e} \quad (t_{1/2} = 30 \text{ years})$$

### 17.2.9   Units of Radiation Dose

The amount of biological damage caused by radiation and the probability of the occurrence of damage are directly related to dosage and so before discussing the effects of radiation it is necessary to consider the units of dose.

*17.2.9.1   Absorbed Dose.*   The amount of energy actually absorbed by tissue or other material from radiation is known as the *absorbed dose* and is expressed in a unit called the *gray* (Gy). One gray is equal to the transfer of one joule of energy to one kilogram of material. Table 17.4 shows the relationship of this and other SI units with the older units they replace.

*17.2.9.2   Dose Equivalent.*   Because the physical and chemical properties of α, β and γ radiations vary, equal absorbed doses of radiation do not necessarily have the same biological effects. In order to equate the biological effects of one type of radiation with another it is therefore necessary to multiply the absorbed dose by a quality factor that accounts for the differences in biological damage caused by radioactive particles having the same energy. This is known as the *dose equivalent* and is expressed in units of *sieverts* (Sv). For γ-rays, X-rays and β particles the quality factor equals 1, whereas for α-particles it is 20. The dose

**Table 17.4** *SI and old radiation units*

| Quantity | SI unit | Old unit | Relationship |
|---|---|---|---|
| Activity | becquerel | curie | 1 Ci = 3.7 × 10^{10} Bq |
| Absorbed dose | gray | rad | 1 rad = 0.01 Gy |
| Dose equivalent | sievert | rem | 1 rem = 0.01 Sv |

**Table 17.5** *ICRP risk weighting factors*

| Tissue or organ | Weighting factor |
|---|---|
| Testes and ovaries | 0.25 |
| Breast | 0.15 |
| Red bone marrow | 0.12 |
| Lung | 0.12 |
| Thyroid | 0.03 |
| Bone surfaces | 0.03 |
| Remainder | 0.30 |
| Whole body total | 1.00 |

equivalent resulting from an absorbed dose of 1 Gy of alpha radiation therefore equals 20 Sv.

*17.2.9.3 Effective Dose Equivalents.* Having established the dose equivalent for each tissue $(H_T)$ it is now necessary to weight this to take account of the differing susceptibilities of different organs and types of tissue to damage. For example the testes and ovaries are more easily damaged than are the lungs or bones. The risk weighting factors $(W_T)$ currently recommended by the International Commission for Radiological Protection (ICRP) are shown in Table 17.5. The effective dose equivalent $(H_E)$ for the body is then expressed as the sum of the weighted dose equivalents:

$$H_E = \sum_T H_T W_T$$

*17.2.9.4 Collective Effective Dose Equivalents.* As well as quantifying the effective dose equivalent (or *dose*) received by individuals it is also important to have a measure of the total radiation dose received by a group of people or population. This is the *collective effective dose equivalent* (or *collective dose*) and is obtained by multiplying the mean effective dose equivalent to the group from a particular source by the number of people in that group, to give units of man Sieverts (man Sv).

## 17.3 BIOLOGICAL EFFECTS OF RADIATION

### 17.3.1 General Effects

The formation of ions and free radicals in tissue by radiation and the subsequent chemical reactions between these reactive species and the tissue molecules causes a range of short-term and long-term biological effects. As a frame of reference, acute exposure can cause death, an instantaneously absorbed dose of 5 Gy probably being lethal. The range of effects include cataracts, gastrointestinal disorders, blood disorders including leukaemia, damage to the central nervous system, impaired fertility, cancer, genetic damage and changes to chromosomes producing mutations in later generations. Of the long-term effects of radiation to the exposed generation, cancer and especially leukaemia is probably the most important. The number of excess (or extra) cancers observed in an exposed group compared with a non-exposed control group divided by the product of the exposed group size and the mean individual dose gives the *risk factor*. This is the risk of the effect occurring per unit dose equivalent.

Risk factors have been estimated by the United Nations Scientific Committee on the Effects of Atomic Radiation (UNSCEAR) by studying various groups of exposed people. The risk factor found for fatal leukaemia, for example, is about $20 \times 10^{-3} \, Sv^{-1}$ or about a 1 in 500 chance of dying of leukaemia following an exposure of 1 Sv. When irradiation of the testes or ovaries occurs before the conception of children there is a risk that damage to the DNA may cause hereditary defects in future generations. The current UNSCEAR estimate of the risk factor for serious hereditary damage to humans is about $2 \times 10^{-2} \, Sv^{-1}$, or about 1 in 50 per Sv, with about half of this damage manifesting itself in the first two generations. When fatal cancers and serious hereditary defects in the first two generations are considered together the current ICRP estimate of the risk factor is $1.65 \times 10^{-2} \, Sv^{-1}$. This is based on the assumption that there is no lower threshold of dose below which the probability of effect is zero. In other words it is assumed that any exposure to radiation carries some risk and that the probability of the effect occurring is proportional to the dose. It is worth noting that the historical trend has been for estimates of risk to increase with time, and hence the estimates of acceptable dose have decreased with time.

### 17.3.2   Biological Availability and Residence Times

The rate of uptake of a radionuclide into the body depends upon its concentrations in the various environmental media (air, water, food, dust, *etc.*), the rates of intake of these media and the efficiency with which the body absorbs the nuclide from the media. This latter parameter largely depends upon the physical and chemical properties of the radioactive substance in question. For example, the chemical properties of strontium are similar to those of another Group II element, calcium, which is of vital biological significance. Calcium enters the body to form bones and to carry on other physiological functions and $^{90}Sr$ can readily follow calcium into the bones where it will remain. In the same way the chemical similarity of caesium to potassium, which is present in all cells in the body, allows caesium to be readily transported throughout the entire body. On the other hand the noble gases krypton, argon, xenon and radon, which are all radioactive air pollutants, are not readily absorbed by the body (although they can still contribute to the exposure of an individual by external irradiation and by forming radioactive daughters with different physico-chemical properties).

In any given organism there exists a balance between the intake of an element and its excretion and this controls the concentration of the element in the organism. For radioactive isotopes this balance will include on the debit side the radioactive decay of the isotope. The effective half-life, $t_{eff}$, of a nuclide in an organism, *i.e.* the time required to reduce the activity in the body by half, is a function of both the biological half-life, $t_{biol}$, and the radioactive half-life, $t_{rad}$:

$$t_{eff} = t_{rad}t_{biol}/(t_{rad} + t_{biol})$$

Whilst $t_{rad}$ is fixed for any given nuclide, $t_{biol}$ will vary from species to species and between individuals with age, sex, physical condition and metabolic rate. For $^{90}$Sr which has $t_{rad}$ = 28 years and $t_{biol}$ = about 35 years the effective half-life for man is about 15.5 years. Obviously those nuclides of most concern as pollutants are those present in the environment in the highest concentrations, with the most energetic emissions and with the longest effective biological half-lives.

## 17.4 NATURAL RADIOACTIVITY

### 17.4.1 Cosmic Rays

The earth's atmosphere is continuously bombarded by highly energetic protons and alpha particles emitted by the sun and of galactic origin. Their energies range from about 1 MeV to about $10^4$ MeV, with a flux rate at the outer edge of the atmosphere of about $2 \times 10^7$ MeV m$^{-2}$ s$^{-1}$. These primary particles have two effects. They cause some radiation exposure directly to man, the magnitude of which varies with altitude and latitude and they interact with stable components of the atmosphere causing the formation of radionuclides. Of these, $^3$H and especially $^{14}$C are important to the biosphere, while several of the shorter-lived nuclides (*e.g.*, $^7$Be, $t_{1/2}$ = 53 days; $^{39}$Cl, $t_{1/2}$ = 55 min) have applications as tracers in the study of atmospheric dispersion and deposition processes.

Carbon-14 is formed from atmospheric nitrogen by:

$$^{14}_{7}\text{N} + ^{1}_{0}\text{n} \rightarrow ^{15}_{7}\text{N}$$
$$^{15}_{7}\text{N} \rightarrow ^{14}_{6}\text{C} + ^{1}_{1}\text{p}$$

The $^{14}$C is then oxidized to $CO_2$ and enters the global biogeochemical cycle of carbon, being incorporated into plants by photosynthesis and into the oceans by absorption. Most natural production of tritium is also by the reaction of cosmic ray neutrons (with energy >4.4 MeV) with nitrogen:

$$^{14}_{7}\text{N} + ^{1}_{0}\text{n} \rightarrow ^{12}_{6}\text{C} + ^{3}_{1}\text{H}$$

Tritium is then oxidized or exchanges with ordinary hydrogen to form tritiated water and enters the global hydrological cycle. The National Radiological Protection Board (NRPB) estimate that the annual effective dose equivalent from cosmic rays in the UK is about 300 $\mu$Sv yr$^{-1}$ on average. However, people that live at high altitudes and high latitudes can receive rather greater doses.

### 17.4.2 Terrestrial Gamma Radiation

The earth's crust contains three elements with radioactive isotopes which contribute significantly to man's exposure to radiation. These are $^{40}$K, with an average concentration in the upper crust of $\sim$ 3 ppm, $^{232}$Th, which is present in granitic rocks at 10–15 ppm and the three isotopes of uranium which total an average 3–4 ppm in granite. These latter nuclides have relative abundances of 99.274% $^{238}$U, 0.7205% $^{235}$U and 0.0056% $^{234}$U in the crust.

The NRPB estimate that the gamma rays emitted by these radionuclides and their daughters in soil, sediments and rocks and in building materials give an average annual effective dose equivalent in the UK of about 400 $\mu$Sv yr$^{-1}$, with the whole body being irradiated more or less equally. The local geology and types of building material used lead to considerable variation about this figure.

### 17.4.3 Radon and its Decay Products

The decay series of $^{238}$U and $^{232}$Th both contain radioactive isotopes of the element radon, $^{222}$Rn in the former case and $^{220}$Rn (sometimes known as thoron) in the latter. Radon, being a noble gas, readily escapes from soil and porous rock and diffuses into the lower atmosphere. There the nuclides decay with half-lives of 3.8 days ($^{222}$Rn) and 55 s ($^{220}$Rn) producing a series of shortlived daughter products. The full $^{238}$U decay series is shown in Figure 17.1. These daughters attach themselves to aerosol particles in the atmosphere. These particles are efficiently deposited in the lungs if inhaled. Their subsequent $\alpha$ and $\beta$ emissions can then irradiate and damage the lung tissue. The mean flux rate of radon from soil to the atmosphere is about 1200 Bq m$^{-2}$ d$^{-1}$ but there are large variations about this value at very small spatial scales.

Ambient outdoor concentrations of radon are low, typically about 3 Bq m$^{-3}$ in the UK. However when these gases enter a building, either through the floor following soil emissions or from the building's construction materials or by desorption from the water supply, the low air exchange rates in modern buildings causes a substantial increase in concentrations. Considerable variations in indoor concentrations have been observed, a positively-skewed distribution with a mean of $\sim$ 21 Bq m$^{-3}$ being typical in the UK. The concentrations in the outside air, in the soil and in soil pore water, the rate of emanation from building materials and the building ventilation rate will all affect the indoor concentration, and in a large scale measurement programme a few individual dwellings with concentrations two or three orders of magnitude above the mean might be observed. Because of the large variations in concentration and differences in people's activity patterns the calculation of an individual's exposure to radiation from radon is rather uncertain. The average annual effective dose equivalent in the UK is estimated by NRPB to be $\sim$ 800 $\mu$Sv yr$^{-1}$, although some individuals may receive very much higher doses than this. In any case radon and its daughters generally give a greater dose to man than any other sources of radiation. In the UK an 'Action Level' has been proposed by NRPB (200 Bq m$^{-3}$) above which remedial action to lower radon concentrations is recommended. Areas with >1% of dwellings with radon levels above the Action Level are designated as 'Affected Areas'.

In the USA, in particular, attempts are made to reduce indoor radon exposure in the worst affected areas and buildings. This can be done by preventing radon entering the building, by increasing the building ventilation rate or by removing the radon decay products from the air by, for example, using an electrostatic precipitator. Since most radon in buildings comes from the ground below the best method is to reduce the ingress of radon into the building by sealing the floor or by creating or increasing underfloor ventilation.

### 17.4.4 Radioactivity in Food and Water

The most important naturally occurring radionuclides in food and water are $^{226}$Ra, formed from $^{238}$U, and its $\alpha$-emitting daughter products $^{222}$Rn, $^{218}$Po and $^{214}$Bi. There is a wide variation in the $^{226}$Ra content of public water supplies, it usually being very low and representing a minor source of intake, but some well and spring water may contain 0.04–0.4 Bq l$^{-1}$. As previously mentioned, the domestic water supply, particularly water used in showers, may act as a source of $^{222}$Rn into the indoor environment. For the population in general the main source of radium intake is from food. Wide variations in concentration are found but a typical daily intake from the average diet is about 0.05 Bq per day. Some foods, for example brazil nuts and pacific salmon, accumulate radium in preference to calcium and can have very much higher concentrations than the average for food. Unbalanced diets based on these foods could lead to enhanced intakes of radium. Both $^{210}$Pb and $^{210}$Po can enter food, either from the soil or by wet and dry deposition from the atmosphere where they are present as daughters of $^{222}$Rn. For the average individual in the UK, the NRPB estimate of the total effective dose equivalent of alpha activity from the diet is about 200 $\mu$Sv yr$^{-1}$.

An additional source of $\alpha$-activity for some individuals which should be mentioned is cigarette smoke. The decay products of radon emitted from the ground under tobacco plants can adsorb onto the growing plant leaves and hence be incorporated into cigarettes. Dose rates have been estimated to be as high as 6–7 mSv per year from this source for some individuals.

Most of the naturally occurring $\beta$ activity in food is due to $^{40}$K. The availability of potassium to plants is subject to wide variations and hence the $^{40}$K activity of foods also varies. However potassium is an essential element for plants and animals, constituting about 0.2% of the soft tissue of the body. This leads to a total $^{40}$K content of $\sim$4 kBq for the average person although the individual value depends upon age, weight, sex and proportion of fat. The NRPB estimate that this gives an average effective dose equivalent of about 170 $\mu$Sv yr$^{-1}$ to individuals in the UK.

## 17.5   MEDICAL APPLICATIONS OF RADIOACTIVITY

Although not generally regarded as being pollutants, radiations used for medical purposes make a significant contribution to man's exposure. Indeed they provide the major artificial route of exposure to man and so warrant mention here. $X$-rays are used for a wide range of diagnostic purposes, a typical chest $X$-ray giving an effective dose equivalent of about 20 $\mu$Sv. Short-lived radionuclides are also used diagnostically. For example, $^{99}$Tc is used for bone and brain scans. The use of radiation for diagnostic purposes gives an average effective dose equivalent in the UK of about 250 $\mu$Sv per year. Of this about 120 $\mu$Sv yr$^{-1}$ is considered to be genetically significant, compared with about 1000 $\mu$Sv yr$^{-1}$ genetically significant dose from natural sources. Restrictions on the use of $X$-rays during pregnancy, limiting the $X$-ray beam area to the minimum needed, reducing radiation leakage from the source and improving the sensitivity and reliability of the measuring methods used are all aimed at reducing the genetically significant dose to reduce the probability of future genetic effects.

Externally administered beams of $X$-rays, gamma rays (from $^{60}$Co sources) and neutrons and radiations from internally administered radionuclides (*e.g.* $^{131}$I in the thyroid) are all used for therapeutic purposes. Some of the doses involved are very high but the potential adverse effects have to be weighed against the benefits accrued to the individual patient. Regular and stringent calibration and testing of medical radiation sources are vital if patients are to receive the correct dose in the correct place.

## 17.6   POLLUTION FROM NUCLEAR WEAPONS EXPLOSIONS

Since 1945 the products of nuclear weapons explosions, both of fission and fusion devices, have caused considerable pollution of the globe. The major bomb testing programmes were held in 1954–8 and 1961–2, with the number of documented test explosions now approaching 1000 (21 by the United Kingdom) and totalling about 1000 megatons of explosive force. Both underground and atmospheric tests have been used, with most of the explosive yield being from explosions in the atmosphere. Following an atmospheric explosion there will be some local deposition (or fallout) of activity but the majority of the products are injected into the upper troposphere and stratosphere, allowing their dispersion and subsequent deposition to the earth's surface on a global scale. About $10^{29}$ Bq of fission products have been injected into the atmosphere by these tests.

Fission bombs, such as those used against Japan, depend upon the rapid formation of a critical mass of $^{235}$U or $^{239}$Pu from several components of subcritical size. The critical mass for $^{235}$U is about 10 kg which on fission produces a very large amount of radioactivity: the total activity released by the 14 kiloton device at Hiroshima was about $8 \times 10^{24}$ Bq. Most of the released activity is in the form of short-lived nuclides, but considerable amounts of environmentally important fission products, including $^{85}$Kr, $^{89}$Sr, $^{90}$Sr, $^{99}$Tc, $^{106}$Ru, $^{137}$Cs, $^{140}$Ba and $^{144}$Ce, are also produced. Of particular biological significance are $^{89,90}$Sr, $^{131}$I and $^{137}$Cs.

As mentioned above, strontium is of importance because it has similar chemical properties to those of calcium. It enters the body by inhalation and ingestion, particularly through milk and vegetables. Iodine differs from strontium in that it is present in the atmosphere in both the gas and aerosol phases. Exposure may be by inhalation and ingestion, but of particular significance is the concentration of iodine in milk caused by cows grazing on contaminated grass and its subsequent consumption by man. Estimates of the amount of $^{137}$Cs released into the atmosphere by bomb tests vary but it is probably of the order of $10^{18}$ Bq, giving rise to human exposure mainly through grain, meat and milk.

Fusion bombs (also known as thermonuclear or hydrogen bombs) use the reaction of lithium hydride with slow neutrons to generate tritium:

$$^{6}_{3}\text{Li} + ^{1}_{0}\text{n} \rightarrow ^{3}_{1}\text{H} + ^{4}_{2}\text{He}$$

which then reacts with deuterium releasing energy:

$$^{2}_{1}\text{H} + ^{3}_{1}\text{H} \rightarrow ^{4}_{2}\text{He} + ^{1}_{0}\text{n}$$

The slow neutrons are initially supplied by the fission of $^{235}$U or $^{239}$Pu and the deuterium by the use of lithium-6 deuteride. Fusion bombs therefore release fission products as well as large amounts of tritium.

Tritium is a pure beta emitter with a half-life of 12.3 years. It is readily oxidized in the environment forming tritiated water and hence enters the hydrological cycle. The amount of tritium injected into the atmosphere by weapons tests probably exceeds $10^{20}$ Bq giving rise to estimated dose commitments over an average lifetime of $2 \times 10^{-5}$ Gy in the Northern Hemisphere and $2 \times 10^{-6}$ Gy in the Southern Hemisphere.

Various other radionuclides are produced by weapons explosions by the neutron activation of other elements in the soil or surface rocks, the air and the bomb casings. Of these $^{14}$C is the most significant pollutant, being formed from stable nitrogen in the air. It is a pure beta emitter of mean energy 49.5 keV and a half-life of 5730 years. The rate of natural production of $^{14}$C by cosmic ray interactions is about $1 \times 10^{15}$ Bq yr$^{-1}$ compared with a mean of $\sim 5 \times 10^{15}$ Bq yr$^{-1}$ produced by weapons testing since 1945.

The other important group of pollutant nuclides produced by weapons testing are the transuranics, including plutonium. Of these $^{239}$Pu is the most significant as it has been produced in large quantities ($\sim 1.5 \times 10^{16}$ Bq) and has a long half-life (24 360 years). It is formed by capture of a neutron by $^{238}$U:

$$^{238}_{92}\text{U} + ^{1}_{0}\text{n} \rightarrow ^{239}_{92}\text{U}^{\beta-} \rightarrow ^{239}_{93}\text{Np}^{\beta-} \rightarrow ^{239}_{94}\text{Pu}$$

The most important route to man for plutonium from weapons tests is by inhalation and the UNSCEAR estimates for the population-weighted dose, up to 2000 AD from tests carried out to 1977, are $1 \times 10^{-5}$ in the Northern Hemisphere and $3 \times 10^{-6}$ Gy in the Southern.

The current NRPB estimate for the average effective dose equivalent in the UK from weapon test activity is about 10 $\mu$Sv per year at present, compared with about 80 $\mu$Sv per year in the early 1960s.

## 17.7  POLLUTION FROM ELECTRIC POWER GENERATION PLANT AND OTHER NUCLEAR REACTORS

### 17.7.1  Emissions Resulting from Normal Reactor Operation

There are a large number of nuclear reactors in operation worldwide of many different sizes, designs and applications. As well as 430 large reactors operational, and 55 under construction (total for 1994), there are many other smaller reactors used for research, isotope production, education, materials testing and military purposes. Included in this latter category are reactors used to power submarines and other naval vessels. However, the greatest use remains the generation of electricity.

The production of electric power by using thermal energy derived from nuclear fission involves a chain of steps from the mining and preparation of fissionable fuel through to the disposal of radioactive wastes. There are actual and potential emissions of radioactivity into the environment at each of these steps.

*17.7.1.1  Uranium Mining and Concentration.*  Uranium ores contain, typically, about 0.15% $U_2O_3$ and require milling, extraction and concentration before shipping to the consumer. At several stages in this process dusts bearing radioactivity are produced which may lead to contamination of the environment in the vicinity of the plant. Large volumes of mill tailings are also produced and as these contain virtually all the radium and thorium isotopes present in the ore they give rise to emissions of radon. Also of environmental significance are liquid releases from the mills which, if uncontrolled, may lead to contamination of surface and ground waters. It should also be mentioned that the mining and milling of uranium ore inevitably leads to the occupational exposure of workers to uranium and its daughters, especially *via* inhaled $^{222}$Rn.

*17.7.1.2  Purification, Enrichment and Fuel Fabrication.*  The concentrated and purified ore extract, known as yellowcake, contains about 75% uranium and is converted to usable forms of uranium metal and uranium oxide by several processes. First uranium tetrafluoride, $UF_4$, is produced. The yellowcake is digested in nitric acid, insoluble impurities removed by filtration and soluble impurities extracted with an organic solvent. The uranyl nitrate solution is concentrated, oxidized to $UO_2$ and reacted with hydrogen fluoride to form uranium tetrafluoride, $UF_4$. The $UF_4$ so produced contains the relative natural abundances of the various uranium isotopes, including about 0.7% of $^{235}$U. This is adequate for some reactors, such as the first generation British Magnox reactors, and unenriched uranium metal is produced for them by reaction of the $UF_4$ with magnesium. The metal is cast and machined into fuel rods, heat treated and then inserted into magnesium aluminium alloy ('Magnox') cans.

Other designs of reactors, including the Advanced Gas-cooled and Pressurized Water Reactors, require fuel in the form of enriched uranium oxide containing 2–3% of $^{235}$U. This is produced by first reacting $UF_4$ with fluorine gas to form uranium hexafluoride, $UF_6$. The $UF_6$ gas is then repeatedly centrifuged or diffused through porous membranes, the small mass differences between the isotopes resulting in their eventual fractionation. Following enrichment the $UF_6$ is hydrolysed and reacted with hydrogen to form uranium oxide, $UO_2$, in powder form. This is pressed into pellets and packed in helium-filled stainless steel or zirconium–tin alloy cans.

Radioactive releases into the environment during all these steps should be minimal, although as with all chemical and mechanical processes some emissions are inevitable. The current NRPB estimates of the maximum annual effective dose equivalents arising from fuel preparation for members of the public living near to the relevant plants in the UK are less than 5 $\mu$Sv from discharges to the air and about 50 $\mu$Sv from discharges to water. These each give estimated *collective* effective dose equivalents for the UK of about 0.1 man Sv per year.

*17.7.1.3 Reactor Operation.* An operational nuclear reactor contains, and potentially may emit, radioactivity from three sources, from the fuel, from the products of fission reactions and from the products of activation reactions. The amount of activity present as fuel itself is of course variable but a Pressurized Water Reactor with 100 tonnes of 3.5% enriched uranium would contain $\sim 0.25 \times 10^{12}$ Bq of $^{235}$U and $1.1 \times 10^{12}$ Bq of $^{238}$U. Unless a catastrophic accident occurs, releases of fuel into the environment should not occur.

Fission of the fuel produces a large number of primary products which in turn decay, producing a large number of secondary decay products. Because of their very wide range of half-lives the relative composition and amounts of fission products present in a reactor varies with time in a complex manner. As each nuclide has a different rate of decay the rate of accumulation will vary. However there reaches a point when the rate of production of a nuclide equals its rate of removal by decay and an equilibrium arises between parent and daughter. If the rate of formation is constant then for all practical purposes equilibrium may be assumed to have been reached after about seven half-lives. For a short-lived nuclide such as $^{85}$Kr, equilibrium will be reached about 10 weeks after the start of the reactor, whereas for $^{137}$Cs with a half-life of 30.1 years it will take more than 200 years of continuous operation to reach equilibrium. As the reactor continues to operate with a given set of fuel rods the long-lived fission products become relatively more important and eventually the reactor might contain $10^{20}$ Bq of activity in total.

Most of the fission products are contained within the fuel cans themselves but total or partial failure of the fuel cladding will allow contamination of the coolant. In addition, contamination of the outer surfaces of the fuel rods by uranium fuel will allow fission products to form in the coolant. Most of the fission product activity is in the form of gaseous elements and although various scrubbing systems are used to remove them from the coolant, leakage into the environment can and does occur. The amount of gaseous fission products released will depend upon the number of fuel cladding failures, the design of the ventilation, cooling and coolant purification systems and the length of operation of the reactor. Of the fission products, $^{85}$Kr, which has a half-life of 10.8 years, makes the greatest contribution to the global dose commitment from reactor operation.

The third source of activity in the reactor results from the neutron activation of elements present in the fuel and its casing, the moderator, the coolant and other components of the reactor itself. Those areas most subjected to neutron activation are those where the incident neutron flux is highest, and include the fuel casing and the coolant system. From an environmental point of view the most important activation product is probably tritium. This is formed by various routes, especially by the activation of deuterium present in the water of Heavy Water Reactors but also by the activation of deuterium present in the water of light water-cooled reactors, the activation of boron added as a regulator to the primary coolant of Pressurized Water Reactors and by activation of boron present in the control rods of Boiling Water Reactors. The small size of the tritium nucleus allows it to diffuse through the cladding materials and so be released into the environment where it will be inhaled and absorbed through the skin. A smaller dose will also be received from drinking and cooking water, following the incorporation of released tritium into the water cycle. Estimates have been made of the annual dose equivalent for individuals living 1, 3 and 10 km from a Canadian CANDU Heavy Water Reactor and are of the order of 8, 3 and 2 $\mu$Sv respectively.

Other important activation products are $^{58,60}$Co, $^{65}$Zn, $^{59}$Fe, $^{14}$C and the actinides. This latter group, which includes the three long lived isotopes of plutonium $^{238}$Pu, $^{239}$Pu and $^{240}$Pu (with half-lives of 86.4, 24 360 and 6580 years respectively), are produced by the neutron activation of fuel uranium. Plutonium has a very low volatility and is not significantly released during normal reactor operation. However it can potentially enter the environment following a reactor accident and as a result of fuel reprocessing.

Various estimates are available for the total dose arising from these various discharges of radioactivity to the environment during the normal operation of power generation reactors. The current NRPB estimate of the maximum effective dose equivalents resulting from discharges to the atmosphere for the most exposed individuals living near to power stations in the UK is about 100 $\mu$Sv per year, giving a collective effective dose equivalent in the UK of about 4 man Sv per year. In addition to discharges to the atmosphere there are also discharges of radioactivity from power stations to natural waters. These result from the temporary storage of used fuel under water in specially constructed ponds. The NRPB estimate that these discharges to natural waters give a maximum effective dose equivalent to local individuals of less than 350 $\mu$Sv per year. The collective effective dose equivalent in the UK, mainly through the eating of contaminated seafood, is about 0.1 man Sv per year.

## 17.7.2 Pollution Following Reactor Accidents

With the large number of reactors operational worldwide it is inevitable that accidents with environmental consequences should occur. To date five major incidents have taken place in the UK, USA and the former USSR.

*17.7.2.1 Windscale, 1957.* In October, 1957, a graphite moderated reactor at Windscale, UK, overheated, rupturing at least one fuel can and causing the release of fission products into the environment. An estimated $740 \times 10^{12}$ Bq of $^{131}$I, $22 \times 10^{12}$ Bq of $^{137}$Cs, $3 \times 10^{12}$ Bq of $^{89}$Sr and $3 \times 10^{12}$ Bq of $^{90}$Sr were released. The cause of the accident was the sudden release of a huge amount of Wigner energy from the graphite in the core. When graphite is bombarded by fast neutrons, as in a reactor, an increase in volume and a decrease in thermal and electrical conductivity results. The graphite can then efficiently store energy and if heated above 300°C, this is released in the form of further heat. Normally this Wigner energy is released by slowly heating the graphite, allowing the stored energy to be released under control and reversing the radiation damage. During 7–8 October, 1957, however, an unsuccessful attempt was made to release the Wigner energy at Windscale, following which the reactor rapidly overheated releasing radioactive material into the atmosphere.

Following the accident the inhalation of aerosol and gaseous nuclides probably gave the largest source of exposure to the general population, although the drinking of milk contaminated by $^{131}$I was also significant, despite a restriction on the sale of milk from farms in the worst affected area. The total committed effective dose equivalent in the UK has been estimated at about 2000 man Sv, giving about 30 excess deaths from cancer in the UK during the first 40 years after the accident.

*17.7.2.2 Kyshtym, 1957.* In 1957 a chemical explosion in a high-level radio-active waste tank caused the release of about $7 \times 10^{17}$ Bq into the environment.

*17.7.2.3 Idaho Falls, 1961.* In January, 1961, during maintenance work on an enriched uranium-fuelled boiling water reactor at the National Reactor Testing Station, Idaho Springs, USA, the central rod was accidentally removed causing a large explosion, killing three men. Most of the coolant water was ejected together with an estimated 5–10% of the total fission products in the core. About $4 \times 10^{12}$ Bq of activity escaped from the reactor building.

*17.7.2.4 Three Mile Island, 1979.* Whilst operating at full power on 28 March, 1979, the cooling water system failed on the pressurized water reactor at Three Mile Island reactor 2 at Harrisburg, Pennsylvania. The reactor automatically shut down and three auxiliary pumps started to provide cooling water. However valves in this circuit had inadvertently been left closed and the resultant increase in temperature and pressure in the primary circuit caused an automatic relief valve to open, allowing about 30% of the primary coolant to escape. Fresh cooling water was injected into the primary circuit by the emergency cooling system but, following an error in their analysis of the accident, this and the main coolant pumps were deactivated by the operators. It was only when cooling water was later added that the overheating of the core ceased.

Despite the reactor core sustaining serious damage, the amount of activity released into the environment was limited to about $10^{17}$ Bq, almost entirely as short-lived noble gases, especially $^{133}$Xe (which has a half-life of 5.2 days). The resultant total committed effective dose equivalent to the population was estimated to be about 20 man Sv with the maximum individual dose equivalent being about 1 mSv.

*17.7.2.5  Chernobyl, 1986.*   The most serious reactor accident to date began on 26 April, 1986, when an explosion and fire occurred at the Chernobyl number 4 reactor near Kiev in the Ukraine. The reactor came into service in 1984 and was one of 14 RBMK boiling water pressure tube reactors operational in the USSR. These are graphite-moderated reactors of 950–1450 MW electrical power generation capacity fuelled with 2% enriched uranium dioxide encased in zirconium alloy tubing. The normal maximum temperature of the graphite is about 700°C and in order to prevent this being oxidized the core is surrounded by a thin-walled steel jacket containing an inert helium/nitrogen mixture.

Prior to the accident the reactor had completed a period of full power operation and was being progressively shut down for maintenance when an experiment was begun to see whether the mechanical inertia of one of the turbogenerators could be used to generate electricity for a short period in the event of a power failure. The reactor core contained water at just below the boiling point but when the experiment began some of the main coolant pumps slowed down causing the core water to boil vigorously. The bubbles of steam so formed displaced the water in the core and, because steam absorbs neutrons much less efficiently than water, the number of neutrons in the core began to rise. This increased the power output of the reactor, so increasing the heat output and the amount of steam in the core, which in turn led to a further rise in the neutron density. This positive feedback mechanism led to a rapid surge in power causing the fuel to melt and disintegrate. As the fuel came into contact with the surrounding water, steam explosions occurred destroying the structure of the core and the pile cap, causing radioactive material to be ejected into the atmosphere. The core fires allowed a continuing release of activity which was slowly reduced by the dumping of clay and other materials onto the core debris. However the core temperature again began to rise and a second peak in activity release occurred on 5 May. After this the core was progressively buried and finally sealed in a concrete sarcophagus.

Estimates of the amount of activity released from the reactor vary but Ukrainian measurements suggest that all of the noble gases, 10–20% of the volatile fission products (mainly iodine and caesium), and 3–4% of the fuel activity, giving a total of $1.85 \times 10^{18}$ Bq, escaped into the environment. On the basis of air concentration and deposition measurements the UK AEA's Harwell Laboratory initially estimated that about $7 \times 10^{16}$ Bq of $^{137}$Cs was released.

The immediate casualties of the accident were 31 killed and about two hundred diagnosed as suffering from acute radiation effects. About 135 000 people and a large number of animals were evacuated from a 30 km radius area surrounding the plant. However, in 1995 the Ukrainian Health Ministry announced that a total of 120 000 fatalities had occurred as a result of the accident but the basis of this claim is not yet clear.

The meteorological conditions prevailing over Europe at the time of the accident were rather complex, leading to the dispersion of activity over a very wide area. In the UK peak air concentrations occurred on 2 May when $\sim 0.5$ Bq m$^{-3}$ of $^{137}$Cs were recorded at Harwell. No Chernobyl radioactivity was immediately detected in the Southern Hemisphere. In the year following the accident the annual mean $^{137}$Cs concentration in the Northern Hemisphere was about the same as that of 1963 when weapons testing activity was at its highest. The amount of $^{137}$Cs deposited on the ground surface obviously varied with air concentration, rainfall and other parameters but close to Chernobyl (within 30 km) as much as $10^4$ kBq m$^{-2}$ of $^{137}$Cs was deposited. The average for Austria was 23 kBq m$^{-2}$, for the UK 1.4 kBq m$^{-2}$ and for the USA 0.04 kBq m$^{-2}$. The very heavy but localized rainfall which occurred in parts of Europe and the UK during the time when the plume was overhead led to a very patchy distribution of deposited activity on the ground. In the UK for example it varied from $> 10$ kBq m$^{-2}$ of $^{137}$Cs in parts of Cumbria to $< 0.3$ kBq m$^{-2}$ in parts of Suffolk.

Following the wet and dry deposition of activity from the atmosphere some contamination of foodstuffs was inevitable and outside of the immediate area around Chernobyl this was the major consequence of the accident. In general the most vulnerable foods in the UK were lamb and milk products. In the UK a ban on the movement and slaughter of lambs was imposed within specified areas until the meat consistently contained less than 1000 Bq kg$^{-1}$ of radiocaesium. No restrictions were placed on the sale of milk.

The main exposure pathways to man following the accident were direct gamma irradiation from the cloud and from activity deposited on the ground (groundshine), inhalation of gaseous and particulate activity from the air and, most importantly, the ingestion of contaminated foods. The Organization for Economic Co-operation and Development Nuclear Energy Agency estimate that the average individual effective dose equivalents received in the first year after the accident range from a few microsieverts or less for Spain, Portugal and most countries outside of Europe to about 0.7 mSv for Austria. However, hidden within these averages are the higher peak doses received by the most exposed individuals, or critical group, in each country. These vary from a few micro-sieverts outside Europe to an upper extreme of 2–3 mSv for the Nordic countries and Italy.

The total collective dose in Eastern and Western Europe has been estimated to be about $1.8 \times 10^5$ man Sv. The NRPB estimate that the average effective dose equivalent received in the UK during the first year was about 40 $\mu$Sv with a total of about 20 $\mu$Sv being received over subsequent years. The total committed effective dose equivalent for the UK population is estimated to be about 2100 man Sv. When compared with the other doses of radioactivity normally received by individuals, for example the 2 mSv per year individual effective dose equivalent received on average in Europe from natural background radiation, these doses are small and probably insignificant. Using the ICRP estimates of risk it is possible to calculate the additional mortality likely to result from Chernobyl. In the 40 years following the accident about 50 excess deaths are likely in the UK (compared with about 145 000 non-radiogenic cancers per year in the UK), indicating that the impact of the accident on future mortality statistics will be small in areas distant from the accident.

The Chernobyl accident presented a unique opportunity for experiments and studies in a wide range of environmental studies. It allowed the validation and refinement of atmospheric dispersion models, the calculation of washout ratios and deposition velocities and the study of the behaviour of caesium and iodine in food chains, natural waters, sediments and in the urban environment. It demonstrated the necessity of better international harmonization of scientific databases and public health protection policies and it provided a vivid example of the long-range transboundary transport of pollutants. In 1994 a very large European Union-funded tracer release experiment (ETEX) was carried out to simulate a nuclear reactor accident on the scale of Chernobyl with the aim of testing the predictive capabilities of long range transport models and forecasts. An inert tracer was released from a site in Brittany, Western France and the resultant plume detected across Europe by a network of monitoring sites and by aircraft.

### 17.7.3   Radioactive Waste Treatments and Disposal

An inevitable consequence of man's use of radioactivity is that radioactive waste material is produced which must then be disposed of. Although there is no universal scheme for the classification of such waste it is usual for it to be categorized in terms of its activity content as low, intermediate and high level.

*17.7.3.1   Low Level Waste.*   Low level wastes are produced in large volumes by all the various medical, industrial, scientific and military applications of radio-activity. They include contaminated solutions and solids, protective, cleaning and decontamination materials, laboratory ware and other equipment. They also include gases and liquids operationally discharged from power stations and other facilities. It has been estimated that during the period 1980–2000 about $3.6 \times 10^6$ m$^3$ of such wastes will be produced, some of which is sufficiently low in activity to be directly discharged into the environment, either with or without prior dilution or chemical treatment. Typical maximum activity concentrations in low level waste are $4 \times 10^9$ Bq t$^{-1}$ (alpha) and $12 \times 10^9$ Bq t$^{-1}$ (beta and gamma). Much low level waste is currently disposed of by shallow burial in landfill sites, often with the co-disposal of other, non-radioactive, controlled wastes or by discharge into surface waters in rivers, lakes, estuaries or coastal seas or by discharge into the atmosphere. If the environmental biophysicochemical behaviour of the radionuclides in question and their possible pathways back to man are well understood then it is possible to make reliable estimates of the likely resultant doses of radioactivity to those most exposed in the population. If these doses are suitably low then the disposal methods may be deemed to be acceptable.

However, at the present time there are many uncertainties in the understanding of such behaviour, pathways and doses, but nevertheless the very large volumes of low level waste being generated will, through lack of economically and environmentally viable alternatives, continue to be disposed of in these relatively uncontrolled ways.

*17.7.3.2 Intermediate Level Waste.* Intermediate level wastes are sufficiently active to prevent their direct discharge into the environment, with maximum specific activities of typically $2 \times 10^{12}$ Bq m$^{-3}$ ($\alpha$) and $2 \times 10^{-14}$ Bq m$^{-3}$ ($\beta$ and $\gamma$). They comprise much of the solid and liquid wastes generated during fuel reprocessing, residues from power station effluent plants and wastes produced by the decommissioning of nuclear facilities. Very large quantities ($\sim 2000$ t yr$^{-1}$) of intermediate and low level wastes have been disposed of by dumping in deep ocean waters in the NE Atlantic. Although some authorities still consider this method of disposal to be the best practicable environmental option for these categories of waste it is no longer practised and intermediate level waste produced in the UK is now stored on land, mainly at Sellafield, awaiting further policy decisions.

*17.7.3.3 High Level Waste.* High level wastes mainly consist of spent fuel and its residues and very active liquids generated during fuel reprocessing. Typical maximum activities are $4 \times 10^{14}$ Bq m$^{-3}$ ($\alpha$) and $8 \times 10^{16}$ Bq m$^{-3}$ ($\beta$ and $\gamma$). At present such wastes generated in the UK are stored at Sellafield in storage ponds where it is proposed they will be vitrified prior to further storage (to allow the decay of shorter lived nuclides) and finally disposal in deep repositories. No such repository yet exists but deep mines and boreholes on land and sea as well as other more exotic solutions including extraterrestrial disposal have all been proposed.

### 17.7.4 Fuel Reprocessing

In most nuclear reactors the economic lifetime of the fuel in the core is determined not by the depletion of fissile material but by the production and accumulation of fission products which progressively reduce the efficiency of the reactor. In a typical reactor the fuel is changed on a three-year cycle, generating large amounts of partially spent fuel contaminated with fission products. Apart from the economic considerations, which may or may not be in favour of recovering the unreacted fissile material from the spent fuel, depending upon the relative costs of reprocessing and importing further uranium ore, there are several reasons why such reprocessing is carried out: it allows the production of plutonium for military purposes, it reduces dependence on imported ores and it reduces the volume of high-level waste produced by a reactor. It is also a very large scale international commercial operation.

In the reprocessing method currently used in the UK the short lived activity is allowed to decay by storage for several months, after which the spent fuel is dissolved in nitric acid and a sequential extraction procedure used successively to remove the uranium, plutonium and fission products. The uranium is then re-enriched and fabricated into fuel rods and the plutonium used in the mixed oxide fuel or for military purposes. About 25 000 t of Magnox fuel has been reprocessed at Sellafield, yielding about 10 000 t of uranium for re-enrichment. The THORP thermal oxide reprocessing plant at Sellafield in NW England processes fuel from Advanced Gas-cooled and Pressurized Water Reactors at a projected rate of about 600 t yr$^{-1}$ and will eventually handle the 2500 t of waste currently being stored at Sellafield and that being produced now and in the future in the UK and overseas. The current NRPB estimates of the annual effective dose equivalents arising from fuel reprocessing to the UK population are 1 mSv yr$^{-1}$ to the most exposed individuals and a collective dose of 80 man Sv yr$^{-1}$. The reprocessing of fuel requires the transport of large quantities of spent and reprocessed fuel around the globe, giving rise to the possibility of further accidental releases into the environment.

## 17.8  POLLUTION FROM NON-NUCLEAR PROCESSES

Two non-nuclear industrial processes, the burning of fossil fuels and the smelting of non-ferrous metals, release non-trivial quantities of radioactivity into the environment and require brief consideration here. Coal contains uranium and thorium in varying concentrations, typically 1–2 ppm of both $^{238}$U and $^{232}$Th but at much higher concentrations (100–300 ppm) in some areas, *e.g.*, the western USA. It also contains significant quantities of $^{14}$C and $^{40}$K. Similarly oil and natural gas both contain members of the naturally-occurring U and Th decay series. When the fuel is burned, as in a conventional power station, these nuclides and any daughters present are either released into the atmosphere in the flue gas and fly ash or retained in the bottom ash. One current estimate of the amount of $^{226}$Ra emitted by a typical 1000 MW coal-fired station is $10^9$–$10^{10}$ Bq yr$^{-1}$ with a larger amount being retained in the ash. These emissions undoubtedly give rise to elevated environmental concentrations with the whole body dose equivalents for those most exposed living in the vicinity of large coal-fired plants being possibly as high as 1 mSv per year.

The second non-nuclear industrial source of radioactivity, the smelting of nonferrous metals, arises because of the natural occurrence of radioactive isotopes of lead. The geochemistry of lead is intimately associated with that of uranium and thorium, there being four radioactive isotopes of lead: $^{210}$Pb and $^{214}$Pb in the $^{238}$U decay chain, $^{211}$Pb in the $^{235}$U decay chain and $^{212}$Pb in the $^{232}$Th decay chain. All except $^{210}$Pb have half-lives of less than 12 hours, but the 22 year half-life of $^{210}$Pb and its subsequent decay to form $^{210}$Po, itself an alpha-emitter of half-life 138 days, makes it of environmental significance. During the primary and secondary smelting of lead and other non-ferrous metals and their ores some lead is released into the atmosphere. A fraction of this will be $^{210}$Pb together with a similar quantity of $^{210}$Po. This aerosol may then be inhaled, giving rise to an exposure of the lung to alpha particles. Investigation and understanding of such 'non-nuclear' pathways of radioactivity to man is at an early stage but it is possible that those living close to large smelters and other sources of $^{210}$Po may receive measurable and non-trivial doses of radioactivity.

*Acknowledgements.* I thank Dr. M. Kelly for his constructive comments and thoughtful assistance and J. Dixon, C. Duckham and M. Heaton for preparing the manuscript.

## 17.9 BIBLIOGRAPHY

R. S. Cambray, *et al.*, 'Observations on Radioactivity from the Chernobyl Accident', *Nucl. Energy*, 1987, **26**, 77–101.

K. D. Cliff, J. C. H. Miles and K. Brown, 'The Incidence and Origin of Radon and its Decay Products in Buildings', NRPB-R159, HMSO, London, 1984.

J. H. Gittus, *et al.*, 'The Chernobyl Accident and its Consequences', United Kingdom Atomic Energy Authority, HMSO, London, 1988.

J. S. Hughes and G. C. Roberts, 'The Radiation Exposure of the UK Population – 1984 Review', NRPB-R173, HMSO, London, 1984.

R. Kathren, 'Radioactivity in the Environment: Sources, Distribution and Surveillance', Harwood, Amsterdam, 1984.

National Radiological Protection Board, 'Living with Radiation', HMSO, London, 1986.

Organization for Economic Co-operation and Development Nuclear Energy Agency, 'The Radiological Impact of the Chernobyl Accident in OECD Countries', OECD, Paris, 1987.

F. Warner and R. M. Harrison, 'Radioecology after Chernobyl', SCOPE 50, John Wiley and Sons, Chichester, 1993.

CHAPTER 18

# Health Effects of Environmental Chemicals

P. T. C. HARRISON

## 18.1 INTRODUCTION

With the established expectation of a long and healthy life, and growing public concern about the impact of man's activities upon the environment, the issue of environmental quality and its relation to human health has come increasingly to the fore. People are becoming more and more concerned about possible effects of environmental chemicals on their health, and the popular press has latched on to this as a newsworthy issue. The apparent increase in the incidence of childhood asthma, for example, is of real public interest and has spawned much popular debate and new research initiatives on possible environmental causes. There continues to be concern about environmental exposure to pesticides, toxic metals and asbestos fibres, and accidents such as those at Bhopal and Seveso only serve to highlight the potential consequences to health of chemicals inadvertently released into the environment.

The public's perception of risk is highly coloured by the origin of the hazard. Chemicals seen as natural (or where exposure is by deliberate choice) are not perceived as having the same danger as those emanating, for example, from an industrial source. Thus natural toxins present in plants, for example, are not regarded in the same light as residues remaining after pesticide treatment.

Industrialization has resulted in increased concentrations of chemicals in air, water and soil. Of course, the presence of a chemical or chemicals in the environment cannot be taken as an indication *per se* that it is harmful to human health; it may have no relevant exposure pathway or be in too low a concentration to pose any real threat to health. However, for the sake of human health and that of the natural environment, particular attention must be paid to those substances which accumulate in the environment and in the tissues of plants and animals and which bioconcentrate in the food chain.

The health outcome resulting from exposure to a toxic substance (either directly or indirectly through the food chain) can range from effects on perceived well-being, through to frank chronic ill-health, cancer and death. There may be

418

reversible physiological changes or more severe pathological consequences to the cardiovascular and respiratory systems, for example. Or there may be adverse reproductive, immunological, neurological and developmental effects.

The factors which will determine the outcome of exposure include the inherent toxicity of the substance and its mode of action, the route of exposure, the concentration and duration of exposure and the susceptibility of the individual exposed. Individuals can be at particular risk because of their genetic makeup, their age, exposure history, current health status or lifestyle.

Different types of exposure to environmental chemicals can be distinguished. There is catastrophic exposure which results from the massive accidental release of material into the environment such as occurred at Seveso or Bhopal. Other localized incidents can occur as a result of heavy contamination of the local environment or of the adulteration of food and water; the outbreaks of mercury poisoning in Iraq and elsewhere as the result of the use of seed grain treated with organic mercurials to make flour are examples of this kind of exposure. Then there is the lower level but more chronic exposure which can occur with air pollutants and food and water contaminants. Sometimes these exposures are inescapable in a society which depends upon the use of chemicals to maintain its way of life, but better understanding of the risks of such exposures can lead to more controlled use of the substances or replacement by proven safer alternative chemicals or processes.

## 18.2 CATASTROPHIC EXPOSURE

The two best known examples of catastrophic exposure to chemicals occurred at Seveso and Bhopal. On 10 July, 1967 a massive release of 2,3,7,8-tetrachlorodibenzo-*p*-dioxin occurred from a chemical plant in Seveso, near Milan in Northern Italy, which was manufacturing 2,4,5-trichlorophenol (2,4,5-TCP). A safety disc in a reaction vessel ruptured and a plume of chemicals containing 2,4,5-TCP blew 30 to 50 m above the factory. As it cooled the material in it was deposited over a cone shaped area downwind from the factory about 2 km long and 700 m wide. In all, an area of 3–4 km$^2$ was contaminated and an estimated 3 to 16 kg of dioxin was released. There were almost 28 000 people living in the vicinity of the factory. Those who lived in the immediate area downwind were evacuated 14 days after the explosion and the area was closed off. About 5000 people in the most heavily contaminated area were allowed to stay in their homes but they were not allowed to cultivate or consume local vegetables or fruit nor to raise or keep poultry or other animals.[1]

Dioxin is both extremely toxic and extremely stable and is known, at sufficient dose levels, to affect fetal development and to have porphyrinogenic effects (often manifested as digestive system and skin disorders), all of which are well documented. It is not used commercially but is found as a contaminant when 2,4,5-TCP is synthesized by the hydrolysis of tetrachlorobenzene at high

---

[1] A. Giovanardi in 'Proceedings of the Expert Meeting on the Problems Raised by TCDD Pollution', Milan, 1976, p. 49.

temperatures. 2,4,5-TCP itself is used to make 2,4,5-T and 2,4-D which are used as herbicides; dioxin is often present in trace amounts in these compounds. Dioxin is also produced as a by-product of waste incineration and continues to be of major concern as a general environmental pollutant. The population at Seveso was screened shortly after the accident happened and a number of positive findings were noted. A few months after the accident, 176 individuals, mostly children, were found to have the skin condition chloracne, 50 of whom came from the most contaminated area. This constituted about 7% of the population estimated to be at risk. A further round of medical screening in February, 1977 revealed a further 137 cases of chloracne but subsequent follow-ups showed that the incidence had decreased and that the individuals already affected had improved.

In addition to the chloracne, some neurological abnormalities were noted. These included polyneuropathy with some symptoms which were due to effects on the central nervous system. These observations were more common amongst the people who lived in the most heavily contaminated zone and the incidence of abnormal nerve conduction tests was significantly increased in those who had chloracne.[2]

Finally, there was evidence of liver enlargement in about 8% of the population and again this was most noticeable amongst the most heavily exposed individuals. Measurement of liver enzyme activity showed some abnormalities which had returned to normal about a year after the explosion. It is interesting and noteworthy that there was no evidence that the immune system had been affected, and there has been no evidence of chromosomal abnormalities or of any damage to the fetus. There were no deaths.[3,4]

A recent analysis of cancer incidence in the exposed population has reported increased hepatobiliary (liver) cancer, elevated incidence of leukaemias and other haematologic neoplasms in men, increased multiple myeloma and myeloid leukaemia (bone marrow cancers) in women and evidence for higher incidences of soft tissue tumours and non-Hodgkin's lymphoma. Interestingly (because dioxin is an anti-oestrogen – see below) breast cancer and endometrial cancer in women were reduced.[5]

The explosion at Seveso excited a great deal of public alarm, particularly because dioxin was involved, but the major harmful effects were directed towards the environment; many farm animals died and the site became a wasteland of dying plants and deserted homes. The incident prompted the European Community to adopt a Directive (the so-called 'Seveso' Directive) aimed at preventing such major chemical accidents.

At Bhopal, the effects were directed almost entirely on the population living around the factory which was involved in the catastrophic release of methyl isocyanate (MIC). The accident at Bhopal, India, occurred on 3 December, 1984

[2]G. Fillipini, B. Bordo, P. Crenna, N. Massetto, M. Musicco and R. Boeri, *Scand J. Work Environ. Health*, 1981, **7**, 257.
[3]F. Pocchiari, V. Silano and A. Zampieri, *Ann. N.Y. Acad. Sci.*, 1979, **320**, 311.
[4]G. Reggiani, *J. Toxicol. Environ. Health*, 1980, **6**, 27.
[5]P. A. Bertazzi, A. C. Pesatori, D. Consomi, A. Tironi, M. J. Landi and C. Zocchetti, *Epidemiology*, 1993, **4**, 398.

at the Union Carbide factory which had been producing the insecticide carbaryl for about eighteen years; MIC was one of the main ingredients. The MIC itself was produced from monomethylamine (MMA) and phosgene, phosgene being produced on site by reacting chlorine and carbon monoxide. The MMA and chlorine were brought by tanker from other plants in India, stored and used when required; chloroform was used as a solvent throughout the process. There were in this plant, then, a variety of extremely toxic materials in use. On the night of the accident it seems that some water inadvertently got into a tank where 41 tonnes of MIC were being stored, causing a runaway chemical reaction. The heat of the reaction, possibly augmented by reactions with other materials present in the tank as contaminants, produced vaporization of such momentum that it could not be contained by the safety systems. The safety valve on the tank remained open for about two hours allowing MIC in liquid and vapour form to escape into the surroundings. The prevailing wind carried the cloud towards the north of the plant, and then towards the west, and affected approximately 100 000 people living in the vicinity. There were at least 2000 deaths, although none of the workers on night duty at the plant was harmed. The most frequent symptoms in those who survived were burning of the eyes, coughing, watering of the eyes and vomiting.[6]

## 18.3 LOCALIZED CONTAMINATION INCIDENTS

Most incidents in which local communities have suffered overt signs of toxicity have involved exposure to food contaminants. Well-known examples include Toxic Oil Syndrome in Spain and methyl mercury poisoning in Iraq. In the UK, incidents have occurred in Epping (Greater London) and in North Cornwall where the water supply was contaminated. There have been other examples where contamination has occurred indirectly as a result of an environmental pollutant entering a food chain, for example at Minimata and Niigata, Japan. The most famous example of health effects being associated with environmental contamination around waste disposal sites is Love Canal in the USA.

### 18.3.1 Toxic Oil Syndrome

Cooking oil was at the heart of an endemic of poisoning in Spain,[7] although the toxic agent responsible has not been identified satisfactorily to this day. The syndrome manifested itself in May, 1981 in an eight year old boy who died with acute respiratory insufficiency. He was one of a family of eight living in Madrid, of whom six eventually became ill. By June, 2000 patients with the condition had been admitted to hospitals in Madrid and a further 600 to hospitals in the provinces. By the end of August, 13 000 people had been treated in hospital and 100 had died. The final toll was about 20 000 persons affected with nearly 400 deaths, a case fatality rate of about 2%.

[6]N. Anderson, M. Kerr, V. Mehra and A. G. Salmon, *Brit. J. Indust. Med.*, 1988, **45**, 469.
[7]World Health Organization, 'Toxic Oil Syndrome', Report of a WHO meeting, Madrid, 21–25 March 1983, WHO, Copenhagen, 1984.

The illness began with a fever which was followed by severe respiratory symptoms and a variety of skin rashes, which led some of the victims to be diagnosed as having measles or German measles. Many of the patients developed signs of cerebral oedema (swelling of the brain) and many had cardiological abnormalities.

The cause of the disease was traced to adulterated cooking oil which was fraudulently sold to the public as pure olive oil. The oil was sold by door to door salesmen in five litre plastic bottles with no labels; since olive oil is an expensive commodity in Spain it was the poorer families in the working class suburbs of Madrid who were visited by the salesmen and it was they who almost entirely bore the brunt of the disease. The composition of the oil varied but rapeseed oil accounted for up to 90%; there were varying amounts of soya oil, castor-oil, olive oil and animal fats. The oil also contained between 1 and 50 ppm of aniline and between 1500 and 2000 ppm of acetanilide.

It seems that those who perpetrated the fraud tried to refine out the aniline and in doing so produced a number of chemical species, one of which was acetanilide which had reacted with fatty acids in the oil to produce oleoanilide which was originally presumed to be the toxic agent.[8] Later work, however, showed the presence of a number of other anilides, the main component of which was a fatty acid diester of 1,2-propanediol-3-aminophenyl. The other aniline compounds were considered to be hydrolysis products of diacyl 1,2-propanediol-3-aminophenyl or positional isomers. It has been suggested that these compounds may have been responsible for the symptoms produced by the toxic oil,[9] but despite much effort the precise toxicant has not been identified.

### 18.3.2   Rice Oil Contamination by Polychlorinated Biphenyls (PCBs)

Episodes of human poisoning with PCBs occurred in Japan (which seems to have had rather more than its share of environmental disasters) and in Taiwan. The disease first made its appearance in Japan in 1968 in the western part, when a number of families were noted to have developed chloracne, the skin condition which affected the victims at Seveso. Chloracne is more severe than the type of acne that occurs in adolescents and it has a rather different distribution on the body. Epidemiological studies brought other cases to light and it was found that the factor which the cases had in common was exposure to a particular batch of one brand of rice oil. Chemical analysis showed that the oil was contaminated with PCBs.

The PCBs were shown to have leaked into the oil from equipment which had been used to process the oil, the PCBs having been used as a heat-transfer fluid as was commonly the case in industry. By the end of 1977, 1665 individuals were considered to have met the diagnostic criteria for what has come to be known as Yusho disease.[10]

[8]J. M. Tabuenca, *Lancet*, 1981, **2**, 567.
[9]A. V. Roncero, C. J. del Valle, R. M. Duran and E. G. Constante, *Lancet*, 1983, **2**, 1025.
[10]H. Urate, K. Koda and M. Asahi, *Ann. N.Y. Acad Sci.*, 1979, **320**, 273.

In addition to chloracne, the patients with Yusho disease had a number of systemic complaints, including loss of appetite, lassitude, nausea and vomiting, weakness and loss of sensation in the extremities. Some also had hyperpigmentation of the face and nails.

The patients were followed up from 1969 to 1975 and in 64% of cases the skin lesions improved. A number of non-specific symptoms persisted, however, including a feeling of fatigue, headache, abdominal pain, cough with sputum, numbness and pain in the extremities and in the females, changes in menstruation. Objective findings included a sensory neuropathy, retarded growth in children and abnormal development of the teeth. Children who had been exposed *in utero* were small for dates and were hyperpigmented.

Some patients were found to be anaemic and some had other abnormalities, but the most striking observation was a marked increase in serum triglyceride levels. The mean in the patients was $134 \pm 60$ mg/100 ml compared with a mean of $74 \pm 29$ mg/100 ml in normal controls. When the PCB concentrations in the serum were measured by gas chromatography it was found that the patients with Yusho had an isomeric pattern which was different from that seen in controls whose only exposure had been from the general environment.[11]

Yusho disease appeared in Taiwan in the spring and summer of 1979 in two prefectures in the middle part of the country. The signs and symptoms were indistinguishable from the Japanese outbreak and by the end of 1980 more than 1800 people had been affected. The source of the PCBs was again contaminated rice oil. The blood levels of those with the disease ranged from 54 to 135 ppb. Later studies showed that polychlorinated dibenzofurans (PCDFs) and polychlorinated quaterphenyls (PCQs) were also present in the blood.[12]

To what extent the symptoms seen in the patients with Yusho disease were *entirely* due to PCBs is difficult to say since it is known that they were also exposed to PCDFs and to PCQs. These compounds are formed when PCBs are heated. In animal models PCDFs and PCQs are more toxic than PCBs and it seems that there may have been some synergistic effects between the different compounds in the oil. The fact that the PCB isomers in the oil were different from those in the general environment may also be of importance since they may have been more toxic than those to which the population at large was exposed.

### 18.3.3 Polybrominated Biphenyls (PBB) in Cattle Feed

PBB is used mainly in plastics as a fire retardant. In May and June of 1973, some ten to twenty bags of PBB were sent in error, instead of a livestock food additive, to a grain elevator. The chemical company which made the PBB normally supplied magnesium oxide to go into the cattle feed but both products were packed in the same colour bag and although the PBB was labelled 'Firemaster' rather than 'Nutrimaster', and although this difference was actually noted by the staff at the grain elevator, it was nevertheless

---

[11] P. H. Chen, J. M. Gaw, C. K. Wong, C. J. Chen, *Bull. Environ. Contam. Toxicol.*, 1980, **25**, 325.
[12] T. Kashimoto, H. Miyata, S. Kunita, T-C. Tung, S-T. Hsu, K-J. Chang, S-Y. Tang, G. Ohi, J. Nakagawa and S-I. Yamamoto, *Arch. Environ. Health*, 1981, **36**, 321.

incorporated into the feed and distributed throughout the state to be fed to the unsuspecting cows.

Reports of sick cows began to surface in August, 1973 and towards the end of the year it was realized that the feed was to blame. Despite this, the contamination continued both because there was cross-contamination of otherwise normal feed from the grain elevator and because the tainted feed was resold at a discount after it had been returned to the suppliers. Not until PBB was identified in the feed in May, 1974 was any attempt made to limit the contamination.

From the time that the feed had become contaminated, dairy products containing PBB had been sold throughout Michigan and cows and other livestock which had been given it had been slaughtered for meat. A representative sample of 2000 individuals was surveyed and more than half had a concentration of PBB in their fatty tissues exceeding 10 ppb.[13] Farmers and others who consumed produce directly from contaminated farms had the highest levels of PBB.

An initial study of 217 farmers concluded that the exposure had caused no deleterious effects to their health, but the study was criticized on the grounds that the control group had also been exposed to PBB.[14] A second study was then carried out in which over 1000 farmers were compared with unexposed farmers from Wisconsin. This second study did find some adverse effects; acne, dry skin, hyperpigmentation and discoloration of the nails were all more common in the exposed group, and they also complained more of headaches, nausea, depression and a number of other non-specific symptoms. Serum levels of hepatic enzymes were higher in the Michigan farmers than their neighbouring controls. Individuals with symptoms were also more likely to have elevated enzyme levels, and it was shown that there were some changes in the immune system[15,16] and that some individuals had enlargement of the liver and a sensory neuropathy.[17]In follow up studies, the PBB levels in the serum were found to decrease, but it was interesting that elevated PCB levels were also found and these were actually higher than the PBB levels. There was no relationship between abnormal liver function tests and serum PBB concentrations but there was a slight (statistically insignificant) negative correlation with serum PBB levels and some tests of thyroid function.[18]

It is of interest that in none of these studies did the subjective or the objective findings correlate with serum or fat PBB concentrations. This may have been because there was another toxic contaminant present which was acting independently of the PBB, or that the levels of PBB had fallen in the interval between ingestion and the beginning of the studies. It may also be the case that levels of PBB in blood and fat are not good indicators of levels in target organs.

[13]I. J. Selikoff, 'A Survey of the General Population of Michigan for Health Effects of PBB Exposure', Final Report submitted to the Michigan Department of Public Health, Michigan, 1979.
[14]H. E. B. Humphrey and N. S. Hayner, 'Polybrominated Biphenyls: An Agricultural Incident and its Consequences: An Epidemiological Investigation of Human Exposure', Michigan Department of Health, Michigan, 1975.
[15]J. G. Bekesi, J. F. Holland and H. A. Anderson, *Science*, 1978, **199**, 1207.
[16]P. J. Landrigan, K. R. Wilcox and J. Silva, *Ann. N.Y. Acad Sci.*, 1979, **320**, 284.
[17]J. E. Stross, R. K. Nixon and M. D. Anderson, *Ann. N.Y. Acad Sci.*, 1979, **320**, 368.
[18]K. Kreiss, C. Roberts and H. H. Humphrey, *Arch. Environ. Health*, 1982, **37**, 141.

### 18.3.4 Mercury Poisoning in Minimata and Niigata

Mercury in its organic form has accounted for many of the episodes of endemic disease resulting from environmental exposure. Probably the best known of these is Minimata Bay disease.[19] This was first noted at the end of 1953 when an unusual neurological disorder began to affect the villagers who lived on Minimata Bay on the south west coast of Kyushu, the most southerly of the main islands of Japan. It was commonly referred to as *kibyo*, that is, the 'mystery illness'. Both sexes and all ages were affected, and those who were affected presented with a mixture of signs relating to the peripheral and central nervous systems. The prognosis of the condition was poor, many patients became disabled and bedridden and about 40% died. The disorder was associated with the consumption of fish and shellfish caught in the bay, but mercury poisoning was not considered in the early investigations of the condition because many of what were then considered the classic symptoms of mercury poisoning were not present. It was only when it was realized that the signs of the disease were similar to those described in a man who had died after being poisoned during the manufacture of alkyl mercury fungicides that the possibility was raised.

The source of the mercury was effluent released into the bay from a factory which was manufacturing vinyl chloride using mercuric chloride as a catalyst. It is claimed that *inorganic* mercury was released by the factory and that this was methylated by microorganisms living in the sediments in the bay. However, the rate of conversion is extremely slow, much too slow to have accounted for the large amounts of methyl mercury which are calculated to have accumulated in the waters of the bay. It is therefore much more likely that the mercury was actually released in the organic form; at the time there were no regulations forbidding this in Japan.

About 700 people were affected at Minimata Bay and there was a second outbreak of methyl mercury poisoning in Japan in 1965 in Niigata, affecting a further 500 or so individuals. This episode followed pollution of the Agano River by industrial effluent and, again, the bioconcentration of mercury in fish subsequently consumed by the local population.[20]

### 18.3.5 Methyl Mercury Poisoning in Iraq

A major poisoning episode occurred in 1971–72 when the Iraqi government imported a large consignment of seed grain treated with an alkyl mercury fungicide, which was then distributed to the largely illiterate rural population. The distribution was accompanied by warnings that the seed was for sowing not for eating and the sacks were marked with warning labels – in English and Spanish! The seed had been treated with a red dye to distinguish it from edible grain but the farmers found that they could remove the dye by washing and equated this with the removal of the poison. The grain was first used to make

[19]M. Katsuna, 'Minimata Disease', Kumamoto University, Japan, 1968.
[20]H. P. Illing in 'General and Applied Toxicology', eds. B. Ballantyne, T. Marrs and P. Turner, Macmillan, London, 1995, p. 1287.

bread in November, 1971 and the first cases of poisoning appeared in December. By the end of March, 1972 there had been 6530 admissions to hospital and 459 (7%) of these had died.[21] These represented only the most severe cases and the true extent of this outbreak will probably never be known, although it has been suggested that the incidence of the disease may have been as high as 73 per 10 000.[22]

### 18.3.6 Aluminium Contamination of Drinking Water in North Cornwall

In July 1988, about 200 tonnes of aluminium sulfate were accidentally deposited into the treated water reservoir at Lowermoor Treatment Works, resulting in major contamination of the drinking water supply to Camelford in Cornwall and the surrounding district.[20] Aluminium levels increased to over 10 mg $l^{-1}$, well above the 0.2 mg $l^{-1}$ limit set by the EC on palatability grounds, and the pH of the water dropped below 5.0.

Despite reassuring messages from local sources, residents and holiday makers reported a large number of acute symptoms and speculation about longer-term effects was rife. The expert assessment, made by the Lowermoor Incident Health Advisory Group, [23] was that the early reported symptoms of gastro-intestinal disturbances, rashes and mouth ulcers were probably due to the incident, but these effects were short-lived. Persistent toxic effects were thought unlikely because of the transient nature of the exposure and because all the known toxic effects of aluminium in man are associated with prolonged exposure.

However, a significant time after the incident, hundreds of people resident in the Lowermoor area continued to attribute health complaints to the incident, including joint and muscle pains, malaise, fatigue and memory problems. Also, various scientific reports of unexpected symptoms and clinical findings were reported. The Advisory Group concluded that, based on all the evidence available to it, some of the symptoms experienced by the residents were probably induced by the sustained anxiety naturally felt by many people, and that others were wrongly attributed to the incident as a result of heightened awareness provoked by the incident and subsequent events. It was however recommended that a developmental follow-up of children exposed *in utero* during the incident should be undertaken, together with formal testing of the possibility of particular individual 'sensitivity' to aluminium.

A complication to this incident was that the low pH of the drinking water resulted in other metals – copper, zinc and lead – being dissolved from domestic plumbing, so that various additive or synergistic effects may have occurred.

[21] F. Bakir, S. F. Damluji, L. Amin-Zaki, M. Murtadha, A. Khalidi, N. Y. al-Rawi, S. Tikviti, H. I. Dhahir, T. W. Clarkson, J. C. Smithy and R. A. Doherty, *Science*, 1973, **181**, 230.

[22] G. Kazantzis, A. W. al-Mufti, A. al-Jawad, Y. al-Shawani, M. A. Majid, R. M. Mahmoud, M. Soufi, K. Tawfiq, M. A. Ibrahim and H. Dabagh, *Bull. WHO*, 1976, **53**, Suppl. 37.

[23] Lowermoor Incident Health Advisory Group, 'Water Pollution at Lowermoor, North Cornwall', 2nd Report, HMSO, London, 1991.

A more recent revisit of the Camelford incident[24] confirmed that a number of individuals showed consistent evidence of impaired information processing and memory, with no obvious relationship with measurement of anxiety and depression. Although the abnormal neuropsychological findings indicated cognitive impairment it was uncertain whether this was caused by an acute episode of brain damage or other causes of stress resulting from the accident.

### 18.3.7 Epping Jaundice – Chemical Contamination of Food During Storage

The ingestion of contaminated wholemeal bread resulted, in February 1965, in an unusual outbreak of jaundice in Epping, UK.[20,25] The outbreak, which affected at least 84 people, was traced to the contamination of flour by 4,4'-diaminophenylmethane. This had spilled from a container onto the floor of a van transporting both the flour and chemicals. The chemicals were absorbed by the flour through the sacking, and the flour was subsequently used to make the bread.

Jaundice and liver enlargement were preceded, in most cases, by severe pains in the upper abdomen and chest. The pains were mostly of acute onset, but in some patients the onset was insidious. Raised levels of serum bilirubin (a blood breakdown product) and the enzymes alkaline phosphatase and aspartate aminotransferase were recorded in most of the 57 patients further investigated. Needle biopsies of the liver showed considerable evidence of inflammation and cholestasis and damage to the liver cells. Experimental studies subsequently showed the liver lesions to be reproducible in mice following administrations of 4,4'-diaminophenylmethane.[26] All the patients eventually recovered.

### 18.3.8 Love Canal

The reporting of chemical odours in the basements of homes in the Love Canal district, USA, led to a toxicological investigation which made this area famous in the history of waste disposal and resulted in much regulatory activity in the USA.

Love Canal was a waste disposal site containing municipal and chemical waste disposed of over a 30 year period up to 1953.[20] Homes were then built on the site during the 1960s and leachates began to be detected in the late 1960s.

Dibenzofurans and dioxins were among the chemicals detected in the organic phase of the leachates. Animal studies indicated possible risks of immunotoxic, carcinogenic and teratogenic effects.[27,28] The episode resulted in significant fears of ill-health and much psychological stress. Limited follow-up of residents identified low birthweights in the offspring of Love Canal residents, [29] but no causal link has been established for cancer incidence in the area. Love Canal

[24]T. M. McMillan, A. J. Freemont, A. Herxheimer, J. Denton, A. P. Taylor, M. Pazianis, A. R. C. Cummin and J. B. Eastwood, *Hum. Exp. Toxicol.*, 1993, **12**, 37.
[25]H. Kopelman, M. H. Robertson, P. G. Saunders and I. Ash, *Br. Med. J.*, 1966, **1**, 514–516.
[26]R. Schoental, *Nature*, 1968, **219**, 1162–1163.
[27]J. B. Silkworth, D. S. Cutler and G. Sack, *Fund. Appl. Toxicol.*, 1989, **12**, 302.
[28]J. B. Silkworth, D. S. Cutler and L. Antrim, *Fund. Appl. Toxicol.*, 1989, **13**, 1.
[29]N. J. Vianna and A. K. Polan, *Science*, 1984, **226**, 1217.

provides one more instance of considerable public anxiety and stress resulting
from the identification of a potential environmental toxic hazard. The psychological and other consequences of such incidents might easily outweigh the actual
toxic effects of chemical exposure.[30] Indeed, having reviewed the literature on
hazardous waste disposal sites, Grishan[31] concluded that 'there are few published
scientific reports of health effects clearly attributable to chemicals from uncontrolled disposal sites'.

## 18.4 GENERALIZED ENVIRONMENTAL CONTAMINATION

While local disasters and incidents such as those described above are often
significant and can tell us much about the consequences of high level, acute or
sub-chronic exposure to environmental contaminants, of greater significance to
public health are the lower level chronic exposures which occur to air pollutants
and other substances more commonly encountered or more widely dispersed in
the environment. This section looks in particular at indoor air pollution, toxic
metals, asbestos and other fibrous materials in the environment, pesticides and
finally an ill-defined group of substances commonly referred to as 'environmental
oestrogens'.

### 18.4.1  Indoor Air Pollution

The presence of noxious substances in the outside air, coming from factories,
domestic fuel combustion and, increasingly, vehicle exhausts continues to be a
major cause of concern. In terms of overall morbidity and mortality, probably
nothing can surpass the damage which has been caused historically by exposure
to the by-products of the burning of coal; certainly there have been remarkable
episodes in which sudden increases in the number of deaths have been noted to
follow exceptionally high pollution, for example, in the Meuse Valley in 1930, in
Donora in 1948 and in London in 1952 when the overall excess in daily deaths
was estimated to be 3500 to 4000.[32] Even though the levels of sulfur dioxide and
smoke seen during these episodes are no longer experienced, there remains
considerable concern that other air pollutants, most notably $PM_{10}$*, continue to
have a real and measurable impact on human health. These concerns about
outdoor air quality are reflected in increasing consideration of the health effects of
vehicle emissions and calls for tighter controls and the setting of standards for the
common air pollutants.

A separate chapter of this book is dedicated to the health effects of air
pollutants, but there is a need here to concentrate on a compartment of the
environment where people spend up to 90% of their time – the indoor
environment. Although outdoor air quality has a considerable influence on

[30]L. H. Roht, S. W. Vernon, F. W. Weir, S. M. Plier, P. Sullivan and L. J. Reed, *Am. J. Epidemiol.*,
1985, **122**, 418
[31]'Health Aspects of the Disposal of Waste Chemicals', ed. J. W. Grishan, Pergamon, Oxford, 1986.
[32]W. P. D. Logan, *Lancet*, 1953, **1**, 336.
*Particulate matter of aerodynamic diameter less than 10 $\mu$m.

indoor air, for most pollutants the greatest proportion of total exposure is determined by exposures indoors. Indoor sources, such as gas cookers, can result in pollutant levels much higher indoors than out, and some pollutants are found only indoors. The impetus over recent years to conserve energy has resulted in 'tighter' buildings with much reduced air exchange and therefore a greater propensity for indoor pollutants to build up.

People are exposed to a wide variety of indoor air pollutants in their work and at home. Concern is directed especially to domestic air quality since the home is where the non-healthy individuals and the very young and old spend much of their time. Pollutants indoors arise from a variety of sources, most notably from the combustion of fuel, emissions from furnishings and from the use of DIY and consumer products.[33-35]

This section briefly reviews the key chemical indoor air pollutants found in homes. Radon (a natural radioactive gas) and house dust mites, bacteria and fungi (biological agents), whilst important, are not included here.

*18.4.1.1 Carbon Monoxide.* Carbon monoxide is one of the most important indoor air pollutants as it continues to kill as many as 100 people a year in the UK through accidental poisoning. It is especially dangerous since it has no colour, smell or taste. Its toxic action is through the displacement of oxygen in haemoglobin in the blood to form carboxyhaemoglobin, thus depriving the tissues of the body of their oxygen supply. Early symptoms of exposure include tiredness, drowsiness, headaches, dizziness, pains in the chest and stomach pains. Excessive exposure can lead to loss of consciousness, coma and death. Exposure *in utero* can also result in adverse effects on the fetus and longer-term effects of chronic exposure in adults are suspected. Carbon monoxide is produced when fuel burns with an inadequate supply of oxygen. High concentrations can be produced by badly installed or inadequately maintained gas and solid fuel appliances, paraffin heaters, *etc.*, or by fumes leaking from a flue into a poorly ventilated room. Most fatal cases of carbon monoxide poisoning result from blockage and/or leakage from flues.

*18.4.1.2 Environmental Tobacco Smoke.* Smoking, where it occurs, is one of the most important sources of indoor pollution. Tobacco smoke contains tar droplets and a cocktail of various other toxic chemicals including carbon monoxide, nitric oxide, ammonia, hydrogen cyanide and acrolein, together with proven animal carcinogens such as *N*-nitrosamines, polycyclic aromatic hydrocarbons and benzene. Environmental tobacco smoke can irritate the eyes, nose and throat, and exposed babies and children are more prone to chest, ear, nose and throat infections. Women exposed during pregnancy tend to have lower birthweight babies and asthmatics may be adversely affected by acute exposures.

[33]Department of the Environment, 'Good Air Quality in Your Home', Department of the Environment, London, 1992.
[34]British Medical Association, 'Living with Risk', John Wiley and Sons, Chichester, 1987.
[35]Institute for Environment and Health, 'Indoor Air Quality in the Home: Nitrogen Dioxide, Formaldehyde, Volatile Organic Compounds, House Dust Mites, Fungi and Bacteria', IEH, Leicester, 1996.

*18.4.1.3   Nitrogen Dioxide.*   This and other oxides of nitrogen, are formed when fuel is burned in air. Thus $NO_2$ is generated indoors by gas, oil and solid fuel appliances. The main sources are unflued appliances such as gas cookers and gas wall heaters. Indoor levels are significantly influenced by outdoor levels, but where an indoor source is present this tends to dominate. Exposure to high levels typically occurs in kitchens during gas cooking. Nitrogen dioxide can irritate the lungs, but the mechanisms of toxic action remain to be fully elucidated and there is continued uncertainty and debate about the actual impact of levels of $NO_2$, as typically found indoors, on the health of occupants. Overall, the weight of published evidence points to a possible hazard of respiratory illness in children, perhaps resulting from increased susceptibility to infections. Interactive effects of $NO_2$, for example with house dust mite allergen, have also been postulated.[36] The possible impact of $NO_2$ on potentially susceptible groups such as asthmatics and people with chronic bronchitis is largely unknown.

*18.4.1.4   Formaldehyde.*   Formaldehyde is a colourless gas with a pungent odour given off from various furnishings and fittings found in the home. One of the most important sources is pressed wood ('chipboard'), made using bonding materials containing urea-formaldehyde resin, which has become increasingly used in furniture items over the last few decades. Another major source is urea-formaldehyde foam insulation (UFFI) installed in wall cavities. Formaldehyde is also generated during the combustion of fuel and is a component of cigarette smoke and vehicle exhaust. Formaldehyde gas can irritate the mucous membranes of those exposed. The odour threshold is in the region of 0.05–1.00 ppm and levels found in some homes may reach the threshold, for some individuals, for transient eye, nose and throat irritation. Although formaldehyde is a sensitizing agent, no studies to date have demonstrated the induction of asthma by domestic exposure.

*18.4.1.5   Volatile Organic Compounds.*   VOCs* in the indoor environment originate from a number of sources including furnishings, furniture and carpet adhesives, building materials, cosmetics, cleaning agents and DIY materials. VOCs also originate from fungi, tobacco smoke and fuel combustion. By far the greatest peak exposure to VOCs occurs during home decorating using solvent-based paints. Glues are another important source of high peak levels. It has been estimated that between 50 and 300 different compounds may occur in a typical non-industrial indoor environment,[37] including aliphatic and aromatic hydrocarbons, halogenated compounds and aldehydes. Because of the diverse range of chemical substances defined as VOCs, determination of health effects is problematic. However, it is known that at levels typically found indoors the major effects are likely to be sensory. Short exposures to high levels of solvent vapours can cause temporary dizziness; lengthy or repeated exposure can irritate the eyes and lungs and may affect the nervous system. Some VOCs, such as benzene, are

[36]W. S. Tunnicliffe, P. S. Burge and J. G. Ayres, *Lancet*, 1994, **344**, 1733.
[37]L. Mølhave, *Indoor Air*, 1991, **1**, 357.

*A wide range of compounds with boiling point between approximately 50 °C and 250 °C and which, at room temperature, produce vapours.

cancer causing agents, but the consequences of exposure to the levels typically found in homes are uncertain.

## 18.4.2 Metals

Episodes or incidents involving exposure to mercury and aluminium, mostly involving exposure through ingestion, have already been described. Added to this is the evidence for more widespread environmental contamination by various metals including mercury, cadmium and lead. For example, lead concentration in polar ice has increased over 20-fold since 1800 and tissue levels of lead and mercury in Greenland Inuit Eskimos is four to eight times higher than preserved ancestors from five centuries ago.[38] Industrial sources of metals include refineries, chemical plants, cement manufacturing, power plants, smelters, incinerators, vehicles (*e.g.* from the use of tetraethyl lead as an anti-knock additive in petrol) and tobacco smoke.[39]

Metals have a variety of effects upon the human body, mostly acting at the cellular level. Some metals disrupt biochemical reactions while others block essential biological processes, including the absorption of nutrients. Some accumulate in the body giving rise to toxic concentrations after many years of exposure and yet others (including arsenic, beryllium, cadmium and chromium) are carcinogens. Exposure to methyl mercury and high levels of lead can cause gross deformities in development.[40]

The health effects of lead have been studied extensively and are well documented.[41,42] Environmental lead exposure has been linked to reduced IQ in children and to elevated blood pressure in adults, although the cause–effect relationship for the latter is not especially robust.[43] Airborne lead exposure has been much reduced in recent years by the introduction in many countries of unleaded gasoline. However, other routes of exposure (most notably ingestion) remain significant. Most human exposure to lead in recent times has arisen from the use of lead in domestic water pipes. The outbreaks of endemic lead poisoning in classical and historical times are well known. Suffice it to say, that the adulteration of food and drink with lead has been a significant contributor to morbidity and perhaps to mortality in the past. The most serious form of endemic lead poisoning arose from the habit of adulterating wine with lead to improve a poor vintage and make it more saleable. During the eighteenth century there were a number of famous outbreaks of lead poisoning, the Devonshire colic being perhaps the most well known. In this case, the adulteration of cider with lead

[38]A. J. McMichael, 'Planetary Overload', Cambridge University Press, Cambridge, 1993.

[39]V. T. Covello and M. W. Merkhofer, 'Risk Assessment Methods', Plenum Press, New York, 1993.

[40]L. Friberg, G. F. Nordberg and V. B. Vouk, 'Handbook on the Toxicology of Metals', Elsevier, Amsterdam, 1986.

[41]D. Krewski, A. Oxman and G. W. Torrance in 'The Risk Assessment of Environmental Hazards', ed. D. J. Paustenbach, Wiley, New York, 1989, p. 1047.

[42]Environmental Protection Agency (EPA), 'Review of the National Ambient Air Quality Standards for Lead: Assessment of Scientific and Technical Information', USEPA, Research Triangle Park, N.C., 1986.

[43]F. K. Hare (Chairman), 'Lead in the Canadian Environment: Science and Regulation, Final Report of the Commission on Lead in the Environment', Royal Society of Canada, Ottawa, 1986.

arose accidentally due to the presence of lead in the pounds and presses used to make the cider. The widespread use of lead in cooking utensils, in glazes and in pewter added to the burden of lead exposure during the eighteenth and nineteenth centuries.[44]

Environmental exposure to cadmium was considered to be the cause of a disease which was first reported in 1955 as occurring in a localized area downstream from a mine on the Juntsu River in the Toyama Prefecture in Japan. It was called *itai-itai* disease meaning 'it hurts'. The condition was almost entirely confined to elderly women who had borne several children and was characterized by severe bone pain, waddling gait, severe osteomalacia (bone softening), pathological fractures and some signs of renal impairment. The water which was used to irrigate crops was frequently contaminated by outpourings from the mine which contained zinc, lead and cadmium. Levels of cadmium in rice samples were shown to be about ten times the amount normally present and the view was gradually formed that it was the cadmium which was responsible for the disease.[45] However, it seems likely that deficiencies of calcium and vitamin D were also at least partially to blame and that cadmium may have been acting only as one factor in what was a multi-factorial aetiology for this disease.[46]

People who depend upon fish as the staple part of their diet may still be at risk of excessive exposure to mercury, even if not on the scale experienced at Minimata and Niigata. For example, blood methyl mercury levels have been found to be almost ten times higher in a Peruvian population who ate on average 10.1 kg of fish per family of (on average) 6.2 persons compared with a control population whose fish intake was considerably more modest (the mean in the high fish eating population was 82 ng ml$^{-1}$ and in the control population, 9.9 ng ml$^{-1}$). Moreover, 29.5% of the heavily exposed population had signs of a sensory neuropathy.[47] In a fish-eating population in New Guinea, hair mercury concentrations were between two and three times that of a control group (6.4 $\mu$g g$^{-1}$ compared with 2.4 $\mu$g g$^{-1}$); there were, however, no demonstrable ill effects in this group.[48]

### 18.4.3   Asbestos and Man-made Mineral Fibres

The industrial exploitation of asbestos began about a hundred years ago, its use probably peaking in the 1960s. Because the health effects of asbestos were unsuspected, workers in the industry were often exposed to massive levels of airborne fibres resulting in serious and often fatal pulmonary interstitial fibrosis, lung cancer and mesothelioma (cancer of the epithelium lining the chest cavity). First reports of these diseases started to appear in the 1900s, 1930s and 1960

[44]T. Waldron in 'Diet and Crafts in Towns', eds. D. Serjeantson and T. Waldron, BAR, Oxford, 1989, 55 pp.
[45]L. Friberg, T. Kjellstrom, G. Nordberg and M. Piscator, 'Cadmium in the Environment, III', USEPA, Washington, 1975.
[46]K. Tsuchiya, *Fed. Proc.*, 1976, **35**, 2412.
[47]M. D. Turner, D. O. Marsh, J. C. Smith, J. B. Inglis, T. W. Clarkson, C. E. Rubio, J. Chiriboga and C. C. Chiriboga, *Arch. Environ. Health*, 1980, **35**, 367.
[48]J. H. Kyle and N. Ghani, *Arch. Environ. Health*, 1982, **37**, 266.

respectively.[49,50] The work of Wagner in South Africa in the 50s and 60s was especially significant in assessing and understanding the health impacts of asbestos fibres. These health concerns resulted in a reduction, at least in the developed world, of asbestos use. The commercial materials marketed as 'asbestos' come from both the amphibole and serpentine groups of silicate minerals. Those from the amphibole group were identified as being particularly hazardous and crocidolite (blue asbestos) use was discontinued in the UK following the voluntary ban on its import in the 1970s. Import of amosite ceased in 1980.[51] Use of chrysotile (white asbestos; serpentine group) has also significantly declined but it is still allowed in many products, most notably asbestos cement and sheeting, and in friction products such as brake linings. For some applications man-made mineral fibres have subsequently been developed as substitute materials, although many of the uses of these fibres are new and unrelated to the historic use of asbestos. The man-made fibres being used in the largest amounts are the insulation wools (glass or fused rock products), which are used for both thermal and acoustic insulation. Energy conservation drives in the last decade or so have resulted in very widespread use of these materials in homes as loft and cavity wall insulation. They also have important applications in horticultural products and smaller amounts of slag wool and ceramic fibre are used in more specialized applications.

Because asbestos is a cheap and effective insulator, fire-retardant and reinforcing material, its use became widespread and it was incorporated into a wide range of products. The demonstrated health risks of asbestos have resulted in considerable concern about environmental exposure to asbestos fibres as well as worries about the legacy of historical occupational exposures. Exposure to asbestos fibres in smokers is especially relevant because of the synergistic interaction that has been demonstrated between asbestos and smoking in the induction of lung cancer.[52] Of particular current interest is the evidence that, contrary to expectations, the age-specific incidence of mesothelioma in men in the UK, instead of falling in parallel with the decrease in asbestos use, is actually increasing.[53] If the projection by Peto *et al.*[53] is accurate, British mesothelioma deaths will approach 3000 per year by 2020. This is in addition to the several hundred deaths per year ascribed to asbestosis and a similar number from lung cancer. Although most of the reported deaths are a result of previous occupational exposure, the level of mortality and morbidity does give rise to concern about environmental exposures, not only to asbestos (especially the amphiboles) but to other fibrous materials with similar characteristics.

The features of importance include physical fibre diameter and shape, and durability in the lung; the chemical composition of fibres, except insofar as it influences these properties, is generally considered less important, although the crystalline habit seems to be relevant, and some scientists believe that the

[49]D. H. K. Lee and I. J. Selikoff, *Environ. Res.*, 1979, **18**, 300–314.

[50]J. C. Wagner, C. A. Sleggs and P. Marchand, *Br. J. Industr. Med.*, 1960, **17**, 260.

[51]Department of the Environment, 'Asbestos Materials in Buildings', HMSO, London, 1986.

[52]P. T. C. Harrison and J. C. Heath in 'Mechanisms in Fibre Carcinogenesis', Plenum Press, New York, 1991, p. 469.

[53]J. Peto, J. T. Hodgson, F. E. Matthews and J. R. Jones, *Lancet*, 1995, **345**, 535.

propensity to give rise to oxygen free radicals is a key factor. Size and shape is important because, to have effect, fibres must be fine enough to penetrate the deep lung; the fibrous shape affects deposition and inhibits removal by the lung's natural defence systems. Many fibrous materials pose little or no risk because they are too coarse to gain access to the lower respiratory tract, or do not liberate fibrous dust during normal handling and use, or are highly soluble and therefore do not persist in the lung. All these factors have to be taken into account when considering the risk to health of fibrous materials, especially of the newer materials, encountered in the environment.

## 18.4.4  Pesticides

As well as their obvious beneficial effects, pesticides do present recognized hazards to human health,[54] mostly through high level occupational exposure and accidental poisoning incidents.

There are a number of ways in which humans can be exposed to pesticides through the environmental route. Pesticides used domestically in wood preservation or as household insecticides may be a particularly important source of exposure for the general public. The effects of pesticide residues in food and water probably cause the greatest public concern, although reports of clinical poisoning by residues seem to be extremely rare. Analysis of reported consumer poisonings by pesticides show that most arise from spillage of pesticides onto food during storage or transport, eating a food article not intended for human consumption (*e.g.* treated grain or seed potatoes) and improper application of pesticides.[55] Much of the concern about pesticide use, however, revolves around the long term accumulation in the environment and the low level but chronic human exposure to these compounds which occurs either directly or indirectly through the food chain.

Pesticides include insecticides, fungicides and herbicides. The main classes of insecticide are organochlorines (*e.g.* DDT, lindane, dieldrin), which are very persistent in the environment, anticholinesterases (the organophosphates – *e.g.* parathion – and the carbamates) pyrethroids and other botanicals and fumigants (*e.g.* ethylene dibromide). Fungicides include organometals, phenols and carbamates. Most herbicides are bipiridinium compounds (*e.g.* Paraquat), phenoxy compounds, organophosphates and substituted anilines. Other compounds are used to kill rodents, mites and ticks, molluscs, bacteria, birds and algae.

The best known pesticides are probably the organochlorines and the organophosphates. The organochlorines act as neurotoxins to the target organisms. As a class, the organochlorine pesticides are less acutely toxic to humans than some other insecticide classes but have greater potential for chronic toxicity. Many are now banned or restricted because of their persistence in the environment and their propensity to accumulate in the tissues of living organisms.[39] They are probably the compounds of most concern with respect to chronic environmental

[54]R. Levine in 'Handbook of Pesticide Toxicology', Vol. 1, Academic Press, New York, 1991, p. 275.
[55]T. C. Marrs in 'General and Applied Toxicology', eds. B. Ballantyne, T. Marrs and P. Turner, Macmillan, London, 1995, p. 1199.

exposure, and recently have also been implicated as possible environmental oestrogens (see below). Organophosphates, on the other hand, which function by blocking the activity of acetylcholinesterase, break down rapidly and do not accumulate in tissues. However, they are often extremely toxic and non-selective, and more instances of acute poisoning have occurred with organophosphates than with any other insecticide class.[56] Recent information on the effects of occupational exposure through their use in sheep dips[57] has given rise to concern in some quarters about possible low level environmental exposure to this group of pesticides. However, as with many other environmental contaminants, the lack of exposure data and the uncertainties regarding low dose extrapolation precludes any firm conclusions being made regarding risks to the general public of environmental exposure to pesticides.

### 18.4.5 Environmental Oestrogens

There is currently much public interest concerning possible adverse consequences arising from the release into the environment of chemicals with oestrogenic properties.

In humans there is an increasing body of evidence for changing trends in reproductive health.[58] In particular, elevated incidence of testicular cancer in men and breast cancer in women have been demonstrated and there is some evidence, albeit incomplete and uncertain, for reduced sperm counts and sperm quality, at least in certain parts of the world. Other effects of concern include cryptorchidism (undescended testes), hypospadias (a congenital malformation of the penis) and male breast cancer. A wide range of environmental pollutants have been implicated as possible xenoestrogens. Of these, synthetic hormones, phytoestrogens, organochlorine pesticides, polychlorinated biphenyls and alkylphenol ethoxylates are perhaps the most widely studied. It is to be expected that, as more and more chemicals are investigated, so the list of those with oestrogenic potential will expand.

Tying up with the concern about possible links between changes in human reproductive health and environmental oestrogens is the demonstrated occurrence of adverse reproductive changes in certain wildlife species linked to pollution. The best documented of these are the findings in Florida alligators. A decline in the population of alligators in lake Apopka in the 1980s was linked to poor reproductive success. Investigations revealed abnormalities in the reproduction organs of both males and females. Subsequent studies showed a possible role for organochlorine pesticides which entered the lake from a major spill of dicofol. Induction of vitellogenin (an egg yolk protein) in male fish and masculinization of females are other effects which have been detected and related to environ-

[56] W. Stopford in 'Industrial Toxicology', eds. P. L. Williams and J. L. Burton, van Nostrand Reinhold, New York, 1985, p. 211.

[57] R. Stephens, A. Spurgeon, I. A. Calvert, J. Beach, L. S. Levy, H. Berry and J. M. Harrington, *Lancet*, 1995, **345**, 1135.

[58] Institute for Environment and Health, 'Environmental Oestrogens: Consequences to Human Health and Wildlife', IEH, Leicester, 1995.

mental pollution. However, because of different physiologies, exposure routes, *etc.*, the relevance of observations in wildlife to effects in humans is very uncertain.

The organochlorine pesticides, because of their propensity to build up in the environment and concerns about their known effects on wildlife, are a particular focus of attention. The *o,p'*-isomers of DDT have been demonstrated to have oestrogenic activity and to bind to the oestrogen receptor, but these isomers are relatively unstable and are rarely found in the environment. The major and persistent DDT metabolite, *p,p'*-DDE, has no significant interaction with the oestrogen receptor and possesses no inherent oestrogenic activity. However, recent studies have shown that *p,p'*-DDE does bind strongly to the androgen receptor and inhibits androgen receptor action. Thus it is possible that DDT may display 'oestrogenic' properties through anti-androgenic activities of its metabolites.

There are many factors to consider when assessing the roles of environmental oestrogens on human health. Firstly, the oestrogenic potency of many of the relevant environmental contaminants is low or very low when compared to endogenous oestrogens such as oestrodiol, although differential protein binding complicates the picture. As with all toxicological reactions, there may be additive or synergistic effects between environmental oestrogens. Also, some environmental pollutants, such as dioxins and PCBs actually have anti-oestrogenic activities. Humans have always been exposed to natural phytoestrogens present in foodstuffs and over the last 50 years have been exposed to high concentrations of hormonally active substances in the form of medical treatments and oral contraceptives. Against this background, and considering the likely importance of the stage of life at which exposure occurs, it is currently very difficult to assess whether or not environmental oestrogens play a significant part in reproduction disorders in the general population. Further research is clearly needed in certain key areas before a firm assessment can be made of the risk to humans of exposure to environmental chemicals with oestrogenic or related activity.[58]

## 18.5  CONCLUSIONS

Following the industrial revolution and especially since the Second World War and the ensuing rapid growth of the chemical industry, an increasing number of foreign substances have been synthesized and either deliberately, accidentally or incidentally released into the environment. In addition, man's activities in utilizing the earth's natural resources has resulted in the release of pollutants, especially combustion products, into the local and global environment. The increasing human population and continuing rapid strides in the industrial development of Third World countries, linked with the inherent desire for improved quality of life, has exacerbated the problem. Against this, however, is the most remarkable trend, beginning in the 1960s, to consider ever more seriously the impacts of man's activities on the environment. This, coupled with increasing concern about how environmental quality can affect human health, has led, for example, to the development of such axioms as 'sustainable development' and the 'precautionary principle', and of 'cradle-to-grave' environmental assessments, 'ecolabelling' and 'product stewardship'.

The link between environmental chemicals and health is thus a topic receiving increasing attention. This review has looked at some of the ways that sudden (catastrophic) or more incipient exposures can result in health effects. In many cases the impacts of low level exposures on human health are suspected rather than proven. This is because of the problems associated with establishing causal links when effects and doses are small, and many other factors confound the situation in human populations. Exposure to contaminants can occur through three major pathways; ingestion, inhalation and dermal contact. The former is of key concern, as demonstrated by the cases reviewed here and by the obvious public interest revolving around the presence of pesticides and other chemical residues in food and water. In particular, the halogenated hydrocarbons, including dioxins and organochlorine pesticides, continue to attract attention. With respect to exposure by inhalation, pollution from road vehicles and the quality of air in the indoor environment are increasingly being seen as important. Clearly, total personal exposure (by whatever route) is fundamentally important, as is individual susceptibility to the effects of toxins determined, for example, by genetic make-up, age, current health status and exposure history. Developments in toxicology including, for example, the application of biomarkers and advances in modelling and risk assessment for low dose exposures, are enabling more accurate estimates to be made of population exposure profiles and health consequences.

One of the biggest issues of the moment relates to the properties of some ubiquitous environmental chemicals to act as oestrogen mimics or to otherwise interfere with the balance of sex hormones and therefore possibly affect human reproductive health and fertility. This, and other facets of the endocrine disruptive activity of chemicals, is likely to be an issue of increasing importance and of scientific and public interest.

CHAPTER 19

# The Legal Control of Pollution

R. MACRORY

## 19.1 FUNCTIONS OF POLLUTION LAW

In a modern industrial society, the control and management of pollution is underpinned by law. Pollution laws* tend to be complex and dynamic, and need to respond both to increased scientific understanding of the nature and causes of pollution, and to public and political perceptions and demands. Britain has a long history of pollution laws, but the last fifteen years have seen substantial changes, with new legislation, reformed institutional structures and far greater interest in the subject being shown by both legal practitioners and academics. At the same time, membership of the European Union and the growth of international environmental treaties has forced Britain to adopt policies and laws which are less isolationist than has sometimes been the case in the past.

The detailed substance of the law, then, may never be static, but the underlying functions of different legal provisions tend to remain constant. A valuable, if not essential, starting point of any analysis of the law is therefore first to consider the main purpose of the provision in question. This section identifies key functions of pollution law, while Section 19.2 examines the various sources of pollution law in this country. Section 19.3 discusses typical features of current British pollution law, illustrating these in more detail by examining the principal legal controls in the fields of waste, water, contaminated land and integrated pollution control. The concluding section deals with a number of significant policy themes that are emerging to influence the future design of effective controls.

Legal principles can, for a start, provide remedies for those whose interests are damaged or threatened by pollution. Such legal actions originally formed the mainstay of British pollution law and continue today to have a significant residual importance. In certain areas such as pollution of inland waters, such actions are fairly frequent with many cases settled by insurers before they ever come to court, and in recent years the growth of private law firms with a specialist

*The law in this chapter largely relates to England and Wales. In the environmental field, the same legislation sometimes covers Scotland and Northern Ireland, but often there are different laws. These usually follow the same policies but reflect the distinct administrative systems in these regions.

interest in environmental law has seen an increase in high profile litigation. But in themselves these types of 'private' remedies are unlikely to provide a wholly satisfactory basis for pollution control. Individuals may have neither the financial resources nor the stamina to enforce their rights through the legal system. In complex or novel cases of pollution, the evidential burden of identifying cause and effect may prove difficult to surmount, and is simply not appropriate to situations where there are many possible or cumulative sources, such as pollution from motor vehicles. Furthermore, in most cases it is not sufficient simply to show that a person or company who is taken to court caused the pollution; the principles of liability in private actions generally require that the defendant is also at fault in some way, or at the very least that the damage caused was reasonably foreseeable at the time of the incident.[1] Again, this may not be easy to prove. The purpose of these types of remedies is to protect individuals, private properties or the rights that are possessed over them. This will encompass many types of pollution damage such as damage to health, trees or crops owned by individuals, or inland waters and fisheries in private hands. But there remain important aspects of the environment which are not explicitly owned or protected by private interests but are subject to common enjoyment – the marine environment, wild flora and fauna and the upper atmosphere are all examples of such 'common' property. In this country we have yet to develop private legal remedies that can adequately protect such interests. Finally, it is generally difficult to enforce such remedies until at least some damage has occurred – this fits uneasily with modern environmental policy demands for taking preventative or anticipatory action in order to protect the environment.

Against this background, legal controls over pollution are now dominated by various forms of 'public' laws initiated by Government and Parliament and which provide a range of mechanisms for controlling pollution, irrespective of possible damage to private interests and with the task of enforcement generally assigned to some official agency or public body. The typical contemporary pollution law which performs this regulatory function will contain, within its provisions, a number of key elements. These will include, first and foremost, the key mechanism of legal control over the type of pollution concerned. This could take the form of a straightforward prohibition of a specified activity, backed by criminal sanctions. In other cases, the law may specify minimum standards for, say, a product that may give rise to pollution – lead in paint, vehicle emission standards and the like. More common, though, particularly when it comes to fixed sources of potential pollution such as industrial sites, is some form of licensing or consent system, ultimately backed by criminal sanctions, but allowing public authorities far greater flexibility to permit discharges or emissions into the environment but under controlled conditions.

A second critical part of a pollution law will be those provisions concerned with the administration of controls. These will identify the particular authority or

---

[1] 'Cambridge Water Co. Ltd *v.* Eastern Counties Leatherwork Plc', [1994] 1 ALL E.R. 53.

body responsible for implementation and enforcement, and will often deal in considerable detail with matters of process – for example, the procedural steps required when considering an application for a licence, the extent to which the general public have rights to be consulted during this process and rights of appeal against decisions, the powers available to enter and search premises and to take samples and the steps that can be taken to enforce the controls. All these elements, though sometimes resulting in dense and complex legal provisions, are in practice vitally important for the effective operation of the controls.

A third important element of any pollution law will be provisions concerned with the overall goals of the controls. Such provisions may be expressed in the broadest and most generalized terms; they may impose express duties on the enforcement body; or, at the other extreme, they may take the form of very specific requirements,expressed in numerate or quantitative terms and perhaps containing time-limits for attainment.

The first two types of provision – those concerning the basic control mechanism and those concerning the administration of the controls – are really necessary elements of any pollution law. It is less essential, though, for the third type to be present. The basic aims or goals of the controls can be considered to be matters that are left to be determined at the discretion of those responsible for administering the controls rather than stated in the body of the law itself. British pollution laws were long characterized by a reluctance to elaborate in legislation detailed environmental standards or goals, an approach encouraged by environmental scientists who argued that the need to take into account the nature of the local receiving environment when deciding what levels of pollutants could be safely released was inimical to national, detailed standards. But it is a feature that no longer can be said to prevail. The nature of European Community legislation, the effect of public sector privatization and growing public demand for more explicit standards has brought about some profound changes in the substantive content of British pollution laws, ushering in an era of more explicit and legally binding goals in many, though not all, areas.

Other important characteristics of pollution law need to be noted. Contemporary legislation may place a great deal of responsibility in the hands of public bodies, but to what extent can they be taken to court if they fail to exercise their powers and duties? Here, principles of what is known as judicial review, which have largely been developed by the courts themselves, will provide the solution. The question will largely turn both on the nature of the duty imposed by the law upon the authority – hence the significance in the way these duties are expressed in the legislation – and issues concerning the standing of those entitled to bring such an action. British courts require that such litigants have 'sufficient interest' in the matter, a deliberately vague phrase, which allows for a considerable amount of discretion and permits the courts to deny action to those they feel are mere busybodies. Generally, though, while still denying that every individual or every amenity group should have a right to bring actions against any government body, the courts in recent years have shown a more liberal approach than would have been apparent in the 1950s or 1960s; for example, environmental interest groups with a history of campaigning in a particular area have been considered to have sufficient interest to bring such actions against public bodies.* At the same time, it must be remembered that time and time again British judges stress that it is not their function to act as a simple appeal body or substitute their own judgement for that of a specialized public agency or official, even if they disagreed with the wisdom of the decision taken by that body – rather they aim to ensure the public bodies follow correct legal procedures, interpret the legislation correctly and do not act totally unreasonably. Actions for judicial review are therefore still rare, but in the environmental field are on the increase, especially as environmental groups and other interested bodies have more ready access to specialized legal advice. With the changing structure of the legislation, those public bodies responsible for its implementation and enforcement are more conscious than ever that they no longer operate in an uncluttered vacuum consisting of just themselves and those that they regulate.

Finally, it is important to emphasize that the efficiency of the type of typical regulatory machinery described above is increasingly being questioned. Such controls still predominate but policy-makers, both in this country and others, are constantly looking for more market-based solutions to replace what are disparagingly described as the traditional 'command and control' mechanisms. Such market-based tools are promoted as a more effective and less bureaucratic means for delivering improved environmental standards, with tax differentials for different forms of motor fuel being a well-publicized example. Improved information on the comparable environmental impacts of products may empower the consumer to influence the market, and persuade manufacturers to reduce their pollution impact. Fiscal charges on pollution discharges may be a greater incentive to reduce pollution than licence conditions which may be unwieldy or difficult to enforce. Markets may be established under which pollution permits may be traded between different industries, ultimately leading to a more efficient use of resources. Free market economists sometimes claim too much for their preferred solutions, and in doing so indulge in an excessive disparaging of the role of the law. Yet it is an important debate, and one that needs to be engaged, even if ultimately it ends up in the rather more prosaic but preferable conclusion that a combination of approaches is required, and that one needs to examine carefully the particular issue at hand to determine the best mix needed.

## 19.2 SOURCES OF LAW

Determining the applicable law to any particular example of pollution is not necessarily straightforward. There is no authoritative codification of law in the United Kingdom, nor is there a single, comprehensive law on pollution and it is therefore necessary to draw on a number of sources. Nearly all relevant legal principles in the field are now contained in Acts of Parliament, passed by Parliament and generally drafted and promoted by Government. These contain at least the main control mechanisms and administrative arrangements, and key

*R *v.* Secretary of State for the Environment, (2) HMIP, (3) MAFF, ex parte (1) Greenpeace, (2) Lancashire County Council; [1993]. See *Journal of Environmental Law* 1994, **6**(2).

recent pieces of legislation include the Environmental Protection Act 1990 (which, *inter alia*, deals with integrated pollution control and waste management), the Water Resources Act 1991 (water pollution) and the Environment Act 1995 (which includes provisions providing for a major restructuring of the institutional arrangements for pollution control). A modern tendency of most Acts of Parliament is to give power to a Government Minister to make regulations or other forms of subsidiary legislation at a later date. These are intended to flesh out legal details that might otherwise overload the body of the main Act of Parliament and since the process of changing and updating such subsidiary legislation is much simpler than for primary Acts, there are obvious attractions in this approach. Increasingly, too, one finds that provisions in environmental laws allow Ministers to provide official Codes of Practice or Guidance in specific areas; such documents are often known as 'quasi-legislation' in that they are not treated as formal laws, but may have considerable influence in the applications of legal controls, and may indeed be given some form of legal status under the main Act.

Statutes and their associated subsidiary legislation are therefore the primary sources of law in the field of pollution control. A secondary, and quite distinct source, derives from the principles developed by the courts when handling individual cases. Case-law has traditionally provided an important and dynamic input into the British system. For a start, the ultimate authority on the interpretation of statutory provisions (which, however well drafted, often contain ambiguities or unanticipated problem areas) rests with the courts. The courts also play a critical role in developing principles of law which are independent from those contained in statutory provisions, though again always in the context of deciding particular cases before them. In case of conflict, legislation will always prevail over such 'common law' principles, but Government and Parliament often consciously leave certain areas exclusively within the purview of the courts. In the context of pollution law, for example, the principles under which individuals may seek compensation for harm caused by polluting activities are almost wholly the creation of the courts, with statute law contributing little. Similarly, the grounds upon which the decisions of government departments or other public bodies may be reviewed by the courts have been almost exclusively developed by the judiciary.

A third source of law derives from international obligations, in the form of Treaties and similar agreements made between countries on a bilateral or multilateral basis. In contrast to the position in some countries, British courts have never recognized that provisions of international treaties can in themselves form part of the national law which can be directly enforced before the national courts. They are essentially treated as political obligations which must be converted by Governments into national legislation if they are to be given true legal status. Failure to comply with international treaties is therefore essentially a matter for international law and the machinery built into treaties to ensure implementation. As such, international treaties should, perhaps, be primarily regarded as an important lever in shaping national pollution laws and policy, although increasingly the courts are prepared to look at the texts of such treaties to assist them where appropriate in the interpretation of national legislation intended to implement their provisions.

The Treaty of Rome and its subsequent amending Treaties, which form the basis of what is now known as the European Union is, in one sense, an international obligation of the type described above. But it possesses such distinctive legal characteristics from other international arrangements that it really should be treated as a fourth source of law, and one that has assumed great significance in the field of pollution control. For a start, the Treaty established permanent European institutions including a body responsible for developing and securing implementation of Community policies, the European Commission, and, especially significant from a legal perspective, an international court, the European Court of Justice, based at Luxembourg. Accession to the Treaty implies an acceptance by the Member States of the role and authority of these institutions, and in the case of the European Court of Justice its rulings on the interpretation and meaning of Community law are treated as final and binding on national courts.

Again, in contrast to most international treaties, the Treaty of Rome contains very broad objectives and gives wide ranging powers to the Community institutions to pass what is in effect secondary legislation. The procedures for doing so are complex, involving the Commission, the European Parliament and representatives of national governments meeting in the Council of Ministers, and will differ according to the subject matter in hand. But, while the Council of Ministers has the final approval of proposed Community legislation, increasingly they may take decisions by a form of majority voting, meaning that sometimes Member States may find themselves bound by an obligation to which they did not agree, a position unique in international law. The original Treaty of Rome made no express reference to pollution or the environment, and it is only since 1987 that express provisions concerning environmental policy were inserted into the Treaty. The absence of such specific provisions did not, however, inhibit the development of Community policy and law in this area and since 1973 both the Commission and the other Community institutions have been highly active, producing a series of Action Programmes and agreeing several hundred individual items of the law relating to environmental protection. These have generally taken the form of 'Directives', implying that strictly they represent obligations on Member States to achieve the aims of the Directive within a specified time-limit (usually two or three years) but leaving each country considerable discretion as to the precise legal or administrative means to do so. Community legislation has had considerable influence on many areas of pollution law within Britain, especially in the fields of air pollution and water pollution. Part of the reason is that they have tended to be much more explicit in terms of quantitative environmental standards than has been the tradition in British legislation, requiring British law to adapt to the new policy demands, and forgo long-prevailing characteristics.

The final, important feature of Community law which distinguishes it from other sources of international law, is that the European Court of Justice has long insisted that where there is a conflict between national and Community law, Community law must prevail. In certain instances, this may require the national law to be interpreted to fit in with the Community law. In other cases, where there is no relevant national law, the European Court of Justice has held that in some instances both provisions of the Treaty itself and subsidiary legislation may be directly invoked before national courts. The British judges have accepted the supremacy of the Community law, some with more enthusiasm than others, and, indeed, the Act of Parliament which governed British accession expressly requires national courts to apply the doctrines of the European Court.[2] The result is that Community law and legislation has a heightened legal significance in the application and interpretation of national law, quite beyond any other international source.

## 19.3    SOME UNDERLYING CHARACTERISTICS OF BRITISH POLLUTION LAW

Before examining in more detail relevant laws in some key areas of pollution control, it is worth emphasizing a number of prevailing characteristics. These are not necessarily common to every aspect of pollution control but are sufficiently consistent to mark a strongly etched pattern, though one that is now subject to substantial change. First, historically, statutory provisions (both Acts of Parliament and subsidiary legislation) have tended to be developed on a specialized basis dealing with different aspects of pollution on a medium by medium basis – air, water, land, *etc.* Inconsistencies in approach between different laws can be readily identified, and these can often be explained by historical factors or sectorial influences operating on the law in question rather than by distinctive issues relevant to the type of pollution in question. Only in 1990 was an integrated system of pollution control regulated by a single agency introduced for certain industries. Under the Environment Act 1995 the different pollution agencies for integrated pollution control, water pollution and waste disposal were combined into a single body, the Environment Agency (EA). For the time being, though, the Agency will still be dealing with different bodies of pollution laws, though the opportunity has been taken to iron out a number of current inconsistencies.

The lack of legally binding environmental standards used to be a prevailing characteristic of British pollution law. Environmental policy makers and scientists resisted the concept of absolute standards relating solely to the property of the substance being emitted or discharged, but insisted that one must take into account the nature of the receiving environment and the target to be protected (whether human, other living organisms or physical features) taking due account of environmental heterogeneity and the resilience of the physical environment to handle and absorb pollutants without causing substantial damage.

Localized enforcement and localized decision-making dominated much of the structure of the law. The last decade, though, has seen substantial changes. Legally binding standards of various sorts, such as quality standards for bathing waters and air emission standards for specific industrial processes have been

[2]European Communities Act, 1972.

introduced into the body of national legislation, largely as a result of European Community legislation. At the same time, as we have seen, there has been a growing centralization of the enforcement bodies, culminating in 1995 with the removal of waste law enforcement from local authorities to the new Environment Agency. By no means are all areas of pollution control covered by legally binding standards – contaminated land, for example, remains an area where despite political demands for the creation of standards, the prevailing philosophy is to resist their introduction. Other fields such as the control of noise pollution from stationary sources remain firmly within the control of local authorities.

Another substantial change that has taken place in British pollution law concerns public information and public participation. Since 1947 most new developments of land have required planning permissions under Town and Country Planning laws, and rights of public consultation and involvement in decision-making have been concentrated at this stage of the procedures. Pollution control legislation, including the licensing of discharges, was considered largely a technical and private matter between the regulator and company or person being regulated. Again, much has changed in the last decade or so. Public registers containing details of licences or discharge consents, and various details of sampling results have been created, and there are more extensive legal provisions concerning public consultation over applications for licences or discharge consents. In 1992, regulations were introduced, implementing a European Community directive on the subject, granting extensive public rights to access to information concerning the environment held by government and other public bodies.[3]

Nearly all statutory pollution controls ultimately rest on criminal offences of various types, enforceable by the regulatory authorities, and normally framed in 'strict' terms meaning that the mere act of non-compliance is sufficient to secure conviction; it is not necessary, as with most other types of crime such as theft, to prove that the defendant actually had a criminal intention. Despite this backdrop, most regulatory authorities, faced with non-compliance, have treated prosecution very much as a last resort, to be pursued only against the recalcitrant or gross violator. In recent years, though, a rather more aggressive approach can be detected. Something of a turning point was marked by the 1989 prosecution of Shell by the newly formed National Rivers Authority for pollution of the Mersey which resulted in an unprecedented £1 million pound fine. Recent legislation has significantly increased the maximum penalties for environmental offences.

## 19.4 WATER POLLUTION (see also Chapter 1)

The Water Act 1989 provided a major shift in contemporary legal controls and brought about the privatization of the responsibilities for the delivery of water supplies and sewage treatment which had previously rested with public sector bodies. The Act established a new government agency, the National Rivers

---

[3]'Environmental Information Regulations', S.I. No. 3240.

Authority (NRA) because it was recognized that environmental protection was a matter that could not rest exclusively in the private sector after privatization, and one of the Authority's primary functions is to set discharge consents and monitor and enforce the applicable pollution controls. The Water Act 1989 also gave the Secretary of State the power to establish statutory water control objectives, which provide an explicit goal for the controls, placing the key responsibilities on the NRA to achieve these objectives.[4] Whilst the 1989 Act was ground-breaking, it was repealed and replaced by new Acts of Parliament in 1991 which were essentially designed to provide a consolidation of the main laws concerning water and sewage. In particular, the Water Resources Act 1991 now provides the key controls concerning the control of water pollution in both ground water, inland waters, estuaries and the sea within three miles of the coast.

### 9.4.1 Discharge Consents

The central control mechanism is a consent or licensing system operated by the EA, and underpinned by criminal offences. In the absence of a discharge consent it is an offence under Section 85 of the Water Resources Act 1991,[5] to either cause or knowingly permit the discharge of any (i) poisonous, noxious or polluting matter, or (ii) any trade or sewage effluent, into controlled waters, whether or not it is potentially harmful. The leading decision of the House of Lords in Alphacell *v* Woodward[6] held that the word 'cause' in the first limb of the offence implied that the prosecution only had to prove a causal connection between the discharge and the defendant's activities – the offence could be committed even though there was no actual intention or negligence present. This approach was reaffirmed in Wrothwell Ltd. *v* Yorkshire Water Authority,[9] where a company director who poured herbicide down a drain connecting to a river was convicted of the offence of causing effluent to enter waters, even though he actually believed the drain only led to a public sewer. But precisely what is meant by causing something to happen has given rise to a considerable body of case-law and more recent cases such as Wychavon District Council *v* NRA[8] and NRA *v* Welsh Development Agency,[9] have suggested there must be some 'positive, deliberate' act on behalf of the defendant if he can be said to cause the discharge – standing by, or even failing to maintain abatement equipment may not be sufficient. If there is no such positive act, it may be more appropriate to prosecute under the 'knowingly permit' limb, though here clearly some knowledge on the part of the defendant is required. Another important recent case, NRA *v* Egger (UK Ltd.),[10] considered what was meant by 'polluting' matter and concluded that it was not necessary for the prosecution to show that harm had actually been caused by the discharge in

[4] Water Act, 1989, Section 106.
[5] Water Act, 1989, Formerly Section 107.
[6] 1972; [3 ALL ER 475]
[7] 1984; [Crim. L.R. 43]
[8] 1993; [ENV. L.R. 330]
[9] 1993; [ENV. L.R. 407]
[10] 1992; [ENDS Rep. 39]

question. The purpose of the controls was to prevent or minimize pollution and in this context 'polluting' implied something that was capable of causing harm.

### 19.4.2   European Community Directives

The development of a number of important European Directives concerning various aspects of water pollution was a central aim of the early years of the Community Environmental Action Programmes. The cumulative effect of the Directives agreed to date has brought about a radical change to the climate of legal regulation in this country by introducing into the system a layer of quantitative numerate standards in respect of water quality.

Many of the EEC Directives relating to water pollution contain a variety of quality standards for various categories of water depending on its use (*e.g.* bathing waters, drinking waters). One of the key Directives (76/464) concerns the discharge of dangerous substances into the aquatic environment and is sometimes known as a 'mother' or framework Directive since it essentially lays down a set of general principles concerning the application of controls over the discharge of effluents, leaving numerate standards for individual substances to be prescribed in subsequent 'daughter' Directives. The framework Directive specifies a number of broad families of chemical substances, dividing them into Lists: List I, containing those considered most toxic (the 'Black list') and List II (the 'Grey list'), containing those that are less dangerous. The stated aims of the Directive are to 'eliminate' pollution caused by List I substances and to 'reduce' pollution caused by List II (see also Chapter 1).

The development of the Directive was particularly controversial because it revealed fundamentally distinct philosophies over the most effective approach towards setting pollution standards, with the United Kingdom pressing for one approach, and other Member States content with a different one. At the time, such Directives had to be agreed by unanimous voting by the Member States and the Directive was only eventually agreed by allowing the option for both approaches for List I substances. These arguments are still apparent in many fields of pollution control beyond the issue of dangerous substances into waters.

Originally both the Commission and other Member States envisaged that in respect of List I substances subsequent daughter Directives relating to the individual substances would specify a minimum emission standard relating to the actual discharge point (known as a 'Limit Value'). Member States could apply stricter standards in particular circumstances but could never go below it, whatever the nature of the receiving waters. This was seen as a clear signal to eliminate the discharges of such substances, an approach that was reasonably straightforward to administer and a system that was transparent, especially to countries with cross-border river systems. In contrast, the UK, with its long standing emphasis on the need to take account of heterogeneous receiving environments in determining individual pollution consents, argued that even a minimum emission standard was sub-optimal for the environment and pressed strongly for the system that prescribed standards of the receiving waters (known in Community law as Quality Objectives) rather than the discharges themselves, allowing considerably more flexibility to determine the most effective means for reaching those standards. Both approaches are now permissible under the Directive for List I substances with subsequent daughter Directives defining both Minimum Emission Limits for discharges and Quality Objectives for the receiving environment, with Member States having the option to choose either route.

Progress in agreeing subsequent daughter directives for the vast quantity of possible chemicals in both List I and List II has proved immensely slow, and in 1994, a draft Directive intended to create a framework for improvements in the ecological quality of all surface waters throughout the EEC was published by the European commission.[11] In many ways the proposed approach reflects the British preferred policy of emphasizing the quality of the receiving environment, and giving considerable discretion as to how to achieve those standards.

## 19.5   CONTAMINATED LAND (see also Chapter 14)

Land contamination is likely to be a hugely complicated problem, focusing on a number of equally complex variables relating to historic uses of a site, the likely mixture of polluting substances on the site, the web of ownership and control over the site and the rights and interests of those who, in terms of proximity, are adversely affected as neighbours or otherwise.[12] The UK's traditional approach to contaminated land has been to promote its restoration, with grants being made available to redevelop sites which have been left in a detrimental condition from past use.[13] This 'end-use' policy is conditional on the particular purpose the land will be put to, so certain developments will require a higher standard of clean-up accordingly.

The United Kingdom possesses little comprehensive legislation directly relating to contaminated land or soil protection as such, but a large number of areas of law are relevant. Key specific provisions can be found in the Environmental Protection Act 1990, the Town and Country Planning Act 1990, the Water Resources Act 1991 and most recently the Environment Act 1995. Crucially, all new development of land generally requires planning permission under the Town and Country Planning legislation, and local authorities have long been encouraged to use the procedures to ensure that land which is contaminated is not developed without sufficient remediation. These provisions will, of course, only have purchase where new developments are taking place. Land contamination can lead to severe pollution of groundwater, and under the Water Resources Act 1991 the NRA (now EA) had the power to carry out anti-pollution works to remove or dispose of polluting matter which might contaminate both surface and ground waters, and to remedy or mitigate its polluting effects in restoring the waters to their previous state.[14] The Authority may also recover any expenses

---

[11]Com. (93) 680, (OJ C222, Vol. 37, 10.8.94)
[12]Hawke, Environmental Health Law, 1995.
[13]'Development of Contaminated Land', Circular 21/87, D.O.E., 1987.
[14]Water Resources Act, 1991, Section 161, HMSO, London, 1991.

incurred from the person who caused or knowingly permitted the polluting matter to be there, but there have been scarcely any instances to date where the NRA recovered its costs under these provisions for any major case of contamination.

The whole issue of contaminated land in the United Kingdom rose on the policy agenda with the publication of the 1990 Report of the House of Commons Environment Committee. The Committee was particularly concerned at the lack of a consistent policy on the subject, and the muddled state of the law, especially when it came to the principles of liability for those who caused contamination or who innocently became owners of land which was contaminated. One of their more controversial recommendations was a proposal that there should be public registers concerning contaminated land. The underlying benefit of these registers would be for developers and others with interests to identify contaminated land, and then take appropriate remedial action. Although the Government initially accepted this recommendation and even provided for establishment of registers in the Environmental Protection Act 1990, the provisions were never brought into effect. The idea of property becoming blighted once it was entered on to the register and the uncertainties as to quite what contamination meant or implied raised immense controversy and the 1995 Environment Act repealed these public register proposals. Instead, a more modest policy is proposed under which local authorities will identify sites in their area which they consider to be contaminated (with guidelines issued by Central Government) and proceed to a programme of restoration. The Act allows for the financial costs of restoration to be borne by landowners and other persons responsible, but with a number of complex provisions which aim to ensure that totally innocent parties do not bear unreasonable costs. One of the principal aims of the Act is to restore confidence and reduce uncertainty in the property market, but it remains to be seen whether this will effectively re-evaluate and improve the law, or whether it will cause greater long term controversy and create a fair amount of significant new law. Certainly, the Act does not deal explicitly with the question of what precisely is meant by 'contaminated' – 'contaminated land' is defined as land which appears to the local authority to be in a condition such that 'significant harm' is being caused or there is a significant possibility of such harm being caused, or that pollution of waters is being or is likely to be caused. Government will be issuing guidance to local authorities of the meaning of these phrases,[15] which will involve questions of risk and hazard assessment. Nor does the legislation try to define what standards of remediation are required. These are questions still argued over by policy makers and technical experts, and perhaps can never be effectively defined in the law.

## 19.6 WASTE (see also Chapter 15)

Nearly all solid wastes arising in the United Kingdom, both domestic and commercial, are disposed of on engineering landfill sites. Specific legislation

---

[15]See Draft Guidance on Determination of Whether Land is Contaminated Land under the provisions of Part IIA of the Environmental Protection Act 1990, Department of the Environment, May 1995.

dealing with the regulation of waste disposal was not introduced until 1974. Part II of the Environmental Protection Act 1990, and a number of European Community Directives have introduced new, more stringent controls that now shape the core of contemporary regulation. The focus is on household, industrial and commercial waste, termed 'controlled waste', and Section 34 of the 1990 Act places a duty of care on all those dealing with waste, from its generation to its disposal. Every person who is subject to that duty of care must ensure not only that they do not commit an offence, but that any person to whom they deliver waste does not either. The aim is to ensure that waste, which will generally have no value to the person trying to get rid of it, remains within legal waste disposal streams from its delivery to transporters to its final resting place. These provisions are backed up by a statutory code of practice,[16] and the breach of the duty of care is in itself an offence, and can lead to criminal proceedings being instigated.[17]

Although the current legislation is far more effective in the enforcement of legal controls it still has its shortcomings. The system may be over optimistic as it makes the duty of care self-enforcing, through waste holders checking on each other primarily to avoid breaking the law.

### 19.6.1   Licensing of Facilities

Section 33 of the Environmental Protection Act 1990 provides the legal backbone for the waste management licensing system, and is based on a classic licensing system backed by criminal penalties. The responsibility for the administration of the controls has traditionally rested with local authorities, generally at County Council level, but these powers have been transferred to the new Environment Agency, bringing the responsibility for nearly all areas of pollution regulation under one body.

The 1990 Act strengthened the previous licensing regime, and in particular attempted to introduce greater professionalism amongst those who were licenced to operate waste disposal facilities by requiring them to be a 'fit and proper' person and relevant considerations will include conviction for certain offences and the technical competence of the applicant. Another key area of reform was to introduce provisions which deal effectively with the position after the actual disposal operations have ceased when technically the operating licence was no longer legally enforceable. Under the 1990 Act licences may only be surrendered if the regulatory authority accepts the surrender, and it must be satisfied that the condition of the land is unlikely to cause pollution of the environment or harm to human health. Clearly, operators of landfill sites will increasingly find the cost of doing so more expensive, and the Government's introduction of landfill tax in 1995 is further intended to increase the cost of landfill in an attempt both to squeeze out less financially secure operators and to reduce the cost disadvantages of other methods of dealing with waste including recycling and incineration. As with contaminated land, the legislation says little in precise detail about the environmental standards to be applied by the authorities when it comes to the

[16]Environmental Protection Act, 1990, Section 34(10).
[17]Environmental Protection Act, 1990, Section 34(6).

licensing of landfill sites, leaving this largely to their discretion but with the advice of various technical guidance notes issued by central government. A proposed European Community Directive on Landfill[18] would introduce more detailed prescriptive rules in this context, but as a result has proved politically controversial. The position can be contrasted with that of one of the other main forms of waste disposal, waste incinerators, where various emissions standards of increasing strictness have been introduced under Community law.

### 19.6.2  Imports and Exports

The main legal controls over the import and export of wastes are currently contained in the Transfrontier Shipment of Waste Regulations 1994.[19] These are intended to implement a European Community Regulation (259/93), itself implementing an international convention on the subject which seeks to control the movement of hazardous wastes worldwide and to prevent illegal movement of waste. This is as a result of growing public and political concern over the whole question of imports and exports. An example of this is the prohibition of exports of hazardous waste to African, Caribbean and Pacific States. The 1994 Regulations also aspire to sanction a free trade area in Western Europe for the free movement of waste. However, there is a deviation from the normal single market rules, which can allow Member States to prohibit imports of waste for disposal from other EEC Member States. The EC Regulation establishes a prior notification system which requires the country to which the waste is destined, consenting before it is shipped. A notification about a shipment is made by a consignment note, and Member States must not accept shipments unless they are accompanied by a consignment note. Additionally, the National Regulatory Authority must be given advance notice of imports, and it has twenty days in which to either acknowledge the consignment note (only then may the holder of the waste proceed with shipment) or to make a written objection.

### 19.7  INTEGRATED POLLUTION CONTROL  (see also Chapter 20)

The genesis of integrated pollution control lay in the 1976 Report of the Royal Commission on Environmental Pollution which stressed the links between air, water and land pollution, and concluded that sub-optimal regulatory decisions could not be reached unless pollution control legislation was applied in a consistent and co-ordinated manner. Fourteen years later, Part I of the Environmental Protection Act 1990 formally introduced the concept of integrated pollution control into British law. The Act applies to specified industrial processes prescribed in regulations by the Secretary of State for the Environment,[20] and these now cover most heavy industrial and chemical industries. The lynch-pin of the controls lies in the requirement of an authorization from Her Majesty's

---

[18]Draft Directive (93/C212/02).
[19]Transfrontier Shipment of Waste Regulations, S.I. No. 1137.
[20]Environmental Protection Act, 1990, Section 2.

Inspectorate of Pollution (now part of the new Environment Agency) before such a process can be operated. At first glance, the new controls were an extension of the long-established requirement to register certain industries with the Inspectorate and its predecessors for the purposes of regulating emissions into the atmosphere, but it is clear that they go far deeper.

Conditions can be attached to an authorization, and the legislation defines the basic objectives of the controls as being to ensure that the operator (a) employs the best available techniques not entailing excessive costs to prevent the release or render harmless substances in so far as they could cause harm if released into the environment and (b) where the process might release substances into more than one environmental medium to employ the 'best practical environmental option' in doing so. Clearly the legislation represents an ambitious approach to adopt a radically different approach towards pollution control, albeit still one rooted in familiar regulatory control methods. The prevention of pollution as opposed to its amelioration or abatement is given greater emphasis. Furthermore, the legislation introduced greater rights of public consultation over applications for authorization, and public registers containing details of authorizations must be maintained. The critical terms, Best Available Technology Not Entailing Excessive Costs (BATNEEC) and the Best Practicable Environmental Option (BPEO) are concepts which receive little in the way of elaborate legal definition, and clearly involve complex economic and technological judgements. With respect to BATNEEC, for example, the Act makes its clear that the term 'techniques' is broad and encompasses the design and layout of buildings, and the numbers and qualifications of staff. But no definition of 'available' or 'not entailing excessive costs' is provided for, raising some interesting questions. For example, to what extent can an advanced pollution abatement technology employed in another country be said to be available here? Are the costs of techniques which might easily be absorbed by a multi-national industry still 'excessive' for a much smaller company which might be undercapitalized due to its own poor financial management? These sorts of questions have initially, at any rate, to be answered by the Inspectorate, both in Guidance Notes they issue and in their own decisions on licences. Eventually, though, it is likely that the courts will be called upon to provide more authoritative interpretations, though not necessarily any more correct from an environmental point of view.

## 19.8   CONCLUSIONS

Pollution law in the United Kingdom has undergone a period of rapid change in recent years. New legislation and institutional structures have seen a shift away from dealing with pollution on a medium by medium basis, and attempting to view the issue on a more holistic basis. Greater policy emphasis on the anticipation and prevention of pollution implies that stronger connections must be made between specialized pollution regulation and other forms of environmental control. In this context, environmental assessment and the linkage between land use planning controls and pollution regulation has a potentially crucial role to play where new developments are concerned, although as yet there remain problems in the effective dove-tailing of the two regimes.[21] The emerging discipline of life-cycle analysis may similarly play a significant factor in assessing and anticipating the pollution implications of products placed on the market. Another connected theme concerns the end-use or decommissioning of facilities and products where as yet pollution law remains at a fairly undeveloped stage, though initiatives have been made in certain areas such as the regulation over the closure of waste disposal sites. At the same time, traditional features of UK pollution regulation emphasizing discretion and flexibility have had to give way to contemporary demands for greater transparency in information and decision-making and greater precision in the body of legislation.

This in itself can bring about fears of excessive legalism and inefficient bureaucracies, and has made all the more appealing the much greater use of market instruments in place of more traditional regulatory techniques. Even the current concern with the principles of civil liability for environmental damage can be seen as part of a concern to internalize environmental costs and shift the balance of legal influence from control by public bodies to private interests. Certainly, in a free-market economy, harnessing the power of the market to assist in improved environmental protection must be an attractive proposition, though it is self-deluding to think that legal regimes can be totally avoided.

Conventional licensing techniques, a long-time central feature of pollution control, tend to focus on point sources of pollution such as discharges from industrial premises. Pollution policy must now also be concerned with a far wider range of disperse sources such as run-offs from agricultural land and, in a broader context, major sources of mobile pollution such as motor vehicles. Here a whole range of legal techniques and policies must be employed, ranging from product standards (such as vehicle emission controls), fiscal incentives and disincentives and various types of land use measures – in an ideal world, all employed in a consistent manner.[22] In effect, the longer-term aim is to modify behavioural patterns and methods of working in a way that goes well beyond the more familiar approach of regulating the results of pollution, and to begin to meet some of the aspirations inherent in the concept of sustainable development. The challenge for pollution law in the first part of the twenty-first century will be its capacity to adapt to these new demands.

*Acknowledgements.* My thanks to Ray Purdy for assistance in the preparation of this chapter.

---

[21]See E.C. Directive on the Assessment of the Effects of Certain Public and Private Projects on the Environment (85/337/EEC) and in particular the Town and Country Planning (Assessment of Environmental Effects) Regulations 1988, S.I. No. 1988/1199.

[22]See the 18th Report of the Royal Commission on Environmental Pollution (1994), HMSO cmnd 2674.

CHAPTER 20

# Integrated Pollution Control and Waste Minimization

D. SLATER

## 20.1 AN INTRODUCTION TO INTEGRATED POLLUTION CONTROL

The introduction of the system of Integrated Pollution Control (IPC), proposed by the Royal Commission on Environmental Pollution (RCEP) in their Fifth Report[1] and embodied in the Environmental Protection Act (EPA 90) 1990, marked an important milestone in the development of the legislative philosophy and framework in the UK. The Act is of major importance since it largely established Britain's strategy for pollution control and waste management for the foreseeable future. The Act itself is divided into nine main parts. However, only Part I is relevant to the theme of this article.

Before the introduction of IPC under the Act in April 1991, emissions from major polluters to the three environmental media of air, water and land were subject to individual and distinct control regimes. IPC provides a mechanism and a legal basis for looking at the impact which a process as a whole has on the environment as a whole. IPC takes a holistic approach, ensuring that substances which are unavoidably released to the environment are released to the medium to which they will cause the least damage. It embodies the precautionary principle: prevention is better than cure. As the saying goes: prevent, minimize and render harmless. Also, the regulatory process, from application, through authorization, to regular returns of monitoring releases to the environment, and, where appropriate, to the enforcement action by Her Majesty's Inspectorate of Pollution (HMIP),* is open to public scrutiny and comment.

The Act, published in 1990, is 'enabling legislation', in which the regulatory framework is established. The specification and, at times, variation of the requirements for authorization of processes are made in Regulations published

[1]'Fifth Report by the Royal Commission on Environmental Pollution', HMSO, London, 1976, Cmnd 6371.

*By the time this chapter is published, both HMIP and NRA will have been subsumed into the Environment Agency.

under the Act, which reflect changes in knowledge and experience allowing continuous improvement in environmental protection without the need to resort to primary legislation. Typical, and perhaps the most important of these Regulations, are the Environmental Protection (Prescribed Processes and Substances) Regulations 1991 (Statutory Instrument SI/1991/472), and it is useful to examine these as an example of the purpose of the Regulations. As described above, Part I of the Act makes provision for integrated pollution control (IPC). It also makes provision for the control of air pollution by local authorities.

Regulations SI/1991/472 also provide a framework for the implementation of a substantial number of EC Directives relating to air pollution from industrial plants. It should be noted that since SI/1991/472 came into force in April 1991 in England and Wales, and in 1992 in Scotland, it has been modified several times. These amended Regulations, in general, extend (or at least redefine) the prescribed processes covered by the original Regulations and the dates for their authorization.

IPC applies to all processes in England, Wales and Scotland falling within any descriptions of processes prescribed for the purpose by the Secretary of State for the Environment. The Act provides that no prescribed process may be operated without an authorization from HMIP after the date specified in the regulations for that description of process.

In setting the conditions within an authorization for a prescribed process, Section 7 of the Environmental Protection Act 1990 places HMIP under a duty to ensure that certain objectives are met. The conditions should ensure:

(i) that the best available techniques not entailing excessive cost (BATNEEC) are used to prevent or, if that is not practicable, to minimize the release of prescribed substances into the medium for which they are prescribed and to render harmless both any prescribed substances which are released and any other substances which might cause harm if released into any environmental medium;

(ii) that releases do not cause or contribute to the breach of any direction given by the Secretary of State to implement European Community or international obligations relating to environmental protection or any statutory environmental quality standards or objectives, or other statutory limits or requirements; and

(iii) that when a process is likely to involve releases into more than one medium (which will probably be the case in many processes prescribed for IPC), the best practicable environmental option (BPEO) is achieved *i.e.* the releases from the process are controlled through the use of BATNEEC so as to have the least effect on the environment as a whole).

HMIP is also charged with delivering the National Plan for reduction of $SO_2$ and $NO_x$ emissions by means of the authorizations which it grants under Part I of the Environmental Protection Act 1990. This translates a blanket concept, which takes no account of the pollution potential of an individual plant, into a site-specific allocation, the use of which can be accounted for by the operator, audited by HMIP, and enforced against if necessary.

The processes covered by the National Plan can broadly be grouped under three headings: first, the electricity supply industry; second, the petrochemical industry, comprising the refineries and third, other industry, which picks up the power-generating combustion processes of 50 MW input or more.

There is often confusion about whether the National Plan takes precedence over BATNEEC and BPEO, or *vice versa?* The answer is simple. All objectives have to be achieved, so effectively it is the most stringent which will prevail. If BATNEEC standards are the tighter, then BATNEEC is pre-eminent. If BATNEEC would allow greater releases than the National Plan allocation, then National Plan prevails.

## 20.2  PROCESS REGULATION

The Environmental Protection Act 1990 requires an 'Authorization' to be issued by the 'Enforcing Authority' which, in the case of IPC, is Her Majesty's Inspectorate of Pollution. 'Authorization' means an authorization for a process (whether on premises or by means of mobile plant) granted under Section 6 of EPA 90.

The main goal, therefore, of an authorization is to specify the limits and conditions which are important to achieving the objectives of IPC in a particular circumstance. These are in the main likely to comprise limits and conditions on feed materials, operating parameters and release levels. In turn, the detail of these, particularly of the latter, such as the period over which they apply, will need to take into account what constitutes BATNEEC, environmental impact, *e.g.* concentration or load-dependent process characteristics, *e.g.* cyclic variations, fluctuations and practical considerations.

The operator must provide a strong, detailed justification of his process. Particularly where it is new or modifications are proposed, then all the options must be explored and justified. He must list the substances which might cause harm, that are used in or result from the process. He must identify the techniques used to prevent, minimize or render harmless such substances. He must assess the environmental consequences of any proposed releases. He must detail the proposed monitoring of the releases. This comprehensive environmental assessment then becomes the corner stone of the assessment of the application by HMIP and available to all on the public register. HMIP can also be judged, by the public, on the way it responds. The public should then have confidence in the system.

The authorization document follows a standardized format, produced within HMIP. The conditions in the authorization have been set to ensure a consistency from one authorization to another. The standard authorization format is set out with standard assumptions. These are that the Applicant should be an expert in his own process, and that the Inspector is the expert on the requirements of the regulations and on assessment. The Applicant will have a much better under-standing of his operations, requirements, process control and variations in conditions than will the Inspector. It is for the Applicant to detail, where relevant to his process, the practicable conditions and assess the environmental implications. It is for the Inspector to assess the validity and acceptability of the proposals in respect of statutory environmental requirements. The application itself forms part of the authorization.

Releases to the various media are dealt with in a consistent manner in separate parts of the authorization. A plan should be included in the application which identifies the position of the various release points. The Inspector must consult with the National Rivers Authority (NRA) with respect to discharges to 'Controlled Waters' from the process, and include as a minimum requirement any conditions which the NRA insist shall be included. HMIP may impose conditions which are more onerous than those required by the NRA (*e.g.* continuous monitoring instead of spot sampling). Inspectors must check that the criteria used by NRA are appropriate. Section 28(3)b of the EPA requires that the inclusion of NRA's requirements are conditions of the authorization.

The Waste Regulatory Authority will be responsible for final disposal of waste to land (including on-site final disposal) in accordance with their licensing system. HMIP is responsible for control of any releases to land not subject to control by the Waste Regulatory Authority.

An authorization includes an Improvement Programme. For a new process the improvement programme may contain requirements for additional equipment/ controls, *etc.* The Applicant must demonstrate that BATNEEC applies during the interim period (*e.g.* cost of lost production *versus* environmental effects if there is a delay or an extended period in delivering equipment). The regulation of releases will be enforced by regular site inspections, checking against authorization conditions. The operator will be required to provide information to HMIP as specified by the authorization conditions. Whenever breaches of authorization conditions are identified, due consideration will be given to enforcement actions that could lead to prosecution in the Courts.

## 20.3 Monitoring

Unfortunately, the word 'monitoring' is open to wide interpretation. In its general sense it includes all HMIP's regulatory functions, including physical inspection of processes and plants, the environment around the sites and the paperwork, *e.g.* operators' returns, *etc.* However, in the context of authorizations we are referring to 'compliance monitoring' involving the measurement and recording of those aspects of a process which are subject to limitation or condition and 'environmental monitoring' when this is required as a condition. Although narrower than the general interpretation, 'compliance monitoring' can still encompass a wide variety of measurements.

When determining compliance monitoring requirements under IPC, Inspectors consider the following points.

(i) *What information is required to provide confidence that the process is being operated within authorized limits and conditions?* These requirements must be capable of providing data for the period over which the limit is expressed. Thus, if the limit is 'instantaneous', then a spot sample (or measurement) will be specified. If the limit is an average concentration expressed over 24 hours, then a 24-hour sample (or measurement) will be specified. If compliance with a load limit is to be monitored, then not only must the sample be taken over the period of the limit but a measure of flow will also be required.

(ii) *How might the information best be obtained?* Could more reliable data be obtained by deduction based on a knowledge of impurity levels in feedstocks and their behaviour during processing, *e.g.* mercury in caustic? Or is measurement the best option? If so, are on-line continuous monitoring systems available – do they constitute BATNEEC for that process? If reliance has to be placed on laboratory analysis of samples, how are they to be taken – by continuous sampling on a flow or time proportional basis, and over what time, or if by periodic sampling at what frequency and duration? Might surrogate measurements be more practicable? For example, for dioxin emissions from incinerators, continuous monitoring is not available; periodic testing should be called for but is too expensive to employ more than perhaps once per month. Measurements of temperature, flow rates, feed composition, *etc.*, would provide valuable information on whether an incinerator of acceptable design has been complying with dioxin emission limits.

EPA 90, Part 1, places no statutory requirement on either operator or regulator to carry out monitoring beyond providing Inspectors with rights of entry and powers to take away samples, *etc.* However, monitoring is fundamental to providing information to the Inspectorate and the general public:

(i)   on whether operators are within authorized limits and conditions;
(ii)  on the levels of releases actually being made to the environment.

In its purest form an authorization need not refer to monitoring at all. The essential point is to set limits and conditions which, if complied with, constitute BATNEEC and provide for satisfactory environmental protection.

The approach that has been adopted under IPC is to place requirements on operators to monitor their releases, to keep the results of their monitoring for inspection and to report key results to the Inspectorate for inclusion on publicly accessible registers. Furthermore, operators are required to use the best available techniques for monitoring, which HMIP regards as on-line instrumentation linked to computer data storage systems wherever practical and not entailing excessive cost.

There are arguments for independent measurement of releases but, with the growing complexity and sophistication of monitoring techniques, there are advantages in requiring the operator to carry out specified measurements of releases and to record them with other details. The regulator must make sure that these are done conscientiously and by way of tamper-proof controls and recording systems. This is part of the inspection task. In addition, confirmation and reassurance that this system is working honestly and effectively is provided by HMIP commissioning check-measurements. The purpose of the latter is not to duplicate or augment the information supplied by the operator but rather to provide information, independently generated, against which the operator's data may be compared.

This approach to monitoring has been portrayed as self-regulation. It is not. Rather, it represents a system which is rigorous, cost-effective for HMIP and operators, requires the operator to take a close interest in the environmental impact of his operation, and will provide increased and better quality data to the public on releases.

## 20.4  BATNEEC

The Environmental Protection Act lays a specific duty upon the Chief Inspector

'to follow developments in technology and techniques for preventing or reducing pollution of the environment due to releases of substances from prescribed processes'.

To discharge this duty, the Department of Environment allocates a sizeable research budget to HMIP, which is used to conduct and commission the research necessary to underpin HMIP's regulatory role. An important area of this research is to find out about available techniques which are in use in the UK and abroad in the process in question, and to evaluate their relative merits. A former Secretary of State for the Environment saw IPC as a driving force for innovation, creating and satisfying demand for new technologies not only in terms of abatement equipment but in terms of the whole process. What IPC seeks to achieve is the right balance between cost to industry of better processes and abatement equipment, and cost to the environment in terms of the damage caused by releases.

As was described earlier in this Article, all IPC processes under Part I of the Environmental Protection Act 1990 are subject to control via the philosophy of BATNEEC, which has been a part of European Community legislation for some time, having been first introduced in 1984 in Directive 84/360/EEC on combating air pollution from industrial plant. This states that:

'... an authorization may be issued only when the competent authority is satisfied (among other things) that all appropriate preventive measures against air pollution have been taken, including the application of best available technology, provided that the application of such measures does not entail excessive costs ...'.

For an overall definition of BATNEEC it is helpful to consider the words separately and together. Thus:

- *'Best'* must be taken to mean most effective preventing, minimizing or rendering polluting releases harmless. There may be more than one set of techniques that achieves comparable effectiveness – that is, there may be more than one set of 'best' techniques. It implies that the technology's effectiveness has been demonstrated.
- *'Available'* should be taken to mean 'procurable by any operator of the class in question'; that is, the technology should be generally accessible (but not necessarily in use) from sources outside as well as within the UK. It does not imply a multiplicity of sources, but if there is a monopoly supplier, the technique will count as being 'available' provided that all operators can procure it. Furthermore, it includes a technique which has been developed (or proven) at a scale which allows its implementation in the relevant industrial context and with the necessary business confidence. Industry often believes that for a technology to be recommended as 'available', it would need to have been in commercial use for at least six to twelve months.
- *'Techniques'* is a term which embraces both the plant in which the process is carried on and how the process is operated. It should be taken to mean the components of which it is made up and the manner in which they are connected together to make the whole. It also includes matters such as numbers and qualifications of staff, working methods, training and supervision and also the design, lay-out and maintenance of buildings, and will affect the concept and design of the process.

The other North Sea States have also adopted (at least in general) the concept of BATNEEC and are in the process of applying it to industrial sectors discharging hazardous substances by reviewing discharge consents. Consents specify BATNEEC in terms of treatment efficiency rather than specifying the actual technology to be used, so as not to hamper innovation, and these are continuously revised to include the latest developments.

As examples of the application of BATNEEC, in Sweden when consents are issued for new or enlarging plants, they also require research and development to further improve treatment technology. The Netherlands and Germany have set two levels of technological requirements. Thus in the Netherlands, Best Technical Means (BTM) applies to all potential EC List I Substances (Black List) and Best Practical Means (BPM) for all List II Substances (Grey List). BTM, in this case, is defined as the most advanced treatment technology which is practically usable, whereas BPM refers to technologies which are affordable for 'averagely profitable companies'. In Germany, Best Available Technology (BAT), similar to BTM in the Netherlands, must be applied to effluents containing dangerous substances, whereas Generally Acceptable Technology (similar to BPM in the Netherlands) must be applied to effluents containing biodegradable substances.

- *'Not entailing excessive cost'* needs to be taken in two contexts, depending on whether it is applied to new processes or existing processes. For new processes the presumption is that the best available techniques will be used. Nevertheless, in all cases BAT can properly be modified by economic considerations where the costs of applying the best available techniques would be excessive in relation to the nature of the industry and to the environmental protection to be achieved.

In relation to existing processes, HMIP seeks from applicants and operators fully justified proposals for the timescale over which their plants will be upgraded to achieve the same standards as would be expected of new plant. This is why for new gas turbine power stations HMIP has not required the fitting of relatively very expensive selective catalytic reduction in most circumstances. Very good control of emissions of nitrogen oxides is achievable by combustor and burner design. It also fits the Best Practicable Environmental Option (BPEO) criteria referred to later.

Reasonableness comes into all of the discussions on BATNEEC, in that a situation is achieved where money is spent on abatement only up to the point where the resulting increase in control of pollution justifies the money spent.

So far as the ground rules are concerned, HMIP has been proceeding on the presumption that coal-burning baseload stations and orimulsion stations would have to be fitted with flue gas desulfurization (FGD) or equivalent, low $NO_x$ burners or equivalent and high efficiency particulate arrestment. Stations on load factors below baseload should be subject to BATNEEC evaluation which balances industry-wide financial considerations against station-specific environmental impacts.

BATNEEC judgements refer to what is achievable by way of pollutant release levels by the application of process/abatement options which are accepted as representing best practice in the relevant context. Ideally, Inspectors' judgements for a specific site should be based on an awareness of financial and economic implications for the affected industry(ies) of what is identified as 'Best Practice', which will enable HMIP to gauge how far and how fast the regulated community can undertake environmental expenditures.

For the choice of production techniques, the options are the alternative ways of producing a specified output either for intermediate use or final consumption. The focus of this analysis is on assessing the costs of each technique against the associated environmental effects, so the options may usefully be identified in terms of the ranked environmental performance of each technique. The analysis can then seek to determine either the total or the incremental net cost of each option.

The method of appraisal outlined here aims to identify for a typical firm the incremental costs (unit or total) associated with using different technologies, and thus different levels of environmental impact, in order to produce the same final output. The comparison could be made in terms of, say, the difference in net present value of annual output of 1000 units for ten years; or the difference in unit costs in a base year, assuming full accounting costs are covered over project life.

A complete cost–benefit analysis would aim to identify and compare the full social costs of two or more different states of the world, so that a decision could be based on the change in Net Social Benefit (*i.e.* the estimated money value of the material gains and losses to all actors) between the different options. Hitherto, it has not been possible for Inspectors' judgements to be based on a full cost–benefit analysis because no reliable, cardinal monetary valuations are available for the environmental outcomes; the ranking is ordinal only. Hence there is no explicit attempt made to equate marginal costs and marginal benefits of environmental improvements. The BATNEEC solution is defined solely in terms of the total or incremental resource cost of implementing that environmental outcome.

Cost effectiveness is the principal economic criterion for identifying the BATNEEC solution. This requires comparison of the relative costs to the operator of alternative environmental quality levels, as determined by the feasible set of production techniques.

## 20.5  PUBLISHED GUIDANCE

The Inspectorate recognizes that the requirements of the Act, its regulations and the procedures to be followed under it are demanding for an applicant for an authorization. Comprehensive guidance has therefore been prepared. General guidance is provided by the publication 'Integrated Pollution Control – A Practical Guide'[2] which

(i)   outlines the origins of IPC;
(ii)  explains authorizations, and the application and consultation procedures;
(iii) explains the meaning of BATNEEC and how it is promulgated through Chief Inspector's Guidance Notes;
(iv)  discusses topics such as authorization variation procedures, fees and charging, the functions of HMIP enforcement, appeals and the interface with other legislation; and
(v)   lists other information such as the prescribed substances and HMIP locations.

In order that all his Inspectors apply the current BAT standards consistently, the Chief Inspector issues guidance to them in the form of published 'Chief Inspector's Guidance Notes' (CIGN). The notes are available to the public and, of course, to operators and developers of prescribed process, so that they have a clear understanding of what are the necessary standards of operation – including the substances released and the emission/release levels that may be stipulated in a specific authorization.

Further general guidance to inspectors on technical matters is published in the form of Technical Guidance Notes, covering such matters as dispersion calculations, chimney height determinations, *etc*. These, like the process guidance notes, are publicly available through Her Majesty's Stationery Office (HMSO). HMIP's published guidance is listed in its bibliography.[3]

The legislation does not contain numerical standards. If the best available techniques enable an operator to prevent a release, and he can employ these

[2]'Integrated Pollution Control: A Practical Guide', HMSO, 1993.
[3]HMIP Bibliography, HMIP, London, May 1994.

techniques without entailing excessive cost, then that is what he will be required to do. Any published guidance is fixed in time, but the yardstick for authorizing his process is dynamic: BAT can change from day to day. It will be up to industry to drive the standards. Industry, whether it be individual companies or their trade associations, must keep abreast of developments.

The CIGNs to Inspectors have no statutory force. They do, however, represent the view of HMIP on appropriate techniques for particular processes, and are therefore a material consideration to be taken into account in every case. HMIP must be prepared to give reasons for departing from the guidance in any particular case. The final decision on a particular application will be taken following consideration of the Applicant's case and of any representations from the public and the statutory consultees.

Similarly, an applicant must not feel constrained by the Guidance Notes: if he deems that a particular technique constitutes 'best available' in the context of his process, it is for him to put that proposition forward, and to justify it in the terms of the Act. He might choose another technique which will deliver the same environmental performance as those mentioned in the Guidance Notes; or he might choose a route which will deliver a better performance, and therefore must be regarded as BATNEEC.

Both HMIP and industry must be constantly re-evaluating the standard, the techniques and the economics relevant to any process. It is this dynamic feature of IPC which will enable sustained and sustainable environmental improvements to be achieved. HMIP is committed to making regular revisions of the CIGNs. As already stated, HMIP conducts and commissions research to find out about available techniques. It is this research which underpins the CIGNs to Inspectors. HMIP reviews and evaluates the techniques available and in use, in the UK and abroad, for the process.

Because BATNEEC is a site-specific concept, CIGNs now provide information relating to achievable release levels for new processes applying the best combination of techniques to limit environmental impact in the context of the processes described. This is aimed at overcoming possible confusion about the expression of release limits in earlier CIGNs which were occasionallyinterpreted as uniform emission standards. This is, of course, contrary to the nature of BATNEEC.

HMIP are currently considering the next steps for the development of CIGNs which aim to incorporate general sectoral advice on economic and market factors which should be taken into account in the assessment of BATNEEC. The programme for the revision of CIGNs commenced with the Fuel and Power sector during 1994/95.

## 20.6  BEST PRACTICABLE ENVIRONMENTAL OPTION (BPEO)

The Act requires that BATNEEC is used to achieve the Best Practicable Environmental Option (BPEO). So what is BPEO? The concept of BPEO was first introduced by the UK's Royal Commission on Environmental Pollution which, in 1988, described BPEO as 'the outcome of a systematic consultative and decision-making procedure which emphasizes the protection and conservation of the environment across land, air and water. The BPEO procedure establishes, for a given set of objectives, the option that provides the most benefit or least damage to the environment as a whole, at acceptable cost, in the long term as well as the short term.'[4]

To authorize an IPC process in accordance with the BPEO (and BATNEEC) objective, it is necessary to compare alternative options for operating a process, and consequent releases, to ensure that one environmental medium is not protected to the detriment of another, and that the impact on the environment as a whole is minimized.

Under IPC, the requirement to satisfy the BPEO objective for such a process has caused HMIP to re-examine the way in which it has traditionally assessed applications for authorization. For example, how does the Inspectorate judge whether or not hydrocarbons should be flared or recycled, or whether or not heavy metals should be disposed of by means of landfill rather than discharged to an estuary? The primary aim must be to identify the maximum concentration of contaminants which can be tolerated in any portion of the environment, considering in particular the most sensitive pollution receptors. For regulatory purposes, this concentration or level can be regarded as an indication of the Environmental Capacity, *i.e.* the ability of a particular portion of the environment to tolerate (an increase in concentration of) a specific contaminant. Furthermore, the proportion of the environmental capacity utilized by a release can be used to indicate the relative harm caused.

Many decisions on IPC applications have been made without the scientific methodology which we all recognize as being so very important. EPA 90 is new legislation which takes a very bold approach and embodies some adventurous objectives. BPEO was, when the Act was passed, a slightly abstract concept which has needed a great deal of development. It is an evolutionary approach, and will bear fruit over many years if not decades. It is a piece of legislation which enables us to apply the very best science and the most up-to-date thinking to the dual challenge of maintaining a healthy business base in this country and of affording the right level of protection to the environment.

Development of an environmental assessment procedure for BPEO began in 1992–93, and was continued strongly during 1993–94, with particular emphasis on consultation. In July 1993 the approach was presented at a seminar attended by nearly 100 delegates from industry, government and environmental organizations, and further discussions have been held with industry groups on other occasions. The procedure has been significantly revised and extended on the basis of comments received, and was circulated as a consultation document to several hundred interested organizations and individuals in 1994. It is intended that following this consultation exercise a Guidance Note on principles for environmental assessment will be produced.

[4]'Best Practicable Environment Option. Twelth Report by the Commission on Environmental Pollution', HMSO, London, February 1988, Cmnd 310.

## 20.7  CLEANER TECHNOLOGY

Cleaner technology is about minimizing the environmental impact of releases from processes. The philosophy behind it is the prevention of waste rather than the cure. Every aspect of a process needs to be optimized to minimize waste in any form.

The basic options to achieve this philosophy are relatively few and can be summarized in order of preference as:

(i) *Reduction at Source* – The most effective way to prevent a material from entering the environment is to stop using or making it.

(ii) *Product Changes* – A process should only be operated if the products cannot be made in a cleaner way. Suitable alternative materials may perform the same function with less environmental consequences.

(iii) *Process Changes* – A process should be designed or changed in such a way that potentially polluting materials are not made or isolated, minimizing the possibility of a release.

(iv) *Re-use* – Re-use of a material is an alternative way of preventing release to the environment.

(v) *On-site Recycling* – Using a by-product of one process as a raw material for another disposes of it without an environmental impact.

(vi) *Off-site Recycling* – Sending a by-product of a process to be used elsewhere is similar to on-site recycling, but the pollution and cost of transport, handling, *etc.* makes this less desirable.

In the event of it not being possible to prevent a pollutant being formed, it must be treated or destroyed to render it harmless. This, of course, is not cleaner technology.

Only after achieving all the possible moves up this scale do you move on to looking at the sort of end-of-pipe methods which are suitable for dealing with those wastes which you have failed to eliminate. And here, the sort of alternatives to be considered might include:

- Incineration;
- Chemical transformation to a less harmful waste;
- Biological treatment;
- Transfer from one environmental medium to another where it might be less harmful;
- Dilution or dispersion.

Cleaner technology is achieved by good engineering design, good management practices and innovative process design. The actual cleaner technology will depend upon the industry or process concerned. For example, water conservation and energy management may be the most significant considerations for a company manufacturing soft drinks, while to chemical manufacturers raw material controls and synthetic routes could be more important. Cleaner technology is thus the most efficient process, ensuring maximum utilization of all raw materials and energy.

Because cleaner technology is about innovative ideas for processes and proper management of people and equipment, it is not necessarily expensive technology. Cleaner processes can be operated by any organization, no matter what the size or type of operation. For many processes the hardware for cleaner operation already exists. What is required is the ingenuity to assemble the appropriate building blocks.

The advantage of using cleaner technology is as much financial as environmental. By designing a process to minimize waste, product yields are usually higher. On some occasions the extra capital costs offset the advantage of increased process efficiency but, by the time disposal costs of waste is taken into account, clean processes are normally economically advantageous.

Organic solvents which give rise to Volatile Organic Compounds (VOCs) are widely used in the manufacture and application of industrial paint systems. Paint manufacturers are developing water-based paint systems which do not contain organic solvents. This is an example of substitution. At present these aqueous systems are generally more expensive to produce. However, being non-flammable they present fewer safety hazards. The operator is not required to take as many safety precautions or fit as much abatement equipment to the paint shop. The cost of application in these circumstances will be less.

The power generation industry is an example where an innovative solution to a difficult problem has given rise to a cleaner technology. Most fossil fuels contain significant quantities of sulfur which is converted to sulfur dioxide during the combustion process. Sulfur dioxide is one of the major gases that gives rise to acid rain. The current practice of removing sulfur dioxide from the flue gas – FGD – prevents air-borne pollution but leaves a solid waste product instead. This meets the current requirements to reduce sulfur dioxide releases but is not clean technology. A new process which removes the sulfur prior to combustion is a much cleaner technology with many potential operational advantages.

This process is called an Integrated Gasification and Combined Cycle (IGCC) system. Coal (or oil) is gasified using one of several proprietary processes. The technology for this stage is well established and has been used in the chemical industry for many years. The synthesis gas generated has the sulfur contaminant present as hydrogen sulfide. The gas volume prior to combustion is about one hundred times less than if the same fuel were burned conventionally. The size of the gas cleaning plant consequently is smaller and less expensive in relation to an FGD plant. The process for removal of hydrogen sulfide from the synthesis gas also is well-established, having been used in the petroleum industry for many years. Sulfur is removed as elemental sulfur which has a ready commercial market. Removal of sulfur as the element is not possible if it has been through the combustion process and been oxidized to sulfur dioxide. Power is generated by burning the clean synthesis gas in a gas turbine working in a combined cycle mode. This part of the overall process is also well established, with many Combined Cycle Gas Turbine stations running on natural gas. The IGCC method of power generation is not only more efficient at converting the fuel to power than a conventional steam-raising power station, but much lower releases of nitrogen oxides (another acid-rain-producing gas and strong irritant) are also possible. Thus, for a given unit of fuel, more power is produced and less of each of the major pollutants is released. This is an illustration of innovative clean technology. Several demonstration plants are being built or planned and this process is the one most likely to be used for power generation in the early part of the next century.

In the case of companies that have made applications for IPC authorization, a striking example of one where a 'dirty' process has been replaced by 'clean' technology is that of a petrochemical company in Southern England. The company required pure butenes for its down-stream processes but the butene supply was heavily contaminated with butadiene. They had a requirement for butadiene and so operated a wet extractive process using cuprous ammonium acetate to separate the butadiene from the butenes. This process typically released annually around 200 tonnes of ammonia and 140 tonnes of VOCs to air, and around 300 tonnes of ammonia and 6 tonnes of copper to the aqueous effluent treatment system. Involving HMIP at the design stage, they resolved to source their butadiene from elsewhere and replace the butene purification process with a new one having essentially zero emissions. Authorization has been given for a catalytic hydrogenation process which selectively reduces the butadiene to the required end-product butene. There are no emissions to air other than those estimated for fugitive releases from flanges, valves, pump glands, *etc.* (7 tonnes annum$^{-1}$ of VOCs), and they estimate that less than 1 kg year$^{-1}$ of hydrocarbons is released to the water course.

Another example of a company changing the process completely to eliminate the release of a prescribed substance – in this case VOCs – is that of an American-owned manufacturer of fluorescent tubes. The company is currently preparing an application for authorization as a 'Mercury process'. The insides of fluorescent tubes are coated with a phosphor that has traditionally been applied as a solution or suspension in a xylene/butanol-based solvent. Following the drying/curing process, around 500 tonnes year$^{-1}$ of solvent is lost to the atmosphere. Having come under near-simultaneous environmental pressures on both sides of the Atlantic, the laboratories of the parent company put around 30 man-years of effort into finding a 'cleaner' process. They have now developed a water-based carrier system for phosphor deposition which is to be installed in the UK factory. The only significant release expected is 1.25 tonnes year$^{-1}$ of ammonia loss to the atmosphere. The capital cost of this change of technology is high so the company is replacing its four existing production units in a four-year rolling programme.

## 20.8 WASTE MINIMIZATION AND IPC

In its widest context, waste can be interpreted as almost any loss or discharge of any material to any medium, and with particular emphasis on specified ranges of substances for the three environmental media. HMIP uses that interpretation in IPC.

In the issues of the Chief Inspector's Guidance Notes covering the Chemical Industry, Inspectors also are advised to encourage applicants to carry out a formal process assessment and have in place a Waste Minimization programme, in advance of submitting their application. In addition to identifying, in a systematic way, those areas in their process where reductions in releases may be accomplished, and thereby providing the foundations for a programme for up-grading the plant, the procedure provides a mechanism whereby applicants can identify deficiencies in the data to be included in their application (which is a problem HMIP has encountered far too frequently). The introduction of the waste minimization concept into those Chief Inspector's Guidance Notes that were issued after the Institute of Chemical Engineers' 'Waste Minimization Guide'[5] was published in 1992, is no coincidence. The appearance of the Guide provided a mechanism (with defined methodology) by which HMIP could encourage operators of IPC processes to take the next step towards the basic principle of IPC – namely, 'Prevention rather than cure'.

Under IPC, strictly it is only releases of substrates prescribed in the Regulations which must be prevented or minimized; other releases need only be rendered harmless. However, the range of prescribed substances, particularly those prescribed for air, is large, so even if waste minimization is restricted to these areas, very significant improvements can be made – and, once the thinking has started, it is unusual for the programme to be restricted solely to prescribed substances.

HMIP, along with the BOC Foundation for the Environment, the NRA and Yorkshire Water Services, also were sponsors of a project on Waste Minimization, co-ordinated by the Centre for Exploitation of Science and Technology. This project concentrated on water treatment for a number of companies in the Aire and Calder catchment area in Yorkshire. The project clearly demonstrated environmental and economic benefits. Most of the benefits were as a result of the reduction in the use of inputs such as water, energy and raw materials.

The Aire and Calder project has proved to us all that to prevent, to minimize and to render harmless our releases to the environment is a readily achievable objective. It is the basic objective of IPC, and seems to me to be a very good one to apply to any of our activities. The Aire and Calder project has already shown us that there are shortcomings in measurement of releases to the environment even at the most modern sites. And as the project has progressed it has become clear that those companies which have wholly embraced monitoring strategies are now in control and can therefore formulate programmes for management, containment and minimization of releases to all three environmental media. So we have a way in to what must be a continuous cycle of measurement, analysis, control and feedback.

The project has demonstrated pay-back at three levels. First order savings are those related to good housekeeping. Second order savings come from an analysis

[5]'Waste Minimization Guide', The Institution of Chemical Engineers, Rugby, 1992.

of product losses and de-bottlenecking. And third order savings will come from a change in ethos from end-of-pipe techniques to inherently 'clean' processes. The conclusion is that pollution prevention pays.

The Centre for the Exploitation of Science and Technology has been commissioned to extend the concept by identifying three suitable areas for further studies. The new studies will be concentrated on industries and regions of the country that have not been subject to this approach with particular reference to those that will soon become subject to IPC regulation.

## 20.9  CONCLUSION

The protection of the environment is our responsibility, whether we are a member of the public as a citizen or consumer or an industrialist producing consumer products, or a part of Government, either in setting environmental policy or in regulating environmental pollution.

Concern for the environment and both national and international regulations will continue to put pressure on industries to minimize their impact on the environment. There are great business opportunities for companies to supply and adopt cleaner technologies and more environmentally friendly techniques. There are challenges to be met in developing new products and processes that are more efficient and produce less waste. There are savings to be made in avoiding the increasing costs of environmentally acceptable waste disposal. There are further opportunities. New technologies can bring together the benefits of greater efficiency, less pollution and a minimizing of impact on the environment as a whole. Maintaining the *status quo* is not acceptable.

Environmental awareness is critical and must be central to every company's activity. It is no longer just a passing phase. The environmental impact of a process must be fully considered both for an existing process and a new process. Changes must be justified against the environmental impact. Waste minimization should be a key part of business strategy in the 1990s and the implementation of IPC has a part to play.

There is a move from a strict regulatory framework, for example as found in the planning system and water and waste regulation, through to the flexible but sophisticated approach of IPC. We are now seeing voluntary, totally integrated, environmental management, as companies appreciate the benefits of environmental commitment.

However, the notion of integration as expressed by the RCEP a number of years ago, embraced not only an overall view of the environment but also advocated a single regulator. It is thinking such as this which led to the establishment of the Environment Agency for England and Wales created in August 1995 under the provision of the Environment Act 1995. The Agency brings together the National Rivers Authority, H.M. Inspectorate of Pollution and local authority Waste Regulation Authorities. The Act also establishes the Scottish Environment Protection Agency. Existing functions, duties and responsibilities of each of the former bodies, including IPC, are transferred to the Agency. Modest legal amendments made by the Environment Act, are designed to enhance harmonization of the formerly separate regulatory regimes. The advent of IPC and the establishment of the Environment Agency has produced pressures and opportunities to forge much closer links and understandings.

# Subject Index